Aquatic Ecosystems in a Changing Climate

Editors

Donat-P. Häder
Department of Biology
Friedrich-Alexander University
Erlangen-Nuremberg, Germany

Kunshan Gao
State Key Laboratory of Marine Environmental Science
Xiamen University
Fujian, People's Republic of China

CRC Press
Taylor & Francis Group
Boca Raton London New York

CRC Press is an imprint of the
Taylor & Francis Group, an **informa** business
A SCIENCE PUBLISHERS BOOK

CRC Press
Taylor & Francis Group
6000 Broken Sound Parkway NW, Suite 300
Boca Raton, FL 33487-2742

Printed on acid-free paper
Version Date: 20180921

International Standard Book Number-13: 978-1-138-35005-2 (Hardback)

Library of Congress Cataloging-in-Publication Data

Names: Häder, H. C. Donat-Peter, editor.
Title: Aquatic ecosystems in a changing climate / editors, H.C. Donat-P. Häder, Department of Biology, Friedrich-Alexander University, Erlangen-Nuremberg, Germany, Kunshan Gao, State Key Laboratory of Marine Environmental Science, Xiamen University, Fujian, People's Republic of China.
Description: Boca Raton, FL : Taylor & Francis Group, [2018] | "A science publishers book." | Includes bibliographical references and index.
Identifiers: LCCN 2018041503 | ISBN 9781138350052 (hardback)
Subjects: LCSH: Aquatic ecology--Climatic factors.
Classification: LCC QH541.5.W3 A6795 2018 | DDC 577.6--dc23
LC record available at https://lccn.loc.gov/2018041503

Visit the Taylor & Francis Web site at
http://www.taylorandfrancis.com

and the CRC Press Web site at
http://www.crcpress.com

Preface

During evolution our planet saw five catastrophic events driven by oxygenation of the atmosphere, carbon dioxide emissions, volcanic eruptions and asteroid impacts, each of which extinguished most of the biota. But after each wipe-out, evolution started triggering more advanced organisms. The Cretaceous-Tertiary extinction eradicated 75% of all plant and animal species about 66 million years ago caused by the impact of a massive asteroid 10 to 15 km in diameter. The most prominent victims were the non-avian dinosaurs. The largest marine extinction has been documented 250–300 million years ago with an event of massive ocean acidification. Today, many scientists are convinced that we are nearing the next mass extinction, this time resulting from the activities of a single species, *Homo sapiens*. It is proven that increasing emissions of carbon dioxide from fossil fuel burning and altered land usage result in global climate change. The question is whether we can stop the global heating at 2° or 3° or 4.5°C. The latter scenario would be catastrophic: the difference between the global mean temperature during the last ice age and today was about 4.5°C. The reason for the uncertainty in the predictions depends on which scenario we will follow. The Intergovernmental Panel on Climate Change has developed a number of models with different inputs, ranging from an immediate stop of CO_2 emissions to business-as-usual. Additional anthropogenic impacts on the biota include stratospheric ozone destruction and pollution. The oceans have absorbed over 90% of the Earth's excessive heat, resulting in ocean warming and deoxygenation with enhanced stratification. The oceans and terrestrial ecosystems are exposed to multiple stresses, but they respond differentially. The ratio of primary production in the oceans vs. terrestrial systems will change, leading to contrasting differences in food supply capacities. Marine ecosystems are responsible for mitigating global climate change and feeding a rapidly growing human population, but they are under a plethora of manmade stress factors including ocean warming, acidification, deoxygenation, and increased exposure to solar UV radiation due to enhanced stratification. These changes alter marine physical and chemical environments to different extents in different regions with wide-range consequences for the food webs up to decreased production for human consumption. This volume reviews the effects of anthropogenic stress factors and their interactions on the aquatic biota.

Donat-P. Häder
Kunshan Gao
Spring 2018

Contents

CHAPTER 1

Introduction

Donat-P. Häder[1,*] and *Kunshan Gao*[2]

More than two-thirds of our planet is covered by water but less than 1% of it is freshwater, with the lion's share being represented by marine habitats. While the biomass in the aquatic ecosystems is only 1% of that of the terrestrial ecosystems, the productivity equals each other in aquatic and terrestrial habitats (Falkowski 2013). Consequently, the aquatic ecosystems sequester the same amount of atmospheric carbon dioxide (estimated as 50–60 Pg carbon) as all terrestrial ecosystems taken together. About 10% of the CO_2 converted into organic biomass in the process of photosynthesis sinks out of the photic zone into the deep sea in the form of dead organisms and fecal pellets which can be detected as marine snow (Laurenceau-Cornec et al. 2015). Eventually less than 0.5% reaches the sea sediment; the rest is mineralized by bacteria in the nutrient cycle. This sequestration of carbon dioxide and removal from the atmosphere is called biological pump (Sigman and Haug 2006). The sinking rate of phytoplankton is about 1 m/per day, therefore, it can take more than 10 years for carbon to reach the ocean floor considering an average depth of 4,000 m in the oceans. In addition, carbon dioxide is incorporated into calcium carbonate in the form of shells by plankton, such as coccolithophorids and foraminifera, animals, such as worms and molluscs and removed from the top layers of the water column, which is called carbonate pump (Hain et al. 2014). Once the carbon has sedimented to the deep sea bottom, it will stay there for millions of years adding to the ca. 36000 Gt which form the largest carbon reservoir on earth. Five major extinction events have occurred on our planet during the evolution of life.

Four of these were mass extinctions, when nearly 75–90% of the marine species vanished, were related to the rapidly increasing levels of carbon dioxide in the atmosphere. The current anthropogenic release of carbon dioxide may initiate a sixth mass extinction (Fischetti and Christiansen 2018). By 1850, the oceans held about 38,000 gigatons of carbon. A recent study indicated that if another 310 gigatons were

[1] Friedrich-Alexander University, Erlangen-Nürnberg, Neue Str. 9, 91096 Möhrendorf, Germany, Europe.
[2] State Key Laboratory of Marine Environmental Science, Xiamen University, Daxue Rd 182, Xiamen, Fujian, 361005, China.
Email: ksgao@xmu.edu.cn
* Corresponding author: donat@dphaeder.de

added to this, it could lead to mass extinction (Rothman 2017). In the recent years, humans are said to have already added about 155 gigatons and may reach up to 400 gigatons before or by 2100—depending on the success or failure in achieving of carbon emission reduction—which could initiate the predicted anthropogenically induced mass extinction.

The Aquatic Food Web

The main primary producers are phytoplankton, which dwell in the photic zone, defined by its lower limit where the surface solar irradiance has been attenuated to 10%. Another definition describes the upper mixed layer (UML) in which most of the phytoplankton live. The lower limit of the UML is the thermocline below which the temperature suddenly drops. This boundary hampers the transition of nutrients from deeper water into the UML (Fischer et al. 2017). Rising temperatures, due to global climate change will increase the stratification and reduce the depth of the UML which exposes the organisms dwelling there to higher solar visible and UV radiation. Phytoplankton is not evenly distributed in the oceans. The highest concentrations are found near the poles while the Northern and Southern mid-latitudes are deserts as seen in pseudocolor images of the chlorophyll distribution (Fig. 1.1).

Macroalgae are mainly limited to the rocky shores of the coasts and continental shelves (Richmond and Stevens 2014). While their productivity is only a fraction of that of the phytoplankton distributed over the whole area of the oceans, they contribute a significant share to the biomass which is of ecological and economic importance: millions of tons of red, brown and green macroalgae are harvested every year for food production, as fertilizers and biofuel and as a basis for many technological applications (Konda et al. 2015, Garcia-Vaquero and Hayes 2016). In addition, the algal beds as well as the seagrass meadows are important refuges for larval fish and other animals.

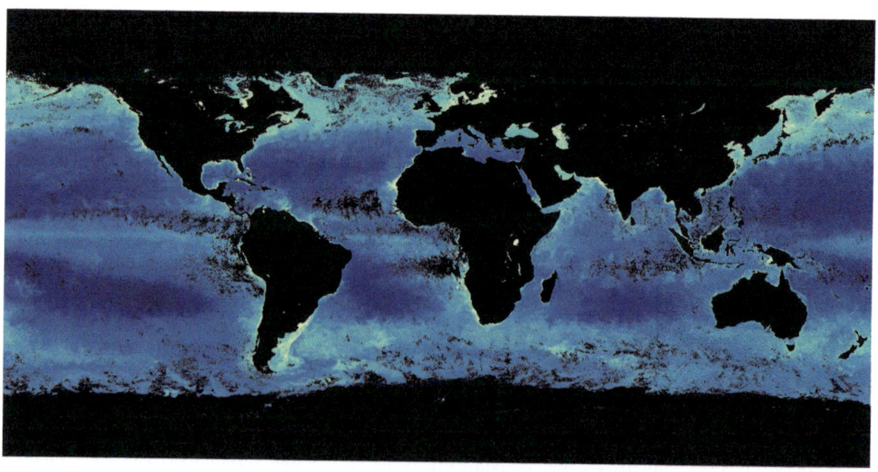

Fig. 1.1. Chlorophyll concentration in the oceans (Aqua/Modis) ranging from 0.01 mg/m^3 (dark blue) to 60 mg/m^3 (bright yellow) as seen by satellite integrated over October 2017. Imagery produced by the Earth Observatory Group in coordination with Gene Feldman and Norman Kuring, NASA Goddard Ocean Color Group.

Changes in temperature and wave patterns affect the macroalgal communities (Harley 2014, Puente et al. 2017). In addition increased dissolved organic matter (DOM) from terrestrial run-off (Clark et al. 2016), changes the transparency of the coastal waters. Increasing pollution including heavy metals, fertilizers, oil products and pharmaceuticals have been found to interfere with the macroalgal populations and change the species composition (Pinto 2015).

The primary producers (prokaryotic and eukaryotic phytoplankton as well as macroalgae and seagrasses) are the basis of the extended marine food webs feeding primary and secondary consumers (Hussey et al. 2014). These food webs, usually starting with herbivorous zooplankton, are highly dynamic and prone to alter rapidly under changing environmental conditions (Quillien et al. 2016). Even subtle changes in the primary elements of the food web can pose significant consequences for the subsequent elements. For example, increased exposure to solar UV radiation affects small phytoplankton more than larger ones. Since zooplankton feed by size and not by taste, organisms adapted to larger food pellets will prosper in contrast to those which prefer smaller prey. This change is relayed though the entire food web. Likewise, increasing temperatures augment the enzymatic repair of UV-induced damages of the DNA and the photosynthetic apparatus (D1 protein in PSII). Therefore, climate change will likely favor organisms with better enzymatic repair systems.

Rising Temperature

While the global atmospheric temperatures have increased by 0.27°C per decade since 1979, ocean surface temperatures have increased by 0.13°C per decade, which is one of the major anthropogenic forcings on the marine biota (Halpern et al. 2008). Increasing temperatures alter habitat selection and change the distribution and species composition of phytoplankton taxa (Thomas et al. 2012), such as dinoflagellates (Hallegraeff 2010, Fu et al. 2012), diatoms, coccolithophorids (Mericoa et al. 2004, Hare et al. 2007) and cyanobacteria (Breitbarth et al. 2007, Pittera et al. 2014). Each organism has a specific temperature window with a minimum, optimum, and maximum. Due to climate change thermal niches of species will move toward the poles and tropical phytoplankton will undergo a significant decline in species diversity (Thomas et al. 2012). In the period between 1960 and 2009, a decline in dinoflagellates was observed in the northeast Atlantic and the North Sea, due to increasing temperatures and resulting stronger stratification as well as ocean acidification and nutrient limitation (Hinder et al. 2012). In contrast, no effect on the density of diatom populations had been noticed.

For corals, the upper permissive temperature of 28°C has exceeded in many tropical habitats which has resulted in the expulsion of the photosynthetic endosymbionts (zooxanthellae) and bleaching of the host animals (Marshall and Baird 2000). In addition, infections with *Vibrio corallilyticus* were more severe at elevated temperatures of 27 and 29°C in contrast to 25°C and resulted in lysis of the corals within two weeks (Ben-Haim et al. 2003). Also the increasing abundance of benthic macroalgae caused alarming declines in reef-building coral species as shown by the effect of green alga *Halimeda opuntia* on the dominant Caribbean coral *Montastraea faveolata* (Nugues et al. 2004). Motile organisms can escape hazardous temperature conditions by moving to a different habitat. For example, tropical *Radiolaria* have been found in Arctic waters (http://earth.columbia.edu, Bjørklund et al. 2012).

Ocean Acidification

The carbon dioxide concentration in the atmosphere has been fairly constant at about 270 ppm over the last thousands of years, but it started to rise noticeably at the beginning of the industrial revolution in the second half of the eighteenth century and has increased at an accelerating pace since. Currently, the CO_2 concentration has exceeded 400 ppm as measured at the Mauna Loa Observatory, Hawaii (Fig. 1.2) (Kaiser 2014, Tans and Keeling 2015). The annual wiggles in the curve are due to the fact that most of the landmass with its vegetation is located on the Northern Hemisphere (Keeling et al. 1996). When the leaves fall and the vegetation decays in the Northern winter the CO_2 concentration rises and falls again in the next summer when more CO_2 is fixed by photosynthesis. It has been estimated that the atmospheric CO_2 concentration will exceed 750 ppm if we continue to utilize fossil fuel in a business-as-usual scenario. A doubling of the concentration measured before the onset of the Industrial Revolution to 650 ppm is regarded as a red line in order to keep the average global temperature increase below the politically pronounced target of 2°C (Knutti et al. 2016).

The CO_2 concentration in the surface layer of the oceans increases in parallel with the atmospheric concentration because both are in an equilibrium (Gill 2016). The oceans take up about 1 million tons of carbon dioxide per hour and remove 25% of the anthropogenically produced CO_2 (Gao et al. 2012b). Therefore, the oceans and the biological pump are a major carbon dioxide sink and play a crucial role in global climate change and limiting the temperature increase (Intergovernmental Panel on Climate Change 2014). The higher CO_2 concentrations in the water result in ocean acidification (Doney et al. 2016). Even though the oceans are well buffered, the pH in surface waters has dropped by about 0.1 units because of the anthropogenically emitted carbon dioxide (Bates et al. 2014, Häder and Gao 2015) which corresponds to a 30% increase in the H^+ concentration. If the anthropogenic CO_2 release increases

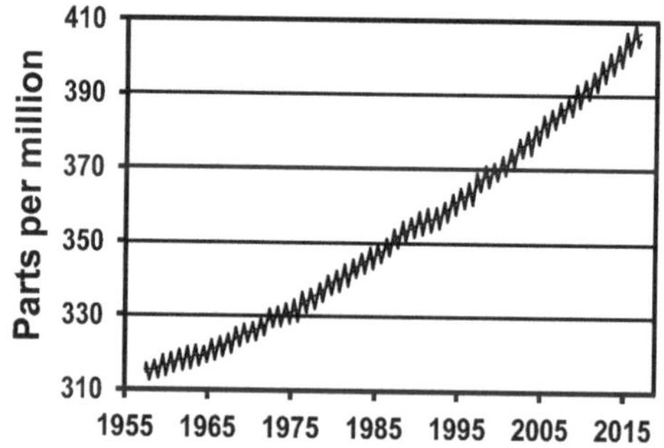

Fig. 1.2. Carbon dioxide concentration in the atmosphere (bold curve) and seasonally corrected data (fine line) measured by the Mauna Loa Observatory. Redrawn after www.esrl.noaa.gov/gmd/ccgg/trends/full.html.

at the current pace, an increase of 100–150% in the H^+ concentration is predicted by the end of the century corresponding to a drop in the pH by 0.3–0.4 units (Feely et al. 2004). In coastal regions decadal changes of up to 0.5 pH units can occur leading to changes in alkalinity and CO_2 fluxes (Feely et al. 2004). Also pelagic waters are affected by ocean acidification (Capone and Hutchins 2014, Waldbusser and Salisbury 2014).

The available CO_2 concentration is a limiting factor for photosynthesis in plants (Farquhar and Sharkey 1982). Therefore, it could be predicted that increasing concentrations in the water could augment productivity in aquatic primary producers. However, investigations at increased acidity showed mixed results depending on species and other environmental stress factors (Gao and Campbell 2014). A diatom-dominated phytoplankton community did not show an increased productivity at 800 µatm pCO_2 during an incubation under about 30% solar irradiance (Tortell 2000). Also in the diatom *Skeletonema* photosynthesis was not stimulated under laboratory conditions at the same CO_2 concentration (Burkhardt and Riebesell 1997, Chen and Gao 2003, Chen and Gao 2004), but in a mesocosm experiment it was enhanced at a CO_2 concentration of 750 µatm (Kim 2006). Also in other diatoms increased CO_2 levels augmented growth and productivity (King et al. 2011, Low-Décarie et al. 2011). In *Phaeodactylum tricornutum* and *Thalassiosira pseudonana* grown at 1,000 µatm CO_2 growth was enhanced at low irradiances but reduced at higher light levels (Gao et al. 2012a). In natural periphyton, communities significantly increased chlorophyll *a* concentrations and diatom abundance were detected under increasing CO_2 concentrations (Johnson et al. 2013). Ocean acidification also affects survival and orientation abilities of larval fishes in the ocean (Siebeck et al. 2015). Growing copepods at 900 µatm CO_2 resulted in a 29% decrease in fecundity (Thor and Dupont 2015). These results indicate that the community structure of phytoplankton assemblages might change with increasing ocean acidification with far-reaching consequences for the food web.

Some phytoplankton, such as coccolithophorids and radiolaria produce an exoskeleton of calcium carbonate. Likewise some macroalgae, such as the green *Acetabularia* and the red *Corallina* produce incrustations of calcium carbonate (Marszalek 1975, Gao et al. 1993). Calcification is found in many zoological taxa, such as worms, corals, mollusks and cephalopods (Ries et al. 2009, Marin et al. 2012). Ocean acidification interferes with the process of calcification (Doney et al. 2016, Jokiel et al. 2016). In addition to protection against predators, calcium carbonate exoskeletons absorb solar UV radiation (Gao et al. 2009, Van Den Broecke et al. 2012). As a consequence of decreased calcification, due to ocean acidification, the organisms are exposed to higher solar visible and short-wavelength irradiation (Gao et al. 2012a, Gao et al. 2012b, Häder et al. 2015).

Even though solar radiation is an essential requirement for all photosynthetic producers, excessive irradiances are stress factors for the biota. In addition to extreme visible irradiances, high-energetic solar UV damages biomolecules, organelles, cells and organisms. UV is divided into UV-C (< 280 nm), UV-B (280–315 nm), and UV-A (315–400 nm). The subsequent wavelength band is described as visible or photosynthetic active radiation (PAR, 400–700 nm) followed by infrared. UV-C radiation is completely absorbed by oxygen and ozone in the atmosphere, and UV-B is strongly attenuated by stratospheric and tropospheric ozone (Häder and Tevini 1987) while most of the UV-A reaches the Earth's surface (Steinbrecht et al. 2009). Since the early 1980s the stratospheric ozone has been found to be catalytically destroyed by anthropogenic release of chlorofluorocarbons (CFCs) and other trace gases paralleled by the increase in solar UV-B radiation (Li et al. 2009, Hoffmann et al. 2014). However, it was a remarkable scientific success story that the production of CFCs was phased out due to the ratification of the Montreal Protocol and its later amendments. The increase in solar UV-B leveled off in the first decade of the current century (Abbasi and Abbasi 2017), but it will take another 50 years to reach pre-1980s ozone values because of the long life of the CFCs in the stratosphere on the order of decades (Steinbrecht et al. 2009, Hoffmann et al. 2014). However, climate change will influence the future ozone depletion and solar UV level (UNEP EEAP 2016). The Antarctic ozone hole is the most prominent feature of ozone depletion (Salby et al. 2012), but also over the Arctic frequent smaller ozone holes have been detected reaching the Northern and mid-Europe (Manney et al. 2011). Smaller but significant ozone losses have been detected over the Northern and Southern mid-latitudes while the tropics did not show an increase over the initially high UV levels which are due to the low zenith angles (Lane 2015).

Solar UV radiation hits many biological targets. It damages cellular DNA in organisms dwelling in the top layers of the water column and impairs photosynthesis in primary producers which have to expose themselves to solar radiation in order to tap the energy for carbon fixation. In addition to direct damage, solar UV-B causes the production of reactive oxygen species upon hitting cellular targets and dissolved organic matter in the water (Pallela 2014, Amado et al. 2015). Even though an increase in solar UV-B is not imminent, organisms in the photic zone are exposed to higher short-wavelength radiation because of the shoaling of the UML and by losing their protective exoskeleton (Gao and Häder 2017, Häder and Gao 2017).

Many exposed organisms have developed strategies to mitigate UV-induced damage by self-shading in crusts or by vertical migration out of the danger zone (Häder et al. 2015, Häder and Horneck 2018). In addition, both prokaryotic and eukaryotic primary producers have developed UV-absorbing substances, such as scytonemin (found only in cyanobacteria) or mycosporine-like amino acids (MAAs, in prokaryotic and eukaryotic phytoplankton and in many macroalgae) which are deposited outside the cell or in outer layers so that damaging radiation is attenuated before it can reach the nucleus and produce damage (Rastogi et al. 2014, Richa et al. 2016, Pandey et al. 2017). In contrast, many consumers do not possess the Shikimate pathway to produce MAAs but can ingest the substances with their diet and utilize them for the same purpose (Ekvall et al. 2015). Extracellularly and intracellularly produced ROS is quenched by enzymatic or non-enzymatic mechanisms to mitigate solar UV-B damage (Häder et al. 2015a, Li et al. 2017). A final method to reduce UV damage is the use of efficient repair mechanisms (Sinha and Häder 2002, Chang et al. 2017). UV-B is known to induce cyclobutane pyrimidine dimers (CPD) which are repaired by photolyases (MacFadyen et al. 2004, Häder and Sinha 2005). However, these repair mechanisms are not foolproof so that mutations and genetic aberrations occur resulting in increased mortality (Lesser and Barry 2003, Häder et al. 2007). The damage of the vital D1 protein in photosystem II, essential for the photosynthetic electron transport, is repaired by a fast and efficient resynthesis—replacing the damaged protein with a new copy (Wong et al. 2015, Wu et al. 2015).

Conclusion

Many investigations into the effects of environmental stress factors in marine organisms have been carried out under controlled laboratory conditions (Häder 2011). This is important to reveal the organismsic and molecular mechanisms of damage and repair. However, often the results of these studies have no bearing for the real ecosystems. For this reason studies in the real world are inevitable to reveal the effects of stress factors on the biota (Gao et al. 2012a, Häder et al. 2015b). Conducting investigations is often difficult because of the vast dimensions of the oceans, changes in irradiation, temperature and other factors and the diluted concentrations of, for instance, plankton in the water column. As a shortcut investigations are carried out in mesocosms with limited volumes or in ship-bound containers using natural communities (Engel et al. 2007, Liu et al. 2017). Different and sometimes contradictory results show that the responses of organisms to certain stress factors depend on other factors as well. For example, the severity of UV damage and its repair depends on temperature and nutrient availability (Wong et al. 2015). Likewise, the effect of ocean acidification depends on the capability of carbon accumulating mechanisms and species composition. Therefore, it is mandatory to analyze the responses of organisms and communities under a wide range of changing environmental factors in order to generate a valid assessment of global climate change on the aquatic biota (Gao and Häder 2017, Häder and Gao 2017).

References

Abbasi, S. and T. Abbasi. 2017. Monitoring ozone loss and its consequences: Past, present, and future. pp. 121–131. *In*: Ozone Hole. Springer, New York, NY.

Amado, A.M., J.B. Cotner, R.M. Cory, B.L. Edhlund and K. McNeill. 2015. Disentangling the interactions between photochemical and bacterial degradation of dissolved organic matter: amino acids play a central role. Microbial Ecology 69: 554–566.

Bates, N., Y. Astor, M. Church, K. Currie, J. Dore, M. Gonaález-Dávila, L. Lorenzoni, F. Muller-Karger, J. Olafsson and M. Santa-Casiano. 2014. A time-series view of changing ocean chemistry due to ocean uptake of anthropogenic CO_2 and ocean acidification. Oceanography 27: 126–141.

Ben-Haim, Y., M. Zicherman-Keren and E. Rosenberg. 2003. Temperature-regulated bleaching and lysis of the coral *Pocillopora damicornis* by the novel pathogen *Vibrio coralliilyticus*. Applied and Environmental Microbiology 69: 4236–4242.

Bjørklund, K.R., S.B. Kruglikova and O.R. Anderson. 2012. Modern incursions of tropical Radiolaria into the Arctic Ocean. Journal of Micropalaeontology 31: 139–158.

Breitbarth, E., A. Oschlies and J. LaRoche. 2007. Physiological constraints on the global distribution of *Trichodesmium*—Effect of temperature on diazotrophy. Biogeosciences 4: 53–61.

Burkhardt, S. and U. Riebesell. 1997. CO_2 availability affects elemental composition (C: N: P) of the marine diatom *Skeletonema costatum*. Marine Ecology-Progress Series 155: 67–76.

Capone, D.G. and D.A. Hutchins. 2014. Microbial biogeochemistry of coastal upwelling regimes in a changing ocean. Nature Geoscience 6: 711–717.

Chang, Y., W.-Y. Lee, Y.-J. Lin and T. Hsu. 2017. Mercury (II) impairs nucleotide excision repair (NER) in zebrafish (*Danio rerio*) embryos by targeting primarily at the stage of DNA incision. Aquatic Toxicology 192: 97–104.

Chen, X. and K. Gao. 2003. Effect of CO_2 concentrations on the activity of photosynthetic CO_2 fixation and extracelluar carbonic anhydrase in the marine diatom *Skeletonema costatum*. Chinese Science Bulletin 48: 2616–2620.

Chen, X. and K. Gao. 2004. Characterization of diurnal photosynthetic rhythms in the marine diatom *Skeletonema costatum* grown in synchronous culture under ambient and elevated CO_2. Functional Plant Biology 31: 399–404.

Clark, C.D., W.J. De Bruyn and P.D. Aiona. 2016. Temporal variation in optical properties of chromophoric dissolved organic matter (CDOM) in Southern California coastal waters with nearshore kelp and seagrass. Limnology and Oceanography 61: 32–46.

Doney, S.C., V.J. Fabry, R.A. Feely and J.A. Kleypas. 2016. Ocean acidification: The other CO_2 problem. Wash. J. Envtl. L. & Pol'y 6: 213.

Ekvall, M.T., S. Hylander, T. Walles, X. Yang and L.-A. Hansson. 2015. Diel vertical migration, size distribution and photoprotection in zooplankton as response to UV-A radiation. Limnology and Oceanography 60: 2048–2058.

Engel, A., K. Schulz, U. Riebesell, R. Bellerby, B. Delille and M. Schartau. 2007. Effects of CO_2 on particle size distribution and phytoplankton abundance during a mesocosm bloom experiment (PeECE II). Biogeosciences Discussions 4: 4101–4133.

Falkowski, P. 2013. Primary productivity in the sea. Springer Science & Business Media.

Farquhar, G.D. and T.D. Sharkey. 1982. Stomatal conductance and photosynthesis. Annual Review of Plant Physiology 33: 317–345.

Feely, R.A., C.L. Sabine, K. Lee, W. Berelson, J. Kleypas, V.J. Fabry and F.J. Millero. 2004. Impact of anthropogenic CO_2 on the $CaCO_3$ system in the oceans. Science 305: 362–366.

Fischer, C.J., J.-Z. Zhang and M.O. Baringer. 2017. An estimate of diapycnal nutrient fluxes to the euphotic zone in the Florida Straits. Scientific Reports 7: 16098.

Fischetti, M. and J. Christiansen. 2018. Killer seas. Scientific American 2018: 76.

Fu, F.X., A.O. Tatters and D.A. Hutchins. 2012. Global change and the future of harmful algal blooms in the ocean. Marine Ecology Progress Series 470: 207–233.

Gao, K., Y. Aruga, K. Asada, T. Ishihara, T. Akano and M. Kiyohara. 1993. Calcification in the articulated coralline alga *Corallina pilulifera*, with special reference to the effect of elevated CO_2 concentration. Marine Biology 117: 129–132.

Gao, K. and D.-P. Häder. 2007. Effects of ocean acidification and UV radiation on marine photosynthetic carbon fixation. pp. 235–250. *In*: Kumar, M. and P.J. Ralph (eds.). Systems Biology of Marine Ecosystems. Springer, Cham, Switzerland.

Gao, K., Z. Ruan, V.E. Villafañe, J.P. Gattuso and E.W. Helbling. 2009. Ocean acidification exacerbates the effect of UV radiation on the calcifying phytoplankter *Emiliania huxleyi*. Limnology and Oceanography 54: 1855–1862.

Gao, K., E.W. Helbling, D.-P. Häder and D.A. Hutchins. 2012a. Responses of marine primary producers to interactions between ocean acidification, solar radiation, and warming. Marine Ecology Progress Series 470: 167–189.

Gao, K., J. Xu, G. Gao, Y. Li, D.A. Hutchins, B. Huang, Y. Zheng, P. Jin, X. Cai, D.-P. Häder, W. Li, K. Xu, N. Liu and U. Riebesell. 2012b. Rising carbon dioxide and increasing light exposure act synergistically to reduce marine primary productivity. Nature Climate Change 2: 519–523.

Gao, K. and D. Campbell. 2014. Photophysiological responses of marine diatoms to elevated CO_2 and decreased pH: A review. Functional Plant Biology 41: 449–459.

Garcia-Vaquero, M. and M. Hayes. 2016. Red and green macroalgae for fish and animal feed and human functional food development. Food Reviews International 32: 15–45.

Gill, A.E. 2016. Atmosphere—Ocean Dynamics. Elsevier.

Häder, D.-P. and M. Tevini. 1987. General Photobiology. Pergamon. Oxford, UK.

Häder, D.-P. and R.P. Sinha. 2005. Solar ultraviolet radiation-induced DNA damage in aquatic organisms: potential environmental imapct. Mutation Research 571: 221–233.

Häder, D.-P., H.D. Kumar, R.C. Smith and R.C. Worrest. 2007. Effects of solar UV radiation on aquatic ecosystems and interactions with climate change. Photochemical & Photobiological Sciences 6: 267–285.

Häder, D.-P. 2011. Does enhanced solar UV-B radiation affect marine primary producers in their natural habitats? Photochemistry and Photobiology 87: 263–266.

Häder, D.-P. and K. Gao. 2015. Interactions of anthropogenic stress factors on marine phytoplankton. Frontiers in Environmental Science 3: 14.

Häder, D.-P., C.E. Williamson, S.-A. Wängberg, M. Rautio, K.C. Rose, K. Gao, E.W. Helbling, R.P. Sinha and R. Worrest. 2015. Effects of UV radiation on aquatic ecosystems and interactions with other environmental factors. Photochemical & Photobiological Sciences 14: 108–126.

Häder, D.-P. and K. Gao. 2017. The impacts of climate change on marine phytoplankton. pp. 901–928. *In*: Phillips, B.F. and M. Pérez-Ramírez (eds.). Climate Change Impacts on Fisheries and Aquaculture. A Global Analysis. Wiley, Hoboken, NJ.

Häder, D.-P. and G. Horneck. 2018. Bioassays for solar UV radiation. pp. 331–346. *In*: Häder, D.-P. and G.S. Erzinger (eds.). Bioassays. Advanced Methods and Applications. Elsevier, Atlanta, GA.

Hain, M.P., D. Sigmal and G.H. Haug. 2014. The biological pump in the past. pp. 485–517. *In*: Reference Module in Earth Systems and Environmental Sciences, Treatise on Geochemistry (Second Edition), The Oceans and Marine Geochemistry. Elsevier, The Netherlands.

Hallegraeff, G.M. 2010. Ocean climate change, phytoplankton community responses, and harmful algal blooms: a formidable predictive challenge. Journal of Phycology 46: 220–235.

Halpern, B.S., S. Walbridge, K.A. Selkoe, C.V. Kappel, F. Micheli, C. D'Agrosa, J.F. Bruno, K.S. Casey, C. Ebert and H.E. Fox. 2008. A global map of human impact on marine ecosystems. Science 319: 948–952.

Hare, C.E., K. Leblanc, G.R. DiTullio, R.M. Kudela, Y. Zhang, P.A. Lee, S. Riseman and D.A. Hutchins. 2007. Consequences of increased temperature and CO_2 for phytoplankton community structure in the Bering Sea. Marine Ecology Progress Series 352: 9–16.

Harley, C. 2014. Seaweed responses to climate change: Predictions, observations, and knowledge gaps. Salish Sea Ecosystem Conference.

Hinder, S.L., G.C. Hays, M. Edwards, E.C. Roberts, A.W. Walne and M.B. Gravenor. 2012. Changes in marine dinoflagellate and diatom abundance under climate change. Nature Climate Change 2: 271–275.

Hoffmann, L., C. Hoppe, R. Müller, G. Dutton, J. Gille, S. Griessbach, A. Jones, C. Meyer, R. Spang and C. Volk. 2014. Stratospheric lifetime ratio of CFC-11 and CFC-12 from satellite and model climatologies. Atmospheric Chemistry and Physics 14: 12479–12497. http://earth.columbia.edu. from http://earth.columbia.edu/articles/view/2993.

Hussey, N.E., M.A. MacNeil, B.C. McMeans, J.A. Olin, S.F. Dudley, G. Cliff, S.P. Wintner, S.T. Fennessy and A.T. Fisk. 2014. Rescaling the trophic structure of marine food webs. Ecology Letters 17: 239–250.

Intergovernmental Panel on Climate Change. 2014. Climate Change 2014–Impacts, Adaptation and Vulnerability: Regional Aspects. Cambridge University Press.

Johnson, Z.I., B.J. Wheeler, S.K. Blinebry, C.M. Carlson, C.S. Ward and D.E. Hunt. 2013. Dramatic variability of the carbonate system at a temperate coastal ocean site (Beaufort, North Carolina, USA) is regulated by physical and biogeochemical processes on multiple timescales. PloS One 8: e85117.

Jokiel, P.L., C.P. Jury and I.B. Kuffner. 2016. Coral calcification and ocean acidification. pp. 7–45. *In*: Hubbard, D.K., C.S. Rogers, J.H. Lipps and G.D. Stanley, Jr. (eds). Coral Reefs at the Crossroads. Coral Reefs of the World. Springer, Dordrecht.

Kaiser, K.L. 2014. The Carbon Cycle and Royal Society Math. Principia Scientific International.

Keeling, C.D., J.F.S. Chin and T.P. Whorf. 1996. Increased activity of northern vegetation inferred from atmospheric CO_2 measurements. Nature 382: 146–149.

Kim, J.-H. 2006. The effect of seawater CO_2 concentration on growth of a natural phytoplankton assemblage in a controlled mesocosm experiment. Limnology and Oceanography 51: 1629–1636.

King, A.L., S.A. Sanudo-Wilhelmy, K. Leblanc, D.A. Hutchins and F. Fu. 2011. CO_2 and vitamin B_{12} interactions determine bioactive trace metal requirements of a subarctic Pacific diatom. The ISME Journal 5: 1388–1396.

Knutti, R., J. Rogelj, J. Sedláček and E.M. Fischer. 2016. A scientific critique of the two-degree climate change target. Nature Geoscience 9: 13–18.

Konda, N.M., S. Singh, B.A. Simmons and D. Klein-Marcuschamer. 2015. An investigation on the economic feasibility of macroalgae as a potential feedstock for biorefineries. BioEnergy Research 8: 1046–1056.

Lane, J.L. 2015. Stratospheric ozone depletion. pp. 51–73. *In*: Hauschild, M. and M. Huijbregts (eds.). Life Cycle Impact Assessment. LCA Compendium – The Complete World of Life Cycle Assessment. Springer, Dordrecht.

Laurenceau-Cornec, E.C., T. Trull, D.M. Davies, S.G. Bray, J. Doran, F. Planchon, F. Carlotti, M. Jouander, A.-J. Cavagna and A. Waite. 2015. The relative importance of phytoplankton aggregates and zooplankton fecal pellets to carbon export: Insights from free-drifting sediment trap deployments in naturally iron-fertilised waters near the Kerguelen Plateau. Biogeosciences 12: 1007–1027.

Lesser, M.P. and T.M. Barry. 2003. Survivorship, development, and DNA damage in echinoderm embryos and larvae exposed to ultraviolet radiation (290–400 nm). Journal of Experimental Marine Biology and Ecology 292: 75–91.

Li, F., R.S. Stolarski and P.A. Newman. 2009. Stratospheric ozone in the post-CFC era. Atmos. Chem. Phys. 9: 2207–2213.

Li, Z.K., G.Z. Dai, P. Juneau and B.S. Qiu. 2017. Different physiological responses of cyanobacteria to ultraviolet-B radiation under iron-replete and iron-deficient conditions: Implications for underestimating the negative effects of UV-B radiation. Journal of Phycology 53: 425–436.

Liu, N., S. Tong, X. Yi, Y. Li, Z. Li, H. Miao, T. Wang, F. Li, D. Yan and R. Huang. 2017. Carbon assimilation and losses during an ocean acidification mesocosm experiment, with special reference to algal blooms. Marine Environmental Research 129: 229–235.

Low-Décarie, E., G.F. Fussmann and G. Bell. 2011. The effect of elevated CO_2 on growth and competition in experimental phytoplankton communities. Global Change Biology 17: 2525–2535.

MacFadyen, E.J., C.E. Williamson, G. Grad, M. Lowery, W.H. Jeffrey and D.L. Mitchell. 2004. Molecular response to climate change: temperature dependence of UV-induced DNA damage and repair in the freshwater crustacean *Daphnia pulicaria*. Global Change Biology 10: 408–416.

Manney, G.L., M.L. Santee, M. Rex, N.J. Livesey, M.C. Pitts, P. Veefkind, E.R. Nash, I. Wohltmann, R. Lehmann, L. Froidevaux, L.R. Poole, M.R. Schoeberl, D.P. Haffner, J. Davies, V. Dorokhov, H. Gernandt, B. Johnson, R. Kivi, E. Kyrö, N. Larsen, P.F. Levelt, A. Makshtas, C.T. McElroy, H. Nakajima, M.C. Parrondo, D.W. Tarasick, P. von der Gathen, K.A. Walker and N.S. Zinoviev. 2011. Unprecedented Arctic ozone loss in 2011. Nature 478(7370): 469–475.

Marin, F., N. Le Roy and B. Marie. 2012. The formation and mineralization of mollusk shell. Front Biosci. 4: 125.

Marshall, P. and A. Baird. 2000. Bleaching of corals on the Great Barrier Reef: Differential susceptibilities among taxa. Coral Reefs 19: 155–163.

Marszalek, D. 1975. Calcisphere ultrastructure and skeletal aragonite from the alga *Acetabularia antillana*. Journal of Sedimentary Research 45.

Mericoa, A., T. Tyrrell, E.J. Lessard, T. Oguz, P.J. Stabeno, S. Zeeman and T.E. Whitledge. 2004. Modelling phytoplankton succession on the Bering Sea shelf: Role of climate influences and trophic interactions in generating *Emiliania huxleyi* blooms 1997–2000. Deep-Sea Research Part A I 51: 1803–1826.

Nugues, M.M., G.W. Smith, R.J. Hooidonk, M.I. Seabra and R.P. Bak. 2004. Algal contact as a trigger for coral disease. Ecology Letters 7: 919–923.

Pallela, R. 2014. Antioxidants from marine organisms and skin care. pp. 3771–3783. *In*: Laher, I. (ed.). Systems Biology of Free Radicals and Antioxidants. Springer, Berlin, Heidelberg.

Pandey, A., S. Pandey, J. Pathak, H. Ahmed, V. Singh, S.P. Singh and R.P. Sinha. 2017. Mycosporine-like amino acids (MAAs) profile of two marine red macroalgae, *Gelidium* sp. and *Ceramium* sp. International Journal of Applied Sciences and Biotechnology 5: 12–21.

Pinto, M.I.A. 2015. Pesticides in water, sediments and biota of semi-closed coastal lagoons: Sources, pathways and impact on aquatic organisms. PhD, Universidade nova.

Pittera, J., F. Humily, M. Thorel, D. Grulois, L. Garczarek and C. Six. 2014. Connecting thermal physiology and latitudinal niche partitioning in marine *Synechococcus*. The ISME Journal 8: 1221–1236.

Puente, A., X. Guinda, J.A. Juanes, E. Ramos, B. Echavarri-Erasun, F. Camino, S. Degraer, F. Kerckhof, N. Bojanić and M. Rousou. 2017. The role of physical variables in biodiversity patterns of intertidal macroalgae along European coasts. Journal of the Marine Biological Association of the United Kingdom 97: 549–560.

Quillien, N., M.C. Nordström, G. Schaal, E. Bonsdorff and J. Grall. 2016. Opportunistic basal resource simplifies food web structure and functioning of a highly dynamic marine environment. Journal of Experimental Marine Biology and Ecology 477: 92–102.

Rastogi, R.P., R.P. Sinha, S.H. Moh, T.K. Lee, S. Kottuparambil, Y.-J. Kim, J.-S. Rhee, E.-M. Choi, M.T. Brown and D.-P. Häder. 2014. Ultraviolet radiation and cyanobacteria. Journal of Photochemistry and Photobiology B: Biology 141: 154–169.

Richa, R.P. Sinha and D.-P. Häder. 2016. Effects of global change, including UV and UV screening compounds. pp. 373–409 *In*: Borowitzka, M., J. Beardall and J. Raven (eds.). The Physiology of Microalgae. Springer, Cham.

Richmond, S. and T. Stevens. 2014. Classifying benthic biotopes on sub-tropical continental shelf reefs: How useful are abiotic surrogates? Estuarine, Coastal and Shelf Science 138: 79–89.

Ries, J.B., A.L. Cohen and D.C. McCorkle. 2009. Marine calcifiers exhibit mixed responses to CO_2-induced ocean acidification. Geology 37: 1131–1134.

Rothman, D.H. 2017. Mathematical Expression of a Global Environmental Catastrophe. Notices of the AMS 64.

Salby, M., E.A. Titova and L. Deschamps. 2012. Changes of the Antarctic ozone hole: Controlling mechanisms, seasonal predictability, and evolution. Journal of Geophysical Research 117: D10111.

Siebeck, U.E., J. O'Connor, C. Braun and J.M. Leis. 2015. Do human activities influence survival and orientation abilities of larval fishes in the ocean? Integrative Zoology 10: 65–82.

Sigman, D.M. and G.H. Haug. 2006. The biological pump in the past. pp. 491–528. *In*: Treatise on Geochemistry. Pergamon Press, New York.

Sinha, R.P. and D.-P. Häder. 2002. UV-induced DNA damage and repair: a review. Photochemical & Photobiological Sciences 1: 225–236.

Steinbrecht, W., H. Claude, F. Schönenborn, I.S. McDermid, T. Leblanc, S. Godin-Beekman, P. Keckhut, A. Hauchecorne, J.A.E. van Gijsel, D.P.J. Swart, G.E. Bodeker, A. Parrish, I.S. Boyd, N. Kämpfer, K. Hocke, R. Stolarski, S.M. Frith, L.W. Thomason, E.E. Remsberg, C. von Savigny, A. Rozanov and J.P. Burrows. 2009. Ozone and temperature trends in the upper stratosphere at five stations of the network for the detection of atmospheric composition change. International Journal of Remote Sensing.

Tans, P. and R. Keeling. 2015. Full Mauna Loa CO_2 record, NOAA/ESRL.

Thomas, M.K., C.T. Kremer, C.A. Klausmeier and E. Litchman. 2012. A global pattern of thermal adaptation in marine phytoplankton. Science 338: 1085–1088.

Thor, P. and S. Dupont. 2015. Transgenerational effects alleviate severe fecundity loss during ocean acidification in a ubiquitous planktonic copepod. Global Change Biology 21: 2261–2271.

Tortell, P.D. 2000. Evolutionary and ecological perspectives on carbon acquisition in phytoplankton. Limnology and Oceanography 45: 744–750.

UNEP EEAP. 2016. Environmental effects of ozone depletion and its interactions with climate change: progress report, 2015. Photochemical & Photobiological Sciences 15: 141–174.

Van Den Broecke, L., K. Martens, V. Pieri and I. Schön. 2012. Ostracod valves as efficient UV protection. Journal of Limnology 71: 12.

Waldbusser, G.G. and J.E. Salisbury. 2014. Ocean acidification in the coastal zone from an organism's perspective: Multiple system parameters, frequency domains, and habitats. Annual Review of Marine Science 6: 221–247.

Wong, C.-Y., M.-L. Teoh, S.-M. Phang, P.-E. Lim and J. Beardall. 2015. Interactive effects of temperature and UV radiation on photosynthesis of *Chlorella* strains from polar, temperate and tropical environments: Differential impacts on damage and repair. PloS One 10: e0139469.

Wu, Y., Z. Li, W. Du and K. Gao. 2015. Physiological response of marine centric diatoms to ultraviolet radiation, with special reference to cell size. Journal of Photochemistry and Photobiology B: Biology 153: 1–6.

CHAPTER 2

Solar UV Radiation and Penetration into Water

Uwe Feister[1,*] and *Donat-P. Häder*[2]

Introduction

Solar UV Radiation

Solar UV Radiation at the Earth's Surface

Total solar irradiance (TSI) of 1361.0 ± 0.5 W m^{-2} varies by about 0.1% (Kopp 2016, Dudok de Wit et al. 2017). Stronger variations occur in the ultraviolet (UV) (100 to 400 nm) region by about 10% at 200 nm and 1% at 300 nm (Pagaran et al. 2009, Ball et al. 2014). An extreme event of sunspots and faculae on October 29, 2003, in solar cycle 23 resulted in a short-time change of TSI by –0.34% (Kopp et al. 2005), –0.5% at 400 nm and + 1.3% at 300 nm (Pagaran et al. 2009). Due to the elliptical path of the Earth around the sun, solar radiation at the top of the Earth's atmosphere varies seasonally by about ± 3.4% (highest at the perihelion around January 3 and smallest at the aphelion around July 6).

On its way to the Earth's surface, solar irradiance is modified by atmospheric extinction (absorption and scattering). Radiation scattered by air molecules and suspended particles is called *diffuse* radiation and the sum of both components *direct* (DIR) and diffuse (DIF) irradiance—both incident at a horizontal surface—is *global* irradiance. The apparent angular diameter of the sun as observed from the earth's surface is 0.52° to 0.54°. Atmospheric refraction, which typically increases from zero with the sun in the zenith to about 0.59° at the horizon, causes the sun to appear slightly higher in the sky than it really is.

Figure 2.1 shows an example of solar spectral irradiance at the Earth's surface for cloudless sky at a solar zenith angle θ of 29° (θ = 0° corresponds to zenith). It reaches

[1] Retired from the German Meteorological Service, Breite Str. 23, 14467 Potsdam, Germany.
[2] Friedrich-Alexander University, Erlangen-Nürnberg, Neue Str. 9, 91096 Möhrendorf, Germany. Email: donat@dphaeder.de
* Corresponding author: uwefeister@web.de

a maximum in the green region around 500 nm at somewhat shorter wavelengths than the maximum of photopic (day) vision of the human eye at 555 nm. *Photosynthetic active radiation* (*PAR*) is defined as radiation in the wavelength region 400 to 700 nm. Another example of biological response shown in Fig. 2.1 is melatonin suppression (Thapan et al. 2001), which is maximum at 450 nm. Melatonin, a hormone of the pineal gland in humans and animals regulates sleep and wakefulness and is also built in plants to act against oxidative stress.

Many biological actions of radiation are maximal in the UV region that is subdivided into UV-C (100–280 nm), UV-B (280–315 nm) and UV-A (315–400 nm) (ISO 2005). It was Johann Wilhelm Ritter (1776–1810) who noted in 1801 invisible 'chemical rays' beyond the violet/blue part of the solar spectrum, which were later given the name 'ultraviolet radiation'. Extraterrestrial UV radiation shows a typical pattern of spectral absorption lines named after the physicist Joseph von Fraunhofer (1787–1826) who described 570 of the thousands of absorption lines in 1814, which had already been noticed in 1802 by William Hyde Wollaston (1766–1828). The Fraunhofer structure is slightly modified on its way through the Earth's atmosphere by Raman scattering from atmospheric N_2 and O_2 (Ring effect) (Lampel et al. 2015) and can beneficially be used for the wavelength alignment of spectroradiometers measuring solar spectral irradiance. Clear-sky ratios UV-B/PAR and UV-A/PAR are highest with small solar zenith angle. While ratios UV-A/PAR decrease by only less than 20% towards high solar zenith angles, ratios UV-B/PAR decrease by more than about 97% of their maximum values and have thus a considerable diurnal and latitudinal variation.

UV irradiance at the Earth's surface mainly varies with solar zenith angle, atmospheric ozone, clouds, aerosols, surface albedo and height above sea level (asl). Far UV radiation (100–200 nm) is completely absorbed by molecular oxygen in the

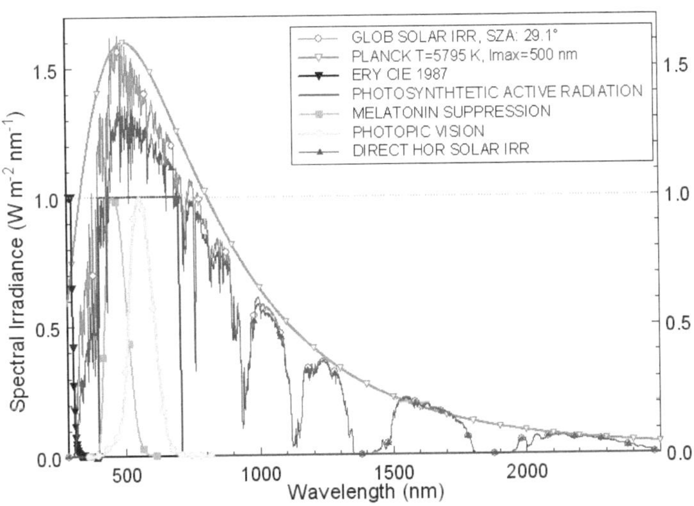

Fig. 2.1. Global and direct solar spectral irradiance at the Earth surface at θ of 29°, Planck black body radiation at 5795 K, and normalized action spectra of erythema (CIE 1987), melatonin suppression (Thapan et al. 2001), photopic vision (Vos 1978) and PAR.

upper atmosphere. UV-C radiation at longer wavelengths is absorbed by stratospheric ozone (O_3) in the Hartley bands detected by Walter Noel Hartley (1845–1913) in 1881. UV-B radiation is partly absorbed in the Huggins bands described by William Huggins (1824–1910) in 1890. A weak radiation absorption by ozone in the visible region between 430 and 750 nm was detected by the French physicist James Chappuis (1854–1934) in 1882. At high solar zenith angles, ozone absorption in the green–orange part of the spectrum contributes to the blue sky color caused by wavelength (λ)-dependent Rayleigh scattering from air molecules according to $\sim \lambda^{-4}$. During twilight at still larger solar zenith angles, it contributes to the red color of the sky (Zagury 2012). Ozone absorption causes a temperature increase with height to its maximum at the stratopause at 50 km, thereby providing energy for atmospheric circulation. Ozone absorption coefficients and their temperature dependences were re-determined in recent years to reduce uncertainties and improve the consistency of ground-based and satellite-based measurements, and model calculations of ozone and solar UV irradiance (Orphal et al. 2016).

Atmospheric *aerosols* from natural (mineral dust from the ground, pollen, sea salt from sea spray, smoke particles from forest fires, volcanic eruptions, etc.), and anthropogenic sources (biomass and fossil fuel burning) absorb and scatter UV irradiance. Large volcanic eruptions emit water vapor and trace gases like sulfur compounds that cross the tropopause (8 km at high latitudes to 18 km in the tropics) and are globally distributed by lower stratospheric circulation. Sulfate aerosols affect the ozone layer as well as radiative and dynamic processes, and reduce solar irradiance in the troposphere down to the surface (Pitari et al. 2016). The portion of absorption to extinction called *single scattering albedo* of aerosols is usually in the range of 0–40% in the UV-B region (Kylling et al. 1998, Mok et al. 2016).

Extinction in *clouds* via suspended water droplets and ice particles modifies spectral UV irradiance. Their optical characteristics, distribution in the sky and movement can result in fast changes of surface spectral UV irradiance that can be observed by fast-measuring instruments having a dynamic range of several orders of magnitude. Optically thick clouds, such as Cumulonimbus (Cb) (WMO 2017) vertically piling up from the lower troposphere to the tropopause or even overshooting it, reduce global UV irradiance at the surface to less than 1% of its cloudless value. Optically very thick clouds can significantly modify spectral irradiance by multiple scattering of radiation at high air pressure and enhanced absorption by ozone in the troposphere (Mayer et al. 1998). Biological radiation effects peaking at shorter wavelengths with stronger ozone absorption, such as vitamin D3 effective irradiance VD3 (CIE 2006) are reduced more by ozone absorption in the cloud than the effects peaking at slightly longer wavelengths, such as erythemal irradiance ERY. Ratios VD3/ERY between vitamin D3 effective and erythemal irradiance depending on column ozone and solar zenith angle (Fioletov et al. 2004) can be reduced to about 50% in optically very thick Cb clouds compared to cloudless sky (Feister et al. 2011).

Clouds modify the smooth cloudless-sky radiance distribution, being maximal close to the sun and minimal about 90° off the sun, to irregular patterns (Blumthaler et al. 1996, Feister and Shields 2005, Shields et al. 2013, Sandmann and Stick 2014, Thosing et al. 2014). Cloud-covered sky parts illuminated by the not-occluded sun can scatter more radiation to the ground than the same cloudless region would scatter.

The Cloud Modification Factor (CMF) defined as the ratio between spectral or broadband irradiance with cloudy conditions (measured) and irradiance with cloudless skies (measured with cloudless sky or taken from cloudless model calculations with the otherwise same conditions) (Schwander et al. 2002, Staiger et al. 2008) has become a useful tool to account for either reduction or enhancement of global spectral irradiance.

Spectral albedo is defined as the ratio between: (i) incident spectral irradiance and (ii) irradiance backscattered and specular reflected (Fresnel reflection called sun glint and glitter in wavy water) from the surface to the atmosphere (Coakley Jr 1914). The albedo of a water surface mainly depends on the purity of water, surface roughness and solar zenith angle (Jin et al. 2004). Specular reflection from water and ice surfaces sharply increases with high solar zenith angles in the morning and evening hours and at high latitudes, while increasing waviness of a water surface reduces sun glint (Haltrin et al. 2001). Due to the small contribution of direct to global irradiance in the UV at high solar zenith angles, the sun glint effect is expected to be smaller in the UV than in the visible region. White caps (foam) on the water surface with high wind speeds mainly increase its albedo. Minimal absorption occurs in the UV-A at 344 nm (Mason et al. 2016) and in the blue region at 418 nm (Pope and Fry 1997). Scattering and absorption of radiation cause pure water of sufficient depth to appear blue (Mobley 2001), though the color impression also depends on the spectral distribution of downward solar irradiance. Spectral extinction of solar radiation by phytoplankton, silt, mineral-like particles, detrital matters, and dissolved organic matter in the water or from microlayers of substances on its surface (Lee et al. 2015) as well as backscattering from the ground and seagrass in shallow water modify backscattered spectral radiance and shift maximum backscattering to longer wavelengths.

Albedo values of water are around 5% in the UV-B (Chadyšiene and Girgždys 2008), 5–9% in the UV-A and about 4% on average in the visible region at 500 nm (Kleipool et al. 2008). High albedo values beyond 10% at 380 nm have been derived over clear water of the Pacific Ocean (Kleipool et al. 2008). Spatial and temporal variations in phytoplankton chlorophyll result in typical seasonal, multi-year (e.g., El Niño Southern Oscillation, ENSO) and long-term variations of albedo patterns (Ye et al. 2016). Enhanced export of UV-absorbing colored dissolved organic matter (DOM) from terrestrial rivers to estuaries and coastal ocean areas as a consequence of more extreme precipitation events and flooding by climate change will affect the sea surface albedo (Sinha et al. 2017). Snow-covered ice shows a higher albedo of 80 to 90% (Zatko and Warren 2015) than bare ice being close to 60% and becoming smaller with melting ice (Fountoulakis et al. 2014). Absorbing impurities including organic carbon, mineral dust, and micro-organisms on ice/snow affect its albedo and accelerate melting (Cook et al. 2017). Expected further melting of Arctic sea ice and Greenland ice sheets, and weakening of the Atlantic Meridional Overturning Circulation (AMOC) are considered policy-relevant 'tipping elements', which by reaching critical 'tipping-points' can flip the Earth's climate into a qualitatively new and lasting state (Lenton 2012).

Height above average sea level (asl) affects the irradiance incident on continental water surfaces a few hundred meters below asl like the Dead Sea (–423 m asl at 31.5° N, 35.5° E) up to mountain lakes and pools a few thousand meters above asl, such as Nevado Ojos del Salada in Argentina/Chile (6,390 m asl at 27.1° S, 68.5° W) and the big lake Titicaca (3,812 m asl at 15.9° S, 69.3° W) on the Altiplano of the Andes region. Air pressure and thus radiation scattering as well as ozone and aerosols overhead decrease with height above asl, while direct and global irradiance increase (Feister et al. 2015b)—depending on atmospheric conditions and solar zenith angle—by rates around 1 (PAR), 2 to 4 (UV-A), and 4 to 9 (UV-B) % per km height (Pfeifer et al. 2006, Dahlback et al. 2007, Cordero et al. 2016). Global UV-B irradiance incident on a mountain lake at 6 km above asl can be about 25 to 50% higher than at asl.

Biologically effective (actinic) irradiance E_B incident on an unobstructed, horizontal surface (WMO 2014) can be derived from spectral irradiance E (λ) in W m^{-2} nm^{-1}, which is the result of integrating spectral radiance in W m^{-2} nm^{-1} sr^{-1} over incidence angles (solid angles in steradian) of the sky and its projection to a horizontal surface, by integration of the product of E(λ) and a dimensionless, normalized spectral biological weighting function ε(λ)

$$E_B = \int_0^{\infty} \varepsilon(\lambda) \cdot E(\lambda) d\lambda \tag{1}$$

If the erythemal action spectrum of human skin is used for ε(λ) (McKinlay and Diffey 1987, Webb et al. 2011) and the wavelength range is limited from 280 to 400 mm, E_B is called *erythemal irradiance*. Erythemal irradiance in W m^{-2} multiplied by 40 is called *UV index*. The UV index was defined as the daily maximum value averaged over a certain time period (30 minutes) as an integer value between 0 and 11+ to serve as a public awareness tool highlighting the risk of sun exposure and the need of sun protection to humans (World Health Organization and International Commission on Non-Ionizing Radiation Protection 2002, Allinson et al. 2012). It has also been used as an instantaneous value extending beyond 11. Integration of biologically effective irradiance E_B over a time period t1 to t2 (hour, day, month, year) provides *radiation (actinic) exposure (dose)* in J m^{-2} (1 J = 1 Ws) as an energy per unit area

$$H_B = \int_{t_1}^{t_2} E_B \; dt \tag{2}$$

If E_B is erythemal irradiance, H_B is *erythemal exposure*. Both E_B and H_B are actinic quantities, though expressed in radiometric units. The specific action spectrum must always be stated to avoid confusion (BIPM 2006). Erythemal exposure H_B is recommended to be given in *Standard Erythemal Dose (SED)* (1 SED = 100 J m^{-2}) (Bouillon et al. 2006) instead of the unit Minimum Erythemal Dose (MED), which depends on the phototype of human skin classified by Thomas B. Fitzpatrick (Fitzpatrick 1986), where for example 1 SED = 0.4 MED_{II} refers to skin type II.

Actinic irradiance in radiometric units can be converted to quantum units. According to the Planck-Einstein relation, the energy of a single photon E_{Photon} increases with decreasing wavelength λ

$$E_{Photon} = \frac{hc}{\lambda} \tag{3}$$

where h = 6.6260755 10^{-34} Js is the Planck constant and c = 2.99792458 · 10^8 m s^{-1} is the speed of light in vacuum (or in the respective medium). The spectral photon rate (number of photons per square meter, seconds and nm, i.e., m^{-2} s^{-1} nm^{-1}) P(λ) is converted from spectral irradiance E(λ) (W m^{-2} nm^{-1}) through

$$P(\lambda) = \frac{E(\lambda)}{E_{Photon} \cdot \eta(\lambda)} \quad (4)$$

where η(λ) is the spectral efficiency of the detector. If the solar spectrum is plotted as a spectral photon rate in dependence of wavelength in analogy to Fig. 2.1 showing E(λ) in radiometric units, the shape of the P(λ) curve differs. The maximum of P(λ) occurs in the red region at longer wavelengths around 635 nm, not in the green region as E(λ), and the decline of P(λ) towards longer wavelengths is more gradual than that of E (λ). Dividing P(λ) by the Avogadro constant $N_A = 6.022140858 \times 10^{17}$ μmol^{-1}, P(λ) has the unit μmol s^{-1} m^{-2} nm^{-1} (Mohr et al. 2015).

Ground-based and Satellite-based UV Radiation Measurements

Networks to measure solar UV irradiance have been closely related to observations of atmospheric ozone (O_3) that started in 1926/1927 with spectrophotometers developed by the British scientist Gordon Miller Bourne Dobson (1889–1976) in Oxford (Brönnimann et al. 2003). Atmospheric column ozone (total ozone) is the amount of ozone overhead at standard conditions of temperature and pressure, and is given in Dobson units (DU), a non SI-unit (1 DU = 1 matm-cm). During and after the International Geophysical Year 1957/1958, ozone research was intensified, ozone measurements at more sites including the Antarctic were opened and World Data Centers inaugurated, such as the World Ozone Data Centre (WODC) in Toronto, Canada in 1960, upgraded in 1992 to the World Ozone and Ultraviolet Radiation Data Centre (WOUDC) (Bojkov 2010).

After the detection of possible effects by nitrogen oxides (NOx) from aircraft (Crutzen 1970) and chlorofluorocarbons (CFC) (Molina and Rowland 1974) to atmospheric ozone, ozone research extended to an environmental issue of global dimension, boosted by the detection of the Antarctic ozone hole from balloon-borne and ground-based ozone measurements (Chubachi 1985, Farman et al. 1985). International organizations, such as the United Nations Environment Program (UNEP) and the World Meteorological Organization (WMO), supported research and monitoring activities (Staehelin 2008). The Vienna Convention for the Protection of the Ozone Layer was adopted in 1985 and entered into force on September 22, 1988, to promote international cooperation by systematic observations, research and information exchange on the atmospheric ozone layer. Legislative and administrative measures to phase out production and release of almost 100 trace gases were taken by the Montreal Protocol on Substances that Deplete the Ozone Layer. First signed by 24 countries in September 1987 and substantially amended in 1990 and 1992, the Montreal Protocol had been signed and ratified by 196 countries by 2009.

An automated instrument, the Brewer spectrometer, which was developed to measure the column ozone, its vertical distribution using the Umkehr effect (Götz 1934) and spectral irradiance in the UV-B and part of the UV-A region, came into operation in the 1980s. Most of the 228 Brewer instruments manufactured are still in use in more than 40 countries around the World (Kerr 2010). International research actions, such as EUBREWNET, were designed to harmonize operations and develop approaches, practices and protocols for consistency in quality control, quality assurance and coordinated operations. Other types of spectroradiometers for spectral UV irradiance and instruments for broad-band UV irradiance were developed and refined for use in networks and at individual sites (De Mazière et al. 2017).

Satellite measurements on polar orbiting satellites about 400 to 800 km above the Earth's surface filled data gaps in between ground-based sites and extended observations to the World's oceans and remote continental areas. The Backscatter UV (BUV) mission on the US National Aeronautics and Space Administration (NASA) satellite Nimbus 4 was started in 1970 (McPeters et al. 2013) by measuring solar spectral UV radiance backscattered from the sunlit part of the earth's surface and atmosphere. Polar night regions were covered by instruments measuring the infrared emission of the earth/atmosphere system in the 9.6 μm ozone band, such as the TIROS Operational Vertical Sounder (TOVS) (Engelen and Stephens 1997). The Total Ozone Mapping Spectrometer (TOMS) on board the NASA satellite Nimbus 7 launched in 1978, also flown from 1991 to 1994 on Russian Meteor satellites, and TOMS Earth Probe in 1996 were next-generation BUV instruments (Arola et al. 2005). The Global Ozone Experiment (GOME) on the ERS-2 satellite in 1990 and a follow-on GOME-2 on METOP in 2006 (Kujanpää and Kalakoski 2015) were the first European satellite missions for global ozone and UV radiation. A Dutch-Finnish Ozone Monitoring Instrument (OMI) was launched on board the NASA Earth Observing System (EOS) Aura spacecraft in 2004 to continue the TOMS ozone record with a higher ground-resolution of 13 x 25 km (McPeters et al. 2015). Satellite-based and ground-based measurements have been analyzed to track long-term homogeneities of records and reduce inconsistencies between data sets (Ialongo et al. 2008, Arola et al. 2009, Loyola et al. 2009, Lee et al. 2013, McPeters et al. 2013, Allaart and Eskes 2015, Bernhard et al. 2015, Dragani 2016).

As polar orbiting satellites provide observations per location only once or twice per day, diurnal and short-time variations of cloudiness, ozone and aerosols are not resolved. Instruments on geostationary satellites at a fixed longitude about 36,000-km above the Earth's equator are capable of providing data with higher resolution in time for latitudes ~ < 65°. As part of the NASA Earth Venture Program, the Tropospheric Emissions Monitoring of Pollution (TEMPO) scheduled for launch in 2018 to 2021 will use a UV/VIS instrument on the GEO satellite focusing at 36.5° N and 100° W to cover large parts of North America including its surrounding ocean areas at a ground-resolution of 2.5 x 4 km (Pierce et al. 2017). Hourly data for column ozone, ozone profiles in the Planetary Boundary Layer (0–2 km) and in the free troposphere, concentrations of other tropospheric trace gases, water vapor, aerosol properties, cloud fraction and cloud top pressure will be derived to calculate downward spectral UV irradiance (Zoogman et al. 2017). TEMPO will be the North American component of a global geostationary constellation for pollution monitoring together with the Korean Geostationary Environment Monitoring Spectrometer (GEMS) on the GEO-KOMPSAT 2B to be positioned at 102.2° E, and the European Space Agency (ESA) Sentinel-4 on the METEOSAT Third Generation Sounder (MTG-S) at 0° longitude (Suleiman et al. 2017). MTG-S will carry a Fourier transform spectrometer measuring the infrared spectral emission from surface and atmosphere, and a UV, Visible and Near-Infrared Sounding (UVN) spectrometer for backscattered solar radiation (Stark et al. 2013). Concurrent low Earth orbit missions will remain indispensable for observations in other regions of the world, particularly the huge oceanic areas outside the fields of views of geostationary satellites.

Global Distribution of Surface UV Irradiance

Figure 2.2 shows an example of the long-term (2004–12) January average of daily solar erythemal exposure H_B integrated by Eq. (2) from modelled hourly cloudless erythemal UV irradiance E_B and hourly CMFs derived from the Deutscher Wetterdienst (DWD) Global Weather Prediction Model at latitude versus longitude (0.5° × 0.75°) grid points (DWD 2014). The database was used in the Personal ERythemal EXposure (PEREX) model for retrospective estimates of personal occupational erythemal UV exposure of seafarers along shipping routes of the world's oceans and validated by ship- and satellite-based measurements (Feister et al. 2015a).

Monthly averages of daily H_B for latitude belts in the Atlantic and Pacific Ocean regions show highest values up to about 90 SED (cloudless) and 80 SED (cloudy) between 30° N and 30° S. A modest average seasonal variation of H_B by 25 to 35% with a double maximum around the equinoxes (March and September) and minimum values around the solstices (June and December) occurs in the equatorial ocean region between 10° N and 10° S (Feister et al. 2015a). In the outer tropical ocean belts from 20° N to 30° N and 20° S to 30°S, H_B shows only one maximum in summer and one minimum in winter. In the Northern hemisphere outer tropics (20° N to 30° N), daily average H_B in summer (June, July) of about 65 SED (cloudless) to 55 SED (cloudy) is about double the minimum values in winter (December, January). In the Southern hemisphere outer tropics (20° S to 30° S), H_B is about 80 (cloudless) to 65 SED (cloudy) in summer (December to January) and 25 (cloudless) to 20 SED (cloudy) in winter

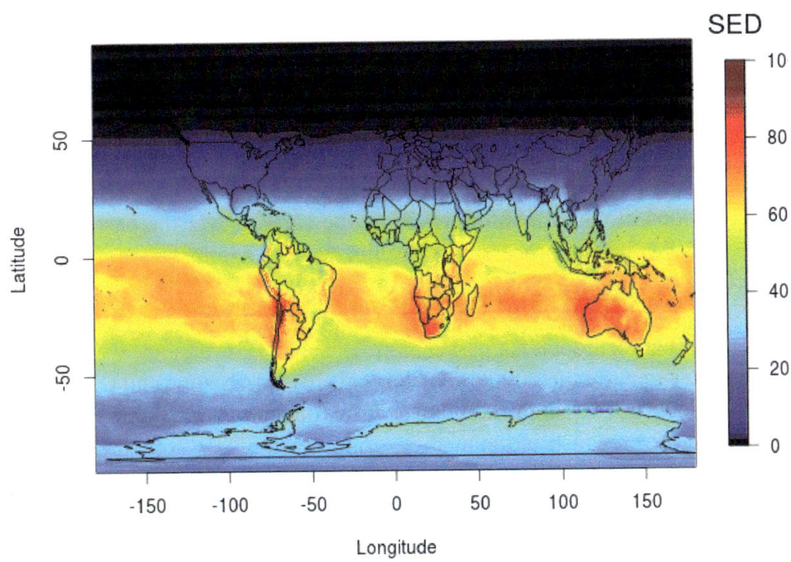

Fig. 2.2. Long-term (2004–2012) January average of daily solar erythemal exposure H_B for cloudy skies in SED (1 SED = 100 Jm^{-2}) (DWD 2014).

(December, January), i.e., H_B in Southern hemisphere summer is three (cloudy) to four times (cloudless) of winter time (June, July) values, and generally higher than in the Northern hemisphere ocean region.

The UV index in the Atlantic and Pacific Ocean can reach high values and show fast short-time variations, as illustrated by ship-based measurements taken on the research vessel METEOR in the eastern tropical Pacific (3.6° S, 84° W) about 420 km off the coast of Peru (Fig. 2.3). Low column ozone and cloud enhancement result in maximum UV index values up to 22, which with a cloudless-sky UV index of 18.4 derived by model calculations using the LibRadtran model (Mayer and Kylling 2005) corresponds to a CMF of 1.19. UV index values on that day are among the highest for the analyzed two-year period of measurements.

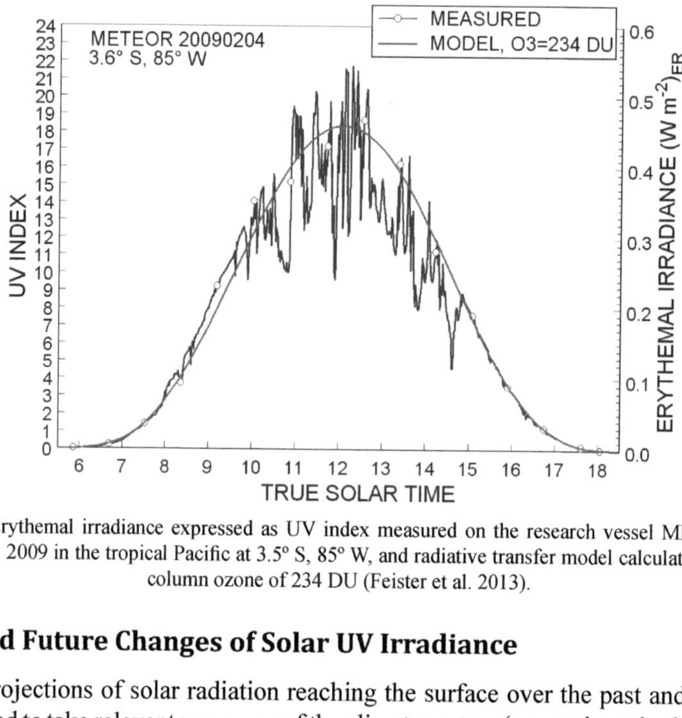

Fig. 2.3. Erythemal irradiance expressed as UV index measured on the research vessel METEOR on February 4, 2009 in the tropical Pacific at 3.5° S, 85° W, and radiative transfer model calculations with a column ozone of 234 DU (Feister et al. 2013).

Past and Future Changes of Solar UV Irradiance

Model projections of solar radiation reaching the surface over the past and into the future need to take relevant processes of the climate system (atmosphere, hydrosphere, cryosphere, biosphere and geosphere), their interactions and changes into account (United Nations Environment Programme Environmental Effects Assessment Panel 2017). Reconstruction models for past changes of solar UV irradiance were developed and compared (Koepke et al. 2006, Litynska et al. 2012) for ozone and UV irradiance, and applied to data of individual sites and geographic regions. Den Outer et al. (2010) compared results of reconstruction methods for erythemal UV exposure at European sites with quality controlled input data in the period 1980 to 2006. They show that erythemal UV radiation exposure increased by 0.2 to 0.6% per year, where about two-thirds of the increase were ascribed to decreasing cloudiness and one-third to decreasing column ozone. Zerefos et al. (2012) studied solar irradiances for two wavelengths (305 and 325 nm) at 12 selected sites in Europe, Canada and Japan between 25° N and 60° N in the period 1990 to 2011. An extended study included sites at high latitudes (Eleftheratos et al. 2015). Increasing anthropogenic emissions, such as sulfur compounds, 4/5 of which are emitted in the Northern hemisphere, had caused a long-term 'global dimming' by the early 1980s. Reduced emissions resulted in a 'global brightening' afterwards (Wild et al. 2017) that was interrupted from about 1991 to 1994 by the volcanic eruption of Mt. Pinatubo (15.14° N, 120.35° E) in June 1991, when about 20 Mt. of sulfur dioxide (SO_2) was emitted high into the atmosphere, and worldwide temperatures subsequently dropped by about 0.5°C (Randel et al. 1995). Solar UV-B irradiance increased in the period 1995 to 2006 due to declining ozone and an aerosol brightening effect by 0.94% at 305 nm and 0.88% at 325 nm per year and showed a leveling afterwards (Zerefos et al. 2012).

Of particular interest have been the potential consequences for human health, agriculture, construction materials and ecosystems, if the Montreal Protocol had not been signed (United Nations Environment Programme Environmental Effects Assessment Panel 2017). Using a fully-coupled radiation-chemical-dynamical model with prescribed tropospheric chemistry, Newman et al. (2009) compared ozone and UV irradiance assuming: (i) no reductions in ozone-depleting gases after 1980 (World-avoided scenario) and (ii) reductions according to the Montreal Protocol. The 'World-avoided scenario' would have resulted in a global ozone decline by about 3% per year, i.e., –17% ozone in the year 2020 and –67% in 2065, corresponding to a column ozone of 100 DU worldwide, and 50 to 80 DU at high latitudes by 2065. The UV index would have increased at Northern hemisphere mid-latitudes (30° to 50° N) from 10 to about 30, and reached still higher values in the tropics and at mountainous regions. For the ship-based measurements in Fig. 2.3, a column ozone of 100 DU would correspond to a hypothetical UV index of 39 (cloudless) and about 47 (with cloud scattering). If tropospheric ozone change and its response to changing UV radiation is included, as was done in another model simulation of 'no Montreal protocol' by Egorova et al. (2013), global ozone would have decreased from 320 DU in 1960 to about 50 DU worldwide and the UV index reached 50 by the year 2100.

Projections of future monthly averaged surface erythemal irradiance E_B at local noon for the time period 1960 to 2100 were derived by radiative transfer calculations applied to results of 14 chemistry climate models (CCM) (Bais et al. 2011) that included chemical processes in the stratosphere, atmospheric circulation, aerosols, surface albedo and cloudiness. E_B first increases by about 3% in the tropics to 25% at high southern latitudes (60° S to 90° S) from 1980 to 2000/2005. E_B decreases by 3–4% at mid-latitudes and 7–12% at high latitudes by the year 2100 due to, both, increasing ozone to levels typical before 1980 and upper-stratospheric cooling by greenhouse gases, which reduces ozone depletion and modifies atmospheric circulation (Table 2.1). In the tropics, E_B increases on average by 1%, in some tropical regions up to 15% (Bais et al. 2011, Meul et al. 2016). Due to melting sea ice and reduced albedo in the Arctic as a result of climate change, UV-B radiation entering the sea is expected to increase by 10 times (Bais et al. 2015).

Table 2.1. Percentage changes in modeled clear-sky and all-sky annually averaged erythemal irradiance from 1975–1984 mean to 2090–2099 mean for different latitude belts from 14 chemistry climate models, after (Bais et al. 2011).

Latitude	Cloudless sky	All-sky
90° N – 60° N	–7.48	–10.72
60° N – 30° N	–4.10	–3.47
30° N – 30° N	0.89	1.10
30° S – 60° S	–4.16	–3.33
60° S – 90° S	–9.80	–12.39

Underwater Light Climate

Terrestrial ecosystems are subject to circadian and annual changes in solar irradiance modulated by the effects of clouds and precipitation. Aquatic ecosystems have to adapt to additional modulations of irradiance, such as tidal rhythms (Gévaert et al. 2003) and changes in transparency. Marine macroalgae and seagrasses growing on the continental shelves experience a modulation in solar light exposure due to the tides independent of the circadian irradiance pattern. Most macroalgae are attached to the rocky bottom in coastal regions and only a few are pelagic and are found floating, such as the species of the genus *Sargassum* in the Sargasso Sea near Bermuda (Carpenter and Cox 1974, Calder 1995).

Attenuation by Dissolved and Particulate Inorganic and Organic Matter

The penetration of solar radiation into the water column is affected by many factors. Depending on the solar zenith angle and the water surface—smooth vs. wavy—more or less light is reflected before entering the water body (Mobley 1999). Radiation entering into the water column is attenuated by dissolved and particulate organic and inorganic absorbers. Particulate inorganic matter (PIM) includes silt and sand and is often found in coastal and freshwater ecosystems (Snyder et al. 2008). Particulate organic material (POM) comprises bacterioplankton, phytoplankton and zooplankton as well as organic debris entering the water by terrestrial runoff (Ittekkot and Laane 1991, Nakatsuka et al. 1992).

The plumes of rivers carrying large amounts of inorganic and organic material can be detected extending dozens or hundreds of kilometres from the mouths into the open sea (Drinkwater 1986) as seen by satellite imaging (Warrick et al. 2004) (Fig. 2.4). Even in large lakes strong gradients in DOM between shore and offshore areas have been found to extend to distances of 20 km or more (Bocaniov et al. 2013). Increasing global temperatures and changing precipitation patterns result in higher terrestrial runoff decreasing transparency in lakes and coastal waters (Larsen et al. 2011, Wilson et al. 2013). In contrast, higher temperatures in polar regions result in increased melting of ice and snow augmenting transparency (Light et al. 2008). This effect is enhanced by a feedback mechanism: polar snow and ice reflect most of the incident solar radiation back into space. After melting, the open sea and land

Fig. 2.4. Landsat 5 satellite image of the plume of the Fraser River west of Vancouver, Canada, carrying sediment from the Columbia Mountains taken 7 September 2011 (reprinted with permission from United States Geological Survey).

surfaces have a lower albedo and absorb more solar radiation augmenting the heating (Bengtsson et al. 2004).

Chromophoric Dissolved Organic Material

Mainly dissolved minerals account for dissolved inorganic matter (DIM) while dissolved organic matter (DOM) is produced by the decay of biological material. In freshwater and coastal habitats DOM is derived mainly from terrestrial vegetation (Mostofa et al. 2013) whereas decaying plankton is the main source of DOM in the open oceans (Steinberg et al. 2004). Alpine lakes above the tree line have considerably lower DOM concentrations than those below and consequently show a higher penetration of solar UV radiation (Rose et al. 2009b). The absorption of solar radiation by DOM is wavelength dependent and shows a higher attenuation at shorter wavelengths especially in the UV wavelength band; for this reason this material is often referred to as CDOM (chromophoric or colored DOM) (Helms et al. 2008). Quantification of CDOM absorption can be used to determine UV transmission as well as to predict water quality, as shown for the shallow Lake Taihu, China, which is an important drinking water reservoir (Zhang et al. 2007).

UV radiation penetrating into the water column has been found to break down CDOM, making it available for bacterial consumption (Feng et al. 2006, Tzortziou et al. 2007, Zhang et al. 2009). This degradation of organic material results in a feedback loop increasing the transmission of solar UV with deleterious consequences for organisms dwelling in the surface layers (Feng et al. 2006). In addition, it is a key process in the carbon cycle and liberates nutrients, such as phosphorous and iron (Shiller et al. 2006, Bastidas Navarro et al. 2009). CDOM has also been found to be a source of reactive oxygen species: UV irradiation of freshwater samples containing CDOM from Antarctica and North America resulted in production of singlet oxygen which contributed to degradation of dissolved free amino acids (Boreen et al. 2008).

Penetration into the Water Column

Water bodies have been classified according to their optical characteristics (Jerlov 1970, Jerlov 1976). Due to the higher concentrations of dissolved and particulate matter attenuation of coastal and freshwater ecosystems is generally higher than that of open oceans (Piazena and Häder 1994). The highly oligotrophic waters of the South Pacific Gyre were found to have extremely low CDOM concentrations; measurements of the diffuse attenuation coefficients indicate that 1% of UV radiation (at 325 nm) incident at the surface reaches a depth of 84 m (Tedetti et al. 2007). Lake Tahoe, California-Nevada, USA is one of the clearest lakes; 1% of UV (320 nm) penetrates to a depth of 27 m (Rose et al. 2009a) and in Crater Lake, Oregon, USA, a penetration of up to 62 m has been determined (Hargreaves et al. 2007).

CDOM concentrations in inland waters have doubled across major areas of northeastern North America and Europe during the past 20 years (Findlay 2005, Evans et al. 2006, Monteith et al. 2007). This has been attributed to the reduction of acid deposition (Evans et al. 2006, Monteith et al. 2007) but climate change may also be responsible (Striegl et al. 2005, Weyhenmeyer and Karlsson 2009). These effects will decrease the spectral irradiance of solar UV radiation in these freshwater ecosystems. Depending on the absorbers in the water, different wavelengths are attenuated differently in freshwater and marine ecosystems (Fig. 2.5).

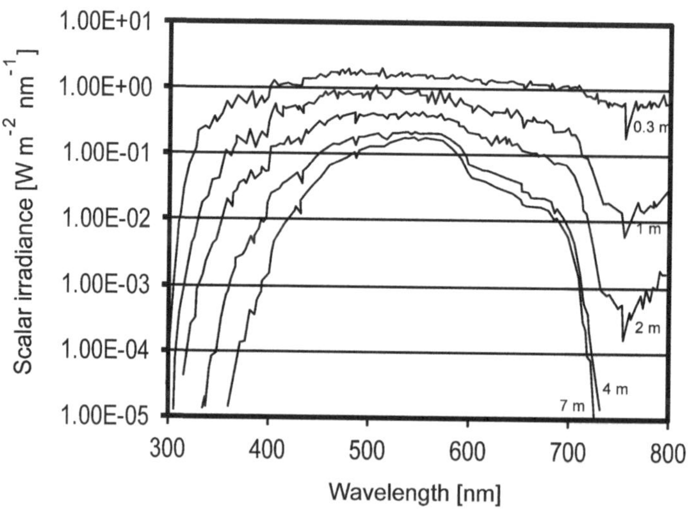

Fig. 2.5. Penetration of solar radiation into marine waters at Gullmarsfjorden, Kattegat, measured with a double monochromator spectroradiometer (Optronic Laboratories, Orlando, FL, USA, model 752) equipped with a 4π sensor attached to a quartz fiber bundle on 27 May 1994. Redrawn from (Piazena and Häder 1997).

Effects of Climate Change on the UV Penetration into Aquatic Ecosystems

The underwater UV radiation level is affected by climate change through a number of different mechanisms. The main aspects are changes in UV transparency and changes in the depth of the upper mixing layer (UML). Essential consequences of climate change are increasing temperature, changing precipitation pattern and enhanced ice melting. Climate models predict reduced precipitation in some regions due to warmer air temperatures while in other regions higher precipitation is expected (IPCC 2007). Higher water temperatures result in stronger stratification of the water column with shoaling of the UML (Jankowski et al. 2006). This phenomenon will expose organisms, such as phytoplankton dwelling in this layer to higher solar visible and UV radiation (Gao et al. 2012). However, this general statement has been questioned by a recent publication showing that stratification is not unequivocally increasing nor is MLD shoaling (Somavilla et al. 2016). It was found that while sea surface temperature increased at three study areas at mid-latitudes, stratification, both, increased and decreased.

References

Allaart, M. and H. Eskes. 2015. Extended and refined multi sensor reanalysis of total ozone for the period 1970–2012. Atmospheric Measurement Techniques 8: 3021–3035.

Allinson, S., M. Asmuss, C. Baldermann, J. Bentzen, D. Buller, N. Gerber, A.C. Green, R. Greinert, M. Kimlin and J. Kunrath. 2012. Validity and use of the UV index: Report from the UVI working group, Schloss Hohenkammer, Germany, 5–7 December 2011. Health Physics 103: 301–306.

Arola, A., S. Kazadzis, N. Krotkov, A. Bais, J. Gröbner and J.R. Herman. 2005. Assessment of TOMS UV bias due to absorbing aerosols. Journal of Geophysical Research: Atmospheres 110.

Arola, A., S. Kazadzis, A. Lindfors, N. Krotkov, J. Kujanpää, J. Tamminen, A. Bais, A. di Sarra, J.M. Villaplana, C. Brogniez, A.M. Siani, M. Janouch, P. Weihs, A. Webb, T. Koskela, N. Kouremeti, D. Meloni, V. Buchard, F. Auriol, I. Ialongo, M. Staneck, S. Simic, A. Smedley and S. Kinne. 2009. A new approach to correct for absorbing aerosols in OMI UV Geophysical Research Letters 36: L22805.

Bais, A., R. McKenzie, G. Bernhard, P. Aucamp, M. Ilyas, S. Madronich and K. Tourpali. 2015. Ozone depletion and climate change: Impacts on UV radiation. Photochemical & Photobiological Sciences 14: 19–52.

Bais, A.F., K. Tourpali, A. Kazantzidis, H. Akiyoshi, S. Bekki, P. Braesicke, M.P. Chipperfield, M. Damerism, V. Eyring, H. Garny, D. Iachetti, P. Jöckel, A. Kubin, U. Langematz, E. Mancini, M. Michou, O. Morgenstern, T. Nakamura, P.A. Newman, G. Pitari, D.A. Plummer, E. Rozanov, T.G. Shepherd, K. Shibata, W. Tian and Y. Yamashita. 2011. Projections of UV radiation changes in the 21st century: impact of ozone recovery and cloud effects. Atmos. Chem. Phys. 11: 7533–7545.

Ball, W.T., N.A. Krivova, Y.C. Unruh, J.D. Haigh and S.K. Solanki. 2014. A new SATIRE-S spectral solar irradiance reconstruction for solar cycles 21–23 and its implications for stratospheric ozone. Journal of the Atmospheric Sciences 71: 4086–4101.

Bastidas Navarro, M., E. Balseiro and B. Modenutti. 2009. Effect of UVR on lake water and macrophyte leachates in shallow Andean-Patagonian lakes: Bacterial response to changes in optical features. Photochemistry and Photobiology 85: 332–340.

Bengtsson, L., V.A. Semenov and O.M. Johannessen. 2004. The early twentieth-century warming in the Arctic—a possible mechanism. Journal of Climate 17: 4045–4057.

Bernhard, G., A. Arola, A. Dahlback, V. Fioletov, A. Heikkilä, B. Johnsen, T. Koskela, K. Lakkala, T. Svendby and J. Tamminen. 2015. Comparison of OMI UV observations with ground-based measurements at high northern latitudes. Atmospheric Chemistry and Physics 15: 7391–7412.

BIPM. 2006. The International System of Units (SI). 8. Appendix 3. Units for photochemical and photobiological quantities, from http://www.bipm.org/utils/common/pdf/si_brochure_8_en.pdf.

Blumthaler, M., J. Gröbner, M. Huber and W. Ambach. 1996. Measuring spectral and spatial variations of UVA and UVB sky radiance. Geophysical Researc Letters 23: 547–550.

Bocaniov, S.A., D.R. Barton, S.L. Schiff and R.E.H. Smith. 2013. Impact of tributary DOM and nutrient inputs on the nearshore ecology of a large, oligotrophic lake (Georgian Bay, Lake Huron, Canada). Aquatic Sciences 75: 321–332.

Bojkov, R.D. 2010. The International Ozone Commission (IO3C): Its history and activities related to atmospheric ozone. Academy of Athens Research Centre for Atmospheric Physics and Climatology Publication.

Boreen, A.L., B.L. Edhlund, J.B. Cotner and K. McNeill. 2008. Indirect photodegradation of dissolved free amino acids: The contribution of singlet oxygen and the differential reactivity of DOM from various sources. Environmental Science & Technology 42: 5492–5498.

Bouillon, R., J. Eisman, M. Garabedian, M.F. Holick, J. Kleinschmidt, T. Suda, I. Terenetskaya and A. Webb. 2006. Action spectrum for the production of previtamin D3 in human skin. CIE, Vienna 174: 2006.

Brönnimann, S., J. Staehelin, S. Farmer, J. Cain, T. Svendby and T. Svenøe. 2003. Total ozone observations prior to the IGY. I: A history. Quarterly Journal of the Royal Meteorological Society 129: 2797–2817.

Calder, D.R. 1995. Hydroid assemblages on holopelagic *Sargassum* from the Sargasso Sea at Bermuda. Bulletin of Marine Science 56: 537–546.

Carpenter, E.J. and J.L. Cox. 1974. Production of pelagic *Sargassum* and a blue-green epiphyte in the western Sargasso Sea. Limnology and Oceanography 19: 429–436.

Chadyšiene, R. and A. Girgždys. 2008. Ultraviolet radiation albedo of natural surfaces. Journal of environmental engineering and landscape management 16: 83–88.

Chubachi, S. 1985. A special ozone observation at Syowa Station, Antarctica from February 1982 to January 1983. pp. 285–289. *In*: Zerefos, C.S. and Ghazi, A. (eds.). Atmospheric Ozone. Springer, Dordrecht.

CIE. 2006. Action spectrum for the production of previtamin D3 in human skin. CIE Technical Report, 174: 1–12.

Coakley Jr, J. 1914. Reflectance and albedo, surface. Encyclopedia of the Atmosphere. Holton, J.R. and Curry, J.A. (eds.). Academic Press.

Cook, J.M., A.J. Hodson, A.S. Gardner, M. Flanner, A.J. Tedstone, C. Williamson, T.D. Irvine-Fynn, J. Nilsson, R. Bryant and M. Tranter. 2017. Quantifying bioalbedo: A new physically-based model and critique of empirical methods for characterizing biological influence on ice and snow albedo. The Cryosphere Discussions: 1–29.

Cordero, R., A. Damiani, G. Seckmeyer, J. Jorquera, M. Caballero, P. Rowe, J. Ferrer, R. Mubarak, J. Carrasco and R. Rondanelli. 2016. The solar spectrum in the Atacama desert. Scientific Reports 6: srep22457.

Crutzen, P.J. 1970. The influence of nitrogen oxides on the atmospheric ozone content. Quarterly Journal of the Royal Meteorological Society 96: 320–325.

Dahlback, A., N. Gelsor, J. Stamnes and Y. Gjessing. 2007. UV measurements in the 3000–5000 m altitude region in Tibet. Journal of Geophysical Research - Atmospheres 112: doi:1029/2006jd007700.

De Mazière, M., A.M. Thompson, M.J. Kurylo, J.D. Wild, G. Bernhard, T. Blumenstock, G.O. Braathen, J.W. Hannigan, J.-C. Lambert, T. Leblanc, T.J. McGee, G. Nedoluha, I. Petropavlovskikh, G. Seckmeyer, P.C. Simon, W. Steinbrecht, and S.E. Strahan. 2018. The network for the detection of atmospheric composition change (NDACC): History, status and perspectives. Atmos. Chem. Phys. Discuss. 18: 4935–4964.

den Outer, P.N., H. Slaper, J. Kaurola, A. Lindfors, A. Kazantzidis, A.F. Bais, U. Feister, J. Junk, M. Janouch and W. Josefsson. 2010. Reconstructing of erythemal ultraviolet radiation levels in Europe for the past 4 decades. Journal of Geophysical Research-Atmospheres 115.

Dragani, R. 2016. A comparative analysis of UV nadir-backscatter and infrared limb-emission ozone data assimilation. Atmospheric Chemistry and Physics 16: 8539–8557.

Drinkwater, K. 1986. On the role of freshwater outflow on coastal marine ecosystems—a workshop summary. pp. 429–438. *In*: Skreslet, S. (ed.). The Role of Freshwater Outflow in Coastal Marine Ecosystems. Springer, Berlin, Heidelberg.

Dudok de Wit, T., G. Kopp, C. Fröhlich and M. Schöll. 2017. Methodology to create a new total solar irradiance record: Making a composite out of multiple data records. Geophysical Research Letters 44: 1196–1203.

DWD. 2014. Deutscher Wetterdienst. Retrieved from https://kunden.dwd.de/uvi/.

Egorova, T., E. Rozanov, J. Gröbner, M. Hauser and W. Schmutz. 2013. Montreal Protocol benefits simulated with CCM SOCOL. Atmospheric Chemistry and Physics 13: 3811–3823.

Eleftheratos, K., S. Kazadzis, C. Zerefos, K. Tourpali, C. Meleti, D. Balis, I. Zyrichidou, K. Lakkala, U. Feister and T. Koskela. 2015. Ozone and spectroradiometric UV changes in the past 20 years over high latitudes. Atmosphere-Ocean 53: 117–125.

Engelen, R.J. and G.L. Stephens. 1997. Infrared radiative transfer in the 9.6-μm band: Application to TIROS operational vertical sounder ozone retrieval. Journal of Geophysical Research: Atmospheres 102: 6929–6939.

Evans, C.D., P.J. Chapman, J.M. Clark, D.T. Monteith and M.S. Cresser. 2006. Alternative explanations for rising dissolved organic carbon export from organic soils. Global Change Biology 12: 2044–2053

Farman, J.C., B.G. Gardiner and J.D. Shanklin. 1985. Large losses of total ozone in Antarctica reveal seasonal ClO_x/NO_x interaction. Nature 315: 207–210.

Feister, U., N. Cabrol and D.-P. Häder. 2015a. UV irradiance enhancements by scattering of solar radiation from clouds. Atmosphere 5: 1211–1228.

Feister, U., G. Laschewski and R.-D. Grewe. 2011. UV index forecasts and measurements of health-effective radiation. Journal of Photochemistry and Photobiology B: Biology 102: 55–68.

Feister, U., G. Meyer and U. Kirst. 2013. Solar UV Exposure of seafarers along subtropical and tropical shipping routes. Photochemistry and Photobiology 89: 1497–1506.

Feister, U., G. Meyer, G. Laschewski and C. Boettcher. 2015b. Validation of modeled daily erythemal exposure along tropical and subtropical shipping routes by ship-based and satellite-based measurements. Journal of Geophysical Research: Atmospheres 120: 4117–4131.

Feister, U. and J. Shields. 2005. Cloud and radiance measurements with the VIS/NIR daylight whole sky imager at Lindenberg (Germany). Meteorologische Zeitschrift 14: 627–639.

Feng, S., Y.L. Zhang and B.Q. Qin. 2006. Photochemical degradation of chromophoric dissolved organic matter in Meiliang Bay of Lake Taihu. China Environmental Science 26: 404–408.

Findlay, S.E.G. 2005. Increased carbon transport in the Hudson River: unexpected consequence of nitrogen deposition? Frontiers in Ecology and the Environment 3: 133–137.

Fioletov, V.E., M.G. Kimlin, N. Krotkov, L.B. McArthur, J.B. Kerr, D.I. Wardle, J.R. Herman, R. Meltzer, T.W. Mathews and J. Kaurola. 2004. UV index climatology over the United States and Canada from ground-based and satellite estimates. Journal of Geophysical Research: Atmospheres 109.

Fitzpatrick, T.B. 1986. Ultraviolet-induced pigmentary changes: Benefits and hazards. pp. 25–38. *In*: Hönigsmann, H. and G. Stingl (eds.). Therapeutic Photomedicine. Karger Publishers.

Fountoulakis, I., A. Bais, K. Tourpali, K. Fragkos and S. Misios. 2014. Projected changes in solar UV radiation in the Arctic and subArctic Oceans: Effects from changes in reflectivity, ice transmittance, clouds, and ozone. Journal of Geophysical Research: Atmospheres 119: 8073–8090.

Gao, K., J. Xu, G. Gao, Y. Li, D.A. Hutchins, B. Huang, Y. Zheng, P. Jin, X. Cai, D.-P. Häder, W. Li, K. Xu, N. Liu and U. Riebesell. 2012. Rising carbon dioxide and increasing light exposure act synergistically to reduce marine primary productivity. Nature Climate Change 2: 519–523.

Gévaert, F., A. Créach, D. Davoult, A. Migné, G. Levavasseur, P. Arzel, A.-C. Holl and Y. Lemoine. 2003. *Laminaria saccharina* photosynthesis measured in situ: photoinhibition and xanthophyll cycle during a tidal cycle. Marine Ecology Progress Series 247: 43–50.

Götz, F. 1934. Über die Deutung des Umkehreffekts bei Messungen des atmosphärischen Ozons. Zeitschrift fur Astrophysik 8: 267.

Haltrin, V., W. McBride III and R. Arnone. 2001. Spectral approach to calculate specular reflection of light from wavy water surface. Proceedings of D. S. Rozhdestvensky Optical Society: International Conference Current Problems in Optics of Natural Waters (ONW'2001), St. Petersburg, Russia, 2001.

Hargreaves, B.R., S.F. Girdner, M.W. Buktenica, R.W. Collier, E. Urbach and G.L. Larson. 2007. Ultraviolet radiation and bio-optics in Crater Lake, Oregon. Hydrobiologia 574: 107–140.

Helms, J.R., A. Stubbins, J.D. Ritchie, E.C. Minor, D.J. Kieber and K. Mopper. 2008. Absorption spectral slopes and slope ratios as indicators of molecular weight, source, and photobleaching of chromophoric dissolved organic matter. Limnology and Oceanography 53: 955–969.

Ialongo, I., G. Casale and A. Siani. 2008. Comparison of total ozone and erythemal UV data from OMI with ground-based measurements at Rome station. Atmospheric Chemistry and Physics 8: 3283–3289.

IPCC. 2007. Summary for Policymakers. pp. 1–18. *In*: Solomon, S., D. Qin, M. Manning, Z. Chen, M. Marquis, K.B. Averyt, M. Tignor and H.L. Miller (eds.). Climate Change 2007: The Physical Science Basis. Contribution of Working Group 1 to the Fourth Assessment report of the Intergovernmental Panel on Climate Change. IPCC, Cambridge University Press, Cambridge, United Kingdom and New York, NY, USA.

ISO. 2005. Definitions of Solar Irradiance Spectral Categories.

Ittekkot, V. and R. Laane. 1991. Fate of riverine particulate organic matter. Biogeochemistry of Major World Rivers 42: 233–242. http://SpaceWx.com. http://www.acttr.com/images/pdf/ISO_DIS_21348.pdf.

Jankowski, T., D.M. Livingstone, H. Buhrer, R. Forster and P. Niederhauser. 2006. Consequences of the 2003 European heat wave for lake temperature profiles, thermal stability, and hypolimnetic oxygen depletion: Implications for a warmer world. Limnology and Oceanography 51: 815–819.

Jerlov, N.G. 1970. General aspects of underwater daylight and definitions of fundamental concepts. pp. 95–102. *In*: Kinne, O. (ed.). Marine Ecology.

Jerlov, N.G. 1976. Marine Optics. Elsevier. Amsterdam.

Jin, Z., T.P. Charlock, W.L. Smith and K. Rutledge. 2004. A parameterization of ocean surface albedo. Geophysical Research Letters 31: L22301.

Kerr, J.B. 2010. The Brewer Spectrophotometer. *In*: Gao, W., Slusser, J.R. and Schmoldt, D.L. (eds.). UV Radiation in Global Climate Change. Springer, Berlin, Heidelberg.

Kleipool, Q., M. Dobber, J. de Haan and P. Levelt. 2008. Earth surface reflectance climatology from 3 years of OMI data. Journal of Geophysical Research: Atmospheres 113: D18308.

Koepke, P., H. De Backer, A. Bais, A. Curylo, K. Eerme, U. Feister, B. Johnsen, J. Junk, A. Kazantzidis, J. Krzyscin, A. Lindfors, J.A. Olseth, P. den Outer, A. Pribullova, A.W. Schmalwieser, H. Slaper, H. Staiger, J. Verdebout, L. Vuilleumier and P. Weihs. 2006. Modelling solar UV radiation in the past: comparison of algorithms and input data. Proc. SPIE 6362, Remote Sensing of Clouds and the Atmosphere XI, 636215 (11 October 2006); doi: 10.1117/12.687682.

Kopp, G. 2016. Magnitudes and timescales of total solar irradiance variability. Journal of Space Weather and Space Climate 6: A30.

Kopp, G., G. Lawrence and G. Rottman. 2005. The total irradiance monitor (TIM): Science results. The Solar Radiation and Climate Experiment (SORCE): 129–139.

Kujanpää, J. and N. Kalakoski. 2015. Operational surface UV radiation product from GOME-2 and AVHRR/3 data. Atmospheric Measurement Techniques 8: 4399–4414.

Kylling, A., A. Bais, M. Blumthaler, J. Schreder, C. Zerefos and E. Kosmidis. 1998. Effect of aerosols on solar UV irradiances during the photochemical activity and solar ultraviolet radiation campaign. Journal of Geophysical Research: Atmospheres 103: 26051–26060.

Lampel, J., U. Frieß and U. Platt. 2015. The impact of vibrational Raman scattering of air on DOAS measurements of atmospheric trace gases. Atmospheric Measurement Techniques 8: 3767–3787.

Larsen, S., T. Andersen and D. Hessen. 2011. Climate change predicted to cause severe increase of ogranic carbon in lakes. Global Change Biology 17: 1186–1192.

Lee, J., W.J. Choi, D.R. Kim, S.-Y. Kim, C.-K. Song, J.S. Hong, Y. Hong and S. Lee. 2013. The effect of ozone and aerosols on the surface erythemal UV radiation estimated from OMI measurements. Asia-Pacific Journal of Atmospheric Sciences 49: 271–278.

Lee, Z., J. Wei, K. Voss, M. Lewis, A. Bricaud and Y. Huot. 2015. Hyperspectral absorption coefficient of "pure" seawater in the range of 350–550 nm inverted from remote sensing reflectance. Applied Optics 54: 546–558.

Lenton, T.M. 2012. Arctic Climate Tipping Points. Ambio 41: 10–22.

Light, B., T.C. Grenfell and D.K. Perovich. 2008. Transmission and absorption of solar radiation by Arctic sea ice during the melt season. Journal of Geophysical Research 113: doi:10.1029/2006JC003977.

Litynska, Z., P. Koepke, H. De Backer, J. Gröbner, A. Schmalwieser, L. Vuilleumier, N. Chubarova, U. Feister, J. Kaurola and A. Kazantzidis. 2012. Long term changes and climatology of UV radiation over Europe. COST Action.

Loyola, D., R. Coldewey-Egbers, M. Dameris, H. Garny, A. Stenke, M. Van Roozendael, C. Lerot, D. Balis and M. Koukouli. 2009. Global long-term monitoring of the ozone layer–a prerequisite for predictions. International Journal of Remote Sensing 30: 4295–4318.

Mason, J.D., M.T. Cone and E.S. Fry. 2016. Ultraviolet (250–550 nm) absorption spectrum of pure water. Applied Optics 55: 7163–7172.

Mayer, B. and A. Kylling. 2005. Technical note: The LibRadtran software package for radiative transfer calculations—description and examples of use. Atmos. Chem. Phys. 5: 1855.

Mayer, B., A. Kylling, S. Madronich and G. Seckmeyer. 1998. Enhanced absorption of UV radiation due to multiple scattering in clouds: Experimental evidence and theoretical explanation. Journal of Geophysical Research: Atmospheres 103: 31241–31254.

McKinlay, A.F. and B.L. Diffey. 1987. A reference action spectrum for ultraviolet induced erythema in human skin. pp. 83–87. *In*: Passchier, W.R. and B.F.M. Bosnjakovic (eds.). Human Exposure to Ultraviolet Radiation: Risks and Regulations. Elsevier, Amsterdam.

McPeters, R., S. Frith and G. Labow. 2015. OMI total column ozone: extending the long-term data record. Atmospheric Measurement Techniques 8: 4845–4850.

McPeters, R.D., P. Bhartia, D. Haffner, G.J. Labow and L. Flynn. 2013. The version 8.6 SBUV ozone data record: An overview. Journal of Geophysical Research: Atmospheres 118: 8032–8039.

Meul, S., M. Dameris, U. Langematz, J. Abalichin, A. Kerschbaumer, A. Kubin and S. Oberländer-Hayn. 2016. Impact of rising greenhouse gas concentrations on future tropical ozone and UV exposure. Geophysical Research Letters 43: 2919–2927.

Mobley, C.D. 1999. Estimation of the remote-sensing reflectance from above-surface measurements. Applied Optics 38: 7442–7455.

Mobley, C.D. 2001. Radiative transfer in the ocean. pp. 2321–2330. *In*: Steele, J.H., S.A. Thorpe and K.K. Turekian (eds.). Encyclopedia of Ocean Sciences. Academic Press.

Mohr, P.J., D.B. Newell and B.N. Taylor. 2015. CODATA Recommended Values of the Fundamental Physical Constants: 2014.

Mok, J., N.A. Krotkov, A. Arola, O. Torres, H. Jethva, M. Andrade, G. Labow, T.F. Eck, Z. Li and R.R. Dickerson. 2016. Impacts of brown carbon from biomass burning on surface UV and ozone photochemistry in the Amazon Basin. Scientific Reports 6: 36940.

Molina, M.J. and F.S. Rowland. 1974. Stratospheric sink for chlorofluoromethanes-Chlorine atom catalyzed destruction of ozone. International Conference on the Environmental Impact of Aerospace Operations in the High Atmosphere, 2nd, San Diego, Calif.

Monteith, D.T., J.L. Stoddard, C.D. Evans, H.A. de Wit, M. Forsius, T. Høgåsen, A. Wilander, B.L. Skjelkvåle, D.S. Jeffries, J. Vuorenmaa, B. Keller, J. Kopácek and J. Vesely. 2007. Dissolved organic carbon trends resulting from changes in atmospheric deposition chemistry. Nature 450: 537–541.

Mostofa, K.M., C.-q. Liu, M.A. Mottaleb, G. Wan, H. Ogawa, D. Vione, T. Yoshioka and F. Wu. 2013. Dissolved organic matter in natural waters. pp. 1–137. *In*: Photobiogeochemistry of Organic Matter. Springer.

Nakatsuka, T., N. Handa, E. Wada and C.S. Wong. 1992. The dynamic changes of stable isotopic ratios of carbon and nitrogen in suspended and sedimented particulate organic matter during a phytoplankton bloom. Journal of Marine Research 50: 267–296.

Newman, P.A., L.D. Oman, A.R. Douglass, E.L. Fleming, S.M. Frith, M.M. Hurwitz, S.R. Kawa, C.H. Jackman, N.A. Krotkov, E.R. Nash, J.E. Nielsen, S. Pawson, R.S. Stolarski and G.J.M. Velders. 2009. What would have happened to the ozone layer if chlorofluorocarbons (CFCs) had not been regulated? Atmos. Chem. Phys. 9: 2113–2128.

Orphal, J., J. Staehelin, J. Tamminen, G. Braathen, M.-R. De Backer, A. Bais, D. Balis, A. Barbe, P.K. Bhartia and M. Birk. 2016. Absorption cross-sections of ozone in the ultraviolet and visible spectral regions: Status report 2015. Journal of Molecular Spectroscopy 327: 105–121.

Pagaran, J., M. Weber and J. Burrows. 2009. Solar variability from 240 to 1750 nm in terms of faculae brightening and sunspot darkening from SCIAMACHY. The Astrophysical Journal 700: 1884.

Pfeifer, M., P. Koepke and J. Reuder. 2006. Effects of altitude and aerosol on UV radiation. Journal of Geophysical Research: Atmospheres 111.

Piazena, H. and D.-P. Häder. 1994. Penetration of solar UV irradiation in coastal lagoons of the southern Baltic Sea and its effect on phytoplankton communities. Photochemistry and Photobiology 60: 463–469.

Piazena, H. and D.-P. Häder. 1997. Penetration of solar UV and PAR into different waters of the baltic sea and remote sensing of phytoplankton. pp. 45–96. *In*: Häder, D.-P. (ed.). The Effects of Ozone Depletion on Aquatic Ecosystems. Acad. Press, R.G. Landes Company, Austin.

Pierce, R.B., M.R. Pippin, A. Saiz-Lopez, R.J.D. Spurr, J.J. Szykman, O. Torres, J.P. Veefkind, B. Veihelmann, H. Wang and J. Wang. 2017. Tropospheric emissions: monitoring of pollution (TEMPO). J. Quant. Spectrosc. Radiat. Transf. 186: 17–39.

Pitari, G., G. Di Genova, E. Mancini, D. Visioni, I. Gandolfi and I. Cionni. 2016. Stratospheric aerosols from major volcanic eruptions: a composition-climate model study of the aerosol cloud dispersal and e folding time. Atmosphere 7: 75.

Pope, R.M. and E.S. Fry. 1997. Absorption spectrum (380–700 nm) of pure water. II. Integrating cavity measurements. Applied Optics 36: 8710–8723.

Randel, W.J., F. Wu, J. Russell, J. Waters and L. Froidevaux. 1995. Ozone and temperature changes in the stratosphere following the eruption of Mount Pinatubo. Journal of Geophysical Research: Atmospheres 100: 16753–16764.

Rose, K.C., C.E. Williamson, J.E. Saros, R. Sommaruga and J.M. Fischer. 2009a. Differences in UV transparency and thermal structure between alpine and subalpine lakes: implications for organisms. Photochemical & Photobiological Sciences 8: 1244–1256.

Rose, K.C., C.E. Williamson, S.G. Schladow, M. Winder and J.T. Oris. 2009b. Patterns of spatial and temporal variability of UV transparency in Lake Tahoe, California-Nevada. Journal of Geophysical Research 114: 1–9.

Sandmann, H. and C. Stick. 2014. Spectral and spatial UV sky radiance measurements at a seaside resort under clear sky and slightly overcast conditions. Photochemistry and Photobiology 90: 225–232.

Schwander, H., P. Koepke, A. Kaifel and G. Seckmeyer. 2002. Modification of spectral UV irradiance by clouds. Journal of Geophysical Research: Atmospheres 107: AAC 7-1–AAC 7-12.

Shields, J.E., M.E. Karr, R.W. Johnson and A.R. Burden. 2013. Day/night whole sky imagers for 24-h cloud and sky assessment: history and overview. Applied Optics 52: 1605–1616.

Shiller, A.M., S. Duan, P. van Erp and T.S. Bianchi. 2006. Photo-oxidation of dissolved organic matter in river water and its effect on trace element speciation. Limnol. Oceanogr. 51: 1716–1728.

Sinha, E., A. Michalak and V. Balaji. 2017. Eutrophication will increase during the 21st century as a result of precipitation changes. Science 357: 405–408.

Snyder, W.A., R.A. Arnone, C.O. Davis, W. Goode, R.W. Gould, S. Ladner, G. Lamela, W.J. Rhea, R. Stavn and M. Sydor. 2008. Optical scattering and backscattering by organic and inorganic particulates in US coastal waters. Applied Optics 47: 666–677.

Somavilla, R., C. González-Pola and J. Fernandez. 2016. The warmer the ocean surface, the shallower the mixed layer: How much of this is true? Journal of Geophysical Research: Oceans.

Staehelin, J. 2008. Global atmospheric ozone monitoring. Bulletin of the World Meteorological Organization 57: 45–54.

Staiger, H., P. Den Outer, A. Bais, U. Feister, B. Johnsen and L. Vuilleumier. 2008. Hourly resolved cloud modification factors in the ultraviolet. Atmospheric Chemistry and Physics 8: 2493–2508.

Stark, H.R., H.L. Moller, G.B. Courreges-Lacoste, R. Koopman, S. Mezzasoma and B. Veihelmann. 2013. The Sentinel-4 Mission and its Implementation. ESA Living Planet Symposium.

Steinberg, D.K., N.B. Nelson, A.C. Craig and A. Prusak. 2004. Production of chromophoric dissolved organic matter (CDOM) in the open ocean by zooplankton and the colonial cyanobacterium *Trichodesmium* spp. Marine Ecology Progress Series 267: 45–56.

Striegl, R.G., G.R. Aiken, M.M. Dornblaser, P.A. Raymond and K.P. Wickland. 2005. A decrease in discharge-normalized DOC export by the Yukon River during summer through autumn. Geophysical Research Letters 32: L21413.

Suleiman, R., K. Chance and X. Liu. 2017. A geostationary air quality monitor for the Middle East. Journal of Physics: Conference Series, IOP Publishing.

Tedetti, M., R. Sempere, A. Vasilkov, B. Charriere, D. Nerini, W.L. Miller, K. Kawamura and P. Raimbault. 2007. High penetration of ultraviolet radiation in the south east Pacific waters. Geophysical Research Letters 34: L126101–L126105.

Thapan, K., J. Arendt and D.J. Skene. 2001. An action spectrum for melatonin suppression: Evidence for a novel non-rod, non-cone photoreceptor system in humans. The Journal of Physiology 535: 261–267.

Thosing, K., M. Schrempf, S. Riechelmann and G. Seckmeyer. 2014. Validation of spectral sky radiance derived from all-sky camera images-a case study. Atmospheric Measurement Techniques 7(2014), Nr. 7.

Tzortziou, M., C.L. Osburn and P.J. Neale. 2007. Photobleaching of dissolved organic material from a tidal marsh-estuarine system of the Chesapeake Bay. Photochemistry and Photobiology 83: 782–792.

United Nations Environment Programme Environmental Effects Assessment Panel. 2017. Environmental effects of ozone depletion and its interactions with climate change: Progress report, 2016. Photochemical & Photobiological Sciences 16: 107–145.

Vos, J.J. 1978. Colorimetric and photometric properties of a 2 fundamental observer. Color Research & Application 3: 125–128.

Warrick, J.A., L.A. Mertes, L. Washburn and D.A. Siegel. 2004. Dispersal forcing of southern California river plumes, based on field and remote sensing observations. Geo-Marine Letters 24: 46–52.

Webb, A.R., H. Slaper, P. Koepke and A.W. Schmalwieser. 2011. Know your standard: clarifying the CIE erythema action spectrum. Photochemistry and Photobiology 87: 483–486.

Weyhenmeyer, G.A. and J. Karlsson. 2009. Nonlinear response of dissolved organic carbon concentrations in boreal lakes to increasing temperatures. Limnology and Oceanography 54: 2513–2519.

Wild, M., A. Ohmura, C. Schär, G. Müller, D. Folini, M. Schwarz, M.Z. Hakuba and A. Sanchez-Lorenzo. 2017. The Global Energy Balance Archive (GEBA) version 2017: a database for worldwide measured surface energy fluxes. Earth System Science Data 9: 601.

Wilson, H.F., J.E. Saiers, P.A. Raymond and W.V. Sobczak. 2013. Hydrologic drivers and seasonality of dissolved organic carbon concentration, nitrogen content, bioavailability, and export in a forested New England stream. Ecosystems 16: 604–616.

WMO. 2014. Guide to Meteorological Instruments and Methods of Observations. WMO-No. 8 (2008, 2014 edition). Part I: Measurement of meteorological variables, Chapter 7: Measurement of radiation. World Meteorological Organization, G. O. R. a. M. P.

WMO. 2017. International cloud atlas. Manual on the observation of clouds and other meteors. WMO-No. 407.

World Health Organization and International Commission on Non-Ionizing Radiation Protection. 2002. Global solar UV index: a practical guide.

Ye, H., J. Li, T. Li, Q. Shen, J. Zhu, X. Wang, F. Zhang, J. Zhang and B. Zhang. 2016. Spectral classification of the Yellow Sea and implications for coastal ocean color remote sensing. Remote Sensing 8: 321.

Zagury, F. 2012. The color of the sky. Atmospheric and Climate Sciences 2: 510–517.

Zatko, M.C. and S.G. Warren. 2015. East Antarctic sea ice in spring: Spectral albedo of snow, nilas, frost flowers and slush, and light-absorbing impurities in snow. Annals of Glaciology 56: 53–64.

Zerefos, C.S., K. Tourpali, K. Eleftheratos, S. Kazadzis, C. Meleti, U. Feister, T. Koskela and A. Heikkilä. 2012. Evidence of a possible turning point in solar UV-B over Canada, Europe and Japan. Atmospheric Chemistry and Physics 12: 2469–2477.

Zhang, Y., M. Liu, B. Qin and S. Feng. 2009. Photochemical degradation of chromophoric-dissolved organic matter exposed to simulated UV-B and natural solar radiation. Hydrobiologia 627: 159–168.

Zhang, Y., B. Qin, G. Zhu, L. Zhang and L. Yang. 2007. Chromophoric dissolved organic matter (CDOM) absorption characteristics in relation to fluorescence in Lake Taihu, China, a large shallow subtropical lake. Hydrobiologia 581: 43–52.

Zoogman, P., X. Liu, R. Suleiman, W. Pennington, D. Flittner, J. Al-Saadi, B. Hilton, D. Nicks, M. Newchurch and J. Carr. 2017. Tropospheric emissions: Monitoring of pollution (TEMPO). Journal of Quantitative Spectroscopy and Radiative Transfer 186: 17–39.

CHAPTER 3

Ocean Climate Changes

Donat-P. Häder[1,*] and *Kunshan Gao*[2]

Introduction

The weather on Earth is nowhere constant. For example, on January 13, 1987, it was –22.2°C in Munich, Southern Germany, and on the same day in 2015 it was + 14.8°C (Deutscher Wetterdienst). In summer it could be sizzling hot with a harvest-threatening drought or it could be very wet with inundations. Snow on the Mediterranean Cote Azure is a rare event, but can happen occasionally as well as tornadoes over central Europe. These exceptional occurrences are characteristic for the day-to-day weather and are no indication for a climate change. In contrast, the average temperature over the Earth integrated over a whole year is a better measure for knowing the climate trend. Even these mean temperatures show a wide range of fluctuations, but when recorded over an extended period of time, such as several decades or even longer, long-term trends can be detected (Fig. 3.1). The strongest increase of about 0.9°C has taken place since 1960 and 16 of the hottest years have been recorded since 2000 (Ripple et al. 2017). The year 2016 was the warmest year since the beginning of precise measurements in 1880 and it was 1.1°C higher than before the industrial revolution and probably the warmest year since the end of the last interglacial period about 115,000 years ago (Dahl-Jensen et al. 2013). The air temperature increases more over land than over the oceans (Morice et al. 2012), though the oceans have absorbed over 90% of the Earth's heat increase (Reid 2016). Between 1970 and 2014 the temperature of the land mass increased by 0.26 K while over the sea it increased by 0.12 K per decade (Met Office). This process is much faster than any other warming during the last 66 million years. If the predictions become true that the global mean temperature will increase by 4–5°C, within the next 100 years this rise will be about 100 times faster than that since the last 10,000 years. Most of the continental mass is located in the Northern Hemisphere and for this reason the temperatures rose stronger than in the Southern

[1] Friedrich-Alexander University, Erlangen-Nürnberg, Neue Str. 9, 91096 Möhrendorf, Germany.
[2] State Key Laboratory of Marine Environmental Science, Xiamen University, Daxue Rd 182, Xiamen, Fujian, 361005, China.
Email: ksgao@xmu.edu.cn
* Corresponding author: donat@dphaeder.de

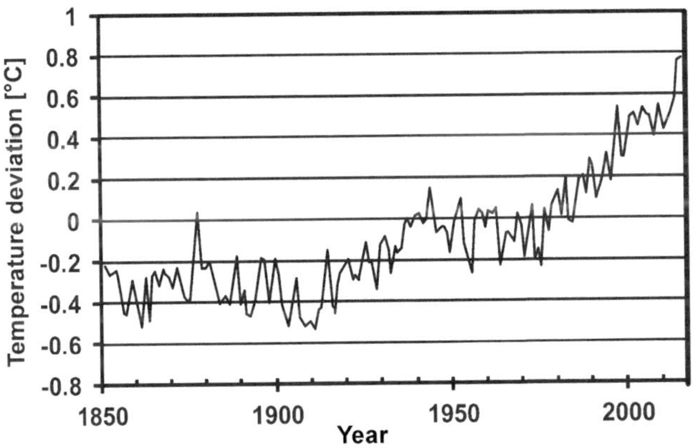

Fig. 3.1. Mean global temperature since 1850 (compared to the global mean value of the years 1961–90) published by the Met Office Hadley Centre and Climatic Research Unit. Similar values have been recorded by the NOAA National Centers for Environmental Information and the NASA Goddard Institute for Space Studies. Redrawn from a press release "Climate breaks multiple records in 2016, with global impacts, published 21 March 2017 by the World Meteorological Organisation.

Hemisphere during the last 100 years. The night and winter temperatures increased stronger than the day and summer temperatures (Vose et al. 2005, Alexander et al. 2006), resulting in the strongest warming during the winter especially over the western part of North America, Scandinavia and Siberia (Hansen et al. 2005). The most dramatic warming occurred over the Arctic where the increase of the mean temperatures was about double that of the global average (IPCC—World Meteorological Organiszation 2007, Milner 2007). In fact the temperature increase and the consequent ice loss has broken one record after the other: in 2017 the winter ice extent was 14.8%, the sea ice loss was 42.5%, and in 2016 the winter air temperature was 8.7°C and the winter water vapour was 41% higher as compared to 1979 (Francis 2018).

This trend is independent of the 11-year cycle of solar activity determined since 1850. It is also independent of the trend in the activity of galactic cosmic rays recorded since the second half of the twentieth century. Rather the global temperatures are linked to the greenhouse gases in the atmosphere including carbon dioxide, methane and dinitrogen monoxide. Most scientists agree that the currently observed climate change is due to the anthropogenic emission of carbon dioxide from fossil fuel burning, tropical deforestation and increased livestock production (Cook et al. 2013).

Currently, the atmosphere holds about 735 Gt carbon (Fig. 3.2) (1 gigatonne = 1 Pg = 10^{15} g). In contrast, the amount bound in fossil fuels is estimated at 7,500 Gt. The terrestrial biomass represents about 600 Gt carbon. Every year it takes up about 52 Gt and releases it again when the vegetation decays so that the net sum is constant. In contrast, humans release about 7 Gt from industry, households and traffic. In addition, 2 Gt carbon are emitted by tropical deforestation and altered land usage. The aquatic ecosystems take up about 50 Gt and release the same amount at the end of the vegetative period. But the biological pump removes about 4 Gt carbon annually by sinking organic and inorganic carbon to the sediments, which contain the largest reservoir on earth, 36,000 Gt. The CO_2 concentration in the atmosphere has risen from

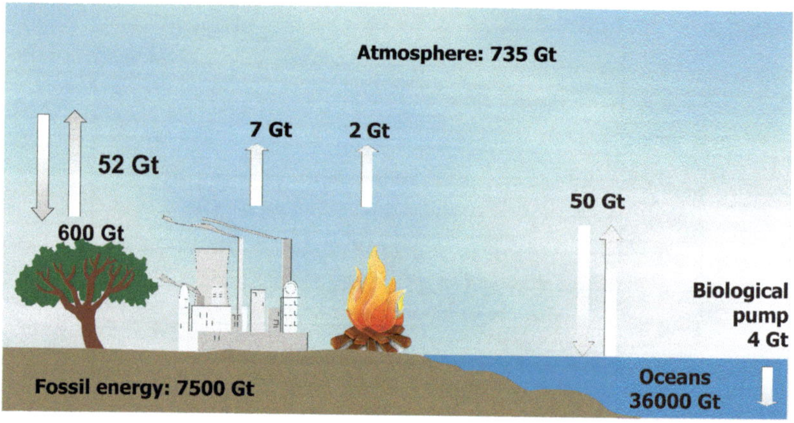

Fig. 3.2. Carbon (not CO_2) fluxes in the atmosphere, fossil fuel deposits and the oceans. Uptake and release equal each other in terrestrial and aquatic biomass even though the biomass is about 100 times larger in the terrestrial than in the aquatic biomass. Net increase in the carbon in the atmosphere is due to anthropogenically released CO_2, net reduction is due to the biological pump in the oceans which sinks organic and inorganic carbon to the deep sea.

280 ppm since the 1880s to more than 400 ppm now. Never in the last 14 million years (mid miocene) was the CO_2 concentration in the atmosphere higher than today. And it is predicted to rise. How much of it will rise up to the end of the current century strongly depends on the scenario: do we phase out of CO_2 emissions, continue the current usage or even increase the emissions?

The methane concentration in the atmosphere has risen from 730 ppb in 1750 to 1800 ppb in 2011, the highest value since almost 800,000 years (Loulergue et al. 2008). Even though the concentrations of CH_4 are much lower than those of CO_2, methane is an important driver of climate change because its greenhouse potential is 25 times higher than that of carbon dioxide (Forster et al. 2007). The increased emission of CH_4 is mainly due to the intensive agriculture and animal husbandry. In addition, methane is released from Arctic Tundra areas where the permafrost soil starts to thaw due to increasing temperatures. Methane is degraded by hydroxyl radicals, so that its lifetime in the atmosphere is about 12 years, while the emitted carbon dioxide has a lifetime of several hundreds to thousands of years (Forster et al. 2007, Blasing 2013).

Ocean Warming

The oceans show a slower increase in temperature than the land masses (Fig. 3.3). New data indicate that there was a steep increase in the observed annual globally averaged sea surface temperature between 1900 and 1945, followed by slow decrease up to 1975. After that the steep increase continued up to today (Yao et al. 2017). However, there are significant regional differences in sea surface temperatures shown by data on a 2° x 2° grid (Huang et al. 2015). The highest increases are seen in the Arctic Sea and around the Antarctic continent, especially near the peninsula pointing towards South America. In contrast, there was a decadal cooling in the tropical Pacific due to intensified trade winds, changes in El Niño activity and increased volcanic activity (Dai et al. 2015).

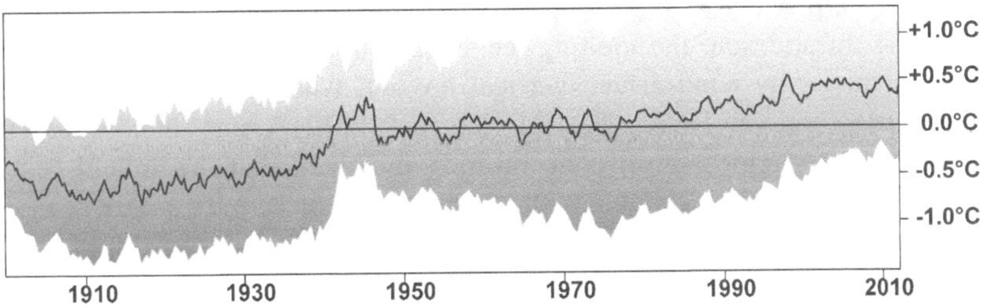

Fig. 3.3. Mean annual temperature of the oceans showing an increase of close to 1°C over the last 100 years, but the error bars are substantial. The values in the 1940s are not reliable, since during World War II several countries did not publish ocean temperature data. Redrawn after (Fischetti 2013).

The strong temperature increase in the Arctic has resulted in a sharp decline of the Arctic sea ice extent (Fig. 3.4) which was on the order of 9.1% per decade between 1979 and 2006 (Stroeve et al. 2007). Most climate models project a continuation of this trend during the current century in response to greenhouse gas forcing (McKenna et al. 2016). Between 1979 and 2017 the Arctic sea ice volume has lost 73.5% when averaged for September of each year (http://climatestate.com/2014/01/26/arctic-death-spiral-1979–2013-sea-ice-decline-deglaciation (ClimateState.com). Recent sea ice models and satellite images suggest that the first ice-free summer will occur before 2020 (Overland and Wang 2013), while the Fifth IPCC report estimates an ice-free summer around 2050 using scenario AR5. The receding ice cover allowed new shipping routes across the Arctic Ocean opening the North East passage which had unsuccessfully been searched for, for several centuries (Howell et al. 2017). The sea ice decline is augmented by a positive feedback called polar amplification: snow and ice cover reflect 50–70% of the incoming solar radiation. When the ice and snow have melted the open water and soil have a much higher absorption (~ 6% reflection) which results in a further heating (Screen and Francis 2016). In addition to a shrinking sea ice cover, the Arctic ice has been found to decrease in thickness as indicated by submarine and ICESAT measurements (Kwok and Rothrock 2009).

The loss of ice over a large area results in a massive increase in phytoplankton biomass production because more sunlight can penetrate into the Arctic Ocean

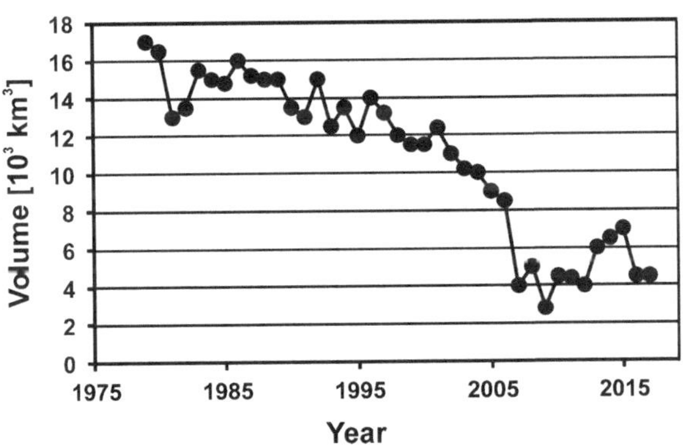

Fig. 3.4. Arctic Sea ice volume averaged over the month of September each year. Redrawn after data from Polar Science Center (University of Washington).

and allowing a pronounced upper ocean stratification (Arrigo et al. 2008, Ardyna et al. 2014). In addition, the melting ice sweeps large amounts of nutrients into the sea enhanced by wind-driven vertical mixing, which further augment rapid growth (Arrigo and van Dijken 2011, Arrigo et al. 2012). Using satellite-measured sea ice extent and chlorophyll concentration indicated that the annual primary production in the Arctic has increased by an average of 27.5 Tg C per year since 2003 and by 35 Tg C per year between 2006 and 2007 (Arrigo et al. 2008). Around 30% of this increase was attributed to a lower ice-covered surface and 70% to a longer growing season. It is estimated that, if the trend continues, Arctic phytoplankton productivity could increase more than threefold above 1998–2002 levels.

It has been widely assumed that increasing water temperatures lead to enhanced stratification both in lakes and in the ocean (Shroyer et al. 2016, Richardson et al. 2017). However, recent analyses challenge this notion and conclude that depth of mixing in lakes and oceans is equally affected by changes in wind strength and the interaction of currents (Kraemer et al. 2015, Somavilla et al. 2017). Simultaneously, the depth of the upper mixed layer (UML) shrinks. This shoaling exposes organisms dwelling in the UML to increased solar visible and ultraviolet radiation (Häder et al. 2015, Häder and Gao 2017). The lower boundary of the UML is the thermocline, which a diver experiences when he swims into deeper waters with a sudden temperature drop (Bruce 2014). Waters of different temperatures or salinities have a tendency not to mix easily (Pickard and Emery 2016). Therefore, stratification or shoaling of the UML hamper the transport of nutrients from deeper water into surface waters and thus limit the growth of phytoplankton (Beardall et al. 2014, Xu et al. 2014). It also reduces the solubility of O_2, leading to oceanic deoxygenation. Increased pCO_2 and decreased pO_2 usually couple inversely well spatio-temporally (Zhai et al. 2009), leading to a decreased respiration index ($Log_{10}\ pO_2/pCO_2$) (Brewer and Peltzer 2009). Another consequence of increased temperatures in the oceans is a poleward shift of ocean currents resulting in an increase in habitat temperatures for some benthic organisms, such as corals and organisms associated with their reefs (Feng et al. 2013, Molinos et al. 2017).

Increasing temperatures lead to melting of ice and snow cover resulting in sea-level rise (Nicholls and Cazenave 2010). During the last 500 years the sea-level has undergone substantial changes of up to 400 m (about 450 million years ago). Between 1870 and 2009 the measured increase of the mean sea-level was about 25 cm. Predictions for the future include melting glaciers and ice sheets, thermal expansion of the water and altered water storage on land. The melting Arctic ice does not significantly contribute to sea-level rise since it is floating. But the glaciers in Greenland, South America and the European Alps contribute a significant amount of water to the oceans (Meier et al. 2007). The velocity of ice loss from glaciers south of 72° N has been found to double since the 1990s resulting in a sea-level rise of more than 0.25 mm per year (Rignot and Kanagaratnam 2006). Model calculations predict that melting of the Greenland Ice Sheet will contribute 0.1–0.3 m to the global sea-level rise (Vasskog et al. 2015). Currently, it contributes about 43% to the sea-level rise (Noël et al. 2017). While the rising temperatures affect the Antarctic Peninsula and lead to break-offs of shelf ice, the ice cover on the continent does not show significant melting. In addition to ice melting, the rising temperatures result in an expansion of the surface waters, contributing significantly to sea-level rise. Both factors—melting ice and thermal expansion resulted in a rise of 3 ± 0.7 mm per year in the period between 1883 and 2010 and a continuation of this trend at the current pace will cause a 30 cm higher water level by the end of the current century (Hay et al. 2015). The predictions have a high degree of uncertainty. Depending on the assumed scenario the mean sea-level will rise by 75, 85 or 130 cm by the year 2150 (Fig. 3.5).

The consequences of the predicted sea-level rise are massive flooding of populated areas. Many human settlements are in low-lying coastal areas. For the biota sea-level rise means changes in coastal wetlands, lagoons as well as macroalga and sea grass habitats (Morris et al. 2002, Carrasco et al. 2016, Takolander et al. 2017).

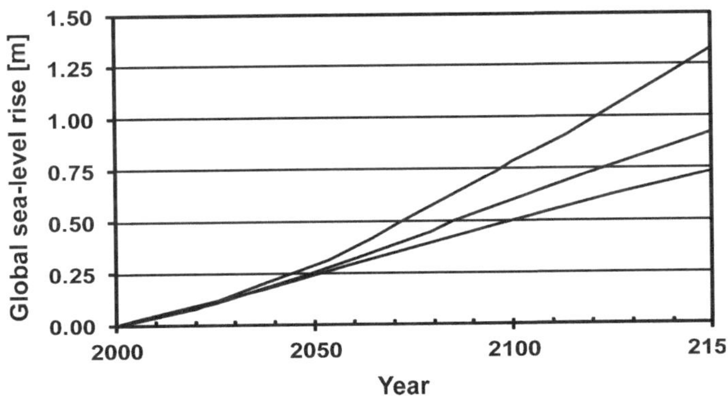

Fig. 3.5. Predicted global sea-level rise based on three different scenarios (representative concentration pathways). Upper line RCP 8.5, medium line RCP 4.5, lower line RCP 2.6. Redrawn after (Kopp et al. 2014).

Ocean Acidification

Increasing CO_2 concentrations in the atmosphere result in higher carbon dioxide concentrations in the surface water since both are in equilibrium by gas transfer (Siegenthaler and Sarmiento 1993), but the dynamics depends on the mixing intensity due to wind and waves. Therefore, the oceans are a major sink for anthropogenic CO_2 emissions (Sabine et al. 2004). If the CO_2 emissions follow a business-as-usual scenario (ICPP, A1F1) (Houghton et al. 2001) the concentration in the atmosphere will reach 800–1000 ppm before the end of the century. In the water CO_2 forms carbonic acid which decreases the pH (Gao and Häder 2017, Häder and Gao 2017).

$$CO_2 + H_2O \rightleftarrows H_2CO_3 \tag{1}$$

This dissociates into a proton and hydrogen carbonate (bicarbonate)

$$H_2CO_3 \rightleftarrows H^+ + HCO_3^- \tag{2}$$

which further dissociates into another proton and a carbonate ion

$$HCO_3^- \rightleftarrows H^+ + CO_3^{2-} \tag{3}$$

These products are in equilibrium which depends on the salinity and temperature of the water. With the increasing H^+ concentration, the capacity for CO_2 to be dissolved decreases (Gattuso et al. 2010), and the CO_3^{2-} and $CaCO_3$ saturation states decline.

General chemical changes associated with ocean acidification include lowered pH, increased bicarbonate ions and pCO_2, decreased CO_3^{2-} but unchanged total alkalinity. The extent of these chemical changes vary in different regions or latitudes due to differences in salinity and temperature.

The increasing CO_2 in the oceans has resulted in a reduction of the pH by 0.1, even though the sea water is well buffered, corresponding to a 30% increase in protons (Caldeira and Wickett 2003). The average global ocean acidification rate is 0.002 pH units per year (Chen et al. 2006). The dissolved carbon dioxide remains in the upper seawater layer for about six years because the mixing with the mesopelagic water is relatively slow on the order of several hundred years. Therefore, most of the carbon absorbed from the atmosphere will remain in the top layer of 0–400 m (Sabine et al. 2004) enhancing upper ocean acidification. Under the A1F1 scenario the pH will be reduced by 0.3–0.4 by the end of the century (Feely et al. 2004) resulting in an increase of protons in the upper water layer by the end of the century by 100–150% (Houghton et al. 2001, Zeebe and Wolf-Gladrow 2001, Gattuso et al. 2015). Judging from the estimated fossil fuel reserves and assuming no drastic reductions in their usage the CO_2 emission into the atmosphere will continue up to a maximum around 2150 and then decline. But the CO_2 has a long lifetime in the atmosphere and therefore the concentrations will remain elevated for thousands of years until the oceans have gradually absorbed the excess carbon dioxide and conveyed it into the deep sea which will simultaneously decrease the pH in the surface ocean waters (Caldeira and Wickett 2003). The atmospheric CO_2 concentrations have changed during the past, but slowly enough so that the concentration in the surface waters was kept more or less constant. In contrast, presently the pH decreases much faster than during the past 300 million years so that the geochemical processes cannot keep up with the mitigation (Hönisch et al. 2012).

In addition to the carbonate chemistry ocean acidification alters other chemical pathways which affect biological processes due to changing environmental conditions (Millero 2007). Ocean acidification interacts with other environmental conditions (Riebesell and Gattuso 2015), such as temperature and solar visible and ultraviolet radiation (Boyd 2011, Brennan and Collins 2015), but these interactions depend on the chemical composition of different waters (Kleypas et al. 1999, Orr et al. 2005).

Decreasing pH values affect the saturation of $CaCO_3$ which can be calculated from $Ca^{2+} \times CO_3^{2-}/K_c$, where K_c is the product of $Ca^{2+} \times CO_3^{2-}$ when the $CaCO_3$ solution is saturated. This is controlled by the crystal type of $CaCO_3$, such as calcite or aragonite. Because the Ca^{2+} concentration in oceanic waters is relatively stable at about 10 mM the $CaCO_3$ saturation depends basically on the CO_3^{2-} concentration. Around 90% of the dissolved inorganic organic carbon (DIC) is represented by HCO_3^-, while CO_3^{2-} accounts for about 9% and CO_2 for less than 1%; but these values can vary between different latitudes and regions (Gao and Häder 2017). If the CO_2 concentration in sea water doubles, the concentration of HCO_3 will increase by about 11%; DIC will increase by 9% and CO_3^{2-} will decrease by about 45% (Kleypas et al. 1999). However, the exact numbers depend on temperature and other factors. For example, the concentration of CO_3^{2-} will decrease to about 149 μmol/L in tropical waters by the end of the current century while it will decrease to 55 μmol/L in polar waters (Orr et al. 2005). Likewise at low temperatures, ocean acidification will decrease the concentration of CO_3^{2-} more significantly than in warmer waters.

Ocean Deoxygenation

Ocean deoxygenation is closely related to climate change (Keeling et al. 2009, Breitburg et al. 2018). It is generally accepted that global warming is the major driver to exacerbate deoxygenation and hypoxia (Levitus et al. 2009). As ocean surface heating reduces gas solubility and enhances stratification of seawater, ocean O_2 content is thought to decline. It has been reported that 94% of dead zones (waters with depleted dissolved O_2) occur in areas where the temperature will increase by at least 2°C by the end of this century (Altieri and Gedan 2015). Over the past 50 years, the degree of hypoxia (< 2 mg L^{-1}) has changed from 400 m to 300 m in the Pacific Ocean and its dissolved oxygen content has also decreased significantly (Whitney et al. 2007). According to the relevant model, dissolved O_2 of seawater will experience a 1–7% decrease in the next few decades, and this decline will continue for the next thousand years (Keeling et al. 2009).

Microbial denitrification and anammox can be promoted by hypoxia, reducing biologically available oxidized nitrogen species back to dinitrogen gas which is unavailable to photosynthetic organisms. Furthermore, nitrous oxide, a greenhouse gas, may be generated during incomplete dissimilatory processes (Gucinski 1994). Sulfide production, another potential threat of hypoxia generated from microbial dissimilatory processes, can jeopardize marine organisms (Capone and Hutchins 2013). Since the physiological performance of most marine organisms requires O_2, their metabolic activities may be affected to an extent that may cause death when dissolved O_2 is below a certain level. Conversely, microbial food webs could overspread taking advantage of ocean deoxygenation, considering some protozoa, such as ciliates which can breathe anaerobically (Capone and Hutchins 2013). For this reason, oceanic carbon and nitrogen cycling and biological productivity as well as the marine biological carbon pump can be influenced by deoxygenation. However, little attention has been given on how deoxygenation or hypoxia may affect biological metabolism and pivotal ecological processes of various especially photosynthetic marine organisms, which has not been highly recognized by the scientific community.

Oxygen consumption is driven mainly by bacterial oxidation of organic matter sinking from the upper water layer as the basis of the biological pump. Due to the decomposition by microorganisms, dissolved oxygen is consumed considerably while large quantities of CO_2 and nutrients are released, forming dead zones at a depth of 500–700 m characterized by low dissolved oxygen and being highly acidified (Brewer and Peltzer 2009) (Fig. 3.6). Thus, hypoxic regions are usually areas of acidification. Because of the eutrophication in coastal waters, the anoxic regions are expanding at an annual rate of 5.5% (Vaquer-Sunyer and Duarte 2008), and the acidification rate in nearshore waters is higher than that in open oceans (Cai et al. 2011).

The interaction between ocean hypoxia and acidification reduce the pO_2/pCO_2 ratio with faster changes in deep oceans (Brewer and Peltzer 2009). The ratio of pO_2 to pCO_2 could also be affected in the euphotic zone when deep seawater is transported to the surface due to physical events, such as upwelling, internal waves or wind forcing. The physical and chemical environment in upwelling regions with high production have also undergone dramatic changes owing to the increase of the atmospheric CO_2

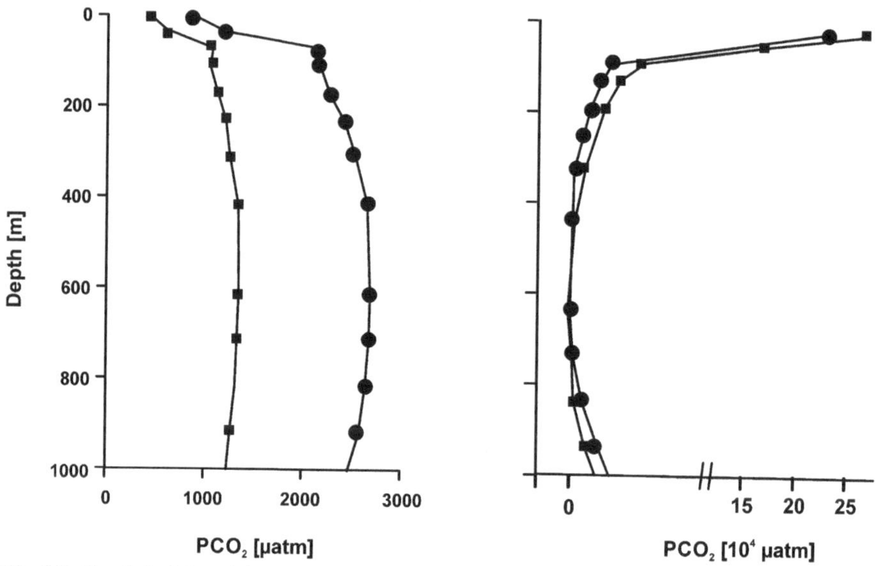

Fig. 3.6. Trends in CO_2 and O_2 concentrations in seawater as a function of depth. Circles indicate values estimated for a doubling of the atmospheric CO_2 concentration. Redrawn after (Brewer and Peltzer 2009).

concentration caused by human activities and the related climate change. It has been observed that the calcium carbonate solubility in upwelling regions of the California coastal ocean is less than 1% due to the interaction of upwelling low-pH seawater and with ocean acidification, which imperils marine organisms and hinders development of marine fisheries. These ecological and economic effects, in addition to the pH changes, are also affected by changes in the seawater pO_2/pCO_2.

Physiological activity of plankton will also be affected by ocean hypoxia and acidification. The declining ratio of pO_2 to pCO_2, while causing huge economic losses to shellfish farming, has also resulted in a decrease in the stocks of zooplankton (Doney et al. 2012). In response to ocean acidification, zooplankton increases its rates of respiration and feeding (Li and Gao 2012). Oxygen content is another factor to affect the respiration of zooplankton, so the coupling of hypoxia with acidification could have an impact on the balance between respiration and energy supply. The reduction of pO_2 or the decrease of pO_2/pCO_2 is favorable to carboxylation (CO_2 fixation) efficiency of ribulose-1,5-bisphosphate carboxylase/oxygenase (Rubisco) and down-regulate oxygenation (decomposing organic carbon and releasing CO_2) of this key enzyme of carbon fixation in photosynthesis (Ogren 1984). Experiments about elevating net photosynthesis through reducing oxygen concentration to 2% found an increase of 45% in C3 plants (Forrester et al. 1966, Reinfelder 2011). In conclusion, ocean deoxygenation and acidification have synergistic or antagonistic effects on photosynthetic carbon fixation, respiration, photorespiration, growth and calcification and we need to pay more attention on this unpredictable issue.

Conclusions

The anthropogenically-induced changes in the environmental factors for aquatic ecosystems spell increased stress which is bound to affect primary productivity with far reaching consequences for primary and secondary consumers and the whole aquatic ecosystems. Rising temperatures may exceed the thermal window of many species. In other cases ocean warming results in poleward movement of habitats and changes in communities. The resulting enhanced stratification exposes organisms in the upper mixed layer to higher solar visible and ultraviolet radiation even though no global increases in solar UV are expected. In addition, it becomes more difficult for nutrients from deeper waters to reach the surface waters where major primary producers and their predators dwell. Melting ice and snow cover create new habitats and extend the vegetative period especially in the Arctic. Their productivity is enhanced by nutrients from the melting ice and terrestrial runoff. Melting glaciers and ice covers enhance sea level rise augmented by thermal expansion which affects coastal habitats, lagoons and wetland biotopes.

Increasing atmospheric carbon dioxide concentrations are reflected by a rising CO_2 partial pressure in surface waters resulting in increasing H^+ concentrations causing ocean acidification. Different organisms respond differently to increasing CO_2 concentrations by either showing enhanced growth, neutral reactions or reduced productivity. Some phytoplankton, macroalgae and many zoological taxa, such as worms, mollusks and corals use calcification for protection, thallus stabilization and UV protection. Ocean acidification interferes with calcification depriving the organisms of their protection.

References

Alexander, L., X. Zhang, T. Peterson, J. Caesar, B. Gleason, A. Klein Tank, M. Haylock, D. Collins, B. Trewin and F. Rahimzadeh. 2006. Global observed changes in daily climate extremes of temperature and precipitation. Journal of Geophysical Research: Atmospheres 111.

Altieri, A.H. and K.B. Gedan. 2015. Climate change and dead zones. Global Change Biology 21: 1395–1406.

Ardyna, M., M. Babin, M. Gosselin, E. Devred, L. Rainville and J.É. Tremblay. 2014. Recent Arctic Ocean sea ice loss triggers novel fall phytoplankton blooms. Geophysical Research Letters 41: 6207–6212.

Arrigo, K.R., G. van Dijken and S. Pabi. 2008. Impact of a shrinking Arctic ice cover on marine primary production. Geophysical Research Letters 35.

Arrigo, K.R. and G.L. van Dijken. 2011. Secular trends in Arctic Ocean net primary production. Journal of Geophysical Research: Oceans (1978–2012) 116: 160–168.

Arrigo, K.R., D.K. Perovich, R.S. Pickart, Z.W. Brown, G.L. van Dijken, K.E. Lowry, M.M. Mills, M.A. Palmer, W.M. Balch, F. Bahr, N.R. Bates, C. Benitez-Nelson, B. Bowler, E. Brownlee, J.K. Ehn, K.E. Frey, R. Garley, S.R. Laney, L. Lubelczyk, J. Mathis, A. Matsuoka, B.G. Mitchell, G.W.K. Moore, E. Ortega-Retuerta, S. Pal, C.M. Polashenski, R.A. Reynolds, B. Schieber, H.M. Sosik, M. Stephens and J.H. Swift. 2012. Massive phytoplankton blooms under Arctic sea ice. Science 336: 1408

Beardall, J., S. Stojkovic and K. Gao. 2014. Interactive effects of nutrient supply and other environmental factors on the sensitivity of marine primary producers to ultraviolet radiation: implications for the impacts of global change. Aquatic Biology 22: 5–23.

Blasing, T. 2013. Recent Greenhouse Gas Concentrations. Oak Ridge National Laboratory: Carbon Dioxide Information Analysis Center.

Boyd, P.W. 2011. Beyond ocean acidification. Nature Geoscience 4: 273–274.

Breitburg, D., L.A. Levin, A. Oschlies, M. Grégoire, F.P. Chavez, D.J. Conley, V. Garçon, D. Gilbert, D. Gutiérrez and K. Isensee. 2018. Declining oxygen in the global ocean and coastal waters. Science 359: eaam7240.

Brennan, G. and S. Collins. 2015. Growth responses of a green alga to multiple environmental drivers. Nature Climate Change 5: 892–897.

Brewer, P.G. and E.T. Peltzer. 2009. Limits to marine life. Science 324: 347–348.

Bruce, J. 2014. Alouette Water Temperature Monitoring: Program No. ALUMON# 5.

Cai, W.-J., X. Hu, W.-J. Huang, M.C. Murrell, J.C. Lehrter, S.E. Lohrenz, W.-C. Chou, W. Zhai, J.T. Hollibaugh and Y. Wang. 2011. Acidification of subsurface coastal waters enhanced by eutrophication. Nature Geoscience 4: 766–770.

Caldeira, K. and M.E. Wickett. 2003. Oceanography: anthropogenic carbon and ocean pH. Nature 425: 365–365.

Capone, D.G. and D.A. Hutchins. 2013. Microbial biogeochemistry of coastal upwelling regimes in a changing ocean. Nature Geoscience 6: 711–717.

Carrasco, A., Ó. Ferreira and D. Roelvink. 2016. Coastal lagoons and rising sea level: A review. Earth-Science Reviews 154: 356–368.

Cook, J., D. Nuccitelli, S.A. Green, M. Richardson, B. Winkler, R. Painting, R. Way, P. Jacobs and A. Skuce. 2013. Quantifying the consensus on anthropogenic global warming in the scientific literature. Environmental Research Letters 8: 024024.

Dahl-Jensen, D., M. Albert, A. Aldahan, N. Azuma, D. Balslev-Clausen, M. Baumgartner, A.-M. Berggren, M. Bigler, T. Binder and T. Blunier. 2013. Eemian interglacial reconstructed from a Greenland folded ice core. Nature 493: 489.

Dai, A., J.C. Fyfe, S.-P. Xie and X. Dai. 2015. Decadal modulation of global surface temperature by internal climate variability. Nature Climate Change 5: 555–559.

Doney, S.C., M. Ruckelshaus, J.E. Duffy, J.P. Barry, F. Chan, C.A. English, H.M. Galindo, J.M. Grebmeier, A.B. Hollowed, N. Knowlton, J. Polovina, N.N. Rabalais, W.J. Sydeman and L.D. Talley. 2012. Climate change impacts on marine ecosystems. Annual Review of Marine Science 4: 11–37.

Feely, R.A., C.L. Sabine, K. Lee, W. Berelson, J. Kleypas, V.J. Fabry and F.J. Millero. 2004. Impact of anthropogenic CO_2 on the $CaCO_3$ system in the oceans. Science 305: 362–366.

Feng, M., M.J. McPhaden, S.-P. Xie and J. Hafner. 2013. La Niña forces unprecedented Leeuwin Current warming in 2011. Scientific Reports 3.

Fischetti, M. 2013. Deep heat threatens marine life. Scientific American 308: 92.

Forrester, M.L., G. Krotkov and C.D. Nelson. 1966. Effect of Oxygen on Photosynthesis, Photorespiration and Respiration in Detached Leaves. I. Soybean. Plant Physiology 41: 422–427.

Forster, P., V. Ramaswamy, P. Artaxo, T. Berntsen, R. Betts, D.W. Fahey, J. Haywood, J. Lean, D.C. Lowe and G. Myhre. Changes in atmospheric constituents and in radiative forcing. Chapter 2. pp. 2007. Climate Change 2007. The Physical Science Basis.

Francis, J.A. 2018. Meltdown. Scientific American 318: 40–45.

Gao, K. and D.-P. Häder. 2007. Effects of ocean acidification and UV radiation on marine photosynthetic carbon fixation. pp. 235–250. *In*: Kumar, M. and P.J. Ralph (eds.). Systems Biology of Marine Ecosystems. Springer, Cham, Switzerland.

Gattuso, J.-P., K. Gao, K. Lee, B. Rost and K.G. Schulz. 2010. Approaches and tools to manipulate the carbonate chemistry. pp. 41–52. *In*: Riebesell, U., V.J. Fabry, L. Hansson and J.P. Gattuso (eds.). Guide to Best Practices Ocean Acidification and Data Reporting. Publications Office of the European Union, Luxembourg.

Gattuso, J.-P., A. Magnan, R. Billé, W. Cheung, E. Howes, F. Joos, D. Allemand, L. Bopp, S. Cooley and C. Eakin. 2015. Contrasting futures for ocean and society from different anthropogenic CO_2 emissions scenarios. Science 349: aac4722.

Gucinski, H. 1994. Microbial production and consumption of greenhouse gases: Methane, nitrogen oxides, and halomethanes. Journal of Environmental Quality 23: 211–212.

Häder, D.-P., C.E. Williamson, S.-A. Wängberg, M. Rautio, K.C. Rose, K. Gao, E.W. Helbling, R.P. Sinha and R. Worrest. 2015. Effects of UV radiation on aquatic ecosystems and interactions with other environmental factors. Photochemical & Photobiological Sciences 14: 108–126.

Häder, D.-P. and K. Gao. 2017. The impacts of climate change on marine phytoplankton. pp. 901–928. *In*: Phillips, B.F. and M. Pérez-Ramírez (eds.). Climate Change Impacts on Fisheries and Aquaculture. A Global Analysis. Wiley, Hoboken, NJ.

Hansen, J., R. Ruedy, M. Sato and K. Lo. 2005. GISS surface temperature analysis global temperature trends: 2005 summation. NASA Goddard Institute for Space Studies, New York, NY. Available at: http://data. giss. nasa. gov/gistemp.

Hay, C.C., E. Morrow, R.E. Kopp and J.X. Mitrovica. 2015. Probabilistic reanalysis of twentieth-century sea-level rise. Nature 517: 481–484.

Hönisch, B., A. Ridgwell, D.N. Schmidt, E. Thomas, S.J. Gibbs, A. Sluijs, R. Zeebe, L. Kump, R.C. Martindale and S.E. Greene. 2012. The geological record of ocean acidification. Science 335: 1058–1063.

Houghton, J.T., Y. Ding, D.J. Griggs, M. Noguer, P.J. van der Linden, X.-. Dai, K. Maskell and C.-A. Johnson. 2001. Climate Change 2001: The Scientific Basis. Cambridge University Press. Cambridge, UK.

Howell, S., L. Pizzolato, J. Dawson and F. Laliberté. 2017. Arctic sea ice decline and its influence on shipping activity. EGU General Assembly Conference Abstracts.

Huang, B., V.F. Banzon, E. Freeman, J. Lawrimore, W. Liu, T.C. Peterson, T.M. Smith, P.W. Thorne, S.D. Woodruff and H.-M. Zhang. 2015. Extended reconstructed sea surface temperature version 4 (ERSST. v4). Part I: upgrades and intercomparisons. Journal of Climate 28: 911–930.

IPCC - World Meteorological Organiszation. 2007. Intergovernmental panel on climate change. World Meteorological Organization. Available at: http://wmo. insomnation. com/sites/default/files/documents/meetings/session20/doc2. pdf.

Keeling, R.F., A. Körtzinger and N. Gruber. 2009. Ocean deoxygenation in a warming world.

Kleypas, J.A., R.W. Buddemeier, D. Archer, J.-P. Gattuso, C. Langdon and B.N. Opdyke. 1999. Geochemical consequences of increased atmospheric carbon dioxide on coral reefs. Science 284: 118–120.

Kopp, R.E., R.M. Horton, C.M. Little, J.X. Mitrovica, M. Oppenheimer, D. Rasmussen, B.H. Strauss and C. Tebaldi. 2014. Probabilistic 21st and 22nd century sea-level projections at a global network of tide-gauge sites. Earth's Future 2: 383–406.

Kraemer, B.M., O. Anneville, S. Chandra, M. Dix, E. Kuusisto, D.M. Livingstone, A. Rimmer, S.G. Schladow, E. Silow and L.M. Sitoki. 2015. Morphometry and average temperature affect lake stratification responses to climate change. Geophysical Research Letters 42: 4981–4988.

Kwok, R. and D. Rothrock. 2009. Decline in Arctic sea ice thickness from submarine and ICESat records: 1958–2008. Geophysical Research Letters 36.

Levitus, S., J.I. Antonov, T.P. Boyer, R.A. Locarnini, H.E. Garcia and A.V. Mishonov. 2009. Global ocean heat content 1955–2008 in light of recently revealed instrumentation problems. Geophysical Research Letters 36.

Li, W. and K.S. Gao. 2012. A marine secondary producer respires and feeds more in a high CO_2 ocean. Marine Pollution Bulletin 64: 699–703.

Loulergue, L., A. Schilt, R. Spahni, V. Masson-Delmotte, T. Blunier, B. Lemieux, J.-M. Barnola, D. Raynaud, T.F. Stocker and J. Chappellaz. 2008. Orbital and millennial-scale features of atmospheric CH_4 over the past 800,000 years. Nature 453: 383–386.

McKenna, C., T. Bracegirdle, E. Shuckburgh and P. Haynes. 2016. The influence of regional Arctic sea-ice decline on stratospheric and tropospheric circulation. EGU General Assembly Conference Abstracts.

Meier, M.F., M.B. Dyurgerov, U.K. Rick, S. O'neel, W.T. Pfeffer, R.S. Anderson, S.P. Anderson and A.F. Glazovsky. 2007. Glaciers dominate eustatic sea-level rise in the 21st century. Science 317: 1064–1067.

Met Office. Observing Changes in the Climate.

Millero, F.J. 2007. The marine inorganic carbon cycle. Chemical Reviews 107: 308–341.

Milner, A. 2007. Arctic climate impact assessment. Cambridge University Press, New York, 2005. ISBN 0521865093. Wiley Online Library.

Molinos, J.G., M. Burrows and E. Poloczanska. 2017. Ocean currents modify the coupling between climate change and biogeographical shifts. Scientific Reports 7.

Morice, C.P., J.J. Kennedy, N.A. Rayner and P.D. Jones. 2012. Quantifying uncertainties in global and regional temperature change using an ensemble of observational estimates: The HadCRUT4 data set. Journal of Geophysical Research: Atmospheres 117.

Morris, J.T., P. Sundareshwar, C.T. Nietch, B. Kjerfve and D.R. Cahoon. 2002. Responses of coastal wetlands to rising sea level. Ecology 83: 2869–2877.

Nicholls, R.J. and A. Cazenave. 2010. Sea-level rise and its impact on coastal zones. Science 328: 1517–1520.

Noël, B., W. van de Berg, S. Lhermitte, B. Wouters, H. Machguth, I. Howat, M. Citterio, G. Moholdt, J. Lenaerts and M.R. van den Broeke. 2017. A tipping point in refreezing accelerates mass loss of Greenland's glaciers and ice caps. Nature Communications 8: 14730.

Ogren, W.L. 1984. Photorespiration: Pathways, regulation, and modification. Annual Review of Plant Physiology 35: 415–442.

Orr, J.C., V.J. Fabry, O. Aumont, L. Bopp, S.C. Doney, R.A. Feely, A. Gnanadesikan, N. Gruber, A. Ishida, F. Joos, R.M. Key, K. Lindsay, E. Maier-Reimer, R. Matear, P. Monfray, A. Mouchet, R.G. Najjar, G.K. Plattner, K.B. Rodgers, C.L. Sabine, J.L. Sarmiento, R. Schlitzer, R.D. Slater, I.J. Totterdell, M.F. Weirig, Y. Yamanaka and A. Yool. 2005. Anthropogenic ocean acidification over the twenty-first century and its impact on calcifying organisms. Nature 437: 681–686.

Overland, J.E. and M. Wang. 2013. When will the summer Arctic be nearly sea ice free? Geophysical Research Letters 40: 2097–2101.

Pickard, G.L. and W.J. Emery. 2016. Descriptive Physical Oceanography: An Introduction. Elsevier.

Reinfelder, J.R. 2011. Carbon concentrating mechanisms in eukaryotic marine phytoplankton. Ann. Rev. Mar. Sci. 3: 291–315.

Richardson, D.C., S.J. Melles, R.M. Pilla, A.L. Hetherington, L.B. Knoll, C.E. Williamson, B.M. Kraemer, J.R. Jackson, E.C. Long and K. Moore. 2017. Transparency, geomorphology and mixing regime explain variability in trends in lake temperature and stratification across Northeastern North America (1975–2014). Water 9: 442.

Riebesell, U. and J.-P. Gattuso. 2015. Lessons learned from ocean acidification research. Nature Climate Change 5: 12–14.

Rignot, E. and P. Kanagaratnam. 2006. Changes in the velocity structure of the Greenland Ice Sheet. Science 311: 986–990.

Ripple, W.J., C. Wolf, T.M. Newsome, M. Mauro Galetti, M. Alamgir, E. Crist, M.I. Mahmoud, W.F. Laurance and and 15.364 bioscientists from 184 countries. World scientists' arning to humanity: a second notice. pp. 1026–1028. *In*: 2017. BioScience.

Sabine, C.L., R.A. Feely, N. Gruber, R.M. Key, K. Lee, J.L. Bullister, R. Wanninkhof, C.S. Won, D.W.R. Wallace, B. Tilbrook, F.J. Millero, T.-H. Peng, A. Kozyr, T. Ono and A.F. Rios. 2004. The oceanic sink for anthropogenic CO_2. Science 305: 367–371.

Screen, J.A. and J.A. Francis. 2016. Contribution of sea-ice loss to Arctic amplification is regulated by Pacific Ocean decadal variability. Nature Climate Change 6: 856–860.

Shroyer, E.L., D.L. Rudnick, J.T. Farrar, B. Lim, S.K. Venayagamoorthy, L.C. St. Laurent, A. Garanaik and J.N. Moum. 2016. Modification of upper-ocean temperature structure by subsurface mixing in the presence of strong salinity stratification. Oceanography 29: 62–71.

Siegenthaler, U. and J. Sarmiento. 1993. Atmospheric carbon dioxide and the ocean. Nature 365: 119–125.

Somavilla, R., C. González-Pola and J. Fernandez. 2017. The warmer the ocean surface, the shallower the mixed layer: How much of this is true? Journal of Geophysical Research: Oceans 122: 7698–7716.

Stroeve, J., M.M. Holland, W. Meier, T. Scambos and M. Serreze. 2007. Arctic sea ice decline: Faster than forecast. Geophysical Research Letters 34.

Takolander, A., M. Cabeza and E. Leskinen. 2017. Climate change can cause complex responses in Baltic Sea macroalgae: A systematic review. Journal of Sea Research 123: 16–29.

Vaquer-Sunyer, R. and C.M. Duarte. 2008. Thresholds of hypoxia for marine biodiversity. Proceedings of the National Academy of Sciences 105: 15452–15457.

Vasskog, K., P.M. Langebroek, J.T. Andrews, J.E.Ø. Nilsen and A. Nesje. 2015. The Greenland Ice Sheet during the last glacial cycle: Current ice loss and contribution to sea-level rise from a palaeoclimatic perspective. Earth-Science Reviews 150: 45–67.

Vose, R.S., D.R. Easterling and B. Gleason. 2005. Maximum and minimum temperature trends for the globe: An update through 2004. Geophysical Research Letters 32.

Whitney, F.A., H.J. Freeland and M. Robert. 2007. Persistently declining oxygen levels in the interior waters of the eastern subarctic Pacific. Progress in Oceanography 75: 179–199.

Xu, J., K. Gao, Y. Li and D.A. Hutchins. 2014. Physiological and biochemical responses of diatoms to projected ocean changes. Marine Ecology Progress Series 515: 73–81.

Yao, S.-L., J.-J. Luo, G. Huang and P. Wang. 2017. Distinct global warming rates tied to multiple ocean surface temperature changes. Nature Climate Change 7: 486–491.

Zeebe, R.E. and D.A. Wolf-Gladrow. 2001. CO_2 in seawater: Equilibrium, kinetics, isotopes. Gulf. Professional Publishing.

CHAPTER 4

Effects of Global Climate Change on Cyanobacteria

Jainendra Pathak,[1] *Haseen Ahmed,*[1] *Rajneesh,*[1] *Shailendra P. Singh,*[2] *Donat-P. Häder*[3] and *Rajeshwar P. Sinha*[1,*]

Introduction

Climate trends have shown a steady rise in global temperature of 0.1–0.2°C per decade over the last 100 years coupled with more extreme weather events (Hansen et al. 2010, NOAA National Centers for Environmental Information published online in June 2017). Reports estimate that the mean global climate has warmed by nearly 1°C since the mid-20th century with some regions experiencing annual temperature anomalies nearly 3°C warmer than average and mean global temperature is expected to rise as much as 4.2°C by the year 2100 owing to anthropogenic carbon emissions (Hansen et al. 2010, Haakonsson et al. 2017, NOAA National Centers for Environmental Information published online in June 2017). These small changes in the environment can severely affect less-tolerant organisms, which can rapidly diminish species diversity at the ecosystem scale (Moorman et al. 2017). More than 70% of the earth surface is covered by water and ~ 99% of this is constituted by the oceans (Charette and Smith 2010).

Cyanobacteria, the ecologically important organisms, are a dominant flora of wetland soils, especially in rice-paddy fields where they serve as natural biofertilizers (Vaishampayan et al. 2001) by virtue of their nitrogen-fixing ability utilizing the enzyme nitrogenase. The prokaryotic and eukaryotic phytoplankton including cyanobacteria, are the major contributors of the biomass in aquatic habitats (Häder et al. 2007). The standing crop of these organisms constitutes only 1% of the biomass of all terrestrial ecosystems combined, but their productivity equals that of all land

[1] Laboratory of Photobiology and Molecular Microbiology, Centre of Advanced Study in Botany, Institute of Science, Banaras Hindu University, Varanasi-221005, India.
[2] Center of Advanced Study in Botany, Institute of Science, Banaras Hindu University, Varanasi-221005, India.
[3] Emeritus from Friedrich-Alexander University, Department of Biology, Neue Str. 9, 91096, Möhrendorf, Germany.
* Corresponding author: r.p.sinha@gmx.net; rpsinhabhu@gmail.com

plants taken together (Gao et al. 2016). Oceans are a major sink for atmospheric CO_2 and hence a major player in the global change in partially mitigating temperature increases (Landschützer et al. 2014). Primary producers, such as cyanobacteria absorb CO_2 in the oceans and fix it *via* photosynthesis to generate organic matter. Global warming has resulted in increasing surface seawater temperatures enforcing stratification and shoaling of the upper mixed layer (UML) above the thermocline (Wang et al. 2015). This exposes the cyanobacteria dwelling in this layer to excessive solar visible (PAR) and UV radiation (UVR) (Gao et al. 2012a, Gao et al. 2012b). It is assumed that UVR will decrease at mid- and high-latitudes relative to the 1960s, but the trend in the tropics may depend on the emission of CO_2, CH_4, and N_2O (Bais et al. 2015). However, simulations based on a chemistry climate model indicated that the total ozone might be lower in the tropics compared to the 1960s (United Nations Environment Programme 2017). For these reasons, the eco-physiological effects of the enhanced UV-B radiation continue to rouse increasing attention (Gao et al. 2017). In addition, the augmented stratification hinders the transport of dissolved inorganic macronutrients from deeper waters into the UML (Behrenfeld et al. 2006).

Ocean warming is known to affect primary productivity both directly and indirectly. The seawater volume-specific primary productivity also decreased with temperature rise due to lower phytoplankton biomass (Van de Poll et al. 2013). CO_2 may be a potentially limiting factor for marine primary productivity because of the low CO_2 level in seawater and the low affinity of the enzyme RuBISCO for dissolved CO_2 (Falkowski and Raven 2013). In addition, CO_2 in seawater diffuses approximately 10,000 times slower than in air, leading to its supply rate being much lower than the demand for photosynthetic carbon fixation (Riebesell et al. 1993). Although phytoplankton have evolved carbon-concentrating mechanisms (CCMs) to cope with these problems (Raven et al. 2012), increased CO_2 concentration may still be beneficial since energy saved due to down-regulation of CCMs under elevated CO_2 can be utilised in other metabolic processes (Gao et al. 2012a). Dissolution of CO_2 increases seawater CO_2 partial pressure and bicarbonate ion levels and decreases pH and carbonate ion concentrations, leading to ocean acidification (OA). By 2100, the global-mean surface pH is projected to decline approximately by 0.065 to 0.31, depending on the representative concentration pathway (RCP) scenario (IPCC 2013). OA can even reduce primary productivity of surface phytoplankton assemblages when exposed to incident solar radiation (Gao et al. 2012b). Frequency and intensity of rainfall events will increase with longer drought periods in between promoting cyanobacterial growth due to a greater nutrient input into water bodies during heavy rainfall events, combined with potentially longer periods of high evaporation and stratification (Reichwaldt and Ghadouani 2012).

Climate change scenarios predict intensified terrestrial storm runoff, providing coastal ecosystems with large nutrient pulses and increased turbidity, with unknown consequences for the phytoplankton community (Deininger et al. 2016). OA, warming, and enhanced exposure to excessive PAR and UVR have positive and negative effects on cyanobacterial productivity and community structure (Wells et al. 2015, Domingues et al. 2017). Phytoplankton communities of lakes are thought to be effected by changes taking place in climate recently and relative abundance of cyanobacteria will increase under the predicted future climate (Elliott 2012). However, testing of such a qualitative prediction is challenging and therefore, lake modeling would serve as an important tool for assessing the impact of climate change upon cyanobacteria. Results indicated an enhancement in relative cyanobacteria abundance with increasing nutrient loads, water temperature and decreased flushing rate. However, models used for such studies are few which can examine the potential effect climate change on cyanobacteria (Table 4.1).

Effects of different environmental factors, such as OA, warming, UV radiation, pollution, eutrophication, etc. in a combined manner on cyanobacteria are still unclear. The following sections will discuss the effects of these ecological stress factors on ecologically and economically important primary producers, cyanobacteria.

Ocean Acidification

Combustion of fossil fuel and tropical deforestation, release of CO_2 and methane from thawing permafrost areas in the Arctic due to rising global temperatures result in an increase of atmospheric CO_2 (Schädel et al. 2016). Hence, CO_2 availability has been augmented in the surface layers of the oceans and thereby enhancing the photosynthetic activity. Consequently, the productivity of the oceanic food webs in total should increase (Riebesell and Tortell 2011).

Early shipboard and laboratory experiments suggested that enhanced CO_2 could increase the phytoplankton growth rates and thus primary productivity (Schippers et al. 2004) but experimental results contradicted this hypothesis as most phytoplankton including cyanobacteria have developed carbon concentrating mechanisms (CCMs) utilizing the enzyme carbonic anhydrase (Bozzo and Colman 2000). As CO_2 serves as main substrate of photosynthesis, any change in the concentration of CO_2 in aquatic ecosystems would cause significant effects on primary producers (Mostofa et al. 2015).

The atmospheric uptake of CO_2 depends on several factors such as temperature, mixing intensity, salinity and chemistry of surface seawater that interacts with biological activities. A decrease of 0.3–0.4 units of pH in oceans is expected by the end of this century as result of OA (Pachauri et al. 2014, Gattuso et al. 2015). OA accelerates with declining capability to take up the CO_2 as more than half of the CO_2 absorbed by the oceans remains in the surface layer (Sabine et al. 2004). However, a decline in pH will be accompanied by the increase of CO_2, resulting in consumption of more energy for maintaining a constant intracellular pH (Bach et al. 2013). Therefore, in view of the changes in substrates for photosynthetic and carbonate chemistry, the impacts of elevated CO_2 on cyanobacteria are complicated.

Atmospheric CO_2 concentrations have changed very slowly during the past 300 million years, which has allowed effective mixing between the deep and surface layers of the oceans, so that OA was moderate (Hönisch et al. 2012). OA will not disappear in the next few hundred years as the deposition of CO_2 into deeper oceanic layers would take thousands of years (Raven et al. 2005). The oceanic carbonate chemistry gets modified by OA (Millero 2007), affecting the organisms and ecosystems to different degrees in different latitudes and waters, considering compounded impacts with multiple environmental drivers (Riebesell and Gattuso 2015).

Table 4.1. Models predicting climate change impacts on cyanobacteria (Adapted from Elliott 2012).

Lake (country) Model(s) used	**Driver**	**Trophic status**	**Volume ($10^6 \cdot m^3$)**	**Depth (m) (mean/max)**	**Regional Climate Model (RCM)**	**Response**	**References**
Bassenthwaite Lake (UK) PROTECH	Higher temperature	Eutrophic	27.9	5.3/19	driven by RCM	No change in overall biomass, earlier growth	Elliott et al. (2005)
Farmoor Reservoir (UK) CLAMM	Reduced short-wave Radiation	Eutrophic	4.5	9.2/11	driven by RCM	None	Howard and Easthope (2002)
Galten basin of Lake Mälaren (Sweden) PROTBAS	Higher temperature	Eutrophic	210	3.4/19	driven by RCM	Increase in dominance (*via* nutrients)	Markensten et al. (2010)
Lake Rotoehu (New Zealand) DYRESM-CAEDYM	Higher temperature /nutrients	Eutrophic	60	8.2/13.5	driven by RCM	Increase in dominance	Trolle et al. (2011)
Ringsjön (Sweden) PROBE & BIOLA	Higher temperature	Eutrophic	184.2	5/17.5	driven by RCM	Increase in overall biomass (*via* nutrients)	Arheimer et al. (2005)
Bassenthwaite Lake (UK) PROTECH	Higher temperature	Eutrophic	27.9	5.3/19	sensitivity method	Increase in dominance	Elliott et al. (2006)
Esthwaite Water (UK) PROTECH	Higher temperature Lower flushing	Eutrophic	6.4	6.4/15.5	sensitivity method	Increase in dominance Increase in dominance	Elliott (2010)
Generic shallow lake PCLake	Higher temperature	Varies	N/A	N/A	sensitivity method	Increase in dominance if nutrients high and/or lake turbid	Mooij et al. (2007)
Loch Leven (UK) PROTECH	Higher temperature	Eutrophic	52.4	3.9/25.5	sensitivity method	None	Elliott and May (2008)
Windermere (UK) PROTECH	Higher temperature	Mesotrophic	314.5	21.3/64	sensitivity method	Increase in dominance	Elliott (in press)

For cyanobacteria, the carbon chemistry in the oceans is very important since they generate organic matter through photosynthesis which depends on the availability of inorganic carbon and changes with altered carbonate chemistry (Britton et al. 2016). Increasing CO_2 concentrations augment photosynthetic carbon fixation and growth of cyanobacteria at low dissolved inorganic carbon (DIC) levels in waters, whereas at high levels these processes are impaired in many species by increasing H^+ concentrations. The dependence of physiological reactions of phytoplankton on the DIC concentration and carbonate chemistry follows an optimum curve, (Fig. 4.1A) for cyanobacteria (Bach et al. 2015, Liu et al. 2017). In heterocystous cyanobacteria, OA can either impair or augment nitrogen fixation (Czerny et al. 2009, Wannicke et al. 2012).

In non-heterocystous cyanobacteria, the rate of nitrogen fixation in response to OA varies depending upon species (Eichner et al. 2014), for example, in *Trichodesmium* and other cyanobacteria, nitrogen fixation and growth are slowed by OA (Shi et al. 2012), whereas in other species little or no effects have been reported (Böttjer et al. 2014, Gradoville et al. 2014). *Trichodesmium* after having been adapted to OA for hundreds of generations showed an enhanced nitrogen fixation rate. Shifts in diel nitrogen fixation patterns were observed in adapted *Trichodesmium*, along with increased activity of a potentially regulatory DNA methyltransferase (Hutchins, Walworth et al. 2015). These differences indicate that the biodiversity and community structure of phytoplankton will change with rising CO_2 (Gradoville et al. 2014). Among the toxic and non-toxic strains of *Microcystis aeruginosa*, the toxic strain was found to be dominant in competition at low CO_2 levels (Van de Waal et al. 2011).

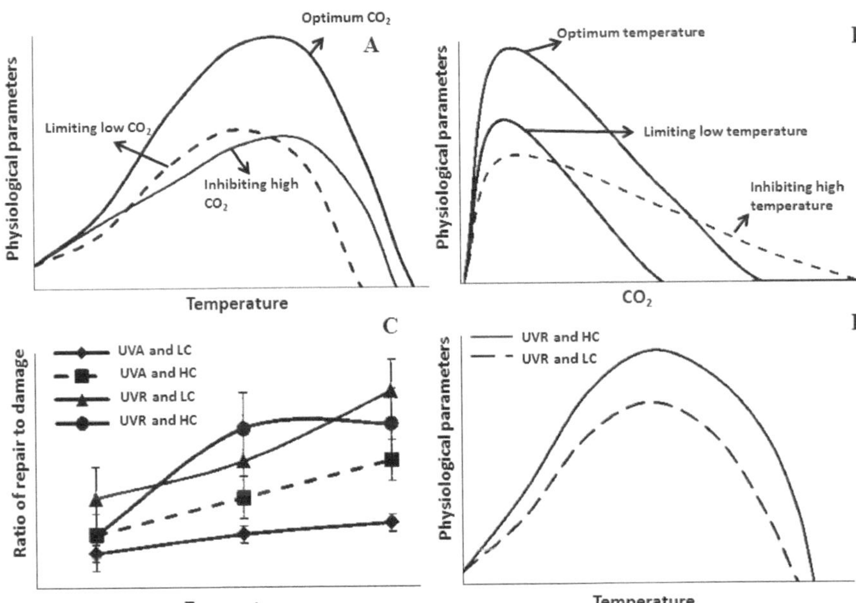

Fig. 4.1. Effects of enhanced CO_2 on cellular reactions depending on temperature (A). CO_2 response curve on increasing temperature (B). Effect of rising temperature and OA on repair to damage ratio induced by UV-A and UV-A + UV-B (C). Response of thermal reaction dependence with rising CO_2 under UVR (D). HC: Phytoplankton grew at 1000 μatm CO_2; LC: Phytoplankton grew at 390 μatm CO_2; UVR: UV-A + UV-B (Based on Li et al. 2012b, Sett et al. 2014, Listmann et al. 2016, Gao et al. 2017).

In marine ecosystems, significant differences in phytoplankton community structures could occur due to species-specific responses in photosynthetic organisms and their grazers to OA. The majority of the earlier investigations have been carried out under controlled conditions inside the laboratory (Grear et al. 2017), hence, it is compulsory to study the effects of OA on growth and photosynthesis of cyanobacteria under natural field conditions for assessment of the influence of multiple environmental factors on physiology and community structure of cyanobacteria.

Global Warming

For phytoplankton productivity, temperature acts as one of the most influential physical variables. In past three decades the surface temperature on earth has increased very fast in comparison to any preceding period since the Industrial Revolution (Hansen et al. 2006). IPCC estimates that surface ocean temperatures will increase by about 1°C (RCP2.6) to 3°C (RCP8.5) by the end of the century (IPCC 2014). Enzymatic efficiency, photosynthetic carbon incorporation, and other metabolic activities are controlled by the ambient temperatures (Beardall and Raven 2004).

As per the thermal performance curves, performance increases initially with increasing temperatures up to an optimum value followed by a sharp decline towards the upper limit of the permissive temperature window (Fig. 4.1B) (Huey and Stevenson 1979). These models depict the optimal temperature, the maximal rate of a response, the width of the permissive thermal window and overall rate across all temperatures (Zhang et al. 2014). Mesocosm experiments also showed that ocean warming can result in decreased phytoplankton biomass but the volume-specific primary productivity remains unaffected (Sommer et al. 2015). Despite having strong influence on cyanobacterial community composition, temperature is unlikely to drive competitive selection of harmful algal blooms forming cyanobacterial (HAB) species over non-HAB species (Wells et al. 2015).

According to satellite records, long-term decreases in globally integrated marine phytoplankton primary production and biomass are linked to global warming (Siegel et al. 2013). Different results from different time scales and over different regions as well as controversial interpretations of satellite observation indicate uncertainties on predicted responses of marine phytoplankton community composition to ocean warming (Henson et al. 2016). Respiration and photorespiration usually further increase with rising temperatures. Because of the large buffer capacity of the oceans this increase is only half of that on land. El Niño, the short-term effects cause warming of the Pacific Ocean and in general higher global temperatures, whereas La Niña events cause the opposite effect. Water column down to 700 m is affected by the increased temperature and this influences the ocean currents and the global weather as well (Fischetti 2013).

Most organisms, which usually dwell in the upper 400 m are affected by this warming. Temperature also governs phytoplankton growth and thus productivity of the aquatic food webs (Halpern et al. 2008). As temperature optima for different species are different, increased temperatures will result in altered species composition affecting the trophic food web (Pörtner and Farrell 2008). *Synechococcus* showed higher productivity at increased temperatures and CO_2, whereas in *Prochlorococcus* this does not happen, hence, the former species is better adapted to increasing temperatures and CO_2 concentrations (Fu et al. 2007). Studies showed that on increasing the temperature by 4°C, growth and nitrogen fixation were significantly augmented in *Trichodesmium* (Hutchins et al. 2007, Levitan et al. 2010). Mesocosm experiment showed that changes in species composition and growth were enhanced in an assemblage of phytoplankton and zooplankton at higher temperatures (Lewandowska and Sommer 2010).

UV Radiation

Cyanobacteria get exposed to lethal doses of ultraviolet radiation (UVR: UV-A 315–400 nm and UV-B 280–315 nm) radiation while harvesting solar energy for photosynthesis and nitrogen fixation in their natural habitats. The highly energetic UVR penetrates a few centimeters in rivers and lakes but deep into the water column up to 20 m in the ocean. This poses a major concern since UV-B radiation not only impairs photo-orientation and motility (Donkor and Häder 1991) but also affects a number of biochemical and physiological processes in cyanobacteria (Singh et al. 2010).

Due to enhanced stratification of the UML in the oceans, exposure of cyanobacteria to solar UV-B radiation is supposed to increase (Singh et al. 2010, Gao et al. 2012b), which is connected with global warming. Due to a thinner UML, cyanobacteria are exposed to higher levels of UVR and visible light resulting in decreased productivity (Häder et al. 2015), but at low irradiances, UV-A also contributes to carbon fixation by photosynthesis in marine phytoplankton (Gao et al. 2007). The light exposure of the phytoplankton is largely affected by the transparency of the water body (Boss et al. 2007).

Dissolved organic matter plays a crucial role in the phytoplankton composition and biomass productivity due to its high UV-B absorbing properties (Frost et al. 2007). The growth of *Oscillatoria priestleyi* (*Phormidium pseudopriestleyi*) and *Phormidium murrayi* (*Wilmottia murrayi*) was suppressed by 100% and 62%, respectively, followed by exposure of UV-B (Quesada and Vincent 1997). Exposure to PAR+UVR inhibited growth of *Anabaena* sp. (Han et al. 2003) while 40% inhibition was observed under solar UVR in *Anabaena* sp. PCC 7120 (Gao et al. 2007). UV-B radiation destroys the cellular constituents which have absorption maxima in the range of 280–315 nm, such as DNA and proteins, hence, affecting the cellular membrane permeability and inducing protein damage which eventually results in the cell death (Vincent and Roy 1993, Richa et al. 2016). Solar UVR suppressed differentiation of heterocysts in *Anabaena* sp. PCC 7120 (Gao et al. 2007) and alteration in the C:N ratio was observed following UV-B exposure suggesting its role for the altered spacing pattern of heterocysts in the filament (Sinha et al. 1996). Under solar UVR, compression and breakage of the spiral filaments of *Arthrospira platensis* have also been reported (Wu et al. 2005, Gao et al. 2008). *Anabaena flos-aquae* showed breakage of the filaments by UVR while PAR had no effect (Singh et al. 2010).

Decrease in phycobiliprotein as well as disassembly of the phycobilisome complex have been observed in several cyanobacteria under UV-B irradiation (Sinha et al. 1997). Photosynthetic parameters, such as O_2 evolution, $^{14}CO_2$ uptake and ribulose-1,5 bisphosphate carboxylase (RUBISCO) activity in cyanobacteria was found to be down-regulated by UVR (Sinha et al. 2008). The major constituents of PS-II reaction centers, i.e., D1 and D2 proteins, are also degraded on exposure to UV-B radiation (Campbell et al. 1998). UV-B exposure reduces photosynthetic activity as evidenced by down-regulation of several groups of mRNAs specifying proteins which were involved in light harvesting and photosynthesis (Huang et al. 2002). It also severely inhibits the process of nitrogen fixation due to the extreme sensitivity of the nitrogenase enzyme to UV-B (Lesser 2008).

Lipid peroxidation of polyunsaturated fatty acids (PUFA) *via* oxidative damage was observed by exposure of UV-B radiation which subsequently affects the integrity of cellular and thylakoid membranes (He and Häder 2002). In *Nostoc calcicola*, complete loss of protein bands between 14.2 and 45 kDa after 90 and 120 min of UV-B exposure was observed by Kumar et al. (Kumar et al. 1996). Similarly, proteome change was monitored following UV-B exposure through two-dimensional gel electrophoresis in *Nostoc commune* and it was observed that out of 1,350 protein spots, 493 were changed by UV-B radiation (Ehling-Schulz et al. 2002). Likewise, in *Synechocystis* sp. PCC 6803 translation activity was suppressed by oxidative stress and oxidation of the elongation factor EF-G played an important role in the inhibited translation activity (Kojima et al. 2007). The DNA lesions were also developed by direct UV-B absorption, such as pyrimidine (6-4) pyrimidone photoproducts (6–4 PPs), *cis-syn* cyclobutane pyrimidine dimers (CPDs) and their Dewar isomers (Singh et al. 2010).

Blotting and the chemiluminescence method were incorporated for examining UVR-induced formation of thymine dimers in *Nostoc, Anabaena* and *Scytonema* sp. (Sinha et al. 2001). Photosynthetic activity of *Scytonema javanicum* declined rapidly after irradiation with UV-B (Chen et al. 2013). Li et al. (2012a) showed that UV-B had a detrimental effect on the phosphatase activity of *Nostoc flagelliforme*. Contrastingly, metabolic functions in cyanobacteria are increased on rising temperatures and may mitigate UV-B-inflicted damage by increasing the repair mechanisms in the cells (Fig. 4.1C–D) but interaction of enhanced temperature, CO_2 and UVR needs to be studied further (Häder et al. 2015).

Terrestrial Runoff and Pollution

All around the globe, enhanced terrestrial runoff of sediments, nutrients and pollutants into the sea and other water bodies is of growing concern and this may lead to inhibitory or stimulatory effects due to nutrient subsidies along with increased light attenuation (GESAMP 2001). Frequent high-intensity rainfall will lead to increased storm runoff and soil erosion, resulting in large pulses of soils and sediments into adjacent coastal marine environments (Nunes et al. 2009), hence, increasing the nutrient availability, but reducing light penetration. Storm runoff and eroded soils contain organic matter and several mineral nutrients, such as P and N. Nutrient subsidy effects are dependent on the extent of light reduction caused by sediment and soil particles entering water bodies along with the storm runoff (Fouilland et al. 2012).

Phytoplankton have developed specific physiological responses in addition to compensatory Chl *a* production with changing light conditions (Goss and Jakob 2010,

Falkowski and Raven 2013). N-limited systems often favour prokaryotic nitrogen-fixing cyanobacteria. P can be the limiting nutrient for cyanobacteria growth, in the absence of N limitation (Martin and Fitzwater 1988) and cyanobacterial blooms are often related to P loadings from the surrounding environment (Seitzinger 1991). However, in case of P availability, biologically available Fe often becomes a major limiting factor in oceans (Martin et al. 1990) as well as estuarine and coastal ecosystems (Hutchins and Bruland 1998). For photosynthesis and nitrogen fixation, cyanobacteria require considerable amounts of Fe and P (Trick et al. 1995, Sañudo-Wilhelmy et al. 2001).

Laboratory studies showed that elevated Fe concentrations increase productivity and phycocyanin production in *Oscillatoria tenius* (Trick et al. 1995), nitrogen fixation in *Trichodesmium* sp. (Rueter et al. 1990) and toxin production by *Microcystis aeruginosa* (Utkilen and Gjølme 1995). Microplastic contamination of the marine environment has been a growing problem. Jambeck et al. (2015) calculated that 4.8 to 12.7 million metric tons of microplastic entered the ocean from 192 coastal countries in 2010 alone. Macroplastic debris as well as tiny plastic fragments, fibres and granules, collectively termed 'microplastics', are pollutants of serious concern (Ryan et al. 2009). They further degrade to be nanoplastic in size, although the smallest microparticle reportedly detected in the oceans at present is 1.6 μm in diameter (Galgani et al. 2010).

Algae/cyanobacteria show different levels of sensitivity to these microparticles (Hund-Rinke and Simon 2006). In marine ecosystems, oil spills are another major cause of concern (González et al. 2009). Usually maritime incidents cause oil contamination and have catastrophic impacts on the marine environment (Celik and Er 2006). Not only gas exchange is limited through oil slicks on the air-sea interface, but it also reduces light penetration into the water column hence, affecting phytoplanktonic photosynthesis. Only few studies have looked into the response of phytoplankton to sporadic oil inputs in natural conditions (Ohwada et al. 2003). Some have reported negative effects on phytoplankton (Sargian et al. 2005) whereas others found between stimulatory effects (Oviatt et al. 1982) and differences in sensitivity varied with different taxonomic groups.

Climate Change and Bloom Formation

Due to climate change effects and eutrophication frequency of cyanobacterial blooms in freshwaters are expected to increase globally. Anthropogenically altered water and land use, and increased eutrophication largely propagate HABs of toxic cyanobacteria (Manning and Nobles 2017, Nguyen et al. 2017). Increase in water temperature and changes in the large-scale hydrological cycle are evident through a greater variability in the temporal and spatial distribution in rainfall (IPCC 2007), but much less information is available regarding the occurrence of cyanobacterial blooms in response to changes in rainfall patterns.

It has been predicted that a warmer world of the late twenty-first century would favor cyanobacteria in terms of biomass and dominance as well (IPCC 2007). Field data of 1997–2015 revealed that *Dolichospermum* (*Anabaena*) was the most frequent freshwater genus which was distributed widely, followed by *Microcystis* (Haakonsson et al. 2017). In lotic ecosystems (freshwater habitat where there is a constant current in one direction), rainfall and temperature explained 27% of the variation in total cyanobacteria biovolumes, while temperature explained 28% and 19% of *Microcystis biovolume* and *Dolichospermum*, respectively. Strains of *Anabaena, Dolichospermum, Oscillatoria, Microcystis, Planktothrix, Lyngbya, and Aphanizonmenon* are common species present in cyanobacterial HABs, which thrive well in nutrient-rich, warm conditions (Manning and Nobles 2017). Heavy rainfall followed by low light conditions might favor non-toxic strains, whereas inorganic nutrient input might promote toxic cyanobacterial HABs (Reichwaldt and Ghadouani 2012).

Consequences of climate change, such as increased vertical stratification, intensification of droughts and storms, salinization, their impacts on nutrient delivery and flushing characteristics of affected water bodies as well, play synergistic roles in bloom formation frequency, geographic distribution, intensity and duration. Shifts in critical nutrient thresholds are caused by rising temperatures which lead to development of cyanobacterial blooms (Paerl et al. 2016), thus for blooms control nutrient reductions need to be more aggressively pursued in response to climatic changes taking place globally.

Conclusion

Cyanobacteria remain exposed to several environmental stresses, such as increasing OA, rising temperatures, pollution and enhanced exposures to UVR. Genetic alterations may occur in cyanobacteria owing to their short lifetime, and it might allow them to adapt to the new environmental conditions with lower pH values in the ocean (Collins et al. 2006). Impacts of global climate change in terms of species composition and phytoplankton productivity still need to be assessed (Tagliabue et al. 2011). After the increase in temperature from 16.5 to 22.5°C in the Kiel Fjord the structure of the phytoplankton community changed (Paul et al. 2016). Additionally, limitation of inorganic macronutrient alters the effect of OA on phytoplankton (Verspagen et al. 2014).

Photosynthetic carbon fixation is reduced in phytoplankton under OA on limiting the dissolved inorganic macronutrients (Matthiessen et al. 2012). Damaging effects of UVR on phytoplankton can be reduced by dissolved inorganic macronutrients and ocean warming (Doyle et al. 2005). During summer, less UV-induced inhibition of photosynthetic carbon fixation was observed in phytoplankton assemblages as compared to winter (Wu et al. 2010). Synthesis of UV-absorbing pigments as well as repair of UV-induced thymine dimers were found to be facilitated by elevated temperatures. Garcia-Corral et al. (Garcia-Corral et al. 2015) showed that an increase of temperature by 3°C worsened the negative effect of UV-B on net community production. Only a few short-term reports are present which investigated the cumulative effects of increasing temperatures, OA, and UVR on phytoplankton (Tong et al. 2017).

Increased temperature and CO_2 synergistically positively nullified the UV-induced inhibition and hence augmented the photosynthetic carbon fixation (Li et al. 2012b); however, increasing temperature or CO_2 concentration may interact neutrally or antagonistically with UVR (Xu and Gao 2015). Increased salinity and drought have shown to promote cyanobacterial blooms and make them one of the dominant groups of microorganisms which are capable of surviving under extreme climatic changes (Crusberg and and Eslamian 2017). It is also suggested that although the climate change can change cyanobacteria abundance and the timing of bloom events in lakes, the amount of phytoplanktonic biomass produced annually is not affected directly by these changes.

Nutrient consumption by the phytoplankton community is increased by warmer waters in the spring which causes nitrogen limitation later in the year, hence posing advantage to the nitrogen-fixing cyanobacteria. It is also predicted that an increase in cyanobacteria dominance will lead to poorer energy flow to higher trophic levels owing to their relatively poor palatability for zooplankton. Conclusively, these environmental factors may interact antagonistically as well as synergistically to alter physiological responses to that of an individual stressor. Our knowledge on how multiple stress factors, such as OA, UVR, ocean warming and salinization might interact positively or negatively in different regions or under different climate scenario is still limited because of lack of thorough understanding of the combined effects of temperature, CO_2, salinity and UVR. Hence further comprehensive field studies are required to assess and predict the effects of global climate change on cyanobacteria.

Acknowledgments

J. Pathak and Rajneesh are thankful to the Council of Scientific and Industrial Research, New Delhi, India, (09/013/0515/2013-EMR-I) and Department of Biotechnology, Goverment of India, (DBT-JRF/13/AL/143/2158), respectively, for the grants in the form of senior research fellowships. H. Ahmed is thankful to UGC (UGC-JRF-21/12/2014 (ii) EU-V) for financial support in the form of senior research fellowship.

References

Bach, L.T., L. Mackinder, K.G. Schulz, G. Wheeler, D.C. Schroeder, C. Brownlee and U. Riebesell. 2013. Dissecting the impact of CO_2 and pH on the mechanisms of photosynthesis and calcification in the coccolithophore *Emiliania huxleyi*. New Phytologist 199: 121–134.

Bach, L.T., U. Riebesell, M.A. Gutowska, L. Federwisch and K.G. Schulz. 2015. A unifying concept of coccolithophore sensitivity to changing carbonate chemistry embedded in an ecological framework. Progress in Oceanography 135: 125–138.

Bais, A., R. McKenzie, G. Bernhard, P. Aucamp, M. Ilyas, S. Madronich and K. Tourpali. 2015. Ozone depletion and climate change: Impacts on UV radiation. Photochemical & Photobiological Sciences 14: 19–52.

Beardall, J. and J.A. Raven. 2004. The potential effects of global climate change on microalgal photosynthesis, growth and ecology. Phycologia 43: 26–40.

Behrenfeld, M., R. O'Malley, D. Siegel, C. McClain, J. Sarmiento, G. Feldman, A. Milligan, P. Falkowski, R. Letelier and E. Boss. 2006. Climate-driven trends in contemporary ocean productivity. Nature 444: 752–755.

Boss, E.S., R. Collier, G. Larson, K. Fennel and W.S. Pegau. 2007. Measurements of spectral optical properties and their relation to biogeochemical variables and processes in Crater Lake, Crater Lake National Park, OR. Hydrobiologia 574: 149–159.

Böttjer, D., D.M. Karl, R.M. Letelier, D.A. Viviani and M.J. Church. 2014. Experimental assessment of diazotroph responses to elevated seawater pCO_2 in the North Pacific Subtropical Gyre. Global Biogeochemical Cycles 28: 601–616.

Bozzo, G.G. and B. Colman. 2000. The induction of inorganic carbon transport and external carbonic anhydrase in *Chlamydomonas reinhardtii* is regulated by external CO_2 concentration. Plant Cell Environment 23: 1137–1144.

Britton, D., C.E. Cornwall, A.T. Revill, C.L. Hurd and C.R. Johnson. 2016. Ocean acidification reverses the positive effects of seawater pH fluctuations on growth and photosynthesis of the habitat-forming kelp, *Ecklonia radiata*. Scientific Reports 6.

Campbell, D., M.J. Eriksson, G. Öquist, P. Gustafsson and A.K. Clarke. 1998. The cyanobacterium *Synechococcus* resists UV-B by exchanging photosystem II reaction-center D1 proteins. Proceedings of the National Academy of Sciences of the Unied States of America 95: 364–369.

Celik, M. and I. Er. 2006. Application requirements of catastrophe theory in maritime transportation industry. International conference on maritime transport.

Charette, M.A. and W.H.F. Smith. 2010. The Volume of Earth's Ocean. Oceanography 23: 112–114.

Chen, L., S. Deng, R. De Philippis, W. Tian, H. Wu and J. Wang. 2013. UV-B resistance as a criterion for the selection of desert microalgae to be utilized for inoculating desert soils. Journal of Applied Phycology 25: 1009–1015.

Collins, S., D. Sültemeyer and G. Bell. 2006. Changes in C uptake in populations of *Chlamydomonas reinhardtii* selected at high CO_2. Plant Cell Environment 29: 1812–1819.

Crusberg, T.C. and S. and S. Eslamian. Drought and water quality. pp. 205-217 *In*: Eslamian, S. and F. A. Eslamian [eds.] 2017. Handbook of Drought and Water Scarcity: Environmental Impacts and Analysis of Drought and Water Scarcity. CRC Press.

Czerny, J., J. Barcelos e Ramos and U. Riebesell. 2009. Influence of elevated CO_2 concentrations on cell division and nitrogen fixation rates in the bloom-forming cyanobacterium *Nodularia spumigena*. Biogeosciences (BG) 6: 1865–1875.

Deininger, A., C.L. Faithfull, K. Lange, T. Bayer, F. Vidussi and A. Liess. 2016. Simulated terrestrial runoff triggered a phytoplankton succession and changed seston stoichiometry in coastal lagoon mesocosms. Marine Environmental Research 119: 40–50.

Domingues, R.B., C.C. Guerra, H.M. Galvão, V. Brotas and A.B. Barbosa. 2017. Short-term interactive effects of ultraviolet radiation, carbon dioxide and nutrient enrichment on phytoplankton in a shallow coastal lagoon. Aquatic Ecology 51: 91–105.

Donkor, V. and D.-P. Häder. 1991. Effects of solar and ultraviolet radiation on motility, photomovement and pigmentation in filamentous gliding cyanobacteria. FEMS Microbiology Ecology 86: 159–168.

Doyle, S.A., J.E. Saros and C.E. Williamson. 2005. Interactive effects of temperature and nutrient limitation on the response of alpine phytoplankton growth to ultraviolet radiation. Limnology and Oceanography 50: 1362–1367.

Ehling-Schulz, M., S. Schulz, R. Wait, A. Görg and S. Scherer. 2002. The UV-B stimulon of the terrestial cyanobacterium *Nostoc commune* comprises early shock proteins and late acclimation proteins. Molecular Microbiology 46: 827–843.

Eichner, M., B. Rost and S.A. Kranz. 2014. Diversity of ocean acidification effects on marine N_2 fixers. Journal of Experimental Marine Biology and Ecology 457: 199–207.

Elliott, J.A. 2012. Is the future blue-green? A review of the current model predictions of how climate change could affect pelagic freshwater cyanobacteria. Water Research 46: 1364–1371.

Falkowski, P.G. and J.A. Raven. 2013. Aquatic Photosynthesis. Princeton University Press.

Fischetti, M. 2013. Deep heat threatens marine life. Scientific American 308: 92.

Fouilland, E., A. Trottet, C. Bancon-Montigny, M. Bouvy, E. Le Floc'h, J.-L. Gonzalez, E. Hatey, S. Mas, B. Mostajir and J. Nouguier. 2012. Impact of a river flash flood on microbial carbon and nitrogen production in a Mediterranean Lagoon (Thau Lagoon, France). Estuarine, Coastal and Shelf Science 113: 192–204.

Frost, P.C., C.T. Cherrier, J.H. Larson, S. Bridgham and G.A. Lamberti. 2007. Effects of dissolved organic matter and ultraviolet radiation on the accrual, stoichiometry and algal taxonomy of stream periphyton. Freshwater Biol. 52: 319–330.

Fu, F.X., M.E. Warner, Y. Zhang, Y. Feng and D.A. Hutchins. 2007. Effects of increased temperature and CO_2 on photosynthesis, growth, and elemental ratios in marine *Synechococcus* and *Prochlorococcus* (Cyanobacteria). Journal of Phycology 43: 485–496.

Galgani, F., D. Fleet, J. Van Franeker, S. Katsanevakis, T. Maes, J. Mouat, L. Oosterbaan, I. Poitou, G. Hanke and R. Thompson. 2010. Marine Strategy Framework directive-Task Group 10 Report marine litter do not cause harm to the coastal and marine environment. Report on the identification of descriptors for the Good Environmental Status of European Seas regarding marine litter under the Marine Strategy Framework Directive. Office for Official Publications of the European Communities.

Gao, G., P. Jin, N. Liu, F. Li, S. Tong, D.A. Hutchins and K. Gao. 2016. Combined effects of elevated pCO_2 and temperature on biomass and carbon fixation of phytoplankton assemblages in the northern South China Sea. Biogeosciences Discuss.

Gao, K., E.W. Helbling, D.-P. Häder and D.A. Hutchins. 2012a. Responses of marine primary producers to interactions between ocean acidification, solar radiation, and warming. Marine Ecology Progress Series 470: 167–189

Gao, K., P. Li, T. Watanabe and E.W. Helbling. 2008. Combined effects of ultraviolet radiation and temperature on morphology, photosynthesis, and DNA of *Arthrospira* (*Spirulina*) *platensis* (Cyanophyta). Journal of Phycology 44: 777–786.

Gao, K., H. Yu and M.T. Brown. 2007. Solar PAR and UV radiation affects the physiology and morphology of the cyanobacterium *Anabaena* sp. PCC 7120. Journal of Photochemistry and Photobiology B: Biology 89: 117–124.

Gao, K., Y. Zhang and D.-P. Häder. 2017. Individual and interactive effects of ocean acidification, global warming, and UV radiation on phytoplankton. Journal of Applied Phycology 30: 1–17.

Gao, K.S., J.T. Xu, G. Gao, Y.H. Li, D.A. Hutchins, B.Q. Huang, L. Wang, Y. Zheng, P. Jin, X.N. Cai, D.-P. Häder, W. Li, K. Xu, N.N. Liu and U. Riebesell. 2012b. Rising CO_2 and increased light exposure synergistically reduce marine primary productivity. Nature Climate Change 2: 519–523.

Garcia-Corral, L.S., J. Martinez-Ayala, C.M. Duarte and S. Agusti. 2015. Experimental assessment of cumulative temperature and UV-B radiation effects on Mediterranean plankton metabolism. Frontiers in Marine Science 2: 48.

Gattuso, J.-P., A. Magnan, R. Billé, W. Cheung, E. Howes, F. Joos, D. Allemand, L. Bopp, S. Cooley and C. Eakin. 2015. Contrasting futures for ocean and society from different anthropogenic CO_2 emissions scenarios. Science 349: aac4722.

GESAMP. 2001. Protecting the oceans from land-based activities. Land-based sources and activities affecting the quality and uses of the marine, coastal and associated freshwater environment. United Nations Environment Program, Nairobi.

González, J., F. Figueiras, M. Aranguren-Gassis, B. Crespo, E. Fernández, X. Morán and M. Nieto-Cid. 2009. Effect of a simulated oil spill on natural assemblages of marine phytoplankton enclosed in microcosms. Estuarine, Coastal and Shelf Science 83: 265–276.

Goss, R. and T. Jakob. 2010. Regulation and function of xanthophyll cycle-dependent photoprotection in algae. Photosynthesis Research 106: 103–122.

Gradoville, M.R., A.E. White, D. Böttjer, M.J. Church and R.M. Letelier. 2014. Diversity trumps acidification: Lack of evidence for carbon dioxide enhancement of *Trichodesmium* community nitrogen or carbon fixation at Station ALOHA. Limnology and Oceanography 59: 645–659.

Grear, J.S., T.A. Rynearson, A.L. Montalbano, B. Govenar and S. Menden-Deuer. 2017. pCO_2 effects on species composition and growth of an estuarine phytoplankton community. Estuarine, Coastal and Shelf Science 190: 40–49.

Haakonsson, S., L. Rodríguez-Gallego, A. Somma and S. Bonilla. 2017. Temperature and precipitation shape the distribution of harmful cyanobacteria in subtropical lotic and lentic ecosystems. Science of the Total Environment 609: 1132–1139.

Häder, D.-P., H. Kumar, R. Smith and R. Worrest. 2007. Effects of solar UV radiation on aquatic ecosystems and interactions with climate change. Photochemical & Photobiological Sciences 6: 267–285.

Häder, D.-P., C.E. Williamson, S.-A. Wängberg, M. Rautio, K.C. Rose, K. Gao, E.W. Helbling, R.P. Sinha and R. Worrest. 2015. Effects of UV radiation on aquatic ecosystems and interactions with other environmental factors. Photochemical & Photobiological Sciences 14: 108–126.

Halpern, B.S., S. Walbridge, K.A. Selkoe, C.V. Kappel, F. Fiorenza Micheli, C. Caterina D'Agrosa, J.F. Bruno, K.S. Casey, C. Ebert, H.E. Fox, R. Fujita, D. Heinemann, H.S. Lenihan, E.M.P. Madin, M.T. Perry, E.R. Selig, M. Spalding, R. Steneck and R. Watson. 2008. A global map of human impact on marine ecosystems. Science 319: 948–952.

Hansen, J., R. Ruedy, M. Sato and K. Lo. 2010. Global surface temperature change. Reviews of Geophysics 48.

Hansen, J., M. Sato, R. Ruedy, K. Lo, D.W. Lea and M. Medina-Elizade. 2006. Global temperature change. Proceedings of the National Academy of Sciences 103: 14288–14293.

He, Y.Y. and D.-P. Häder. 2002. Reactive oxygen species and UV-B: effect on cyanobacteria. Photochemical & Photobiological Sciences 1: 729–736.

Henson, S.A., C. Beaulieu and R. Lampitt. 2016. Observing climate change trends in ocean biogeochemistry: when and where. Global Change Biology 22: 1561–1571.

Hönisch, B., A. Ridgwell, D.N. Schmidt, E. Thomas, S.J. Gibbs, A. Sluijs, R. Zeebe, L. Kump, R.C. Martindale and S.E. Greene. 2012. The geological record of ocean acidification. Science 335: 1058–1063.

Huang, L., M.P. McCluskey, H. Ni and R.A. LaRossa. 2002. Global gene expression profiles of the cyanobacterium *Synechocystis* sp. strain PCC 6803 in response to irradiation with UV-B and white light. Journal of Bacteriology 184: 6845–6858.

Huey, R.B. and R. Stevenson. 1979. Integrating thermal physiology and ecology of ectotherms: a discussion of approaches. American Zoologist 19: 357–366.

Hund-Rinke, K. and M. Simon. 2006. Ecotoxic effect of photocatalytic active nanoparticles (TiO_2) on algae and daphnids (8 pp). Environmental Science and Pollution Research 13: 225–232.

Hutchins, D., F.-X. Fu, Y. Zhang, M.E. Warner, Y. Feng, K. Portune, P.W. Bernhardt and M.R. Mulholland. 2007. CO_2 control of *Trichodesmium* N_2 fixation, photosynthesis, growth rates, and elemental ratios: implications for past, present, and future ocean biogeochemistry. Limnol Oceanogr 52: 1293–1304.

Hutchins, D.A. and K.W. Bruland. 1998. Iron-limited diatom growth and Si: N uptake ratios in a coastal upwelling regime. Nature 393: 561.

IPCC. 2007. Summary for Policymakers. In: *Climate Change 2007:* The Physical Science Basis. *Contribution of Working Group 1 to the Fourth Assessment report of the Intergovernmental Panel on Climate Change* [Solomon, S., D. Qin, M. Manning, Z. Chen, M. Marquis, K.B Averyt, M. Tignor and H.L. Miller (eds.)]. IPCC, Cambridge University Press, Cambridge, United Kingdom and New York, NY, USA: 1–18.

IPCC. 2013. Summary for Policymakers: Climate change 2013 - The physical science basis. Working Group 1 Contribution to the IPCC Fifth Assessment Report. Stocker, T. F. S. T. G., D. Qin, G.-K. Plattner, M. Tignor, S. K. Allen, J. Boschung, A. Nauels, Y. Xia, V. Bex and P. M. Midgley: 1–38.

IPCC. 2014. Climate Change 2014: Impacts, Adaptation, and Vulnerability. Part B: Regional Aspects. Contribution of Working Group II to the Fifth Assessment Report of the Intergovernmental Panel on Climate Change, 2014.

Jambeck, J.R., R. Geyer, C. Wilcox, T.R. Siegler, M. Perryman, A. Andrady, R. Narayan and K.L. Law. 2015. Plastic waste inputs from land into the ocean. Science 347: 768–771.

Kojima, K., M. Oshita, Y. Nanjo, K. Kasai, Y. Tozawa, H. Hayashi and Y. Nishiyama. 2007. Oxidation of elongation factor G inhibits the synthesis of the D1 protein of photosystem II. Mol. Microbiol. 65: 936–947.

Kumar, A., R.P. Sinha and D.-P. Häder. 1996. Effect of UV-B on enzymes of nitrogen metabolism in the cyanobacterium *Nostoc calcicola*. Journal of Plant Physiology 148: 86–91.

Landschützer, P., N. Gruber, D. Bakker and U. Schuster. 2014. Recent variability of the global ocean carbon sink. Global Biogeochemical Cycles 28: 927–949.

Lesser, M.P. 2008. Effects of ultraviolet radiation on productivity and nitrogen fixation in the cyanobacterium, *Anabaena* sp. (Newton's strain). Hydrobiologia 598: 1–9.

Levitan, O., S.A. Kranz, D. Spungin, O. Prasil, B. Rost and I. Berman-Frank. 2010. Combined effects of CO_2 and light on the N_2-fixing cyanobacterium *Trichodesmium* IMS101: a mechanistic view. Plant Physiology 154: 346–356.

Lewandowska, A. and U. Sommer. 2010. Climate change and the spring bloom: a mesocosm study on the influence of light and temperature on phytoplankton and mesozooplankton. Marine Ecology Progress Series 405: 101–111.

Li, P., W. Liu and K. Gao. 2012a. Effects of temperature, pH, and UV radiation on alkaline phosphatase activity in the terrestrial cyanobacterium *Nostoc flagelliforme*. Journal of Applied Phycology 25: 1031–1038.

Li, Y., K. Gao, V. Villafañe and E. Helbling. 2012b. Ocean acidification mediates photosynthetic response to UV radiation and temperature increase in the diatom *Phaeodactylum tricornutum*. Biogeosciences 9: 3931–3942.

Liu, N., J. Beardall and K. Gao. 2017. Elevated CO_2 and associated seawater chemistry do not benefit a model diatom grown with increased availability of light. Aquatic Microbial Ecology 79: 137–147.

Manning, S.R. and D.R. Nobles. 2017. Impact of global warming on water toxicity: Cyanotoxins. Current Opinion in Food Science 18: 14–20.

Martin, J.H. and S.E. Fitzwater. 1988. Iron deficiency limits phytoplankton growth in the north-east Pacific subarctic. Nature 331: 341–343.

Martin, J.H., S.E. Fitzwater and R.M. Gordon. 1990. Iron deficiency limits phytoplankton growth in Antarctic waters. Global Biogeochemical Cycles 4: 5–12.

Matthiessen, B., S.L. Eggers and S. Krug. 2012. High nitrate to phosphorus regime attenuates negative effects of rising pCO_2 on total population carbon accumulation. Biogeosciences (BG) 9: 1195–1203.

Millero, F.J. 2007. The marine inorganic carbon cycle. Chemical Reviews 107: 308-341.

Moorman, M.C., T. Augspurger, J. Stanton and S. A.D. 2017. Where's the grass? Disappearing submerged aquatic vegetation and declining water quality in Lake Mattamuskeet. Fish Wildlife Management 8: 401–417.

Mostofa, K., C. Liu, W. Zhai, M. Minella, D. Vione, K. Gao, D. Minakata, T. Arakaki, T. Yoshioka and K. Hayakawa. 2015. Reviews and Syntheses: Ocean acidification and its potential impacts on marine ecosystems. Biogeosciences Discussions 12.

Nguyen, H.H., F. Recknagel, W. Meyer, J. Frizenschaf and M.K. Shrestha. 2017. Modelling the impacts of altered management practices, land use and climate changes on the water quality of the Millbrook catchment-reservoir system in South Australia. Journal of Environmental Management 202: 1–11.

NOAA National Centers for Environmental Information. published online June 2017. State of the Climate: Global Climate Report for May 2017.

Nunes, J., J. Seixas, J. Keizer and A. Ferreira. 2009. Sensitivity of runoff and soil erosion to climate change in two Mediterranean watersheds. Part I: model parameterization and evaluation. Hydrological Processes 23: 1202–1211.

Ohwada, K., M. Nishimura, M. Wada, H. Nomura, A. Shibata, K. Okamoto, K. Toyoda, A. Yoshida, H. Takada and M. Yamada. 2003. Study of the effect of water-soluble fractions of heavy-oil on coastal marine organisms using enclosed ecosystems, mesocosms. Marine Pollution Bulletin 47: 78–84.

Oviatt, C., J. Frithsen, J. Gearing and P. Gearing. 1982. Low chronic additions of No. 2 fuel oil: chemical behavior, biological impact and recovery in a simulated estuarine environment. Marine Ecology Progress Series: 121–136.

Pachauri, R.K., M. Allen, V. Barros, J. Broome, W. Cramer, R. Christ, J. Church, L. Clarke, Q. Dahe and P. Dasgupta. 2014. Climate Change 2014: Synthesis Report. Contribution of Working Groups I, II and III to the Fifth Assessment Report of the Intergovernmental Panel on Climate Change.

Paerl, H.W., W.S. Gardner, K.E. Havens, A.R. Joyner, M.J. McCarthy, S.E. Newell, B. Qin and J.T. Scott. 2016. Mitigating cyanobacterial harmful algal blooms in aquatic ecosystems impacted by climate change and anthropogenic nutrients. Harmful Algae 54: 213–222.

Paul, C., U. Sommer, J. Garzke, M. Moustaka-Gouni, A. Paul and B. Matthiessen. 2016. Effects of increased CO_2 concentration on nutrient limited coastal summer plankton depend on temperature. Limnology and Oceanography.

Pörtner, H.O. and A.P. Farrell. 2008. Physiology and climate change. Science 322: 690–692.

Quesada, A. and W.F. Vincent. 1997. Strategies of adaptation by Antarctic cyanobacteria to ultraviolet radiation. European Journal of Phycology 32: 335–342.

Raven, J., K. Caldeira, H. Elderfield, O. Hoegh-Guldberg, P. Liss, U. Riebesell, J. Shepherd, C. Turley and A. Watson. 2005. Ocean acidification due to increasing atmospheric carbon dioxide. The Royal Society. London.

Raven, J.A., M. Giordano, J. Beardall and S.C. Maberly. 2012. Algal evolution in relation to atmospheric CO_2: carboxylases, carbon-concentrating mechanisms and carbon oxidation cycles. Phil. Trans. R. Soc. B 367: 493–507.

Reichwaldt, E.S. and A. Ghadouani. 2012. Effects of rainfall patterns on toxic cyanobacterial blooms in a changing climate: between simplistic scenarios and complex dynamics. Water Research 46: 1372–1393.

Richa, R.P. Sinha and D.-P. Häder. 2016. Effects of global change, including UV and UV screening compounds. pp. 373–409. *In*: Borowitzka, M., J. Beardall and J. Raven (eds.). The Physiology of Microalgae. Springer, Cham.

Riebesell, U. and J.-P. Gattuso. 2015. Lessons learned from ocean acidification research. Nature Climate Change 5: 12–14.

Riebesell, U. and P.D. Tortell. 2011. Effects of ocean acidification on pelagic organisms and ecosystems. pp. 99–116. *In*: Gattuso, J. P. and L. Hansson (eds.). Ocean Acidification. Oxford University Press, Oxford, UK.

Riebesell, U., D.A. Wolf-Gladrow and V. Smetacek. 1993. Carbon dioxide limitation of marine phytoplankton growth rates. Nature 361: 249–251.

Rueter, J.G., K. Ohki and Y. Fujita. 1990. The effect of iron nutrition on photosynthesis and nitrogen fixation in cultures of *Trichodesmium* (Cyanophyceae). Journal of Phycology 26: 30–35.

Ryan, P.G., C.J. Moore, J.A. van Franeker and C.L. Moloney. 2009. Monitoring the abundance of plastic debris in the marine environment. Philosophical Transactions of the Royal Society B: Biological Sciences 364: 1999–2012.

Sabine, C.L., R.A. Feely, N. Gruber, R.M. Key, K. Lee, J.L. Bullister, R. Wanninkhof, C.S. Won, D.W.R. Wallace, B. Tilbrook, F.J. Millero, T.-H. Peng, A. Kozyr, T. Ono and A.F. Rios. 2004. The oceanic sink for anthropogenic CO_2. Science 305: 367–371.

Sañudo-Wilhelmy, S.A., A.B. Kustka, C.J. Gobler, D.A. Hutchins, M. Yang, K. Lwiza, J. Burns, D.G. Capone, J.A. Raven and E.J. Carpenter. 2001. Phosphorus limitation of nitrogen fixation by *Trichodesmium* in the central Atlantic Ocean. Nature 411: 66.

Sargian, P., B. Mostajir, K. Chatila, G.A. Ferreyra, É. Pelletier and S. Demers. 2005. Non-synergistic effects of water-soluble crude oil and enhanced ultraviolet-B radiation on a natural plankton assemblage. Marine Ecology Progress Series 294: 63–77.

Schädel, C., M.K.-F. Bader, E.A. Schuur, C. Biasi, R. Bracho, P. Čapek, S. De Baets, K. Diáková, J. Ernakovich and C. Estop-Aragones. 2016. Potential carbon emissions dominated by carbon dioxide from thawed permafrost soils. Nature Climate Change 6: 950–953.

Schippers, P., M. Lürling and M. Scheffer. 2004. Increase of atmospheric CO_2 promotes phytoplankton productivity. Ecology Letters 7: 446–451.

Seitzinger, S.P. 1991. The effect of pH on the release of phosphorus from Potomac Estuary sediments: Implications for blue-green algal blooms. Estuarine, Coastal and Shelf Science 33: 409–418.

Shi, D., S.A. Kranz, J.-M. Kim and F.M. Morel. 2012. Ocean acidification slows nitrogen fixation and growth in the dominant diazotroph *Trichodesmium* under low-iron conditions. Proceedings of the National Academy of Sciences 109: E3094–E3100.

Siegel, D.A., M.J. Behrenfeld, S. Maritorena, C.R. McClain, D. Antoine, S.W. Bailey, P.S. Bontempi, E.S. Boss, H.M. Dierssen and S.C. Doney. 2013. Regional to global assessments of phytoplankton dynamics from the SeaWiFS mission. Remote Sensing of Environment 135: 77–91.

Singh, S.P., D.-P. Häder and R.P. Sinha. 2010. Cyanobacteria and ultraviolet radiation (UVR) stress: mitigation strategies. Ageing Research Reviews 9: 79–90.

Sinha, R.P., M. Dautz and D.-P. Häder. 2001. A simple and efficient method for the quantitative analysis of thymine dimers in cyanobacteria, phytoplankton and macroalgae. Acta Protozoologica 40: 187–195.

Sinha, R.P., R.P. Rastogi, N.K. Ambasht and D.-P. Häder. 2008. Live of wetland cyanobacteria under enhancing solar UV-B radiation. Proceedings of the National Academy of Science India B 78: 53–65.

Sinha, R.P., N. Singh, A. Kumar, H.D. Kumar and D.-P. Häder. 1997. Impacts of ultraviolet-B irradiation on nitrogen-fixing cyanobacteria of rice paddy fields. Journal of Plant Physiology 150: 188–193.

Sinha, R.P., N. Singh, A. Kumar, H.D. Kumar, M. Häder and D.-P. Häder. 1996. Effects of UV irradiation on certain physiological and biochemical processes in cyanobacteria. Journal of Photochemistry and Photobiology B: Biology 32: 107–113.

Sommer, U., C. Paul and M. Moustaka-Gouni. 2015. Warming and ocean acidification effects on phytoplankton—from species shifts to size shifts within species in a mesocosm experiment. PLoS One 10: e0125239.

Tagliabue, A., L. Bopp and M. Gehlen. 2011. The response of marine carbon and nutrient cycles to ocean acidification: Large uncertainties related to phytoplankton physiological assumptions. Global Biogeochemical Cycles 25: GB3017.

Tong, S., D. Hutchins and K. Gao. 2017. Physiological and biochemical responses of *Emiliania huxleyi* to ocean acidification and warming are modulated by UV radiation. Biogeosciences Discussions.

Trick, C.G., S.W. Wilhelm and C.M. Brown. 1995. Alterations in cell pigmentation, protein expression, and photosynthetic capacity of the cyanobacterium *Oscillatoria tenuis* grown under low iron conditions. Canadian Journal of Microbiology 41: 1117–1123.

United Nations Environment Programme, E.E.A.P. 2017. Environmental Effects of Ozone Depletion and its Interactions with Climate Change. Progress Report, 2016. Photochem. Photobiol. Sci. 16: 107–145.

Utkilen, H. and N. Gjølme. 1995. Iron-stimulated toxin production in *Microcystis aeruginosa*. Applied and Environmental Microbiology 61: 797–800.

Vaishampayan, A., R.P. Sinha, D.-P. Häder, T. Dey, A.K. Gupta, U. Bhan and A.L. Rao. 2001. Cyanobacterial biofertilizers in rice agriculture. Botanical Review 67: 453–516.

Van de Poll, W., G. Kulk, K. Timmermans, C. Brussaard, H. Van Der Woerd, M. Kehoe, K. Mojica, R. Visser, P. Rozema and A. Buma. 2013. Phytoplankton chlorophyll a biomass, composition, and productivity along a temperature and stratification gradient in the northeast Atlantic Ocean. Biogeosciences 10: 4227–4240.

Van de Waal, D.B., J.M. Verspagen, J.F. Finke, V. Vournazou, A.K. Immers, W.E.A. Kardinaal, L. Tonk, S. Becker, E. Van Donk and P.M. Visser. 2011. Reversal in competitive dominance of a toxic versus non-toxic cyanobacterium in response to rising CO_2. The ISME Journal 5: 1438.

Verspagen, J.M., D.B. Van de Waal, J.F. Finke, P.M. Visser and J. Huisman. 2014. Contrasting effects of rising CO_2 on primary production and ecological stoichiometry at different nutrient levels. Ecology Letters 17: 951–960.

Vincent, W.F. and S. Roy. 1993. Solar ultraviolet-B radiation and aquatic primary production: damage, protection, and recovery. Environmental Reviews 1: 1–12.

Wang, G., S.-P. Xie, R.X. Huang and C. Chen. 2015. Robust warming pattern of global subtropical oceans and its mechanism. Journal of Climate 28: 8574–8584.

Wannicke, N., S. Endres, A. Engel, H.-P. Grossart, M. Nausch, J. Unger and M. Voss. 2012. Response of *Nodularia spumigena* to pCO_2–Part 1: Growth, production and nitrogen cycling. Biogeosciences 9: 2973–2988.

Wells, M.L., V.L. Trainer, T.J. Smayda, B.S. Karlson, C.G. Trick, R.M. Kudela, A. Ishikawa, S. Bernard, A. Wulff and D.M. Anderson. 2015. Harmful algal blooms and climate change: Learning from the past and present to forecast the future. Harmful Algae 49: 68–93.

Wu, H., K. Gao, V.E. Villafane, T. Watanabe and E.W. Helbling. 2005. Effects of solar UV radiation on morphology and photosynthesis of filamentous cyanobacterium *Arthrospira platensis*. Applied Environental Microbiology 71: 5004–5013.

Wu, Y., K. Gao and U. Riebesell. 2010. CO_2-induced seawater acidification affects physiological performance of the marine diatom *Phaeodactylum tricornutum*. Biogeosciences 7: 2915–2923.

Xu, K. and K. Gao. 2015. Solar UV irradiances modulate effects of ocean acidification on the Coccolithophorid *Emiliania huxleyi*. Photochemistry and Photobiology 91: 92–101.

Zhang, Y., R. Klapper, K.T. Lohbeck, L.T. Bach, K.G. Schulz, T.B. Reusch and U. Riebesell. 2014. Between- and within-population variations in thermal reaction norms of the coccolithophore *Emiliania huxleyi*. Limnology and Oceanography 59: 1570–1580.

CHAPTER 5

Phytoplankton Responses to Ocean Climate Change Drivers

Interaction of Ocean Warming, Ocean Acidification and UV Exposure

Donat-P. Häder[1,*] and *Kunshan Gao*[2]

Introduction

Aquatic primary producers constitute only 1% of the global biomass, but their productivity equals that of all terrestrial ecosystems taken together (cf. Chapter 3, this volume). They take up a large fraction of anthropogenically emitted CO_2, thereby mitigating climate change. In addition, the oceans are a major CO_2 sink, because part of the absorbed carbon sinks to the deep sea as oceanic snow consisting of dead organisms and excretion products. In the sediments its lifetime is millions of years so that this fraction of carbon is effectively removed from the atmosphere.

Macroalgae and sea grasses are important constituents of the aquatic primary producers, but they are limited to coastal areas (cf. Chapters 12, 13 and 14). In contrast, phytoplankton is responsible for the lion's share of aquatic productivity. These organisms range from prokaryotic cyanobacteria (pico-, nano- and microplankton) to unicellular and colonial eukaryotic species in numerous taxa, such as chlorophytes, dinoflagellates, chrysophytes, diatoms, cryptophytes and haptophytes (Mignot et al. 2014, Uriel Hernandez-Becerril and Pasten-Miranda 2015). Because of their small

[1] Friedrich-Alexander University, Erlangen-Nürnberg, Neue Str. 9, 91096 Möhrendorf, Germany.

[2] State Key Laboratory of Marine Environmental Science, Xiamen University (Xiang-An campus, ZhouLongQuan A1-211, Xiang-An Nanlu 4221, Xiang-An Qu, Xiamen, Fujian, 361102 China. Email: ksgao@xmu.edu.cn

* Corresponding author: donat@dphaeder.de

size, picoplankton have long been ignored in quantitative analyses, but modern technological developments, such as flow cytometry revealed that they constitute a significant share of the autotrophic plankton (Tamm et al. 2018). For example, in the West-Estonian Archipelago Sea picoplankton contribute about 20% of the total chlorophyll *a*. In the southern Gulf of Mexico 26% of the total picoplankton biomass was found to be composed of three groups, the cyanobacteria *Prochlorococcus* and *Synechococcus* in addition to picoplankton eukaryotes, of which *Prochlorococcus* was the most abundant with > 80% productivity and > 60% of the total picoplankton biomass (Linacre et al. 2015).

Phytoplankton biomass productivity is governed by a number of external factors, such as availability of solar radiation which restricts the organisms to the photic zone (also called euphotic zone) defined by its lower limit where the surface irradiance is attenuated to 1%. Depending on the transparency of the water column this zone may extent from a few centimeters in turbid eutrophic lakes to about 200 m in the open ocean. Most of the photosynthesis occurs in this zone and about 90% of the marine organisms are found here (Morel and Berthon 1989). However, the organisms dwelling in this zone are simultaneously exposed to damaging solar UV radiation.

Protein biosynthesis, growth and cell division are dependent on the availability of nutrients. Nitrate is often a limiting factor which is necessary for protein synthesis (Read et al. 2000). However, many cyanobacteria are capable of converting atmospheric nitrogen into nitrate and nitrite (Falkowski 1997). Other often limiting factors for phytoplankton growth are iron and phosphorous (Martin 1992, Mills et al. 2004). Nutrients are provided by mixing of the surface layer with deep sea waters, terrestrial runoff and/or deposited particles from the atmosphere. However, thermally-induced stratification limits this nutrient penetration into the upper mixing layer (UML) since the UML and deeper waters are separated by the thermocline (or pycnocline), a boundary which prevents mixing. Stratification is enhanced by ocean warming due to climate change (Häder and Gao 2015, Häder and Gao 2017). Other sources of mineral fertilization are wind-blown nutrient transport from land (Hsu et al. 2007).

Phytoplankton productivity is affected by numerous environmental factors augmented by climate change including temperature increase, ocean acidification and excessive solar visible and ultraviolet radiation (Gao et al. 2018). These anthropogenically induced rapid changes alter habitats and phytoplankton communities with far reaching consequences for the extended oceanic food webs affecting the sustainable development of resources for a rapidly increasing human population (Walker and Salt 2012).

Effects of Ocean Warming

Fossil fuel consumption for industry, traffic and households results in fast increasing carbon dioxide accumulation in the atmosphere (cf. Chapter 3). The resulting greenhouse effect heats the atmosphere and—to a lower extent—also the surface of the water column (Fischetti 2013). During the last 30 years the atmospheric temperature has increased faster than any time period since the late 1880s (Hansen et al. 2006). During the past 40 years, the oceans have absorbed over 90% of the greenhouse heat

(Reid 2016); and, by the end of this century, the global mean sea surface temperature is predicted to rise by 1.1–6.4°C (Meehl et al. 2007, Huertas et al. 2011, IPCC 2014).

Elevated temperatures may enhance growth and productivity of phytoplankton because enzyme activity, photosynthetic carbon fixation and other biochemical activities are augmented by increasing temperatures (Beardall and Raven 2004). Higher temperatures also favor bacterial breakdown of organic matter resulting in an increased nutrient availability for phytoplankton growth mitigating ocean oligotrophication associated with the stabilization of the thermocline (Arandia-Gorostidi et al. 2017, Ayo et al. 2017). However, increased respiratory carbon loss and energy cost also increase with increasing temperature.

The dependence of phytoplankton growth on the ambient temperature can be quantified by thermal performance curves (TPC). The details of these curves differ in different species or for the same species in different seasons, but they follow a common pattern: above a lower limit (which in temperate phytoplankton is often close to freezing) the growth rate shows an increase up to an optimal temperature followed by a sharp decline with further increase of temperature which still allows net positive growth. The lower and upper limit define a permissive temperature window (Huey and Stevenson 1979). These curves can be determined experimentally by evaluating the growth rate at different temperatures or they can be extracted from a mathematical model (Boyd et al. 2013); they are useful to determine the permissive thermal window and the response to all temperatures inside this window (Zhang et al. 2014). However closer analysis of relevant growth factors has indicated that the response to a single factor is modified by other external factors. For example, the width of the thermal window and the growth rate within the TPC is affected by other environmental drivers, such as ocean acidification, deoxygenation and pollutants (Dupont and Pörtner 2013, Listmann et al. 2016, Leung et al. 2017).

Laboratory experiments showed that the optimal growth temperature for several *Skeletonema menselii* and five *S. ardens* strains was between 30° and 35°C (Table 5.1). In eight diatom species isolated from the Arctic to mid-latitudes the optimal temperature was found close to the upper limit of a wide temperature window (Suzuki and Takahashi 1995). And it is interesting to note that the optimal temperature is usually higher than the ambient temperature in the habitat of the organisms, with higher optimal temperatures during the hotter season in view of photosynthetic performance (Gao 1990). Even within the same genus the optimal growth temperature can vary significantly as found in seven *Skeletonema* species grown between 10° to 40°C (Kaeriyama et al. 2011). In 11 strains of the Noelaerhabdaceae *Emiliania huxleyi* the growth rates have been measured between 8° and 28°C (Zhang et al. 2014). While these strains originated from different regions, their optimum growth temperature and temperature window was found to be related to the temperature regimes of their original habitat. Also the minimum growth temperature was related to the original ecosystem conditions in five coccolithophorid species analyzed between 6° and 25°C (Buitenhuis et al. 2008). When grown at 15° or 26.3°C for two and a half years resulted in an adaptation to the growth temperature in *E. huxleyi* isolated from Norwegian coastal waters: the strain grown at the higher temperature was found to have a 0.7°C higher optimum than that grown at the lower temperature and the maximal permissive temperature was 1° to 3°C higher in the strain grown at the higher temperature (Buitenhuis et al. 2008).

Table 5.1. Optimal temperature (T_{opt}), maximal growth rate (μ_{max}) and temperature window width (w) of different phytoplankton species isolated from various habitats. NS indicates no source was found (Gao et al. 2018).

Species/Strain	Habitat	*In situ* Tem. (°C)	T_{opt} (°C)	μ_{max} (d^{-1})	w (°C)	Incubation light intensity	Reference
Asterionella formosa	36° N, 140° E	8	20	1.3	2–30	200 µmol m^{-2} s^{-1}	(Suzuki and Takahashi 1995)
Chaetoceros pseudocurvisetus	35° N, 139° E	25	25	1.6	15–35	200 µmol m^{-2} s^{-1}	(Suzuki and Takahashi 1995)
Nitzschia frigida	75° N, 95° W	−1.8	2	0.4	−1.8–5	100 µmol m^{-2} s^{-1}	(Suzuki and Takahashi 1995)
Skeletonema costatum	36° N, 140° E	15	25	1.5	2–35	200 µmol m^{-2} s^{-1}	(Suzuki and Takahashi 1995)
Thalassiosira nordenskioeldii	44° N, 144° E	−1.8	15	1.0	−1.8–20	100 µmol m^{-2} s^{-1}	(Suzuki and Takahashi 1995)
Ditylum brightwelii	NS	NS	20	0.7	NS	50 µmol m^{-2} s^{-1}	(Montagnes and Franklin 2001)
Phaeodactylum tricornutum	NS	NS	20	1.0	NS	50 µmol m^{-2} s^{-1}	(Montagnes and Franklin 2001)
S. costatum	NS	NS	20	0.9	NS	50 µmol m^{-2} s^{-1}	(Montagnes and Franklin 2001)
T. pseudonana	NS	NS	25	1.4	5–32	130 µmol m^{-2} s^{-1}	(Claquin et al. 2008)
S. marinoi	NS	NS	23	1.2	0–31	130 µmol m^{-2} s^{-1}	(Claquin et al. 2008)
Pseudo-nitzschia fraudulenta	NS	NS	21	0.8	5–24	130 µmol m^{-2} s^{-1}	(Claquin et al. 2008)
S. ardens	33° N, 130° E	28	35	2.3	10–40	150 µmol m^{-2} s^{-1}	(Kaeriyama et al. 2011)
S. costatum	33° N, 130° E	20	30	1.3	10–35	150 µmol m^{-2} s^{-1}	(Kaeriyama et al. 2011)
S. marinio-dohrnii	33° N, 130° E	12	30	1.7	5–35	150 µmol m^{-2} s^{-1}	(Kaeriyama et al. 2011)
S. japonicum	33° N, 130° E	25	25	1.6	5–30	150 µmol m^{-2} s^{-1}	(Kaeriyama et al. 2011)
S. menzelii	33° N, 130° E	28	30	2.6	10–40	150 µmol m^{-2} s^{-1}	(Kaeriyama et al. 2011)
S. tropicum	33° N, 130° E	20	25	1.6	10–35	150 µmol m^{-2} s^{-1}	(Kaeriyama et al. 2011)

Table 5.1 contd. ...

... *Table 5.1 contd.*

Species/Strain	**Habitat**	***In situ* Tem. (°C)**	**T_{opt} (°C)**	**μ_{max} (d^{-1})**	**w (°C)**	**Incubation light intensity**	**Reference**
T. pseudonana CCMP 1011	18° N, 65° E	NS	25	1.1	6–32.5	Saturating light levels	(Boyd et al. 2013)
T. pseudonana CCMP 1012	32° N, 116° W	NS	25	1.3	6–32.5	Saturating light levels	(Boyd et al. 2013)
T. pseudonana CCMP 1013	53° N, 4° W	NS	25	1.4	6–35	Saturating light levels	(Boyd et al. 2013)
T. pseudonana CCMP 1014	28° N, 155° E	NS	25	1.4	6–35	Saturating light levels	(Boyd et al. 2013)
T. pseudonana CCMP 1015	49° N, 123° E	NS	25	1.4	6–35	Saturating light levels	(Boyd et al. 2013)
T. pseudonana CCMP 1335	41° N, 73° E	NS	30	1.4	6–35	Saturating light levels	(Boyd et al. 2013)
Trichodesmium erythraeun KO4-20	15° S, 155° E	NS	28	0.3	16–35	Saturating light levels	(Boyd et al. 2013)
T. erythraeun 2175	7° N, 49° W	NS	26	0.3	18–35	Saturating light levels	(Boyd et al. 2013)
Crocosphaera watsonii WH 3A	7° N, 49° W	NS	28	0.3	22–35	Saturating light levels	(Boyd et al. 2013)
C. watsonii WH84	11° S, 32° W	NS	28	0.4	22–35	Saturating light levels	(Boyd et al. 2013)
C. watsonii WH0005	21° N, 157° W	NS	28	0.4	22–35	Saturating light levels	(Boyd et al. 2013)
T. pseudonana	NS	NS	25	1.3	NS	NS	(Leung et al. 2017)
Emiliania huxleyi B92/21	60° N, 5° E	NS	18	1.3	NS	100 µmol m^{-2} s^{-1}	(Conte et al. 1998)
E. huxleyi G1779Ga	60° N, 20° W	NS	21	1.7	NS	100 µmol m^{-2} s^{-1}	(Conte et al. 1998)
E. huxleyi M181	32° N, 62° W	NS	24	1.5	NS	100 µmol m^{-2} s^{-1}	(Conte et al. 1998)
E. huxleyi S. Africa	29° S, 31° E	NS	24	1.3	NS	100 µmol m^{-2} s^{-1}	(Conte et al. 1998)

E. huxleyi Van556	49° N, 144° W	NS	18	1.1	NS	100 µmol m^{-2} s^{-1}	(Conte et al. 1998)
Gephyrocapsa oceanica AB1	36° S, 174° E	NS	24	1.3	NS	100 µmol m^{-2} s^{-1}	(Conte et al. 1998)
E. huxleyi TQ26DIP	NS	NS	20	1.4	4–30	180 µmol m^{-2} s^{-1}	(Buitenhuis et al. 2008)
G. oceanica NS6-2	NS	NS	25	0.9	NS	180 µmol m^{-2} s^{-1}	(Buitenhuis et al. 2008)
Calcidiscus leptoporus NS10-2	NS	NS	12	0.6	9–25	180 µmol m^{-2} s^{-1}	(Buitenhuis et al. 2008)
C. leptoporus N482–1	NS	NS	20	0.4	9–25	180 µmol m^{-2} s^{-1}	(Buitenhuis et al. 2008)
Coccolithus braarudii N476-2	NS	NS	15	0.4	9–25	180 µmol m^{-2} s^{-1}	(Buitenhuis et al. 2008)
E. huxleyi M23	38° N, 28° W	17	24	1.6	4–30	160 µmol m^{-2} s^{-1}	(Zhang et al. 2013)
E. huxleyi M22	38° N, 28° W	17	24	1.6	4–30	160 µmol m^{-2} s^{-1}	(Zhang et al. 2013)
E. huxleyi M21	38° N, 28° W	17	24	1.6	4–30	160 µmol m^{-2} s^{-1}	(Zhang et al. 2013)
E. huxleyi M19	38° N, 28° W	17	23	1.6	4–30	160 µmol m^{-2} s^{-1}	(Zhang et al. 2013)
E. huxleyi M13	38° N, 28° W	17	25	1.5	4–30	160 µmol m^{-2} s^{-1}	(Zhang et al. 2014)
E. huxleyi M10	38° N, 28° W	17	24	1.6	4–30	160 µmol m^{-2} s^{-1}	(Zhang et al. 2013)
E. huxleyi 85	60° N, 5° E	10	23	1.6	4–28	160 µmol m^{-2} s^{-1}	(Zhang et al. 2013)
E. huxleyi 63	60° N, 5° E	10	23	1.6	4–28	160 µmol m^{-2} s^{-1}	(Zhang et al. 2013)
E. huxleyi 62	60° N, 5° E	10	23	1.7	4–28	160 µmol m^{-2} s^{-1}	(Zhang et al. 2013)
E. huxleyi 41	60° N, 5° E	10	23	1.8	4–28	160 µmol m^{-2} s^{-1}	(Zhang et al. 2013)
E. huxleyi 17	60° N, 5° E	10	23	1.6	4–28	160 µmol m^{-2} s^{-1}	(Zhang et al. 2013)
E. huxleyi RCC1710	34° N, 129° E	NS	25	1.2	6–30	300 µmol m^{-2} s^{-1}	(Rosas-Navarro et al. 2016)
E. huxleyi RCC1252	41° N, 140° E	NS	25	1.2	6–30	300 µmol m^{-2} s^{-1}	(Rosas-Navarro et al. 2016)
E. huxleyi IAN01	NS	NS	25	1.3	6–30	300 µmol m^{-2} s^{-1}	(Rosas-Navarro et al. 2016)

Generally, strains isolated from tropical waters have higher values for the thermal window than those originating from mid-latitude or polar regions. Optimum growth temperatures are usually much higher than the ambient temperature in cool waters. In contrast, strains isolated from warm waters have optima closer to the ambient temperature of their habitat (Thomas et al. 2012). Even though thermal windows and optimal growth temperatures are species- or strain-specific, and they can be modified by different culture conditions, irradiances or nutrient availability (Boyd et al. 2013).

From the results detailed above it can be concluded that climate change-related temperature increases will have significant impacts on phytoplankton communities. The most prominent ocean warming is found in polar waters (Hansen et al. 2006). The effects of increased temperatures have been demonstrated: when a Ross Sea plankton community was incubated for one week at 4°C higher than the local ambient temperature the microzooplankton abundance increased by 43% (Rose et al. 2009). This effect was synergistically augmented by the availability of iron. In the East China Sea, warming is predicted to decrease the abundance of diatoms and increase that of dinoflagelates (Xiao et al. 2018).

Warming has a more pronounced effect on heterotrophic processes than on photosynthesis as shown in a mesocosm experiment; so less biomass and energy is transferred to higher trophic levels. Increased temperatures also shift the seasonal growth and reproduction cycles of copepods to earlier in the year which also decreases the energy transfer to higher trophic levels (Edward and Richardson 2004).

Increases in abundance of toxic harmful alga blooms (HAB) have been associated with increased temperatures (Hallegraeff 1993, Hallegraeff 2010, Hallegraeff 2014). Studies of the fossil record of dinoflagellate cysts, long-term monitoring programs and short-term phytoplankton community responses to El Niño phenomena and North Atlantic Oscillation periods indicate that increasing temperatures, enhanced stratification, shifts in ocean currents as well as micronutrient availability result in more frequent and longer blooms of harmful algae. The ecological results are range expansion of warm-water species, poleward movement, species-specific changes in seasonal windows of harmful algal taxa and earlier occurrences of blooms of tropical benthic dinoflagellates. The produced and excreted algal biotoxins pose serious problems for filter feeders, such as shells, the food web and for human consumption (Stephanie et al. 2008). Climate change is also made responsible for the global expansion of harmful cyanobacterial blooms which produce neurotoxins and hepatotoxins such as microcystin, cylindropermopsin and antoxin and other poisonous substances which threaten humans and animals (Metcalf et al. 2008, Paerl and Huisman 2009). Increasing temperatures are an important factor for the increasing abundance of HAB species and provides a competitive advantage, but it is probably not the only environmental clue (Halac et al. 2013, Wells et al. 2015).

Satellite imaging monitoring chlorophyll *a* abundance have shown a significant decrease in globally integrated oceanic biomass and primary production over a long time period which was associated with climate change-induced increased temperature (Siegel et al. 2013). This result was confirmed by century-long monitoring of water transparency (Boyce et al. 2010) and carbon-climate models (Steinacher et al. 2010). Elevated temperatures reduced the growth rates of the diatom *Cylindrotheca closterium* and, in contrast, increased the abundance of small chlorophytes in a reef lagoon in the Caribbean (Halac et al. 2013, Wells et al. 2015). Elevated water temperatures, in addition to enhanced solar radiation and reduced wind speed result in lengthening of the bloom period. However, there was a decreased bloom intensity in the Baltic Sea during the period 2000–14, which was associated with a decreased nutrient availability (Groetsch et al. 2016). Elevated temperature results in movement of communities: diatoms, dinoflagellates and copepods in the North-East Atlantic and the North Sea were found to undergo a fast poleward migration between 1954 and 2013 which followed the isotherm movement, while in contrast, diatoms did not migrate significantly (Chivers et al. 2017). Satellite imaging of chlorophyll *a* concentrations during 1998–2014 indicated that phytoplankton blooms lasted longer during open water episodes in the North Water polynya, while the periods were shorter during years with persistent ice cover (Marchese et al. 2017). This decrease in bloom length was augmented by a massive input of melt water. In summary, these results indicate changes in marine phytoplankton communities due to ocean warming. However, regional differences and controversial conclusions from satellite monitoring are the reasons for divergent interpretations of the current situation and future scenarios (Chavez et al. 2011, Siegel et al. 2013).

The Effects of Ocean Acidification

Increasing CO_2 concentrations in the atmosphere resulting from fossil fuel burning, tropical deforestation and altered land use are mirrored by higher carbon dioxide in oceanic surface waters and its increased dissolution into deeper waters. The absorption of fossil fuel CO_2 is mainly driven by the biologicial CO_2 pump and the solubility pump. The uptake depends on temperature, salinity, mixing intensity, surface water chemistry and hydrological conditions. Increased dissolution of CO_2 into the oceans result in decreasing pH, leading to ocean acidification, even though seawater has high buffering capacity (Caldeira and Wickett 2003). Even if the anthropogenic emission could be stopped immediately, OA would not be reversed within a thousand years (Raven et al. 2005). If the emissions continue in a business-as-usual scenario (RCP8.5) OA will result in a pH decrease of 0.4 by the end of the century (Zeebe and Wolf-Gladrow 2001, Caldeira and Wickett 2003, Gattuso et al. 2015) (cf. Chapter 3, this volume).

The modified carbonate chemistry resulting from OA affects other chemical processes (Millero 2007), and influences phytoplankton and their ecosystems differently depending on their habitat and latitude because of the other environmental drivers, such as solar UV radiation, temperature and nutrient concentrations (Riebesell and Gattuso 2015). These interacting external factors control the phytoplankton physiology and ecological processes in the marine environment (Boyd 2011, Brennan and Collins 2015, Gattuso et al. 2015). The presence of carbon is the basis for the photosynthetic processes in prokaryotic and eukaryotic photosynthetic organisms, but the availability of CO_2 depends on the carbonate chemistry in the UML (Raymont 2014, Britton et al. 2016) and the thickness of the diffusion layer around the organisms. It goes without saying that the biomass productivity by the primary producers is a profound prerequisite for the primary and secondary consumers as well as the entire food web. OA is known to affect food quality of phytoplankton (Jin et al. 2015) and shrimps (Dupont et al. 2014).

Usually, temporary pCO_2 levels in seawaters are a limiting factor for photosynthetic carbon fixation, therefore, increasing CO_2 availability was thought to play a role as a fertilizing driver. This is usually true for some large diatoms and macroalgae (Gao et al. 1991, Wu et al. 2014). However, elevated CO_2 concentrations usually exacerbate an acidic stress due to increased H^+ concentrations in the water. Quantitatively, the biochemical reactions in phytoplankton controlled by dissolved inorganic carbon and its chemistry follows an optimum curve as found in ecologically important phytoplankton taxa, such as cyanobacteria, coccolithophorids and diatoms (Bach et al. 2015). This findings may explain the recent contradicting results that ocean acidification can either enhance phytoplankton growth (Riebesell and Tortell 2011), have no marked effect (Tortell et al. 2000, Kim 2006, Gao and Campbell 2014) or decrease growth (Wu et al. 2010, Gao et al. 2012a, Mackey et al. 2015). The latter effect has been found to be due to augmented photorespiration and mitochondrial respiration (Wu et al. 2010, Gao et al. 2012a, Mackey et al. 2015). High photosynthetic radiation levels (PAR) in the presence of decreased pH have been found to lower carbon fixation in diatoms, while at low PAR levels growth of three diatoms was found to be stimulated (Gao et al. 2012b). This is due to the fact that many phytoplankton organisms possess a CO_2-concentrating mechanism based on the activity of an carbonic anhydrase enzyme (Aizawa and Miyachi 1986) and active transport of CO_2 and bicarbonate. In the presence of high CO_2 concentrations this mechanism is down-regulated which saves energy otherwise used for the active accumulation mechanism so that growth at low irradiances is enhanced (Gao et al. 2012b). The intracellular concentration of dissolved inorganic carbon in the model diatom *Phaeodactylum tricornatum* was shown to be 2–3 fold lower in the high CO_2-grown cells than in the ambient CO_2-grown ones (Liu et al. 2017) indicating a direct evidence of OA-related down-regulation of carbon concentrating mechanisms.

At high light intensities OA reduced the growth rates in the coccolithophorid *Gephyrocapsa oceanica* during laboratory growth (Zhang et al. 2015). In contrast, OA augmented the growth rate in *Emiliania huxleyi* at both high and low irradiances when grown under sunlight (Jin et al. 2017a), indicating that his coccolithophorid may have the capability to counteract acidic stress at low pH because of the energy requirement for calcification and maintaining cellular homeostasis. Diatoms and coccolithophorids have been found to synthesize and accumulate toxic phenolics when exposed to OA. These substances can be metabolized to obtain extra energy using mitochondrial respiration under acidic stress (Jin et al. 2015). For example, ocean acidification increases β-oxidation of lipids, the tricarboxylic acid cycle and glycolysis (Jin et al. 2015). In harmful algal species the synthesis of biotoxins is also linked to the autotrophic metabolism, and higher temperatures in the presence of ocean acidification may augment toxin synthesis (Fu et al. 2012). This has also been demonstrated in the HAB diatom *Pseudo-nitzschia* (Sun et al. 2011) and the toxic dinoflagellate *Alexandrium fundyense* (Hattenrath-Lehmann et al. 2015). However, the toxin concentrations in phytoplankton represent only a very small amount of carbon in the cell so that no direct effect of the pH can be deduced (Wells et al. 2015). Ocean acidification increased the exopolymer particle concentration in an algal bloom in the Baltic Sea (Engel et al. 2014) indicating that increasing CO_2 concentrations result in production and excretion of carbon-containing particles.

Ocean acidification induces different effects in different phytoplankton taxa. The abundance of *Phaeocystis* increased at low CO_2 concentrations by about 60% as measured off the Peruvian coast, while that of diatoms increased at high CO_2 concentrations (Tortell et al. 2002). In Raunefjorden, Norway, the carbon dioxide consumption of a phytoplankton assemblage increased under higher CO_2 concentrations, but the nutrient uptake did not change (Riebesell et al. 2007). Elevated carbon dioxide concentrations decreased the concentration of viruses specific for *E. huxleyi*, so that the phytoplankton concentrations increased (Larsen et al. 2008). In contrast, under ocean acidification a virus infection enhanced photoinhibition and decreased photosynthetic electron transport in *Phaeocystis globosa* (Chen et al. 2015).

Cyanobacterial diazotrophs are the only phytoplankton organisms which can convert atmospheric nitrogen into nitrate and nitrite thereby creating nitrogen fertilizers which can be used by other phytoplankton for growth and carbon incorporation (Eichner et al. 2014). Simultaneously this mechanism mitigates global warming (Michaels et al. 2001, Berthelot et al. 2015). Ocean acidification can either enhance or decrease nitrogen fixation in heterocyst-containing cyanobacteria (Czerny et al. 2009, Wannicke et al. 2012). In nitrogen-fixing cyanobacteria without heterocysts, the responses to ocean acidification were reported to vary among species (Eichner et al. 2014) or among different studies with the same strain. E.g., in *Trichodesmium* ocean acidification decreases nitrogen fixation and growth; however after adaption for several hundred generations to higher CO_2 concentrations nitrogen fixation was augmented. These adapted cyanobacteria showed shifts in the circadian nitrogen fixation pattern which could be due to an increased activity of methyltransferase which potentially regulates DNA activity (Hutchins et al. 2015). On the other hand, a recent study showed that OA decreases nitrogen fixation in the same *Trichodesmium* strain (Hong et al. 2017). Such controversial findings open further oppertunities to look into the relationship of nitrogen fixation under OA in *Trichodesmium*. In other species little or no change was observed (Böttjer et al. 2014, Gradoville et al. 2014). These puzzling results may be due to physiological responses of specific species or strains to ocean acidification or different culture conditions and length of adaptation. Due to the different responses to ocean acidification, shifts in biodiversity and phytoplankton community structure will occur under continuing ocean acidification with consequences for the grazers and the food web (Hutchins et al. 2013, Gradoville et al. 2014). For example, at low CO_2 concentration toxic strains of the freshwater *Microcystis aeruginosa* became dominant in competition with non-toxic strains (Van de Waal et al. 2011).

One important aspect of ocean acidification is the interference of increased H^+ concentrations with calcification. Many plankton, such as foraminifera and coccolithophorids, macroalgae, such as *Acetabularia*, *Halimeda* and coralline algae and many benthic taxa including worms, bryozoa, mollusks, echinoderms, crustaceans and corals produce exo- or endoskeletons for stability and protection against predators and solar UV radiation (Andersson et al. 2008, Ries et al. 2009, Monteiro et al. 2016). Even though this mechanism requires large amounts of energy, it is very successful as indicated by the wide distribution and large biomass production of coccolithophorids in the oceans. In coccolithophorids the production of the outer shell is affected by ocean acidification (Beaufort et al. 2011), increased temperatures adds to the negative effects on *E. huxleyi* (Milner et al. 2016), and exposure to UV radiation synergistically acted with OA to reduce the calcification (Gao et al. 2009). In coralline algae, when grown under solar radiation with or without UV irradiances, OA treatment led to less calcification, lower contents of photosynthetic pigments, reduced photosynthetic O_2 evolution but increased UV-screening compounds (Gao and Zheng 2010).

Responses of pelagic and coastal phytoplankton may be different to increased CO_2 concentrations because of their adaptive acclimation to different environmental conditions. *Thalassiosira weissflogii* (a coastal diatom) and *Thalassiosira oceanica* (a pelagic diatom) showed differential physical performance under constant or diel fluctuation of pH under ambient or elevated CO_2 levels, with the former's growth being enhanced by OA and the latter inhibited (Li et al. 2016). Most of the earlier studies have been carried out in the laboratory under controlled conditions. However, because of the interaction with other environmental factors it is important to conduct studies on the effect of ocean acidification under natural conditions and in large scales (Riebesell and Gattuso 2015, Tilstone et al. 2016, Grear et al. 2017).

Responses of Phytoplankton to Solar UV Radiation

The current scenario of solar UV radiation, its projected future development and factors affecting the penetration into aquatic ecosystems are detailed in Chapter 3 of this volume. Because of the Montreal protocol and its amendments the maximum of stratospheric ozone depletion with the accompanying increased levels of solar UV has passed during the first decade of the 21st century and is expected to return to pre-1970 levels after 2050 (United Nations Environment Programme Environmental Effects Assessment Panel 2012, Caron 2014). However, the penetration of solar radiation into the water column will change with climate change (Zepp et al. 2007, Häder et al. 2011). Dissolved organic carbon (DOC) is a major absorber of UV-A and UV-B in the water column attenuating the penetration of damaging solar short-wavelength radiation. UV-B breaks down the DOC molecules making the fragments more palatable for bacteria. As a consequence the turbidity decreases and UV can penetrate to greater depths. In addition, ocean warming results in increased shoaling of the upper mixed layer (UML) exposing the organisms dwelling in this layer to higher solar visible and ultraviolet radiation (Häder et al. 2015, Häder and Gao 2017).

Even though UV-B accounts for less than 1% of the total solar radiation in terms of energy this high energetic radiation per wavelength is very effective in damaging biological molecules or processes. There is a plethora of targets for solar UV radiation in the cell. One important effect is the damage of nuclear, mitochondrial and plastid DNA (Sinha and Häder 2002, Meador et al. 2009, Häder and Gao 2015). The most pronounced damage by UV in DNA is the formation of cyclobutane pyrimidine dimers (CPD). Cells have efficient repair mechanisms to split the dimers using the energy of UV-A or blue photons with the help of a photolyase (Häder and Sinha 2005). Being an enzymatic repair, it is augmented by increasing temperatures. However, the repair mechanism itself is inhibited by excessive solar UV radiation (Gao et al. 2008, Rastogi et al. 2014). UV-B is absorbed by DNA, proteins and other biomolecules of the cell, which are damaged by the radiation. In addition, UV can induce the production of free oxygen radicals which results in oxidative stress to the cells which in turn damages photosynthetic pigments and decreases biomass production (Häder and Gao 2015). For example, diatoms were found to show decreased growth rates in a coastal lagoon in southern Portugal (Domingues et al. 2017). In contrast, cryptophytes were found to have increased growth rates.

UV-B effects on phytoplankton have been measured in the lab under constant conditions which was useful to determine the mechanisms of damage. In their natural habitats, phytoplankton distributed in the UML are moved vertically by the action of wind and waves and/or by seawater gravity changes induced by diel temperature changes. On bright days, the organisms near the surface are exposed to strong solar visible and ultraviolet radiation which may cause damage to the DNA, the photosynthetic apparatus and other biologically important components (Häder et al. 2015). When they are moved passively to the bottom of the UML they have time to repair the damage. Therefore, the net damage depends on the mixing velocity and the depth of the UML which decreases with increasing temperatures (Helbling et al. 2003, Villafañe et al. 2007). In addition, the damage by excessive PAR or UV depends on the light history of the phytoplankton community (Guan and Gao 2008, Häder et al. 2014). Excessive UV-A radiation is damaging as well as UV-B, but at low or moderate levels of irradiances on cloudy days UV-A radiation can even contribute to photosynthetic carbon fixation as shown in a plankton assembly irradiated by light in which PAR was removed by a filter (Gao et al. 2007a). In contrast, UV-B always reduced carbon fixation in a manner dependent on the irradiance. Larger phytoplankton are more tolerant to UV-A than smaller ones (Li et al. 2011, Li and Gao 2013).

Cyanobacteria are affected by UV radiation through a number of mechanisms. In addition to DNA damage, UV bleaches phycobilins, the photosynthetic accessory pigments (Sinha et al. 1995a, Sinha et al. 1995b), impairs motility and photoorientation (Donkor and Häder 1991) and other vital biological functions, such as nitrogen fixation (Sinha et al. 2008, Singh et al. 2010, Cai et al. 2017). But also excessive PAR can bleach the phycobilins phycocyanin, phycoerythrin and allophycocyanin (Bhandari and Sharma 2006). UV-B breaks down the phycobilisome complexes disassembling the protein structure (Sinha et al. 1995b, Sinha et al. 1997). In addition, it induces ROS increasing the oxidative stress in the cells (He and Häder 2002a, He and Häder 2002c). Exposure to 120–180 min UV-B killed several cyanobacterial species, such as *Oscillatoria priestleyi* and *Phormidium murrayi* quantitatively (Quesada and Vincent 1997). Under natural conditions growth was inhibited up to 40% in the rice-field cyanobacterium *Anabaena* sp. (Gao et al. 2007b). Cyanobacteria which are embedded in a heavy mucilaginous sheath, such as *Scytonema* sp. and *Nostoc commune* tolerate exposure to UV better than *Anabaena* sp. and *Nostoc* sp. which have no or only thin sheaths (Sinha et al. 1995a). The terrestrial cyanobacterium *Nostoc flagelliforme* showed insensitive photosynthetic response to incident solar UV radiation (Gao and Ye 2007). The conversion of vegetative cells into heterocysts is impaired by UV as shown in several *Anabaena* species (Blakefield and Harris 1994, Gao et al. 2007b). Even the morphology can be altered by exposure to UV: the spiral filaments of *Arthrospira platensis* are broken and compressed by solar UV (Wu et al. 2005, Gao et al. 2008), while PAR effects on its morphology depends on the presence of UV and growth temperature (Ma and Gao 2009, Singh et al. 2010, Häder et al. 2011). Furthermore, UV-B results in lipid peroxidation of polyunsaturated fatty acids thus affecting the integrity of cellular and thylakoid membranes (He and Häder 2002b).

Another major target of UV-B radiation is the D1 protein in the photosystem II (PS II) complex which is responsible for the electron transport from the excited chlorophyll *a* to the primary acceptor (Sass et al. 1997, Campbell et al. 1998). This damage is repaired by degrading the damaged protein and substituting by freshly synthesized protein. However, this repair mechanism is also affected by UV-B since it down-regulates mRNA synthesis coding for proteins involved in light harvesting and electron transport exposure (Huang et al. 2002).

UV radiation controls the uptake of nutrients by phytoplankton. At replete phosphate concentration UV increased the uptake of phosphate and augmented the activity of alkaline phosphatase which enhanced the uptake of phosphorous and decreased the Chl *a* concentration (Villar-Argaiz et al. 2017). A ratio of nitrogen to phosphorous of 16:1 (Redfield ratio) has been found to be optimal for photophosphorylation and growth in the dinoflagellate *Karenia mikimotoi* (Guan and Li 2017).

Because phytoplankton have constantly been exposed to solar UV radiation they have developed effective mitigating strategies mechanisms (Vincent and Roy 1993, Sinha et al. 1998, Cockell and Knowland 1999). One mechanism is downward migration in the water column which is instrumentalized by reducing the volume of internal gas vacuoles thus reducing the buoyancy (Häder 1987, Quesada and Vincent 1997). Nevertheless, under isoenergetic levels of PAR or UVR, *Arthrospira platensis* migrated downward under PAR but remained buoyant under UVR (Ma and Gao 2009). Other organisms form crusts and biofilms for mutual shading (Moisan et al. 2009). Effective repair mechanisms are utilized to reactivate photosynthesis and repair damage in DNA (Britt 1995, Kim and Sancar 1995, Chen et al. 2013). Reactive oxygen stress is mitigated by quenching ROS with enzymatic and non-enzymatic antioxidants (He and Häder 2002b, Singh et al. 2013, Häder et al. 2015, Richa et al. 2016).

One effective mechanism is the production of UV-screening pigments which are concentrated in the outer cell layers or even outside the cell in order to protect the central nuclear DNA. Scytonemin is produced only in some cyanobacteria and not in eukaryotic phytoplankton (Garcia-Pichel and Castenholz 1991, Garcia-Pichel and Castenholz 1993). This pigment absorbs in the UV-B and UV-A and interestingly also in the UV-C (260 nm). While today UV-C radiation does not reach the surface of the Earth because of the absorption in the stratosphere by oxygen and ozone, it was an additional stress factor during the early evolution of cyanobacteria before oxygen accumulated in the atmosphere. Thus scytonemin can be regarded as a relic from the early evolution which still has retained the feature of UV-C absorption. Scytonemin is a yellow-brown hydrophobic pigment with a molecular mass of 544 Da which is excreted into the extracellular polysaccharide sheath of some cyanobacteria (Geitler 1932, Desikachary 1959, Garcia-Pichel and Castenholz 1991, Garcia-Pichel and Belnap 1996). It is a dimeric molecule with indolic and phenolic subunits connected by an olefinic carbon atom. Recently, some derivatives of scytonemin have been identified, such as fuscochlorin, dimethoxyscytonemin, thetramethoxyscytonemin and scytonin (Garcia-Pichel and Castenholz 1991, Bultel-Poncé et al. 2004, Varnali and Edwards 2013).

The major UV-absorbing pigments are mycosporine-like amino acids (MAAs) found in cyanobacteria, phytoplankton and many macroalgae (Gröniger et al. 2000, Sinha et al. 2007). They are small (< 400 Da) hydrophilic molecules with a cyclohexenone or cyclohexenimine chromophore conjugated with the nitrogen group of an amino acid or its imino alcohol (Singh et al. 2008). They contain a glycine moiety at the third carbon of the ring. Some MAAs possess sulfate esters or glycosidic groups at the imine group. Interestingly, zooplankton and other heterotrophic taxa cannot produce MAAs, because they lack the shikimate pathway via gadosols (Bandaranayake 1998, Shick and Dunlap 2002, Singh et al. 2008) which constitutes the first part of the biosynthesis of these UV-absorbing pigments, but are not without protection since they can take MAAs up with their diet, transport them to external cell layers and tissues and utilize them for the same purpose. The most common MAAs are shinorine and porphyra-334 found in both freshwater and marine organisms (Karsten and Garcia-Pichel 1996, Sinha et al. 2001a, Sinha et al. 2001b, Sinha et al. 2003a, Sinha et al. 2003b, Volkmann et al. 2006, Sinha et al. 2007, Sinha and Häder 2008, Singh et al. 2010).

MAAs absorb in the UV-A and UV-B region with high molar extinction coefficients between 28,100 and 50,000 M^{-1} cm^{-1}. Due to these properties three out of ten UV photons are absorbed before they can hit cytoplasmic targets (Garcia-Pichel and Castenholz 1993). The mechanism of UV protection is based on the absorption of the incident UV photon, the energy of which is dissipated as heat. They are highly stable at high temperatures and extreme pH values as well as in different solvents (Richa and Sinha 2013). More than 22 different MAAs have been identified in terrestrial, marine and freshwater photosynthetic organisms (Fig. 5.1). Identification is made according to their absorption maxima and their retention times in high performance liquid chromatography (Garcia-Pichel and Castenholz 1993, Sinha et al. 1998).

In several cases it could be shown that MAAs are induced only when they are needed. Switching from indoor low PAR without UV to outdoor solar radiation with UVR modulates significantly the contents of MAAs in a red alga, being much higher under the solar exposure (Zheng and Gao 2009). Synthesis is induced by exposure to UV-B irradiation (Klisch and Häder 2000, Klisch 2002, Klisch and Häder 2002, Sinha et al. 2003b).

Interactive Effects of Climate Change Stress Factors

Assessment of phytoplankton responses to individual stress factors, such as elevated temperature, ocean acidification or ultraviolet radiation has yielded different, sometimes contradictory results. This may be due to different experimental protocols, different environmental conditions, pre-acclimation or species-specific differences (Domingues et al. 2017, Jin et al. 2017b). Therefore, it is important to study the interactive effects of multiple environmental stress factors under as much as possible natural conditions (Beardall et al. 2014, Riebesell and Gattuso 2015). This is not an easy task because of the wide oceanic spaces, low organisms densities in the water column and seasonal changes in the phytoplankton communities. In some cases attempts have been made to use mesocosms with manageable volumes (Kim et al. 2006, Riebesell et al. 2007, Berthelot et al. 2015, Liu et al. 2017).

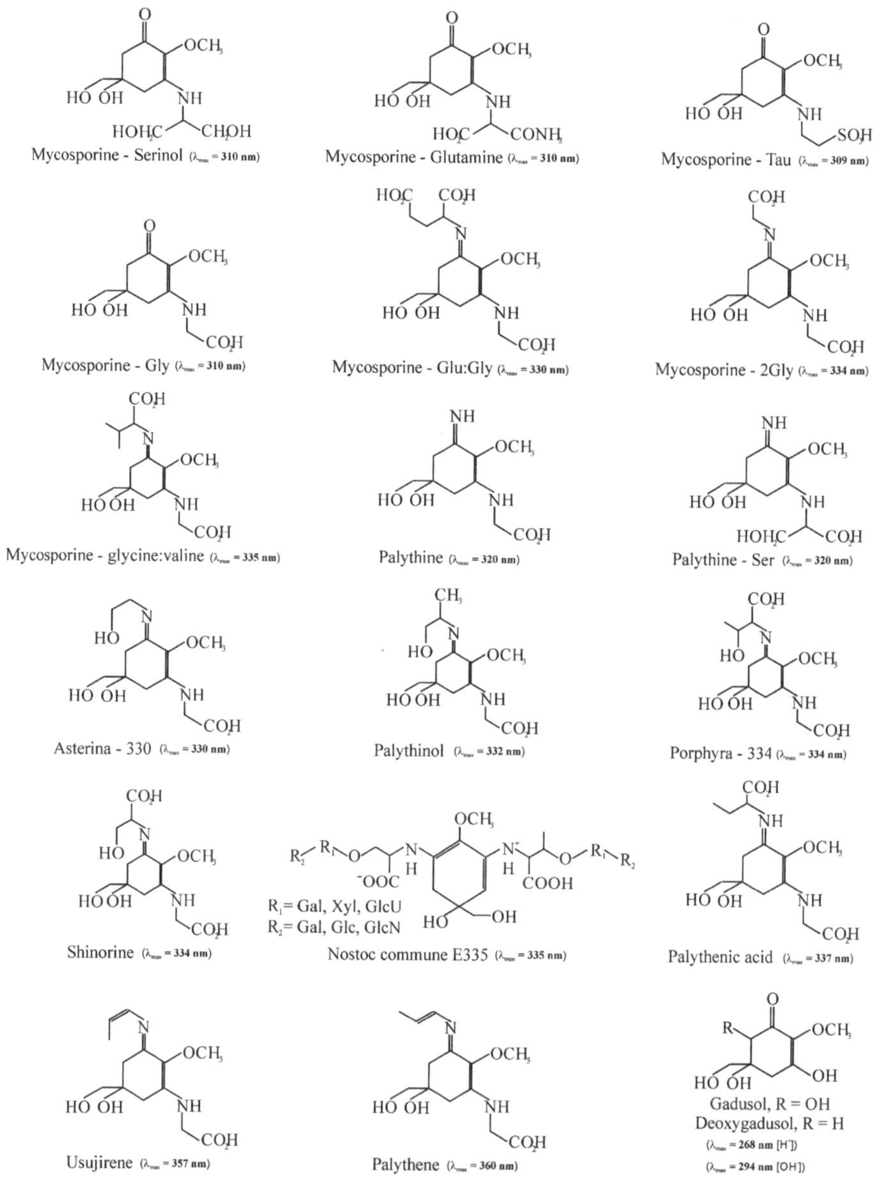

Fig. 5.1. Molecular structures and corresponding absorption maximum (λ_{max}) of mycosporine-like amino acids (MAAs). Redrawn after (Sinha and Häder 2003).

In the diatom *Cylindrotheca closterium* the photochemical efficiency of PSII decreased substantially when grown under UV radiation in the presence of 1,000 μatm CO_2 even after an acclimation for 9 days (Wu et al. 2012); however, it did not reduce the growth rate. A similar result was found for the Prymnesiophyte *Phyaeocystis globosa* (Chen and Gao 2011). Ocean acidification in conjunction with UV exposure decreased photosynthetic carbon fixation in *E. huxleyi* at low PAR, but at higher irradiances carbon fixation increased (Xu and Gao 2015). Later it was found that higher PAR levels abolished the negative effects of ocean acidification on calcification in this phytoplankton (Jin et al. 2017a). Ocean acidification appears to remediate the damaging effects of UV-B in diatoms (Li et al. 2012a); it acts synergistically with fluctuation of solar radiation or mixing to lower carbon fixation in the coccolithophorid *Gephyrocapsa oceanica*, although it appears to decrease ultraviolet B-related photochemical inhibition (Jin et al. 2013).

Ocean acidification thins the protective calcified outer skeleton in calcifying algae, which allows more UV to penetrate thus worsening the damaging impacts of solar short-wavelength radiation (Gao et al. 2009, Gao and Zheng 2010, Xu et al. 2014). Ocean warming with simultaneous exposure to enhanced CO_2 concentration increased the growth rate in several *Skeletonema marinoi* strains isolated from the North Sea but not in strains isolated from the Adriatic Sea (Kremp et al. 2012). Growth rates of *S. costatum, Chaetoceros debilis, Thalassiosira nordenskioeldii* and *C. didymus* responded differently to rising temperatures, but ocean acidification alone or in combination with higher temperatures had no effect (Hyun et al. 2014). In contrast, in *E. huxleyi* the growth rate decreased when the cells were transferred from 15° to 26°C and this decrease was even higher at elevated CO_2 concentrations (Schlüter et al. 2014, Listmann et al. 2016). However, after adaptation to the elevated temperature for one year carbon fixation and growth rate increased substantially when exposed to higher CO_2 concentrations. In Arctic phytoplankton communities the growth rate decreased when the temperature was increased from 1° to 6° or 10°C, while increased CO_2 concentrations slightly increased productivity at 1° and 6°C (Coello-Camba et al. 2014). Simultaneously, the community structure changed with the temperature increase: large centric diatoms dominating the assembly at 1°C were increasingly replaced by smaller diatoms and flagellates. The gross primary production in the Arctic Ocean increased with rising temperatures and enhanced ocean acidification (Holding et al. 2015). Further studies revealed that large diatoms and cryptophytes showed the opposite effects (Halac et al. 2013). Nutrient limitation modifies the effect of ocean acidification on phytoplankton (Li et al. 2012a, Verspagen et al. 2014): both synergistically reduce the cell size, carbon fixation and food value (Li et al. 2012a, Matthiessen et al. 2012).

This puzzling results could be explained by the following hypothesis: Low CO_2 concentrations limit photosynthetic carbon fixation; rising temperatures increase the growth rate so slowly that the temperature optimum is not revealed. At high CO_2 levels the high H^+ concentration inhibits physiological reactions even at low temperatures. Near the lower limit of the thermal window physiological reactions are controlled by the temperature so that higher CO_2 concentrations are of no advantage.

Ocean warming can mitigate the damage inflicted by UV radiation because it augments the synthesis of UV-absorbing pigments and the enzymatic repair mechanisms for DNA and photosynthesis; increased nutrient concentrations augment this process (Doyle et al. 2005). For example, at low temperatures and limiting nutrient concentrations UV impaired growth in the diatoms *Fragilaria crotonensis* and *Asterionella formosa*, the dinoflagellate *Gymnodinium* sp. and the chrysophyte *Dinobryon* sp., while at 14°C the damaging effect of UVR was much lower. Elevated temperatures also mitigated the UV-induced inhibition of the photochemical efficiency in the model diatom *Phaeodactylum tricornutum* under either low or high CO_2 concentrations (Li et al. 2012a); it also extenuated photochemical inhibition induced by UV in *D. salina*, *I. galbana*, *C. gracilis* and *T. weissflogii*, but not in *P. micans* (Halac et al. 2010, Halac et al. 2014). In *Gymnodinium chlorophorum*, the UV-induced decrease in photosynthetic efficiency was mitigated when the temperature was raised from 15° to 20°C but decreased again at 25°C (Häder et al. 2014). UV-inhibition of calcification and carbon fixation were also mitigated at higher temperatures in *E. huxleyi* (Xu et al. 2011). Consequently, UV-inflicted inhibition of carbon fixation in phytoplankton is more pronounced during the winter than during the summer (Wu et al. 2010).

Studies on the interactive effects of ocean acidification, elevated temperatures and UV radiation in phytoplankton are rare (Tong et al. 2017). Higher CO_2 concentrations partially mitigated the UV damage of PSII; but the effect was less obvious at elevated temperatures (Li et al. 2012b). This can be explained by assuming that the repair outweighed the damage with increasing carbon dioxide concentrations and temperature (Li et al. 2012a). These results indicate that individual stress factors may interact synergistically, antagonistically or neutrally and the physiological response of phytoplankton depends on the independent level of each factor (Li et al. 2012a, Xu and Gao 2015).

Conclusions and Future Research

In the past responses of phytoplankton to individual stress factors have been studied in the laboratory under controlled conditions. This approach was useful to reveal the molecular and biochemical mechanisms of physiological stress. But soon it was obvious that the results of these investigations did not correlate well with the behavior of phytoplankton communities in their natural habitat. For example, it became clear that phytoplankton were not eliminated by solar UV radiation even under the Antarctic ozone hole. Obviously, the reactions had to be investigated holistically under the prevailing environmental parameters to reveal the interactions of the stress parameters which may operate either synergistically, antagonistically or neutrally. The increasing impact of climate change and its manifestations in elevated temperatures, ocean stratification, ocean acidification, and nutrient supply need to be determined under as natural conditions as possible (Lohbeck et al. 2012, Schlüter et al. 2014, Gaitán-Espitia et al. 2017). It is obvious that this is a Herculean task because of the vast areas, distinctly different environmental conditions and the usually low phytoplankton concentrations. However, modern technological advances, such as fluorometry and pseudocolor monitoring of Chl *a* concentrations allow insight into the long-term development of phytoplankton communities which is mandatory to understand the impact of climate change stress factors on primary and secondary consumers and the whole intricate food web. These responses are key to the biogeochemical cycles in marine ecosystems and the carbon storage capacity of the oceans which determine future climate change of the earth.

References

Aizawa, K. and S. Miyachi. 1986. Carbonic anhydrase and CO_2 concentrating mechanisms in microalgae and cyanobacteria. FEMS Microbiol. Lett. 39: 215–233.

Andersson, A.J., F.T. Mackenzie and N.R. Bates. 2008. Life on the margin: implications of ocean acidification on Mg-calcite, high latitude and cold-water marine calcifiers. Mar. Ecol. Prog. Ser. 373: 265–273.

Arandia-Gorostidi, N., P.K. Weber, L. Alonso-Sáez, X.A.G. Morán and X. Mayali. 2017. Elevated temperature increases carbon and nitrogen fluxes between phytoplankton and heterotrophic bacteria through physical attachment. The ISME journal 11: 641.

Ayo, B., N. Abad, I. Artolozaga, I. Azua, Z. Baña, M. Unanue, J.M. Gasol, C.M. Duarte and J. Iriberri. 2017. Imbalanced nutrient recycling in a warmer ocean driven by differential response of extracellular enzymatic activities. Global Change Biology, doi 10.1111/gcb.13779.

Bach, L.T., U. Riebesell, M.A. Gutowska, L. Federwisch and K.G. Schulz. 2015. A unifying concept of coccolithophore sensitivity to changing carbonate chemistry embedded in an ecological framework. Progress in Oceanography 135: 125–138.

Bandaranayake, W.M. 1998. Mycosporines: are they nature's sunscreens? Natural Product Reports 15: 159–172.

Beardall, J. and J.A. Raven. 2004. The potential effects of global climate change on microalgal photosynthesis, growth and ecology. Phycologia 43: 26–40.

Beardall, J., S. Stojkovic and K. Gao. 2014. Interactive effects of nutrient supply and other environmental factors on the sensitivity of marine primary producers to ultraviolet radiation: implications for the impacts of global change. Aquatic Biology 22: 5–23.

Beaufort, L., I. Probert, T. De Garidel-Thoron, E.M. Bendif, D. Ruiz-Pino, N. Metzl, C. Goyet, N. Buchet, P. Coupel, M. Grelaud, B. Rost, R.E.M. Rickaby and C. de Vargas. 2011. Sensitivity of coccolithophores to carbonate chemistry and ocean acidification. Nature 476: 80–83.

Berthelot, H., T. Moutin, S. L'Helguen, K. Leblanc, S. Hélias, O. Grosso, N. Leblond, B. Charrière and S. Bonnet. 2015. Dinitrogen fixation and dissolved organic nitrogen fueled primary production and particulate export during the VAHINE mesocosm experiment (New Caledonia lagoon). Biogeosciences 12: 4099–4112.

Bhandari, R. and P.K. Sharma. 2006. High-light-induced changes on photosynthesis, pigments, sugars, lipids and antioxidant enzymes in freshwater (*Nostoc spongiaeforme*) and marine (*Phormidium corium*) cyanobacteria. Photochemistry and Photobiology 82: 702–710.

Blakefield, M.K. and D.O. Harris. 1994. Delay of cell differentiation in *Anabaena aequalis* caused by UV-B radiation and the role of photoreactivation and excision repair. Photochemistry and Photobiology 59: 204–208.

Böttjer, D., D.M. Karl, R.M. Letelier, D.A. Viviani and M.J. Church. 2014. Experimental assessment of diazotroph responses to elevated seawater pCO_2 in the North Pacific Subtropical Gyre. Global Biogeochemical Cycles 28: 601–616.

Boyce, D.G., M.R. Lewis and B. Worm. 2010. Global phytoplankton decline over the past century. Nature 466: 591–596.

Boyd, P.W. 2011. Beyond ocean acidification. Nature Geoscience 4: 273–274.

Boyd, P.W., T.A. Rynearson, E.A. Armstrong, F. Fu, K. Hayashi, Z. Hu, D.A. Hutchins, R.M. Kudela, E. Litchman, M.R. Mulholland, U. Passow, R.F. Strzepek, K.S. Whittaker, E. Yu and M.K. Thomas. 2013. Marine phytoplankton temperature versus growth responses from polar to tropical waters - Outcome of a scientific community-wide study PLoS ONE 8: 8: e63091.

Brennan, G. and S. Collins. 2015. Growth responses of a green alga to multiple environmental drivers. Nature Climate Change 5: 892–897.

Britt, A.B. 1995. Repair of DNA damage induced by ultraviolet radiation. Plant Physiology 108: 891–896.

Britton, D., C.E. Cornwall, A.T. Revill, C.L. Hurd and C.R. Johnson. 2016. Ocean acidification reverses the positive effects of seawater pH fluctuations on growth and photosynthesis of the habitat-forming kelp, *Ecklonia radiata*. Scientific Reports 6: 26036.

Buitenhuis, E.T., T. Pangerc, D.J. Franklin, C. Le Quéré and G. Malin. 2008. Growth rates of six coccolithophorid strains as a function of temperature. Limnology and Oceanography 53: 1181–1185.

Bultel-Poncé, V., F. Felix-Theodore, C. Sarthon, J.F. Ponge and B. Bodo. 2004. New pigments from the terrestrial cyanobacterium *Scytonema* sp. collected on the Mitaraka Inselberg, French Guyana. Journal of Natural Products 67: 678–681.

Cai, X., D.A. Hutchins, F. Fu and K. Gao. 2017. Effects of ultraviolet radiation on photosynthetic performance and N_2 fixation in *Trichodesmium erythraeum* IMS 101. Biogeosciences 14: 4455.

Caldeira, K. and M.E. Wickett. 2003. Oceanography: anthropogenic carbon and ocean pH. Nature 425(6956):365.

Campbell, D., M.J. Eriksson, G. Öquist, P. Gustafsson and A.K. Clarke. 1998. The cyanobacterium *Synechococcus* resists UV-B by exchanging photosystem II reaction-center D1 proteins. Proceedings of the National Academy of Sciences of the Unied States of America 95: 364–369.

Caron, D.D. 2014. Protection of the Stratospheric Ozone Layer and the Structure of International Environmental Law Making.

Chavez, F.P., M. Messié and J.T. Pennington. 2011. Marine primary production in relation to climate variability and change. Annual Review of Marine Science 3: 227–260.

Chen, L., S. Deng, R. De Philippis, W. Tian, H. Wu and J. Wang. 2013. UV-B resistance as a criterion for the selection of desert microalgae to be utilized for inoculating desert soils. Journal of Applied Phycology 25: 1009–1015.

Chen, S. and K. Gao. 2011. Solar ultraviolet radiation and CO_2-induced ocean acidification interacts to influence the photosynthetic performance of the red tide alga *Phaeocystis globosa* (Prymnesiophyceae). Hydrobiologia 675: 105–117.

Chen, S., K. Gao and J. Beardall. 2015. Viral attack exacerbates the susceptibility of a bloom-forming alga to ocean acidification. Global Change Biology 21: 629–636.

Chivers, W.J., A.W. Walne and G.C. Hays. 2017. Mismatch between marine plankton range movements and the velocity of climate change. Nature Communications 8: 14434.

Claquin, P., I. Probert, S. Lefebvre and B. Veron. 2008. Effects of temperature on photosynthetic parameters and TEP production in eight species of marine microalgae. Aquatic Microbial Ecology 51: 1–11.

Cockell, C.S. and J. Knowland. 1999. Ultraviolet radiation screening compounds. Biological Reviews 74: 311–345.

Coello-Camba, A., S. Agustí, J. Holding, J.M. Arrieta and C.M. Duarte. 2014. Interactive effect of temperature and CO_2 increase in Arctic phytoplankton. Frontiers in Marine Science 1: 49.

Conte, M.H., A. Thompson, D. Lesley and R.P. Harris. 1998. Genetic and physiological influences on the alkenone/alkenoate versus growth temperature relationship in *Emiliania huxleyi* and *Gephyrocapsa oceanica*. Geochimica et Cosmochimica Acta 62: 51–68.

Czerny, J., J. Barcelos e Ramos and U. Riebesell. 2009. Influence of elevated CO_2 concentrations on cell division and nitrogen fixation rates in the bloom-forming cyanobacterium *Nodularia spumigena*. Biogeosciences (BG) 6: 1865–1875.

Desikachary, T.V. 1959. Cyanophyta. Indian Council of Agriculture Research, New Delhi, India, 1–686.

Domingues, R.B., C.C. Guerra, H.M. Galvão, V. Brotas and A.B. Barbosa. 2017. Short-term interactive effects of ultraviolet radiation, carbon dioxide and nutrient enrichment on phytoplankton in a shallow coastal lagoon. Aquatic Ecology 51: 91–105.

Donkor, V. and D.-P. Häder. 1991. Effects of solar and ultraviolet radiation on motility, photomovement and pigmentation in filamentous gliding cyanobacteria. FEMS Microbiology Ecology 86: 159–168.

Doyle, S.A., J.E. Saros and C.E. Williamson. 2005. Interactive effects of temperature and nutrient limitation on the response of alpine phytoplankton growth to ultraviolet radiation. Limnology and Oceanography 50: 1362–1367.

Dupont, S., E. Hall, P. Calosi and B. Lundve. 2014. First evidence of altered sensory quality in a shellfish exposed to decreased pH relevant to ocean acidification. Journal of Shellfish Research 33: 857–861.

Dupont, S. and H. Pörtner. 2013. Marine science: get ready for ocean acidification. Nature 498: 429.

Edward, M. and A.J. Richardson. 2004. Impact of climate change on marine pelagic phenology and trophic mismatch. Nature 430: 881–884.

Eichner, M., B. Rost and S.A. Kranz. 2014. Diversity of ocean acidification effects on marine N_2 fixers. Journal of Experimental Marine Biology and Ecology 457: 199–207.

Engel, A., J. Piontek, H.-P. Grossart, U. Riebesell, K.G. Schulz and M. Sperling. 2014. Impact of CO_2 enrichment on organic matter dynamics during nutrient induced coastal phytoplankton blooms. Journal of Plankton Research 36: 641–657.

Falkowski, P.G. 1997. Evolution of the nitrogen cycle and its influence on the biological sequestration of CO_2 in the ocean. Nature 387: 272.

Fischetti, M. 2013. Deep heat threatens marine life. Scientific American 308: 92.

Fu, F.X., A.O. Tatters and D.A. Hutchins. 2012. Global change and the future of harmful algal blooms in the ocean. Marine Ecology Progress Series 470: 207–233.

Gaitán-Espitia, J.D., D. Marshall, S. Dupont, L.D. Bacigalupe, L. Bodrossy and A.J. Hobday. 2017. Geographical gradients in selection can reveal genetic constraints for evolutionary responses to ocean acidification. Biology Letters 13: 20160784.

Gao, K. 1990. Seasonal variation of photosynthetic capacity in *Sargassum horneri*. Jpn. J. Phycol. 38: 25–33.

Gao, K., Y. Aruga, K. Asada, T. Ishihara, T. Akano and M. Kiyohara. 1991. Enhanced growth of the red alga *Porphyra yezoensis* Ueda in high CO_2 concentrations. Journal of Applied Phycology 3: 355–362.

Gao, K. and D. Campbell. 2014. Photophysiological responses of marine diatoms to elevated CO_2 and decreased pH: a review. Functional Plant Biology 41: 449–459.

Gao, K., E.W. Helbling, D.-P. Häder and D.A. Hutchins. 2012a. Responses of marine primary producers to interactions between ocean acidification, solar radiation, and warming. Marine Ecology Progress Series 470: 167–189

Gao, K., P. Li, T. Watanabe and E.W. Helbling. 2008. Combined effects of ultraviolet radiation and temperature on morphology, photosynthesis, and DNA of *Arthrospira* (*Spirulina*) *platensis* (Cyanophyta). Journal of Phycology 44: 777–786.

Gao, K., Z. Ruan, V.E. Villafañe, J.P. Gattuso and E.W. Helbling. 2009. Ocean acidification exacerbates the effect of UV radiation on the calcifying phytoplankter *Emiliania huxleyi*. Limnology and Oceanography 54: 1855–1862.

Gao, K., Y. Wu, G. Li, H. Wu, V.E. Villafañe and E.W. Helbling. 2007a. Solar UV radiation drives CO_2 fixation in marine phytoplankton: a double-edged sword. Plant Physiology 144: 54–59.

Gao, K. and C. Ye. 2007. Photosynthetic insensitivity of the terrestrial cyanobacterium *Nostoc flagelliforme* to solar UV radiation while rehydrated or dessiccated. Journal of Phycology 43: 628–635.

Gao, K., H. Yu and M.T. Brown. 2007b. Solar PAR and UV radiation affects the physiology and morphology of the cyanobacterium *Anabaena* sp. PCC 7120. Journal of Photochemistry and Photobiology B: Biology 89: 117–124.

Gao, K., Y. Zhang and D.-P. Häder. 2018. Individual and interactive effects of ocean acidification, global warming, and UV radiation on phytoplankton. Journal of Applied Phycology.

Gao, K. and Y. Zheng. 2010. Combined effects of ocean acidification and solar UV radiation on photosynthesis, growth, pigmentation and calcification of the coralline alga *Corallina sessilis* (Rhodophyta). Global Change Biology 16: 2388–2398.

Gao, K.S., J.T. Xu, G. Gao, Y.H. Li, D.A. Hutchins, B.Q. Huang, L. Wang, Y. Zheng, P. Jin, X.N. Cai, D.P. Häder, W. Li, K. Xu, N.N. Liu and U. Riebesell. 2012b. Rising CO_2 and increased light exposure synergistically reduce marine primary productivity. Nature Climate Change 2: 519–523.

Garcia-Pichel, F. and J. Belnap. 1996. Microenvironments and microscale productivity of cyanobacterial desert crusts. Journal of Phycology 32: 774–782.

Garcia-Pichel, F. and R.W. Castenholz. 1991. Characterization and biological implications of scytonemin, a cyanobacterial sheath pigment. Journal of Phycology 27: 395–409.

Garcia-Pichel, F. and R.W. Castenholz. 1993. Occurrence of UV-absorbing, mycosporine-like compounds among cyanobacterial isolates and an estimate of their screening capacity. Applied and Environmental Microbiology 59: 163–169.

Gattuso, J.-P., A. Magnan, R. Billé, W. Cheung, E. Howes, F. Joos, D. Allemand, L. Bopp, S. Cooley and C. Eakin. 2015. Contrasting futures for ocean and society from different anthropogenic CO_2 emissions scenarios. Science 349: aac4722.

Geitler, L. 1932. Cyanophyceae (Blaualgen). pp. 1–119. *In*: Rabenhorst, L. (ed.). Kryptogamen-Flora von Deutschland, Österreich und der Schweiz. Akademische Verlagsgesellschaft, Leipzig.

Gradoville, M.R., A.E. White, D. Böttjer, M.J. Church and R.M. Letelier. 2014. Diversity trumps acidification: Lack of evidence for carbon dioxide enhancement of *Trichodesmium* community nitrogen or carbon fixation at Station ALOHA. Limnology and Oceanography 59: 645–659.

Grear, J.S., T.A. Rynearson, A.L. Montalbano, B. Govenar and S. Menden-Deuer. 2017. pCO_2 effects on species composition and growth of an estuarine phytoplankton community. Estuarine, Coastal and Shelf Science 190: 40–49.

Groetsch, P.M., S.G. Simis, M.A. Eleveld and S.W. Peters. 2016. Spring blooms in the Baltic Sea have weakened but lengthened from 2000 to 2014. Biogeosciences 13: 4959.

Gröniger, A., R.P. Sinha, M. Klisch and D.-P. Häder. 2000. Photoprotective compounds in cyanobacteria, phytoplankton and macroalgae—a database. Journal of Photochemistry and Photobiology B: Biology 58: 115–122.

Guan, W. and K. Gao. 2008. Light histories influence the impacts of solar ultraviolet radiation on photosynthesis and growth in a marine diatom, *Skeletonema costatum*. Journal of Photochemistry and Photobiology B: Biology 91: 151–156.

Guan, W. and P. Li. 2017. Dependency of UVR-induced photoinhibition on atomic ratio of N to P in the dinoflagellate *Karenia mikimotoi*. Marine Biology 164: 31.

Häder, D.-P. 1987. Photomovement. pp. 325–345. *In*: Fay, P. and C. Van Baalen (eds.). The Cyanobacteria. Elsevier Amsterdam, New York, Oxford.

Häder, D.-P. and K. Gao. 2015. Interactions of anthropogenic stress factors on marine phytoplankton. Frontiers in Environmental Science 3: 14.

Häder, D.-P. and K. Gao. 2017. The impacts of climate change on marine phytoplankton. pp. 901–928. *In*: Phillips, B.F. and M. Pérez-Ramírez (eds.). Climate Change Impacts on Fisheries and Aquaculture. A Global Analysis. Wiley, Hoboken, NJ.

Häder, D.-P., E.W. Helbling, C.E. Williamson and R.C. Worrest. 2011. Effects of UV radiation on aquatic ecosystems and interactions with climate change. Photochememical and Photobiological Sciences 10: 242–260.

Häder, D.-P., P. Richter, V.E. Villafañe and E.W. Helbling. 2014. Influence of light history on the photosynthetic and motility responses of *Gymnodinium chlorophorum* exposed to UVR and different temperatures. Journal of Photochemistry & Photobiology, B: Biology 138: 273–281.

Häder, D.-P. and R.P. Sinha. 2005. Solar ultraviolet radiation-induced DNA damage in aquatic organisms: potential environmental imapct. Mutation Research 571: 221–233.

Häder, D.-P., C.E. Williamson, S.-A. Wängberg, M. Rautio, K.C. Rose, K. Gao, E.W. Helbling, R.P. Sinha and R. Worrest. 2015. Effects of UV radiation on aquatic ecosystems and interactions with other environmental factors. Photochemical & Photobiological Sciences 14: 108–126.

Halac, S., V. Villafañe, R. Gonçalves and E. Helbling. 2014. Photochemical responses of three marine phytoplankton species exposed to ultraviolet radiation and increased temperature: Role of photoprotective mechanisms. Journal of Photochemistry and Photobiology B: Biology 141: 217–227.

Halac, S.R., S.D. Guendulain-García, V.E. Villafañe, E.W. Helbling and A.T. Banaszak. 2013. Responses of tropical plankton communities from the Mexican Caribbean to solar ultraviolet radiation exposure and increased temperature. Journal of Experimental Marine Biology and Ecology 445: 99–107.

Halac, S.R., V.E. Villafañe and E.W. Helbling. 2010. Temperature benefits the photosynthetic performance of the diatoms *Chaetoceros gracilis* and *Thalassiosira weissflogii* when exposed to UVR. Journal of Photochemistry and Photobiology B: Biology 101: 196–205.

Hallegraeff, G.M. 1993. A review of harmful algal blooms and their apparent global increase. Phycologia 32: 79–99.

Hallegraeff, G.M. 2010. Ocean climate change, phytoplankton community responses, and harmful algal blooms: a formidable predictive challenge. Journal of Phycology 46: 220–235.

Hallegraeff, G.M. 2014. Harmful algal blooms in the Australian region: changes between the 1980s and 2010s. The 9th International Conference on Molluscan Shellfish Safety.

Hansen, J., M. Sato, R. Ruedy, K. Lo, D.W. Lea and M. Medina-Elizade. 2006. Global temperature change. Proceedings of the National Academy of Sciences 103: 14288–14293.

Hattenrath-Lehmann, T.K., J.L. Smith, R.B. Wallace, L.R. Merlo, F. Koch, H. Mittelsdorf, J.A. Goleski, D.M. Anderson and C.J. Gobler. 2015. The effects of elevated CO_2 on the growth and toxicity of field populations and cultures of the saxitoxin—producing dinoflagellate, *Alexandrium fundyense*. Limnology and Oceanography 60: 198–214.

He, Y.Y. and D.-P. Häder. 2002a. Involvement of reactive oxygen species in the UV-B damage to the cyanobacterium *Anabaena* sp. Journal of Photochemistry and Photobiology B: Biology 66: 73–80.

He, Y.Y. and D.-P. Häder. 2002b. Reactive oxygen species and UV-B: effect on cyanobacteria. Photochemical & Photobiological Sciences 1: 729–736.

He, Y.Y. and D.-P. Häder. 2002c. UV-B-induced formation of reactive oxygen species and oxidative damage of the cyanobacterium *Anabaena* sp.: protective effects of ascorbic acid and *N*-acetyl-L-cysteine. Journal of Photochemistry and Photobiology B: Biology 66: 115–124.

Helbling, E.W., K. Gao, R.J. Gonçalves, H. Wu and V.E. Villafañe. 2003. Utilization of solar UV radiation by coastal phytoplankton assemblages off SE China when exposed to fast mixing. Marine Ecology Progress Series 259: 59–66.

Hernandez-Becerril, D.U. and N. Pasten-Miranda. 2015. Abundance and distribution of the picoplanktonic cyanobacteria *Synechococcus* in La Paz bay and Carmen basin, Gulf of California (June, 2001). Hidrobiológica 25: 357–364.

Holding, J.M., C.M. Duarte, M. Sanz-Martín, E. Mesa, J.M. Arrieta, M. Chierici, I. Hendriks, L. Garcia-Corral, A. Regaudie-de-Gioux and A. Delgado. 2015. Temperature dependence of CO_2-enhanced primary production in the European Arctic Ocean. Nature Climate Change 5: 1079–1082.

Hong, H., R. Shen, F. Zhang, Z. Wen, S. Chang, W. Lin, S.A. Kranz, Y.-W. Luo, S.-J. Kao and F.M. Morel. 2017. The complex effects of ocean acidification on the prominent N_2-fixing cyanobacterium *Trichodesmium*. Science 356: 527–531.

Hsu, N.C., T. Si-Chee, M.D. King and D.J. Diner. 2007. Dust in the wind. pp. 56–60. *In*: King, M.D., C.L. Parkinson, K.C. Partington and R.G. Williams (eds.). Our Changing Planet. The View from Space. Cambridge University Press, Cambridge.

Huang, L., M.P. McCluskey, H. Ni and R.A. LaRossa. 2002. Global gene expression profiles of the cyanobacterium *Synechocystis* sp. strain PCC 6803 in response to irradiation with UV-B and white light. Journal of Bacteriology 184: 6845–6858.

Huertas, I.E., M. Rouco, V. López-Rodas and E. Costas. 2011. Warming will affect phytoplankton differently: evidence through a mechanistic approach. Proceedings of the Royal Society B: Biological Sciences 278: 3534–3543.

Huey, R.B. and R. Stevenson. 1979. Integrating thermal physiology and ecology of ectotherms: a discussion of approaches. American Zoologist 19: 357–366.

Hutchins, D.A., F.-X. Fu, E.A. Webb, N. Walworth and A. Tagliabue. 2013. Taxon-specific response of marine nitrogen fixers to elevated carbon dioxide concentrations. Nature Geoscience 6: 790–795.

Hutchins, D.A., N.G. Walworth, E.A. Webb, M.A. Saito, D. Moran, M.R. McIlvin, J. Gale and F.-X. Fu. 2015. Irreversibly increased nitrogen fixation in *Trichodesmium* experimentally adapted to elevated carbon dioxide. Nature Communications 6: 8155.

Hyun, B., K.-H. Choi, P.-G. Jang, M.-C. Jang, W.-J. Lee, C.-H. Moon and K. Shin. 2014. Effects of increased CO_2 and temperature on the growth of four diatom species (*Chaetoceros debilis, Chaetoceros didymus, Skeletonema costatum* and *Thalassiosira nordenskioeldii*) in laboratory experiments. Journal of Environtal Science International 23: 1003–1012.

IPCC. 2014. Climate Change 2014: Impacts, Adaptation, and Vulnerability. Part B: Regional Aspects. Contribution of Working Group II to the Fifth Assessment Report of the Intergovernmental Panel on Climate Change, 2014.

Jin, P., J. Ding, T. Xing, U. Riebesell and K. Gao. 2017a. High levels of solar radiation offset impacts of ocean acidification on *Emiliania huxleyi*, with special reference to calcifying and non-calcifying strains. Marine Ecology Progress Series 568: 47–58.

Jin, P., C.M. Duarte and S. Agustí. 2017b. Contrasting responses of marine and freshwater photosynthetic organisms to UVB radiation: a meta-analysis. Frontiers in Marine Science 4: 45.

Jin, P., K. Gao, V. Villafañe, D. Campbell and W. Helbling. 2013. Ocean acidification alters the photosynthetic responses of a coccolithophorid to fluctuating ultraviolet and visible radiation. Plant Physiology 162: 2084–2094.

Jin, P., T. Wang, N. Liu, S. Dupont, J. Beardall, P.W. Boyd, U. Riebesell and K. Gao. 2015. Ocean acidification increases the accumulation of toxic phenolic compounds across trophic levels. Nature Communications 6: 8714.

Kaeriyama, H., E. Katsuki, M. Otsubo, M. Yamada, K. Ichimi, K. Tada and P.J. Harrison. 2011. Effects of temperature and irradiance on growth of strains belonging to seven *Skeletonema* species isolated from Dokai Bay, southern Japan. European Journal of Phycology 46: 113–124.

Karsten, U. and F. Garcia-Pichel. 1996. Carotenoids and mycosporine-like amino acid compounds in members of the genus *Microcoleus* (Cyanobacteria): a chemosystematic study. Systematic and Applied Microbiology 19: 285–294.

Kim, J.-H. 2006. The effect of seawater CO_2 concentration on growth of a natural phytoplankton assemblage in a controlled mesocosm experiment. Limnology and Oceanography 51: 1629–1636.

Kim, J.-M., K. Lee, K. Shin, J.-H. Kang, H.-W. Lee, M. Kim, P.-G. Jang and M.-C. Jang. 2006. The effect of seawater CO_2 concentration on growth of a natural phytoplankton assemblage in a controlled mesocosm experiment. Limnology and Oceanography 51: 1629–1636.

Kim, S.T. and A. Sancar. 1995. Photorepair of nonadjacent pyrimidine dimers by DNA photolyase. Photochemistry and Photobiology 61: 171–174.

Klisch, M. 2002. Induktion von UV-Schirmpigmenten in Marinen Dinoflagellaten Dissertation, Friedrich-Alexander University Erlangen-Nürnberg, Germany.

Klisch, M. and D.-P. Häder. 2000. Mycosporine-like amino acids in the marine dinoflagellate *Gyrodinium dorsum*: induction by ultraviolet irradiation. Journal of Photochemistry and Photobiology B: Biology 55: 178–182.

Klisch, M. and D.-P. Häder. 2002. Wavelength dependence of mycosporine-like amino acid synthesis in *Gyrodinium dorsum*. Journal of Photochemistry and Photobiology B: Biology 66: 60–66.

Kremp, A., A. Godhe, J. Egardt, S. Dupont, S. Suikkanen, S. Casabianca and A. Penna. 2012. Intraspecific variability in the response of bloom—forming marine microalgae to changed climate conditions. Ecology and Evolution 2: 1195–1207.

Larsen, J., A. Larsen, R. Thyrhaug, G. Bratbak and R.-A. Sandaa. 2008. Response of marine viral populations to a nutrient induced phytoplankton bloom at different pCO_2 levels. Biogeosciences 5: 523–533.

Leung, P.T., A.X. Yi, J.C. Ip, S.S. Mak and K.M. Leung. 2017. Photosynthetic and transcriptional responses of the marine diatom *Thalassiosira pseudonana* to the combined effect of temperature stress and copper exposure. Marine Pollution Bulletin 124: 938–945.

Li, G. and K. Gao. 2013. Cell size-dependent effects of solar UV radiation on primary production in coastal waters of the South China Sea. Estuaries and Coasts 36: 728–736.

Li, G., K. Gao and G. Gao. 2011. Differential impacts of solar UV radiation on photosynthetic carbon fixation from the coastal to offshore surface waters in the South China Sea. Photochemistry and Photobiology 87: 329–334.

Li, W., K. Gao and J. Beardall. 2012a. Interactive effects of ocean acidification and nitrogen-limitation on the diatom *Phaeodactylum tricornutum*. Plos One 7: e51590.

Li, W., Y. Yang, Z. Li, J. Xu and K. Gao. 2017. Effects of seawater acidification on the growth rates of the diatom *Thalassiosira* (Conticribra) *weissflogii* under different nutrient, light, and UV radiation regimes. Journal of Applied Phycology 29: 133–142.

Li, Y., K. Gao, V. Villafañe and E. Helbling. 2012b. Ocean acidification mediates photosynthetic response to UV radiation and temperature increase in the diatom *Phaeodactylum tricornutum*. Biogeosciences 9: 3931–3942.

Linacre, L., R. Lara-Lara, V. Camacho-Ibar, J.C. Herguera, C. Bazán-Guzmán and V. Ferreira-Bartrina. 2015. Distribution pattern of picoplankton carbon biomass linked to mesoscale dynamics in the southern gulf of Mexico during winter conditions. Deep Sea Research Part I: Oceanographic Research Papers 106: 55–67.

Listmann, L., M. LeRoch, L. Schlüter, M.K. Thomas and T.B. Reusch. 2016. Swift thermal reaction norm evolution in a key marine phytoplankton species. Evolutionary Applications 9: 1156–1164.

Liu, N., J. Beardall and K. Gao. 2017. Elevated CO_2 and associated seawater chemistry do not benefit a model diatom grown with increased availability of light. Aquatic Microbial Ecology 79: 137–147.

Lohbeck, K., U. Riebesell and T. Reusch. 2012. Adaptive evolution of a key phytoplankton species to ocean acidification, Nat. Geosci., 5, 346–351.

Ma, Z. and K. Gao. 2009. Photoregulation of morphological structure and its physiological relevance in the cyanobacterium *Arthrospira* (*Spirulina*) *platensis*. Planta 230: 329–337.

Mackey, K.R., J.J. Morris, F.M. Morel and S.A. Kranz. 2015. Response of photosynthesis to ocean acidification. Oceanography 28.

Marchese, C., C. Albouy, J.-É. Tremblay, D. Dumont, F. D'Ortenzio, S. Vissault and S. Bélanger. 2017. Changes in phytoplankton bloom phenology over the North Water (NOW) polynya: a response to changing environmental conditions. Polar Biology: 1–17.

Martin, J.H. 1992. Iron as a limiting factor in oceanic productivity. pp. 123–137. *In*: Falkowski, P.G., A.D. Woodhead and K. Vivirito (eds.). Primary Productivity and Biogeochemical Cycles in the Sea. Environmental Science Research, vol 43. Springer, Boston, MA.

Matthiessen, B., S.L. Eggers and S. Krug. 2012. High nitrate to phosphorus regime attenuates negative effects of rising pCO_2 on total population carbon accumulation. Biogeosciences (BG) 9: 1195–1203.

Meador, J.A., A.J. Baldwin, P. Catala, W.H. Jeffrey, F. Joux, J.A. Moss, J.D. Pakulski, R. Stevens and D.L. Mitchell. 2009. Sunlight-induced DNA damage in marine micro-organisms collected along a latitudinal gradient from 70 degrees N to 68 degrees S. Photochemistry and Photobiology 85: 412–421.

Meehl, G.A., T.F. Stocker, W.D. Collins, P. Friedlingstein, T. Gaye, J.M. Gregory, A. Kitoh, R. Knutti, J.M. Murphy and A. Noda. 2007. Global climate projections. pp. 747–846. *In*: IPCC. Climate Change 2007: The Physical Science Basis. Contribution of Working Group I to the Fourth Assessment Report of the Intergovernmental Panel on Climate Change. Cambridge University Press, Cambridge, U.K.

Metcalf, J.S., S.A. Banack, J. Lindsay, L.F. Morrison, P.A. Cox and G.A. Codd. 2008. Co-occurrence of β-N-methylamino-l-alanine, a neurotoxic amino acid with other cyanobacterial toxins in British waterbodies, 1990–2004. Environmental Microbiology 10: 702–708.

Michaels, A.F., D.M. Karl and D.G. Capone. 2001. Element stoichiometry, new production and nitrogen fixation. Oceanography - Washington DC-Oceanography Society 14: 68–77.

Mignot, A., H. Claustre, J. Uitz, A. Poteau, F. D'Ortenzio and X. Xing. 2014. Understanding the seasonal dynamics of phytoplankton biomass and the deep chlorophyll maximum in oligotrophic environments: A bio-argo float investigation. Global Biogeochemical Cycles 28: 856–876.

Millero, F.J. 2007. The marine inorganic carbon cycle. Chemical Reviews 107: 308–341.

Mills, M.M., C. Ridame, M. Davey, J. La Roche and R.J. Geider. 2004. Iron and phosphorus co-limit nitrogen fixation in the eastern tropical North Atlantic. Nature 429: 292–294.

Milner, S., G. Langer, M. Grelaud and P. Ziveri. 2016. Ocean warming modulates the effects of acidification on *Emiliania huxleyi* calcification and sinking. Limnology and Oceanography 61: 1322–1336.

Moisan, T.A., J. Goes and P.J. Neale. 2009. Mycosporine-like amino acids in phytoplankton: biochemistry, physiology and optics. pp. 119–143. *In*: Kersey, W.T. and S.P. Munger (eds.). Marine Phytoplankton. chap. 4, Nova Science Publishers, New York.

Montagnes, D.J. and M. Franklin. 2001. Effect of temperature on diatom volume, growth rate, and carbon and nitrogen content: reconsidering some paradigms. Limnology and Oceanography 46: 2008–2018.

Monteiro, F.M., L.T. Bach, C. Brownlee, P. Bown, R.E. Rickaby, A.J. Poulton, T. Tyrrell, L. Beaufort, S. Dutkiewicz and S. Gibbs. 2016. Why marine phytoplankton calcify. Science Advances 2: e1501822.

Morel, A. and J.F. Berthon. 1989. Surface pigments, algal biomass profiles, and potential production of the euphotic layer: Relationships reinvestigated in view of remote-sensing applications. Limnology and Oceanography 34: 1545–1562.

Paerl, H.W. and J. Huisman. 2009. Climate change: a catalyst for global expansion of harmful cyanobacterial blooms. Environmental Microbiology Reports 1: 27–37.

Quesada, A. and W.F. Vincent. 1997. Strategies of adaptation by Antarctic cyanobacteria to ultraviolet radiation. European Journal of Phycology 32: 335–342.

Rastogi, R.P., R.P. Sinha, S.H. Moh, T.K. Lee, S. Kottuparambil, Y.-J. Kim, J.-S. Rhee, E.-M. Choi, M.T. Brown and D.-P. Häder. 2014. Ultraviolet radiation and cyanobacteria. Journal of Photochemistry and Photobiology B: Biology 141: 154–169.

Raven, J., K. Caldeira, H. Elderfield, O. Hoegh-Guldberg, P. Liss, U. Riebesell, J. Shepherd, C. Turley and A. Watson. 2005. Ocean acidification due to increasing atmospheric carbon dioxide. The Royal Society. London.

Raymont, J.E. 2014. Plankton & Productivity in the Oceans: Volume 1: Phytoplankton. Elsevier.

Read, J., M. Lucas, S. Holley and R. Pollard. 2000. Phytoplankton, nutrients and hydrography in the frontal zone between the Southwest Indian Subtropical gyre and the Southern Ocean. Deep Sea Research Part I: Oceanographic Research Papers 47: 2341–2367.

Reid, P.C. 2016. Ocean warming: setting the scene. pp. 17–45. *In*: Laffoley, D. and J.M. Baxter (eds.). Explaining Ocean Warming: Causes, Scale, Effects and Consequences. International Union for Conservation of Nature (IUCN), Gland, Switzerland.

Richa and R.P. Sinha. 2013. Biomedical applications of mycosporine-like amino acids. pp. 509–534. *In*: Kim, S.K. (ed.). Marine Microbiology, Bioactive Compounds and Biotechnological Applications. Wiley-VCH Publishers, Germany.

Richa, R.P. Sinha and D.-P. Häder. 2016. Effects of global change, including UV and UV screening compounds. pp. 373–409. *In*: Borowitzka, M., J. Beardall and J. Raven (eds.). The Physiology of Microalgae. Springer, Cham.

Riebesell, U. and J.-P. Gattuso. 2015. Lessons learned from ocean acidification research. Nature Climate Change 5: 12–14.

Riebesell, U., K.G. Schulz, R.G.J. Bellerby, M. Botros, P. Fritsche, M. Meyerhöfer, C. Neill, G. Nondal, A. Oschlies, J. Wohlers and E. Zöllner. 2007. Enhanced biological carbon consumption in a high CO_2 ocean. Nature 450: 545–548.

Riebesell, U. and P.D. Tortell. 2011. Effects of ocean acidification on pelagic organisms and ecosystems. pp. 99–116. *In*: Gattuso, J.P. and L. Hansson (eds.). Ocean Acidification. Oxford University Press, Oxford, UK.

Ries, J.B., A.L. Cohen and D.C. McCorkle. 2009. Marine calcifiers exhibit mixed responses to CO_2-induced ocean acidification. Geology 37: 1131–1134.

Rosas-Navarro, A., G. Langer and P. Ziveri. 2016. Temperature affects the morphology and calcification of *Emiliania huxleyi* strains. Biogeosciences 13: 2913–2926.

Rose, J.M., Y. Feng, G.R. DiTullio, R.B. Dunbar, C.E. Hare, P.A. Lee, M. Lohan, M. Long, W.O. Smith Jr., B. Sohst, S. Tozzi, Y. Zhang and D.A. Hutchins. 2009. Synergistic effects of iron and temperature on Antarctic phytoplankton and microzooplankton assemblages. Biogeosciences 6: 3131–3147.

Sass, L., C. Spetea, Z. Mate, F. Nagy and I. Vass. 1997. Repair of UV-B induced damage of Photosystem II via *de novo* synthesis of the D1 and D2 reaction centre subunits in *Synechocystis* sp. PCC 6803. Photosynthesis Research 54: 55–62.

Schlüter, L., K.T. Lohbeck, M.A. Gutowska, J.P. Gröger, U. Riebesell and T.B. Reusch. 2014. Adaptation of a globally important coccolithophore to ocean warming and acidification. Nature Climate Change 4: 1024–1030.

Shick, J.M. and W.C. Dunlap. 2002. Mycosporine-like amino acids and related gadusols: biosynthesis, accumulation, and UV-protective functions in aquatic organisms. Annual Review of Physiology 64: 223–262.

Siegel, D.A., M.J. Behrenfeld, S. Maritorena, C.R. McClain, D. Antoine, S.W. Bailey, P.S. Bontempi, E.S. Boss, H.M. Dierssen and S.C. Doney. 2013. Regional to global assessments of phytoplankton dynamics from the SeaWiFS mission. Remote Sensing of Environment 135: 77–91.

Singh, G., P.K. Babele, R.P. Sinha, M.B. Tyagi and A. Kumar. 2013. Enymatic and non-enzymatic defense mechanisms against ultraviolet-B radiation in two *Anabaena* species. Process Biochemistry 48: 796–802.

Singh, S.P., D.-P. Häder and R.P. Sinha. 2010. Cyanobacteria and ultraviolet radiation (UVR) stress: Mitigation strategies. Ageing Research Reviews 9: 79–90.

Singh, S.P., S. Kumari, R.P. Rastogi, K.L. Singh and R.P. Sinha. 2008. Mycosporine-like amino acids (MAAs): Chemical structure, biosynthesis and significance as UV-absorbing/screening compounds. Indian Journal of Experimental Biology 46: 7–17.

Sinha, R.P., N.K. Ambasht, J.P. Sinha and D.-P. Häder. 2003a. Wavelength-dependent induction of a mycosporine-like amino acid in a rice-field cyanobacterium, *Nostoc commune*: Role of inhibitors and salt stress. Photochemical & Photobiological Sciences 2: 171–176.

Sinha, R.P., N.K. Ambasht, J.P. Sinha, M. Klisch and D.-P. Häder. 2003b. UV-B-induced synthesis of mycosporine-like amino acids in three strains of *Nodularia* (cyanobacteria). Journal of Photochemistry and Photobiology B: Biology 71: 51–58.

Sinha, R.P. and D.-P. Häder. 2002. UV-induced DNA damage and repair: a review. Photochemical & Photobiological Sciences 1: 225–236.

Sinha, R.P. and D.-P. Häder. 2003. Biochemistry of mycosporine-like amino acids (MAAs) synthesis: Role in photoprotection. Recent Research in Developments in Biochemistry 4: 971–983.

Sinha, R.P. and D.-P. Häder. 2008. UV-protectants in cyanobacteria. Plant Science 174: 278–289.

Sinha, R.P., M. Klisch, A. Gröniger and D.-P. Häder. 1998. Ultraviolet-absorbing/screening substances in cyanobacteria, phytoplankton and macroalgae. Journal of Photochemistry and Photobiology B: Biology 47: 83–94.

Sinha, R.P., M. Klisch, A. Gröniger and D.-P. Häder. 2001a. Responses of aquatic algae and cyanobacteria to solar UV-B. Plant Ecology 154: 221–236.

Sinha, R.P., M. Klisch, E.W. Helbling and D.-P. Häder. 2001b. Induction of mycosporine-like amino acids (MAAs) in cyanobacteria by solar ultraviolet-B radiation. Journal of Photochemistry and Photobiology B: Biology 60: 129–135.

Sinha, R.P., H.D. Kumar, A. Kumar and D.-P. Häder. 1995a. Effects of UV-B irradiation on growth, survival, pigmentation and nitrogen metabolism enzymes in cyanobacteria. Acta Protozoologica 34: 187–192.

Sinha, R.P., S. Kumari and R.P. Rastogi. 2008. Impacts of ultraviolet-B radiation on cyanobacteria: photoprotection and repair. Journal of Science and Reserach 52: 125–142.

Sinha, R.P., M. Lebert, A. Kumar, H.D. Kumar and D.-P. Häder. 1995b. Disintegration of phycobilisomes in a rice field cyanobacterium *Nostoc* sp. following UV irradiation. Biochemistry and Molecular Biology International 37: 697–706.

Sinha, R.P., N. Singh, A. Kumar, H.D. Kumar and D.-P. Häder. 1997. Impacts of ultraviolet-B irradiation on nitrogen-fixing cyanobacteria of rice paddy fields. Journal of Plant Physiology 150: 188–193.

Sinha, R.P., S.P. Singh and D.-P. Häder. 2007. Database on mycosporines and mycosporine-like amino acids (MAAs) in fungi, cyanobacteria, macroalgae, phytoplankton and animals. Journal of Photochemistry and Photobiology B: Biology 89: 29–35.

Steinacher, M., F. Joos, T.L. Froelicher, L. Bopp, P. Cadule, V. Cocco, S.C. Doney, M. Gehlen, K. Lindsay, J.K. Moore, B. Schneider and J. Segschneider. 2010. Projected 21st century decrease in marine productivity: a multi-model analysis. Biogeosciences 7: 979–1005.

Stephanie, M., T. Vera, M. Nathan, P. Micaela, L. Edward, B. Lorraine and F. Lora. 2008. Impacts of climate variability and future climate change on harmful algal blooms and human health.

Sun, J., D.A. Hutchins, Y. Feng, E.L. Seubert, D.A. Caron and F.-X. Fu. 2011. Effects of changing pCO_2 and phosphate availability on domoic acid production and physiology of the marine harmful bloom diatom *Pseudo-nitzschia multiseries*. Limnology and Oceanography 56: 829–840.

Suzuki, Y. and M. Takahashi. 1995. Growth responses of several diatom species isolated from various environments to temperature. Journal of Phycology 31: 880–888.

Tamm, M., P. Laas, R. Freiberg, P. Nõges and T. Nõges. 2018. Parallel assessment of marine autotrophic picoplankton using flow cytometry and chemotaxonomy. Science of The Total Environment 625: 185–193.

Thomas, M.K., C.T. Kremer, C.A. Klausmeier and E. Litchman. 2012. A global pattern of thermal adaptation in marine phytoplankton. Science 338: 1085–1088.

Tilstone, G., B. Šedivá, G. Tarran, R. Kaňa and O. Prášil. 2017. Effect of CO_2 enrichment on phytoplankton photosynthesis in the North Atlantic sub-tropical gyre. Progress in Oceanography 158: 76–89.

Tong, S., D. Hutchins and K. Gao. 2017. Physiological and biochemical responses of *Emiliania huxleyi* to ocean acidification and warming are modulated by UV radiation. Biogeosciences Discussions.

Tortell, P.D., G.R. DiTullio, D.M. Sigman and F.M. Morel. 2002. CO_2 effects on taxonomic composition and nutrient utilization in an equatorial Pacific phytoplankton assemblage. Marine Ecology Progress Series 236: 37–43.

Tortell, P.D., G.H. Rau and F.M. Morel. 2000. Inorganic carbon acquisition in coastal Pacific phytoplankton communities. Limnology and Oceanography 45: 1485–1500.

United Nations Environment Programme Environmental Effects Assessment Panel. 2012. Environmental effects of ozone depletion and its interactions with climate change: Progress report, 2011. Photochemical & Photobiological Sciences 11: 13–27.

Van de Waal, D.B., J.M. Verspagen, J.F. Finke, V. Vournazou, A.K. Immers, W.E.A. Kardinaal, L. Tonk, S. Becker, E. Van Donk and P.M. Visser. 2011. Reversal in competitive dominance of a toxic versus non-toxic cyanobacterium in response to rising CO_2. The ISME Journal 5: 1438.

Varnali, T. and H.G. Edwards. 2013. Theoretical study of novel complexed structures for methoxy derivatives of scytonemin: potential biomarkers in iron-rich stressed environments. Astrobiology 13: 861–869.

Verspagen, J.M., D.B. Van de Waal, J.F. Finke, P.M. Visser and J. Huisman. 2014. Contrasting effects of rising CO_2 on primary production and ecological stoichiometry at different nutrient levels. Ecology Letters 17: 951–960.

Villafañe, V.E., K. Gao, P. Li and E.W. Helbling. 2007. Vertical mixing within the epilimnion modulates UVR-induced photoinhibition in tropical freshwater phytoplankton from southern China. Freshwater Biology 52: 1260–1270.

Villar-Argaiz, M., E. Balseiro, B. Modenutti, M. Souza, F. Bullejos, J. Medina-Sánchez and P. Carrillo. 2017. Resource versus consumer regulation of phytoplankton: Testing the role of UVR in a Southern and Northern hemisphere lake. Hydrobiologia 816: 107–120.

Vincent, W.F. and S. Roy. 1993. Solar ultraviolet-B radiation and aquatic primary production: damage, protection, and recovery. Environmental Reviews 1: 1–12.

Volkmann, M., A.A. Gorbushina, L. Kedar and A. Oren. 2006. Structure of euhalothece-362, a novel red-shifted mycosporine-like amino acid, from a halophilic cyanobacterium (*Euhalothece* sp.). FEMS Microbiological Letters 258: 50–54.

Walker, B. and D. Salt. 2012. Resilience Thinking: Sustaining Ecosystems and People in a Changing World. Island Press. St. Louis.

Wannicke, N., S. Endres, A. Engel, H.-P. Grossart, M. Nausch, J. Unger and M. Voss. 2012. Response of *Nodularia spumigena* to pCO_2–Part 1: Growth, production and nitrogen cycling. Biogeosciences 9: 2973–2988.

Wells, M.L., V.L. Trainer, T.J. Smayda, B.S. Karlson, C.G. Trick, R.M. Kudela, A. Ishikawa, S. Bernard, A. Wulff and D.M. Anderson. 2015. Harmful algal blooms and climate change: Learning from the past and present to forecast the future. Harmful Algae 49: 68–93.

Wu, H., K. Gao, V.E. Villafañe, T. Watanabe and E.W. Helbling. 2005. Effects of solar UV radiation on morphology and photosynthesis of filamentous cyanobacterium *Arthrospira platensis*. Applied and Environmental Microbiology 71: 5004–5013.

Wu, X., G. Gao, M. Giordano and K. Gao. 2012. Growth and photosynthesis of a diatom grown under elevated CO_2 in the presence of solar UV radiation. Fundamental and Applied Limnology/Archiv für Hydrobiologie 180: 279–290.

Wu, Y., D.A. Campbell, A.J. Irwin, D.J. Suggett and Z.V. Finkel. 2014. Ocean acidification enhances the growth rate of larger diatoms. Limnology and Oceanography 59: 1027–1034.

Wu, Y., K. Gao and U. Riebesell. 2010. CO_2-induced seawater acidification affects physiological performance of the marine diatom *Phaeodactylum tricornutum*. Biogeosciences 7: 2915–2923.

Xiao, W., X. Liu, A.J. Irwin, E.A. Laws, L. Wang, B. Chen, Y. Zeng and B. Huang. 2018. Warming and eutrophication combine to restructure diatoms and dinoflagellates. Water Research 128: 206–216.

Xu, J., K. Gao, Y. Li and D.A. Hutchins. 2014. Multiple future ocean changes interactively alter physiological and biochemical processes of diatoms. Marine Ecology Progress Series 515: 73–81.

Xu, K. and K. Gao. 2015. Solar UV irradiances modulate effects of ocean acidification on the Coccolithophorid *Emiliania huxleyi*. Photochemistry and Photobiology 91: 92–101.

Xu, K., K. Gao, V. Villafañe and E. Helbling. 2011. Photosynthetic responses of *Emiliania huxleyi* to UV radiation and elevated temperature: roles of calcified coccoliths. Biogeosciences 8: 1441–1452.

Zeebe, R.E. and D.A. Wolf-Gladrow. 2001. CO_2 in seawater: equilibrium, kinetics, isotopes. Gulf Professional Publishing.

Zepp, R.G., D.J. Erickson, N.D. Paul and B. Sulzberger. 2007. Interactive effects of solar UV radiation and climate change on biogochemical cycling. Photochemical & Photobiolological Sciences 6: 286–700.

Zhang, Y., L.T. Bach, K.G. Schulz and U. Riebesell. 2015. The modulating effect of light intensity on the response of the coccolithophore *Gephyrocapsa oceanica* to ocean acidification. Limnology and Oceanography 60: 2145–2157.

Zhang, Y., H.B. Jiang and B.S. Qiu. 2013. Effects of UVB Radiation on competition between the bloom-forming cyanobacterium *Microcystis aeruginosa* and the Chlorophyceae *Chlamydomonas microsphaera*. Journal of Phycology 49: 318–328.

Zhang, Y., R. Klapper, K.T. Lohbeck, L.T. Bach, K.G. Schulz, T.B. Reusch and U. Riebesell. 2014. Between- and within-population variations in thermal reaction norms of the coccolithophore *Emiliania huxleyi*. Limnology and Oceanography 59: 1570–1580.

Zheng, Y. and K. Gao. 2009. Impacts of solar UV radiation on the photosynthesis, growth and UV-absorbing compounds in *Gracilaria lemaneiformis* (Rhodophyta) grown at different nitrate concentrations. Journal of Phycology 45: 314–323.

CHAPTER 6

Are Warmer Waters, Brighter Waters?

An Examination of the Irradiance Environment of Lakes and Oceans in a Changing Climate

Patrick Neale[1,*] and *Robyn Smyth* [2]

Introduction

Climate change is warming surface waters globally. According to the IPCC, the surface ocean (upper 75 m) has warmed on average 0.11°C per decade since 1971 (Rhein et al. 2013) while summer surface water temperatures in over 200 globally-distributed lakes have increased on average 0.3°C per decade since the mid-1980s (O'Reilly et al. 2015). With warming, the density differences between surface and 'deep' waters are increasing in both lakes and the ocean (Rhein et al. 2013, Kraemer et al. 2015). Ocean model projections suggest decreased nutrient upwelling and primary productivity in, for example, expanding oligotrophic subtropical gyres while other regions, such as the Southern Ocean where productivity is light-limited due to deep mixing, shoaling mixed layers are expected to increase mean irradiance and enhance productivity (Sarmiento et al. 2004, Boyce et al. 2010, Steinacher et al. 2010, Bopp et al. 2013). Decreasing mixing depths can also increase exposure to inhibitory irradiance and reduce productivity (Neale et al. 2003). In lakes, the effects of temperature and mixed layer depth on productivity are complex and mediated by nutrient availability, lake size, and transparency (Lewis 2011).

[1] Patrick Neale, Smithsonian Environmental Research Center, Edgewater MD, USA 21027.

[2] Robyn Smyth, Bard College, Annandale-on-Hudson, NY USA 12504.
Email: rsmyth@bard.edu

* Corresponding author: nealep@si.edu

Rising air temperature is the most commonly identified driver of surface water heating but there are many variables influencing the net energy flux at the air-water interface that ultimately determine the mixed layer depth (MLD) and pelagic light environment. Changes in wind forcing, solar radiation, ice cover, and transparency can also alter dynamics of stratification and mixing in ways that are both synergistic and antagonistic with the overall warming trend (Denman and Gargett 1983, Rhein et al. 2013, Fink et al. 2014). Therefore, the net effect of climate change-driven warming of surface waters on mixing depth and primary productivity is not easy to predict and likely to vary widely with local conditions. In this chapter, we focus on the effects of changing mixed layer depths on the light environment for photosynthesis and other irradiance-dependent processes in both the ocean and lakes. First, we briefly describe the major drivers of mixed layer depth in pelagic environments and how they are heterogeneously impacted by climate change. Then we examine trends in mixed layer depth and light climate using time series measurements of temperature and attenuation spanning two decades or more and compare those to model predictions. Finally, we consider the implications of these findings on future investigations of climate change effects on aquatic ecosystems.

Climate Change Impacts on Mixed Layer Depth and Irradiance Environment

The depth and intensity of mixing directly determines light exposure for photosynthesis and other irradiance-dependent processes in open water environments. MLDs are dynamic and difficult to determine with continuity over space and time (see Box 6.1). The depth of mixing increases and decreases in response to atmospheric forcing at subdaily, seasonal, and interannual timescales. The overall response to climate change is not simple to predict because the MLD is the net result of surface energy transfers by multiple co-occurring processes that are all susceptible to climate change. Solar heating stabilizes the water column while a combination of atmospherically-driven processes generates turbulence and mixing that redistribute heat and other constituents into a well-mixed surface layer (Boehrer and Schultze 2008, Holte and Talley 2009). The depth of mixing reflects a balance between the potential energy of the density gradient resulting from the net surface heat flux and the mechanical energy from wind, waves, and currents that generate mixing. While warming alone will tend to increase surface stratification and reduce the MLD, coincident changes in net radiation, wind, currents, and transparency can either counteract or reinforce the effects of warming, resulting in MLDs that are deeper or unchanged as well as more shallow (Table 6.1). In North America and Europe, some lake warming has been linked to increased incident irradiance due to reductions in air pollution (Fink et al. 2014). Altering water inflows through changes in precipitation and ice cover can also affect stratification dynamics and mixed layer depth (Rhein et al. 2013). The most recent analyses of thermal stratification trends in lakes and the ocean consider how trends in these other factors create varying outcomes in how climate change influences the mixed layer (e.g., Richardson et al. 2017, Somavilla et al. 2017).

Water transparency is an important mediator of changes in stratification also subject to climate change effects. It can decline from increases in biomass (eutrophication),

Box 6.1. Defining mixed layer depth.

Mixed layer depth (MLD) is an important but challenging parameter to define for understanding and modeling pelagic processes. Due to a general scarcity of turbulence measurements, MLD is commonly determined from temperature or density profiles using a specified gradient or step change from the surface, sometimes employing interpolation or more advanced curve fitting procedures to account for data collection at discrete depth intervals (e.g., Fee et al. 1996, Read et al. 2011, Somavilla et al. 2017). The criteria used should correspond to the process under study (Brainerd and Gregg 1995). There are three main categories of depths identified in the literature and illustrated in the figure with density profiles from lakes Lacawac and Giles. The deepest depth, the **thermocline**, is commonly defined as the depth of maximum density gradient (black squares). It varies on seasonal to interannual timescales and is appropriate for describing long-term changes in thermal structure. The most shallow and dynamic is the near-surface **mixing depth,** identified by density gradients less than 0.05 kg m^{-3} and corresponding to the zone of active turbulence and exchange with the atmosphere. This depth is appropriate for studies of gas exchange and active fluorometry and can be quite shallow in small lakes, as shown by the triangles in the figure defined with the density step criterion of 0.03 kg m^{-3} (equivalent to 0.15°C temperature change at 20°C) used in Somavilla et al. (2017) and here for oceanic profiles. Intermediary between the thermocline and the active mixing depth is the **mixed layer depth (MLD)**, defined as the zone of recent mixing as determined by atmospheric forcing and background stratification on a daily to weekly timescale. Defined here with a 0.2 kg m^{-3} per meter density gradient (equivalent to 1°C per meter), the MLD corresponds to the epilimnion as traditionally defined for lakes (dashed lines in Fig. B1).

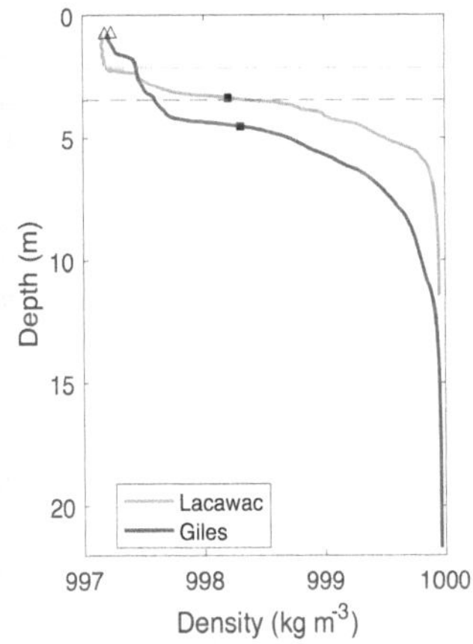

Fig. B1. Density profiles from temperature profiles show in Fig. 1 with active mixing depth (triangles), MLD (dashed lines), and thermocline (squares).

Table 6.1. Expected effect of increasing (↑) and decreasing (↓) trends in climate-related drivers of water column stability (mixed layer depth).

Drivers	Stabilizing (shoaling)	Destabilizing (deepening)
External		
Air temperature	↑	↓
Net radiation	↑	↓
Wind speed	↓	↑
Internal		
Transparency	↑	↓
Currents	↓	↑
Albedo	↓	↑

increases in chromophoric dissolved organic matter (CDOM), known as browning, or increases in turbidity (sediment erosion and suspension). Runoff events can also transiently affect transparency in coastal marine waters (Baum et al. 2007, Moore et al. 2011, Fichot et al. 2014). Melting of the Arctic permafrost is mobilizing dissolved organic carbon (DOC) into rivers, which is ultimately transported into the Arctic ocean substantially increasing the CDOM and decreasing UV transparency (Fichot et al. 2013, Spencer et al. 2015). A class of lakes in which surface warming is particularly rapid is small lakes with increasing DOC (Box 6.2). These trends are expected to continue at least into the near future (de Wit et al. 2016, Weyhenmeyer et al. 2016). Thus, when it comes to predicting how climate change will impact pelagic primary productivity, it is not appropriate to assume MLD will shoal in all cases. Changes in all components of the surface energy budget and water transparency should be considered when projecting pelagic productivity and other irradiance-dependent processes in a changing climate.

Climate-driven changes in the MLD can impact productivity by altering access to nutrients and light (Boyd et al. 2010). Irradiance in the mixed layer acts on phytoplankton and other organisms on a range of time scales, from seconds in terms of the instantaneous rates of photosynthesis to days or weeks for changes in biochemical composition that accompany acclimation to the light environment (MacIntyre et al. 2000, Neale et al. 2003). From an ecological standpoint, the surface layer is considered mixed if biochemical and taxonomic characteristics are uniform over the depth of the layer in question, meaning that vertical exchange is occurring fast enough and often enough to forestall the acclimation processes that create depth gradients in these characteristics in a non-mixing environment (Cullen and Lewis 1988). Nevertheless, there are fast (time-scale seconds to minutes) responses, such as fluorescence quenching, that show depth gradients even in 'mixed layers' (e.g., Neale et al. 2012). More details on the relationship between vertical mixing and

Box 6.2. Lake size effects on MLD.

Many of the physical processes capable of generating turbulent mixing in stratified lakes are dependent upon lake size (Boyce 1974, Yeates and Imberger 2003) which spans eight orders of magnitude from 0.001 km^2 ponds to the Caspian Sea at 371,000 km^2 (Verpoorter et al. 2014). The largest lakes, like the ocean, have geostrophic currents from the effects of the Earth's rotation (Gorham and Boyce 1989, Wüest and Lorke 2003). The basin size at which rotational effects become important varies with latitude and stratification (parameterized internal Rossby radius). For mid-latitude lakes, rotational effects apply to basins greater than 3 km long (Gorham and Boyce 1989). Below this threshold, physical mechanisms continue to scale with lake size and other lake characteristics. For example, the transfer of momentum from wind forcing is dependent upon the height of the surface waves which, in turn, scales with lake fetch (Gorham and Boyce 1989; Wuest and Lorke 2003). The most efficient transfer of wind energy to the water column occurs when the surface waves reach steady state. At wind speeds of 2 and 6 m s^{-1}, it takes lateral distances of 10 km and 100 km, respectively, for surface waves to reach steady state (Wuest and Lorke 2003). Thus, in lakes with shorter basins, a greater proportion of the wind shear is lost maintaining unsteady surface waves and, therefore, is not available to mix deeper than the wave-affected layer (Wüest and Lorke 2003). Of the eight classes of internal motion that occur in stratified lakes, six occur only in lakes with length scales greater than 10 km (Boyce 1974). The limited mixing in small lakes results in greater diurnal temperature variation and dominance by convective over wind-driven mixing in lakes less than 1 km^2 (Read et al. 2012, Woolway et al. 2016).

various photobiological responses including photoinhibition can be found in reviews by Gargett and Marra (2002) and Neale et al. (2003).

Here we focus on average mixed layer irradiance as the best measure of the planktonic light environment. This measure assumes that dissolved constituents and suspended particles spend equal time at all depths in the mixed layer. For comparison among different mixing conditions, the effect of differences in mixed layer depth is highlighted by considering the average irradiance in proportion to surface irradiance, i.e., the average relative irradiance over the MLD, $Avg(E^*) = \int_{z=0}^{MLD} E(z)dz/(E_0 \times MLD)$, where $E(z)$ is irradiance at depth, z, and E_0 is irradiance at the surface. The asterisk (*) in this expression denotes that this is a non-dimensional metric (fraction of surface irradiance). $E(z)$ is a function of water transparency, commonly characterized by the diffuse attenuation coefficient defined as $K_d(z) = -\frac{1}{E_d(z)}\frac{dE_d(z)}{dz}$ so that $E_d(z) = E_0 exp(-\int_0^z K_d(z)dz)$. The inclusion of subscript '*d*' in these expressions refers to 'diffuse' irradiance which is typically measured using an upward looking sensor that captures downwelling irradiance coming from all angles but it is also known to be a good estimate of the water column transparency to 'scalar' irradiance [$E(z)$]—the sum of both downwelling and (the small amount of) upwelling irradiance. Given this assumption, subsequent expressions for irradiance omit the subscript '*d*'. The attenuation coefficient is a function of the absorbance and scattering properties of water as well as dissolved and particulate constituents. The latter should be relatively homogenous over the mixed layer. Thus, K_d(z) can usually be approximated by an average K_d over the mixed layer, so that the relative irradiance profile reduces to $exp(-K_d z)$. The average relative irradiance in a layer of depth MLD is thus the integral of this expression divided by MLD, i.e.,

$$Avg(E^*) = \frac{(1 - \exp(-K_d \times MLD))}{K_d \times MLD} \tag{1}$$

However, K_d does vary considerably with wavelength which is related to the apparent color of a water body. The color indicates the wavelength of greatest transparency (least absorbed and thus most backscattered)—blue in clear open ocean waters going to green and brown in the presence of high phytoplankton and chromophoric dissolved organic matter (CDOM) absorbance, respectively (Kirk 1994). The most important light dependent process in the mixed layer is phytoplankton photosynthesis which derives energy primarily from irradiance with wavelengths from 400 to 700 nm, known as photosynthetically available radiation (PAR). Since individual wavelengths composing PAR irradiance are differentially attenuated through the water column, the attenuation coefficient for PAR, K_{dPAR}, is, by definition, not constant with depth—even over a mixed layer with homogeneous optical properties. This depth dependence is most evident in oceanic waters and very oligotrophic lakes where strongly wavelength-dependent optical properties of water itself is the main determinant of transparency. The depth dependence of K_{dPAR} is moderated in most lakes where particles and CDOM dominate the optics. Thus, in lakes and many coastal areas, PAR transparency is typically approximated by a single attenuation coefficient (K_{dPAR}).

Climate Change and the Ocean Surface Mixed Layer

Concerns about the effects of climate change on the physics and biology of the ocean were expressed soon after the first IPCC consensus scenarios were established in the early to mid-1990s (e.g., IPCC 1995). These scenarios initially assumed that climate change had no effect on the oceanic physics or biology. This motivated the development of fully Coupled Ocean Atmosphere Simulations (COAS) which allow the general circulation of atmosphere and the ocean to interact. The initial results from such simulations showed that IPCC scenarios with CO_2 increasing above 500 ppm by 2050 (the IS92a scenario) would result in substantial surface warming and an increase in the vertical density gradient (Sarmiento et al. 1998). The stronger gradient was related to greater warming at the surface than at depth except at high latitudes where surface freshening due to ice melt was the main contributor to the density difference. These results were followed up by the development of a range of COAS, also termed Atmosphere Ocean Global Circulation Models. Although some of the details differed, all of these models predicted that global climate change would increase the strength of stratification and expand the shallowly-stratified oligotrophic regions in subtropical gyres, a biome that covers almost 50% of the ocean surface (Sarmiento et al. 2004). The winter time mixed layer depth was predicted to shoal, particularly poleward of 40°N. Later coupled model experiments also predicted an increased density difference between 0 and 200 m for much of the ocean (Capotondi et al. 2012, Schmidtko et al. 2017).

Although Sarmiento and co-workers emphasized the considerable uncertainty in these predictions related to an oversimplified and/or incomplete understanding of ocean/atmosphere physics, the predictions of COAS models have been widely used by biological oceanographers to establish conditions to test for the possible effects of changes in the surface layer stratification on ocean biology (Boyd and Doney 2002). A trend of decreasing estimates of global phytoplankton productivity was inversely proportional to the trend in stratification index based on the density difference between 0 and 200 m (Behrenfeld et al. 2006, Martinez et al. 2009). In predicting how different open-ocean phytoplankton groups may respond to climate change, Boyd and Doney (2002) assumed, based on COAS results, that "climate change will increase stratification, shoal the surface mixed layer, and warm the upper ocean." Moreover, several authors have suggested that enhanced stratification should lead to shoaling of the surface mixed layer and greater exposure of surface layer plankton to PAR and UV radiation (Beardall et al. 2009, Gao et al. 2012, Häder et al. 2015). The most recent set of model predictions comes from the fifth Coupled Modeled Intercomparison Project (CMIP5). Several of these models predict that increasing stratification will result in an overall decrease in phytoplankton productivity in the future to at least 2100, though much of that effect is due to a reduction in nutrient transport into the surface layer (Bopp et al. 2013, Cabré et al. 2015, Fu et al. 2016). After 2100, models predict widespread changes in ocean biology due to sequestration or 'trapping' of nutrients in the polar and deep ocean after the disappearance of sea-ice (Moore et al. 2018).

The actual response of pelagic productivity to climate change is likely more nuanced and variable than the conclusions drawn from the current generation of COAS models. In all these models, stratification is measured on a very coarse scale, i.e., the density difference between surface and 200 or 500 m. This is a reasonable measure of what is happening with the deep, permanent thermocline, but not the more dynamic seasonal thermocline that determines the light climate for photosynthesis (Box 6.1). Indeed, 'mixed layer depth' when defined by COAS models is limited to the maximum, winter time mixed layer. It is important to account for deep, winter mixing to determine nutrient transport to the surface layer but it is only one of several possible influences of MLD on productivity. Even for this maximum mixed layer depth, the CMIP5 model predictions are biased shallow compared to observations in the Southern Ocean (Sallée et al. 2015) where productivity is predicted to increase with less light limitation. Other deficiencies have been identified in the predicted thermohaline circulation by CMIP5 models which will be used to refine model experiments conducted for CMIP6. Moreover, recent analyses using the latest observations from ships and Argo floats found no overall correlation between interannual variations in primary productivity and the density difference between 0 and 200 m (Lozier et al. 2011, Dave and Lozier 2013, Dave 2014). Instead, these studies show a mosaic of trend correlations throughout the ocean, some positive and others negative (Dave and Lozier 2013, Dave 2014), further suggesting a complex biophysical response to warming.

Somavilla et al. (2017) contrasted the COAS model predictions of the surface mixed layer with long-term hydrographic records. They found that while sea surface temperature (SST) is increasing in mid-latitudes and sub-tropical gyres worldwide, stratification is not uniformly increasing nor is the mixed layer uniformly shoaling. In many areas there has been enhanced deepening of the winter MLD at rates > 10 m decade^{-1}, a trend that is contrary to model predictions. This deepening is attributed to changes in the extent of winter time cooling and changes in wind patterns, mechanisms that are not adequately represented in COAS models. Similar trends were estimated for broad ocean regions based on data from Argo floats and long-term oceanographic time-series acquired at ocean observatories. The long-term changes in average relative mixed layer irradiance (Avg (E*)) in the subtropical gyres have been examined by combining the Somavilla et al. (2017) estimates of MLD with measurements of spectral transparency at the ocean observatories of the Hawaii Ocean Time Series (HOTS) in the North Pacific (sg-NPac, 22.8° N, 158° W) and the Bermuda Atlantic Time-series Study (BATS) in the North Atlantic (sg-Natl, 32.2° N, 64.5° W). These sites are of particular interest because decreases in productivity and increased deep water anoxia in subtropical gyres have been attributed to increased stratification and shoaling mixed layers (Behrenfeld et al. 2006, Polovina et al. 2008, Boyce et al. 2010, Talley et al. 2016).

To examine surface layer transparency and irradiance trends, we applied procedures similar to those used by Somavilla et al. (2017). This analysis used the MLD estimates calculated by Somavilla et al. which were graciously provided by the authors. They calculated MLD using three different methods: (i) depth at which density gradient exceeds 0.03 kg m^{-3} ('Δsigma'), (ii) based on the fit of the density profile to an 'ideal' functional form as described by Gonzalez-Pola et al. (2007) ('GP'), and (iii) a hybrid algorithm which uses various metrics to select the optimum MLD estimate for a given profile by selecting between threshold, gradient and curve-fitting approaches (Holte and Talley 2009) ('HT'). Box 6.1 discusses how these methods compare. These MLD estimates were matched with contemporaneous spectral irradiance profiles. At

BATS, irradiance was profiled at 12 wavelengths from 410 to 683 nm from 1992 to 1999 and at eight wavelengths from 380 nm to 683 nm from 1999 to 2012 (Allen et al. 2017). At HOTS, irradiance profiles were measured at six wavelengths from 412 to 665 nm from 1998 to 2009, after which profiles were made with a hyperspectral profile covering 350 to 800 nm (10 nm bandwidth) (Letelier et al. 2017).

In these oligotrophic open ocean, 'blue', surface mixed layers there are sharp differences in attenuation between wavelengths. As mentioned previously, this is related to the strong wavelength dependence of attenuation by water molecules. Thus, the effective attenuation coefficient for broadband PAR varies by depth, with stronger attenuation near the surface as the red, green and violet components of sunlight are absorbed, leading to weaker attenuation of the remaining blue irradiance at depth. Moreover, the response of phytoplankton to irradiance is actually dependent on the variable spectral absorption of light by phytoplankton pigments (Neori et al. 1984). Thus, the efficiency of phytoplankton light utilization in subtropical gyres also varies with depth (Letelier et al. 2017). These depth dependencies make it difficult to characterize changes in average mixed layer irradiance due to variation in mixed layer depth. Here, we use changes in average irradiance at 490 nm as an index of PAR availability. Attenuation is usually at the minimum at this wavelength which is widely-used as a reference for ocean optics and remote sensing and is available over the full time-period in both the HOTS and BATS data. Trends in other wavelengths measured were similar but are not presented here for brevity. In the BATS data, we also considered the time series of attenuation at 380 nm as an indicator of UV transparency. No trend analysis of UV transparency was conducted for HOTS since data were only available from 2012.

Attenuation coefficients (K_d) as function of depth (5 m resolution) for BATS profiles were calculated by Allen et al. (2017) and were kindly provided by the authors. Since these profiles were acquired at different times of the day over the period of the series, K_d was normalized to remove the influence of the consequent differences in the angular distribution of irradiance. For HOTS, irradiance profiles were downloaded from public data archives at Oregon state (Ocean Microbial Ecology Laboratory, 2018). Attenuation coefficient profiles ($K_d(z)$) were estimated from irradiance ($E_d(z)$) using a similar procedure as followed by Allen et al. (2017). $K_d(z)$ at 1 m intervals was taken as the slope of a robust fit of the logarithm $E_d(z)$ normalized to surface irradiance in a 3-m depth interval above and below the depth of interest. The resulting profiles were then smoothed using a 15 m moving median filter. Since HOTS profiles were all measured at mid-day, no normalization was performed.

Average relative irradiance in the surface layer was estimated using all three of the Somavilla et al. (2017) MLDs (GP, HT and Δsigma). Each attenuation profile was paired with the MLD from profiles conducted on the same day or on the nearest date available. We averaged MLDs when multiple profiles were conducted on the same day. While the average K_d should be a good approximation of K_d at any depth in the mixed layer, we retained the full depth resolution in the database to keep as close as possible to the original data. The deepest K_d was extrapolated if needed to get K_d at MLD and the K_d at 3 m was extrapolated to the surface. K_d estimates in the range of the

MLD were interpolated as necessary to 1 m resolution. The integrated attenuation for relative irradiance (i.e., $exp(-\int_0^z K_d(z)\,dz)$) was approximated as the summed continued product of relative irradiance over successive one meter increments. Thus average irradiance relative to surface irradiance is obtained as:

$$Avg(E^*) = \sum_{z=1}^{MLD} \prod_{x=1}^{z} \exp(-K_d(x))/MLD$$

Average K_d in the mixed layer was also calculated.

Trends in average MLD, mixed layer K_d and irradiances were calculated using the same procedure used by Somavilla et al. (2017). A third-order polynomial was fitted to the full data set as a function of day of year (Matlab *polyfit*). An anomaly series was calculated by subtracting the seasonal averages and temporal trend determined by linear regression (Matlab *regress*). The results are shown ± an estimated 95% confidence interval for the estimated slope (Tables 6.2 and 6.3).

At both the Pacific and Atlantic subtropical gyre sites the annual mean SST has increased by about 0.1°C per decade, nevertheless the trend in annual average MLD has been a strong deepening in the Pacific and moderate deepening in the Atlantic (Somavilla et al. 2017). The trends were of similar magnitude and sign no matter which approach was used to estimate MLD. Using their own independent calculations, Allen et al. (2017) reported similar trends for full year average temperature and MLD at BATS. At both sites, the trend in annual average MLD is strongly influenced by a significant deepening of the winter-time mixed layer (Somavilla et al. 2017). The deepening is particularly strong at BATS, where the frequency and strength of winter storms has been increasing (Lomas et al. 2010). The maximum depth of winter-time overturn is important because it determines the nutrient availability in the surface layer for the following growing season (Somavilla et al. 2017). One consequence of increased winter storm activity (itself a possible result of climate change) is an enhancement of winter-spring diatom blooms at BATS (Lomas et al. 2013).

Table 6.2. Long-term trends in mixed layer depth (MLD, m decade^{-1}), attenuation coefficient for the PAR wavelength of 490 (Kd, m^{-1} decade^{-1}) and average mixed layer irradiance as a percent of surface irradiance Avg(E*), % decade^{-1}) for the ocean observatory sites HOTS in the N. Pacific subtropical gyre (sg-Npac). MLD was estimated by Somavilla et al. (2017) who used three different techniques (details in text). Trends were estimated as the linear regression slope on anomalies from seasonal values as described in the text and are given ± the 95% confidence interval. Seasonally adjusted trends are shown for May through November (n = 203), i.e., excluding winter time mixed layers, and June through August (n = 85), i.e., summer.

Period	MLD type	sg-Npac	PAR	Avg(E*) (% decade^{-1})
		MLD (m decade^{-1})	Mean K_{d490nm} (m^{-1} decade^{-1})	
May–Nov	GP	–0.32 ± 3.15	0.0002 ± 0.0027	1.0 ± 4.7
	HT	–1.84 ± 3.27	0.0001 ± 0.0032	2.2 ± 5.7
	Δsigma	–1.02 ± 3.67	0.0001 ± 0.0032	1.2 ± 5.7
Jun–Aug	GP	0.88 ± 4.29	–0.0025 ± 0.0056	1.6 ± 7.6
	HT	–1.29 ± 4.3	–0.0028 ± 0.0070	2.8 ± 8.1
	Δsigma	–1.44 ± 4.67	–0.0025 ± 0.0088	1.7 ± 9.0

Table 6.3. Long-term trends in mixed layer depth (MLD, m decade^{-1}), attenuation coefficient for the UV wavelength of 380 nm and PAR wavelength of 490 nm (Kd, m^{-1} decade^{-1}) and average mixed layer irradiance as a percent of surface irradiance (Avg(E*), % decade^{-1}) for the ocean observatory site BATS in the N. Atlantic subtropical gyre (sg-Natl). MLD was estimated by Somavilla et al. (2017) who used three different techniques (details in text). Seasonally adjusted trends are shown for the PAR wavelength of 490 nm and for the UV wavelength of 380 nm at sg-Natl for May through Nov. (n = 330), i.e., excluding winter time mixed layers, and June through August (n = 176), i.e., summer.

Period	MLD type	sg-Natl	UV	Avg(E*) (% decade^{-1})	PAR	Avg(E*) (% decade^{-1})
		MLD (m decade^{-1})	Mean K_{d380nm} (m^{-1} decade^{-1})		Mean K_{d490nm} (m^{-1} decade^{-1})	
May–Nov	GP	−0.30 ± 1.38	**0.0040 ± 0.0032**	0.7 ± 3.8	**0.0017 ± 0.0012**	−0.9 ± 1.7
	HT	0.05 ± 1.63	0.0031 ± 0.0032	4.0 ± 4.3	**0.0014 ± 0.0012**	1.0 ± 1.9
	Δsigma	−0.09 ± 1.84	**0.0047 ± 0.0036**	0.2 ± 4.6	**0.0023 ± 0.0013**	−0.8 ± 2.1
Jun–Aug	GP	−0.42 ± 1.44	**0.0068 ± 0.0042**	−0.9 ± 4.2	0.0015 ± 0.0018	0.3 ± 1.8
	HT	−0.80 ± 1.26	**0.0052 ± 0.0042**	1.5 ± 4.5	0.0010 ± 0.0017	1.0 ± 1.9
	Δsigma	−0.30 ± 1.23	**0.0084 ± 0.0044**	**−4.6 ± 4.1**	**0.0023 ± 0.002**	−1.6 ± 1.8

The analysis presented here, however, excludes the winter period with its highly-variable MLDs. Somavilla et al. (2017) addressed the variability of the MLD in the winter by only using the maximum depth in their analysis. This is not informative as far as estimating trends in mixed layer irradiance. Instead, we conducted our analysis on the periods when MLD is more stable. The first period considered was the overall ‘growing season’ of May through November. This period brackets the summertime peak in productivity at HOTS and most of the productive period at BATS, but excludes the spring bloom at BATS which generally peaks in March. Different from the full-year average, there was no significant trend in MLD at either site for the May to November period (Tables 6.2 and 6.3), though most of the estimated slopes were negative (shoaling) or very slight deepening (MLD-HT at BATS). Trends in MLD were also not significant for the summer time (June through August) mixed layer, again with negative slopes (shoaling) except for the GP estimate of MLD at HOTS. Summer is important as this is when irradiance and growth are maximal. Our estimates of summer time trends are statistically identical to those of Somavilla et al. (2017) however, the actual values are slightly different because our analysis uses the MLDs that match up with K_d estimates.

In the Atlantic, average UV and PAR attenuation both increased with a significant trend for the May to November period (Table 6.3). This result is consistent with the findings of Allen et al. (2017) who concluded that the increased attenuation was due to increased phytoplankton pigment absorbance, indicating more biomass. This observation contrasts with earlier suggestions that phytoplankton abundance was decreasing globally (Boyce et al. 2010). The combined effect of the increased attenuation with a (slightly) shoaling mixed layer was a non-significant trend in mixed layer irradiance, for UV and PAR. For HT estimates, neither MLD nor attenuation had a significant trend. At the Pacific (HOTS) site there has been no increase in attenuation coefficient so the trend in shoaling over the May to November period does lead to an increase in average relative irradiance, though, like the MLD trend, the trend in Avg(E*) was not significant.

The strongest trend found in the analysis was UV attenuation for the summer time period at BATS (Table 6.3). This contrasts with a small or non-significant trend of increased PAR attenuation. The preferential increase of attenuation in the UV suggests an increase in CDOM. Allen et al. (2017) found that there was no significant long-term increase in CDOM at BATS, but only considered full-year period. Interestingly, the summer time period at BATS also has the most negative trends in MLD (Table 6.3). However, the shoaling was not enough to overcome the effect of the increased attenuation, such that Avg(E*) in the UV either had no significant trend or significantly decreased for the summer time. In contrast at HOTS, there has been a slight decrease in attenuation in the summer (not significant). The result was a small, not significant, increase in Avg(E*) for PAR (490 nm).

The foregoing analysis focused on just two wavelengths but trends at other wavelengths were also examined with similar results. In particular, significant trends of increased attenuation were found for most wavelengths measured at BATS. We also analyzed independent estimates of MLD at HOTS (downloaded from http://hahana.soest.hawaii.edu) and BATS (Allen et al. 2017) and the estimates of long-term trends were within the confidence intervals for the estimates in Tables 6.2 and 6.3. This, together with the generally consistent estimates of trends using the three different approaches implemented by Somavilla et al. (2017), suggests that our estimates are methodologically robust. The overall conclusion is that ocean warming, to the extent that it has already occurred, has been insufficient to significantly shoal the mixed layer and increase average mixed layer irradiance, either in the UV or PAR. Consistent with this conclusion is that neither site has observed a long-term change in productivity for the periods examined (Church et al. 2013). Annual productivity has increased at BATS, but this is due to higher productivity in winter-spring blooms associated with more frequent mixing episodes (Lomas et al. 2010, Lomas et al. 2013). Productivity at HOTS increased transiently from 1999 to 2002 then decreased to previous levels (Luo et al. 2012). Sensitivity analyses of the effect of different factors on modeled productivity showed that the primary cause for this transient elevated production was most probably an episode of increased vertical transport of nitrate and not a change in mixed layer depth (Luo et al. 2012).

There are no clear results on the effect of climate change on mixed layer irradiance to date for marine mixed layers. This still leaves open the question of what might be expected as climate change continues in the future. Some clues to how marine mixed layers may respond in the future might be inferred from analogous observations of MLD and attenuation in lakes. Many lakes are warming at rates far exceeding those observed for the upper ocean (O'Reilly et al. 2015). While each lake has features (size, morphology, residence time) that influence how mixing processes respond to climate change, general relationships can be developed using observations in large data sets. Such relationships are considered in the next section.

Climate Change and the Irradiance Environment in Lakes

As in the ocean, the irradiance environment in lakes is controlled by water transparency and mixing depth. Transparency of surface waters to UV (290–400 nm) is primarily controlled by the amount of chromophoric dissolved organic matter (CDOM) which, together with algal biomass and other constituents, regulates the availability of PAR. In small lakes, there can be a strong feedback between transparency and mixing depth which is discussed first in this section. Subsequently we consider larger lakes in which CDOM is a secondary determinant of mixed layer depths. Instead, like the ocean, MLD are more related to wind forcing, currents, and in cold regions, presence or absence of ice.

The Irradiance Environment and Lake 'Browning'

Over the last couple of decades, DOC has been increasing in many temperate and boreal waters of the northern hemisphere and associated with this is an increase in the colored component of DOC, CDOM. This long-term trend, termed lake 'browning', is attributed to several global change trends (Monteith et al. 2007). The first is the imposition of pollution controls on coal burning power plants, particularly the reduction in SO_2 emission into the atmosphere and the consequent acid deposition into watersheds ('acid rain'). As the pH in soils increases, this increases the solubility of DOC and its mobilization into ground water and subsequent runoff into lakes. This mobilization is further augmented in regions where climate change has increased precipitation and associated runoff from watersheds into lakes (Williamson et al. 2014).

The physical effect of increased DOC, and thus CDOM, in the surface layer of lakes is to absorb more quanta of downwelling irradiance closer to the surface. Most of the energy in these quanta is transformed into heat raising the water temperature which decreases the density. The greater warming near the surface steepens the density gradient and increases the mechanical work needed to form a mixed layer. Thus, given constant surface forcing, browning should (and does in some cases) lead to shallower surface mixed layers (Fig. 6.1).

While the shoaling by itself increases average irradiance and UV exposure in the mixed layer (Eqn. 1), the greater light attenuation from CDOM decreases it. The following analysis considers what is the net result of these opposing effects on mean mixed layer irradiance.

Although the overall principle of CDOM enhanced shoaling of the mixed layer is straightforward (Fig. 6.1), the many factors influencing mixing make it a challenge to model in actual lakes. Heiskanen et al. (2015) required a computationally expensive 20-layer model and detailed surface observations to accurately predict the time course of stratification in a small mesotrophic lake in Finland. After tuning the model to fit lake conditions, they simulated the effect of increased attenuation coefficient for visible radiation (K_d), as would occur with higher amounts of CDOM. The results clearly showed that higher K_d leads to consistently shallower mixed layers over the course of the open water period (Fig. 6.2).

Overall, the simulations of Heiskanen et al. (2015) show that darker waters lead to earlier and stronger stratification, and shallower mixed layers. Overall heat storage in the lake declines with increasing K_d leading to earlier autumn overturns. Importantly, the increase in attenuation had a much greater effect on the depth of penetration of irradiance (horizontal lines in Fig. 6.2) than on the mixed layer (approximated by the 16°C isotherm during the summer stratified period). For example, in mid-July

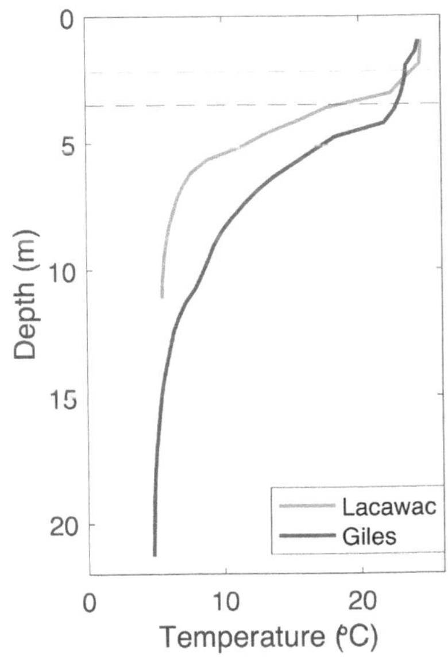

Fig. 6.1. The effect of transparency on thermal structure in two small lakes shown with temperature profiles from low DOC Giles (black) vs high DOC Lacawac (gray) lakes located in NE Pennsylvania taken on consecutive afternoons in June 2015. In Lacawac, light is attenuated closer to the surface leading to warmer temperatures near the surface, colder temperature at depth and a more shallow MLD. Broken horizontal lines indicate the MLD.

increasing K_d from 0.6 to 1.9 m^{-1} (bottom left panels in Fig. 6.2) decreased thermocline depth from 5 to 3 m, a factor of 0.6, while optical depth decreased twice as much, a factor 0.6/1.9 = 0.32. Therefore, despite the mixed layer being shallower, the average irradiance in the mixed layer is lower (cf. Eq. 1).

Based on this, the average irradiance experienced by constituents or organisms in the surface layer associated with the previous darkening example (K_d increasing from 0.6 to 1.9 m^{-1}) decreases from 32% of incident in the 5 m mixed layer to 17% of incident in the 3 m mixed layer. Similar results were obtained in another lake by specific simulation conducted for Trout Bog, a small lake in Wisconsin although the question was framed in terms of a DOC decrease (Read and Rose 2013). Using a model calibrated to the lake, they considered the depth of the mixed layer at ambient and 50% decreased DOC. The decrease in DOC was projected to deepen the mixed layer from 0.85 to 1.4 m. Average PAR was lower in the shallow, high DOC, layer compared to the deep, low DOC, layer, i.e. 28% vs 33%, respectively (using $K_{dPAR} = 0.22 \times$ DOC, [Morris et al. 1995]).

These model results suggest that increases in DOC ('browning') should decrease average irradiance in the surface layer in these specific lakes. However, the lake modeled by Heiskanen et al. (2015) had a relatively long fetch (> 2 km) so that mixing was particularly sensitive to wind stress despite the small size (0.62 km^2). Otherwise lakes of < 1 km^2 tend to be sheltered from wind effects (Box 6.2) making it easier to model the annual evolution of the thermal profile and its response to changing attenuation (Read and Rose 2013). Nevertheless, models that can be easily

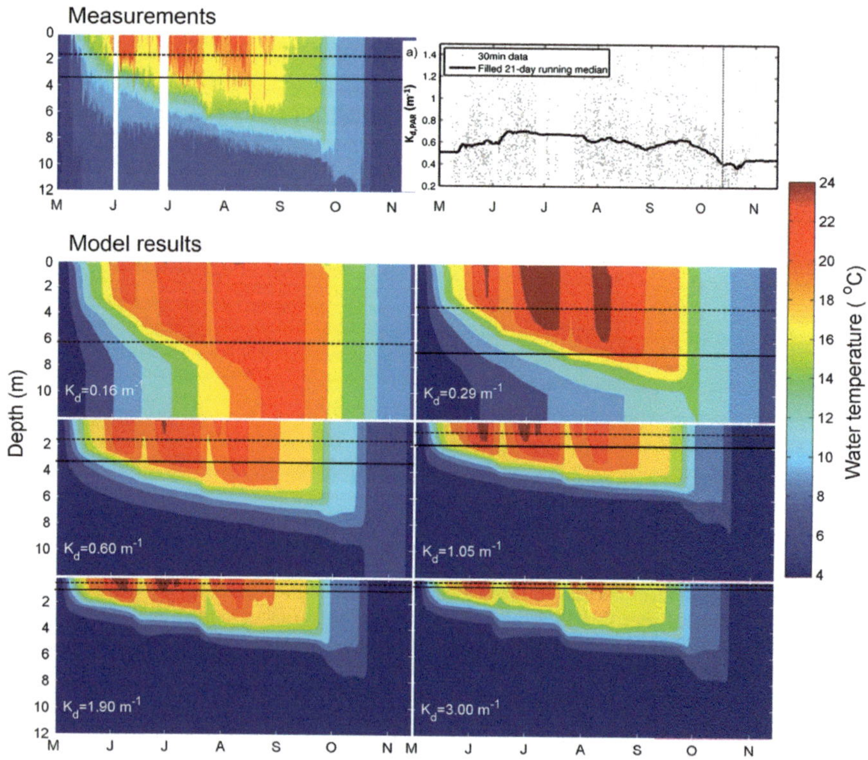

Fig. 6.2. Isotherms of Lake Kuivajärvi as observed in 2013 or as predicted by the Lake model using K_d that are either lower or higher than the observed K_d (panel in upper right). Horizontal lines indicate optical depths of $1/K_d$ and $2/K_d$. Adapted from Heiskanen et al. (2015).

implemented for a broader range of lake conditions, like the General Lake Model (GLM), are not good predictors of even thermocline depth—not to mention mixed layer depth (Bruce et al. 2018).

Another approach to define the dependence of the MLD and irradiance environment on DOC and transparency is to consider the relationship between parameters within long-term data sets which have been compiled for many well-studied lakes, for example the lakes of the Canadian Experimental Lakes Area (ELA). Fee et al. (1996) defined relationships for ELA lakes which estimate summer MLD as a function of lake size and % transparency of PAR ($T = 100*\exp(-K_d)$) based on 186 observations in lakes less than 500 ha (= 5 km^2) in size. The relationship is illustrated in Fig. 6.3 that shows MLDs are deeper than the average for a given size when transparency is high and more shallow when it is low.

Fee et al. (1996) fitted these observations to a series of equations, here combined for conciseness to predict MLD as:

$$MLD = -0.08 + 0.0607\, A^{0.25} + 0.8\, e^{-0.022\, T} \quad (2)$$

To test the effect of changing transparency and lake size, we used MLD so defined to estimate average relative mixed layer irradiance (Eq. 1) for the ELA lakes [taking

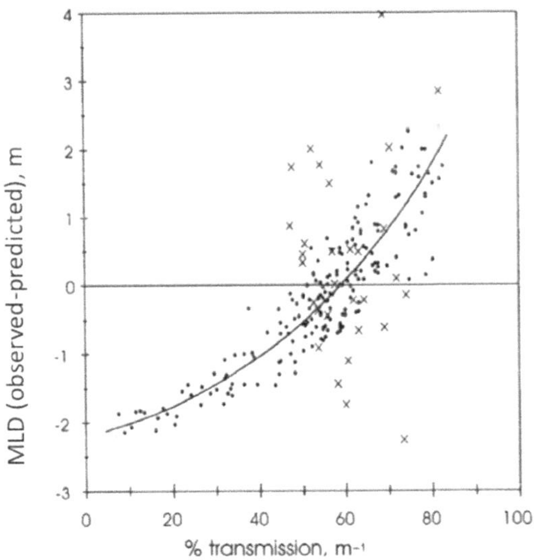

Fig. 6.3. Variation in the depth of the Mixed Layer (MLD) in Lakes of the Canadian Experimental Lakes Area as a function of the % Transmission of PAR at 1 m depth. Y axis is the deviation in mixed layer depth relative to that expected for lakes with the same area and 60% PAR transmission (=0 deviation). Dots are lakes with area < 500 ha, x-lakes with area > 500 ha which do not follow the relationship. From Fee et al. (1996—their Fig. 5).

K_{dPAR} as -ln(T/100)]. The basic relationship is shown by the contour-surface plot in Fig. 6.4.

For small lakes following the relationship shown in Fig. 6.4, as % transparency decreases (K_{dPAR} increases) average PAR irradiance in the surface layer (vertical z-axis) decreases. With respect to area, larger lakes have deeper mixed layers and lower average irradiance, but decreasing transparency is related to decreased Avg(E*) for all size lakes considered in their analysis (lake area: 4–500 ha).

The implication of the Fee et al. analysis is that increased DOC reduces the light available in the surface layer through its synergistic control on transparency and mixed layer depth. However, Fee et al. did not include DOC directly in their analysis. The effect of DOC on the surface irradiance environment of lakes can be tested directly using the long-term data set of lake characteristics compiled by Craig Williamson and co-workers (cf. Williamson et al. 2015). The data set mainly includes lakes in the N.E. Pennsylvania Poconos region and is distinctive in including not only K_{dPAR} and mixed layer depth, but also K_d for UV irradiance (320 nm) and measurements of surface DOC. Here, we used this database to visualize directly the relationship between DOC and mixed layer depth (Fig. 6.5).

This data set includes all ice-free months of the year (generally April to November), so there are some high outliers associated with deep mixing during overturn in the late fall and early spring when surface energy inputs for mixing (wind and heat loss) exceed the stabilizing effects of solar radiation irrespective of DOC and transparency. Otherwise, DOC accounts for much of the variation in mixed layer depth ($R^2 = 0.55$ for fitted line, n = 148). As DOC increases, transparency decreases and heating is

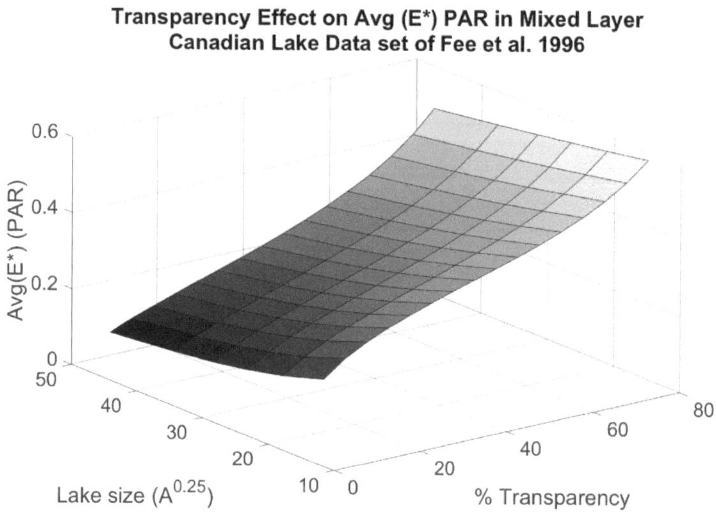

Fig. 6.4. Average PAR irradiance in the surface layer as a function of lake size and % transparency using the surface depth relationships for Canadian Lakes reported by Fee et al. 1996.

confined to a shallower depth range, confirming the model results shown in Fig. 6.2. This is also consistent with the relationship shown in Fig. 6.3, keeping in mind that as DOC increases, PAR transparency decreases so the trend of the curve is the reverse of the trend shown in Fig. 6.3.

Using observed mixed layer depth and attenuation coefficients for both PAR (K_{dPAR}) and a representative UV wavelength of 320 nm (K_{d320}), average mixed layer irradiance was estimated using Eq. 1. The relationship of average PAR and UV in the surface layer to DOC is shown in Fig. 6.6.

CDOM absorbs UV more strongly than PAR, so average E^* in the UV is less than half of E^* for PAR. As for the Canadian lakes, as DOC increases, Avg(E^*) both PAR and UV decreases. Average irradiance in the UV has a stronger relationship with DOC since CDOM absorbs more in the UV than in the visible range which is most important for PAR. The decrease in average irradiance with increasing DOC is most pronounced for DOC < 4 mg L^{-1}. At higher DOC concentrations, average irradiance is already so low that incremental increases have little additional effect. Consistent with this decrease in overall irradiance, Seekell et al. (2015) showed that, in general, increased DOC is associated with lower lake primary productivity.

Overall, both modeling results and observations suggest that increased DOC in small lakes will lead to overall lower irradiance for PAR and, especially, UV in the surface mixed layer. To evaluate these relationships in the context of climate change effects on MLD, the trends in mixed layer depth, lake transparency and inferred average irradiance in the epilimnia of small lakes in different regions of N. America were examined. Lakes with at least 18 years of temperature profiles and light transmission measuring as either K_d or secchi depth were selected. In the former category are the aforementioned PA lakes and data from the North Temperate Lake Long Term Ecological Research (LTER) site in Wisconsin (Magnuson et al. 2018a, Magnuson et al. 2018b). All these lakes have < 500 ha surface areas. A larger

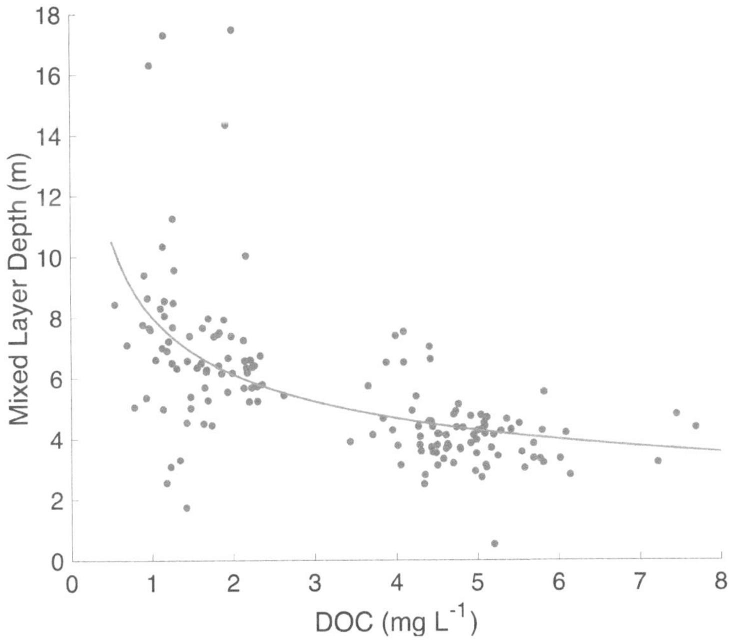

Fig. 6.5. Mixed Layer Depth vs. DOC for the small lakes primarily in the Poconos region of NE Pennsylvania (cf. Williamson et al. 2015). Line is shows fitted power function (MLD = 8.0$DOC^{-0.39}$).

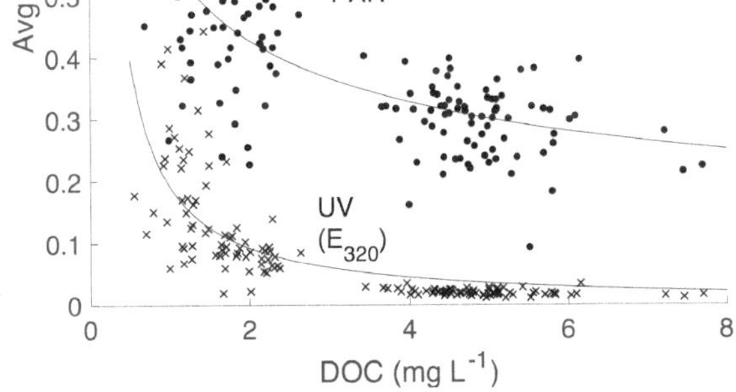

Fig. 6.6. Average mixed layer irradiance vs DOC for the small lakes primarily in the Poconos region of NE Pennsylvania (cf. Williamson et al. 2015). Circles – average for PAR, x-average for UV at 320 nm. Lines fit to a power function.

data set in which transparency was just measured by secchi depth is composed of lakes monitored for the effects of acid deposition in the Adirondack mountains of New York (n = 28, [Leach et al. 2018]) and lakes in Maine monitored as part of the

Maine Volunteer Lake Monitoring Program (n = 140, VLMP – www.lakesofmaine.org). The maximum area in the Adirondack set is 512 ha. Most of the Maine lakes are small (median area 186 ha) but 37 have areas > 500 ha, maximum is 6,600 ha. MLD was estimated using the Lake Analyzer Matlab code for top of metalimnion with the density gradient criterion set to 0.2 kg m^{-4} (Read et al. 2011) (see Box 6.1 for further discussion of MLD determination in lakes). To estimate long-term trends, monthly data from the ice-free period (generally May–November) was seasonally adjusted by subtracting monthly means and a linear rate of change estimated using regression, similar to the approach previously used to analyze trends in the ocean time series.

In general, MLDs in these lakes are shoaling (Table 6.4). Consistent with the relationships previously demonstrated between mixing depth and transparency, the rate of shoaling is also positively correlated with the rate at which K_{dPAR} is increasing or secchi depth is decreasing (Fig. 6.7). Trends in shoaling and loss of transparency are both significant in Lake Giles in PA and Crystal Lake in WI (Table 6.4). These two trends have opposing effects on the average irradiance in the mixed layer and yet only in these two lakes was there a significant decrease in average PAR. In other lakes with significant shoaling but insignificant change in K_d there was an increase in average irradiance in the mixed layer of < 2% per decade (Table 6.4). In Big Muskellunge– the largest lake in the set (396 ha, arrow in Fig. 6.7a), MLD is shoaling without a change transparency (probably because of decreased wind mixing). Lacawac had a significant increase in *K*d but an insignificant shoaling of MLD. Lacawac is a small humic lake with already high DOC and shallow MLD. In agreement with Pilla et al. (2018), these results suggest the thermal structure in Lacawac is less responsive to increasing DOC than nearby Giles which is larger and more transparent. There was, however, a significant decreasing trend in mixed layer UV for both the PA lakes due to increasing attenuation with increasing DOC (see previous section).

In the larger sets of lake data where transparency was measured using a secchi disk, there is more scatter in the relationship between trends in transparency and mixed layer (=epilimnetic) depth (Fig. 6.7b). However, generally lakes with shoaling mixed layers also tended to have trends of decreasing secchi depth. This is especially apparent in the Adirondacks (solid symbols in Fig. 6.7b) where MLD is decreasing by more than a half metre a decade in lakes where secchi depth is significantly decreasing. Another useful data set of long-term observations are those acquired by the Maine Volunteer Lake Monitoring Program, which included 140 lakes that seasonally stratify and had at least 18 years of temperature profiles. While a broad range of trends occur in these lakes, most of the significant trends in epilimnetic depths are negative while secchi depths are moderately increasing, decreasing or not changing (Fig. 6.7b). The overall average trend in MLD (weighted by series length) is significantly negative for both Adirondack and Maine Lakes: –0.54 and –0.22 m $decade^{-1}$, respectively (Table 6.4). The overall weighted average trend in secchi depth for Adirondack lakes is significantly declining at 0.5 m $decade^{-1}$ (less transparency) but for the Maine lakes is slightly positive 0.04 m decade. Overall, the trend analysis shows the important but complex effects of transparency changes on MLDs in small lakes.

To estimate the overall effect of these trends on the mixed layer irradiance environment, K_d was approximated using the relationship K_d=1.7/Z_{sd}, where Z_{sd} is the

Table 6.4. Linear trends of anomalies from monthly means of MLD (m decade^{-1}), attenuation coefficient for PAR (K_{dPAR}, m^{-1} decade^{-1}), attenuation coefficient for UV at 320 nm (K_{dUV}), and mixed layer average PAR (Avg(E*) PAR, % decade^{-1}) and average UV at 320 nm (Avg(E*) UV, % decade^{-1}).

For Lakes with K_d measurements			Trends decade^{-1}(linear slope±95% confidence interval, bold p<0.05)				
						Avg(E*)	
Region	**Lake**	**Years**	**MLD (m)**	K_{dPAR} **(m^{-1})**	K_{dUV} **(m^{-1})**	**PAR (%)**	**UV (%)**
Poconos PA	Giles	1989–2017	**–0.79 ± 0.29**				
		1993–2017	**–0.94 ± 0.39**	**0.09 ± 0.02**	**0.79 ± 0.11**	**–7.8 ± 2.5**	**–8.1 ± 2.3**
	Lacawac	1988–2017	–0.13 ± 0.14				
		1993–2017	–0.16 ± 0.19	**0.07 ± 0.04**	**1.74 ± 0.89**	–1 ± 1.5	**–0.2 ± 0.2**
LTER lakes	Allequash	1981–2016	0.1 ± 0.14	–0.02 ± 0.02		–0.7 ± 1.6	
	Big Muskellunge	"	**–0.35 ± 0.3**	–0.01 ± 0.01		1.9 ± 1.5	
	Crystal	"	**–0.31 ± 0.24**	**0.03 ± 0**		**–1.5 ± 1.3**	
	Sparkling	"	–0.03 ± 0.22	0 ± 0.01		0.3 ± 1.3	
Cascade Project	Paul	1984–2007	**–0.1 ± 0.09**	**0.03 ± 0.02**		1.8 ± 2.3	
	Peter	"	**–0.41 ± 0.1**	0.03 ± 0.07		1.8 ± 2.2	
	Tuesday	1984–2002	–0.03 ± 0.13	0.02 ± 0.07		0.3 ± 6.1	
Lakes with Secchi Disk Measurements			**Average in Set**				
			MLD (m)	**Secchi (m)**		PAR (%)	
Adirondacks+		1994–2012	**–0.54 ± 0.13**	**–0.48 ± 0.12**		1.24 ± 1.34	
Maine*		1974–2016	**–0.22 ± 0.03**	**0.04 ± 0.01**		**1.18 ± 0.19**	

\+ Average trends in 28 lakes monitored by the Adirondack Effects Assessment Program Aquatic Biota Study and the Adirondack Long Term Monitoring Program see Leach et al. (2018).

* Average trends in 140 lakes monitored by the Maine Volunteer Lake Monitoring Program www.lakesofmaine.org.

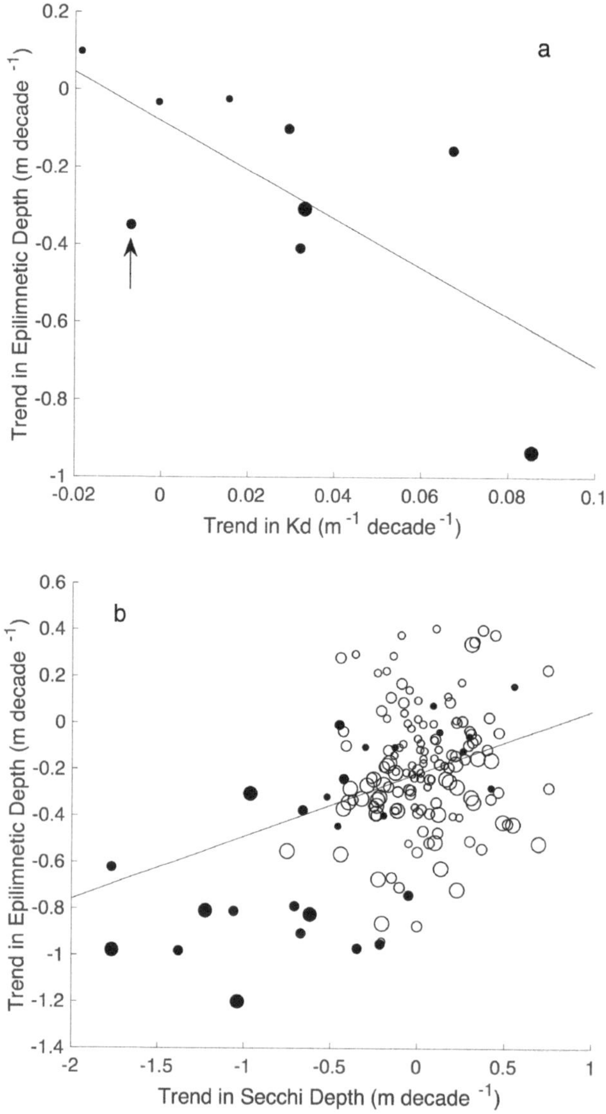

Fig. 6.7. Trends in epilimnetic depth vs. trends in transparency, as either (a) attenuation coefficient (Kd) or (b) secchi depth. In the latter, filled symbols are for a set of 28 lakes in the Adirondacks, open symbols 140 lakes in Maine monitored by the Maine Volunteer Lake Monitoring Program. Symbol size conveys whether the trend determined by linear regression was significantly different from zero ($p < 0.05$). Small circles – neither trend was significant, medium circles – only one of the trends was significant, large circles – both trends were significant. Lines show fitted regression of epilimnetic depth weighted by the inverse standard error of each trend.

secchi depth (Poole and Atkins 1929). Average irradiance was then calculated using Eq. 1, and an anomaly time series produced from monthly time series observations minus long-term monthly means. The weighted average trend in irradiance was slightly positive (~ 1% per decade) for both Adirondack and Maine lakes. This average was not significant for the Adirondack lakes but was significantly positive for the Maine lakes. In the 59 lakes (of 140 total) with both shoaling mixed layers and increasing transparency over the monitored time period, average change in mixed layer irradiance was 2.5% per decade, still small compared to the average transparency of 56% in these lakes. Nevertheless, it would be interesting to better understand what factors are operating to bring about this combination of trends that affect the light environment.

The mixed layer irradiance estimates for lakes in which transparency was monitored using secchi depths are subject to considerable uncertainty because it implicitly assumed that the product of the attenuation coefficient and secchi depth is a constant (1.7). In reality, the constant can vary from less than one to more than three depending on the optical characteristics of the lake with the highest values associated with humic lakes (Koenings and Edmundson 2003). To test the sensitivity of the results to the choice of constant, lake trends were also calculated with a constant of 2.8. This resulted in slightly different estimates of trends, but no changes in the sign or significance. For the case of increasing CDOM, it is probable that the K_d x Z_{sd} product would effectively increase with time, which would add a positive component, i.e., decreasing transparency, to the K_d trends. Thus, trends estimated using a fixed K_d x Z_{sd} product are, if anything, probably underestimating the decrease or overestimating the increase in average irradiance for these lakes. Overall we find, as expected, that there is little or no change in average mixed layer irradiance in lakes where shoaling MLDs are driven by the heat trapping effect of CDOM.

The Irradiance Environment in Lakes not Affected by 'Browning'

Not all lakes are experiencing changes in DOC so we now consider the more general case of lakes in which surface temperature trends are not dominated by the 'browning' phenomenon. This includes small lakes that have not experienced increases in DOC, such as those in alpine environments or drier climates. Also included are larger lakes, i.e., surface area > 500 ha (= 5 km^2), where wind, waves, and currents prevent the formation of near surface density gradients that initiate the heat trapping feedback that is shoaling MLDs where DOC is increasing (Kirillin and Shatwell 2016). As previously mentioned, observations in many of these lakes show long-term increases in surface temperatures and more mixed results for deep-water temperatures. Kraemer et al. (2015) reports that surface temperatures warmed by 0.84°C over the period from 1970 to 2010 across 26 globally distributed lakes, compared to only 0.05°C warming in the bottom temperatures. Their data set refers to mostly large lakes ranging in size up to 68,800 km^2, which together comprise 42% of world's freshwater. Only three lakes in their study are small lakes sensitive to browning, so global increase in air temperature is the most probable cause of the warming in this set of lakes. In another analysis, lakes showed surface warming of summertime temperatures of 0.34°C per decade across 235 globally distributed lakes between 1985 and 2009 (O'Reilly et al. 2015). In a more regional analysis of lakes in northeastern North America which does include lakes affected by browning, surface water temperatures have increased more than bottom temperatures (Richardson et al. 2017).

Putting together the two trends of warming surface waters and less warming (and some cases, cooling) of deep waters, a necessary consequence is stronger stratification between the surface layer (epilimnion) and deep layer (hypolimnion). Indeed, in the previously cited studies various measures of stratification, including top-bottom temperature difference, density difference and maximum buoyancy frequency are increasing in concert with surface temperature. In cases where stronger stratification occurs due to transparency-related heat trapping (previous section), the outcome tends to be a shallower mixed layer. What about the more general case in which warming is externally forced by atmospheric warming or more incoming radiation? Overall, this remains an open question because mixed layer and thermocline depths in large lakes have not been tracked with the same frequency and duration as surface temperatures (i.e., requires more detailed profiling to estimate MLD as described in Box 6.1). However, thermocline depth was tracked in the lake records analyzed by Kraemer et al. (2015) who found no change or some deepening on average in thermocline depth over their set of 26 lakes despite significant long-term warming at the surface. Conspicuous in their absence from the Kraemer et al. analysis are the Laurentian Great Lakes. King et al. (1997) found significantly warming epilimnetic temperatures and shoaling thermoclines in weekly-monthly temperature profiles collected in South Bay of Lake Huron from 1955–92 that correlated with warmer air temperatures and higher radiation. More recent studies show some slow warming of surface temperatures (Austin and Colman 2008, Mason et al. 2016), but little attention has been given to trends in mixed layer depth and transparency in the Great Lakes overall. This certainly should be an objective for future study.

Conclusions

In today's world, climate change is no longer an uncertain possibility for the future, but rather something that is well underway, currently modifying bio-physical relationships in aquatic ecosystems in complex ways. From the outset, it has been obvious that better appreciation of the effects of climate change will derive from both more intensive environmental observations of what is changing as well as more sophisticated models to provide robust predictions of future change. This has been the general trend over the last 20 years in predicting climate change overall and is specifically the case for questions about climate change effects on mixed layer irradiance. Initial modeling of the ocean indicated the possibility of stratification change and shoaling of the mixed layer, but much was uncertain. With the realization that lack of data was a major reason for this uncertainty, there has been a massive increase in observational data from ocean and lakes enabled by advances in instrumentation, particularly that deployed on moorings and autonomous vehicles. The resulting data sets have motivated a new round of analyses on how climate change is affecting stratification and mixing and is enabling the development of modeling approaches that will more accurately predict these changes. These efforts are ongoing, but what is emerging is a mosaic of differing effects on mixed layer depth and irradiance created by spatial variation in the changes in the determinants of these two variables.

Our objective in this chapter has been to provide some insight into how and why different trends are occurring in the ocean vs. lakes, and among different regions of the ocean and lakes of varying size and clarity. It is clear that with climate change, there will be both shallower and deeper mixed layers and that phytoplankton and other constituents of surface layer of aquatic ecosystems will be exposed to both brighter and dimmer average irradiance. As more regional data and model output becomes available, area specific trend determinations and predictions can be made. Such an approach has been recently taken for the Southern Ocean, where higher resolution models of climate change effects now predict that the mixed layer is expected to shoal in some areas and deepen in other areas (see review Deppeler and Davidson 2017). This is due to zonal differences in the predictions of how surface winds and ice cover respond to climate change. Finally, both trend analysis and fine resolution model results in turn inform experimental design as to what parameter range and what types of parameter interactions should be considered in studies of how the chemistry and biology of aquatic systems will respond to ongoing climate change.

Acknowledgements

The authors are grateful to Raquel Somavilla (Spanish Institute of Oceanography, Gijón) and James Allen (University of California, Santa Barbara) for sharing their estimates of mixed layer depths and attenuation coefficients, respectively. Kevin Rose, Jordan Read, and Thomas Shatwell assisted in identifying long-term data archives for lakes. Financial support was provided to Neale by the Smithsonian Institution, the National Aeronautics and Space Administration (NNX14AJ88G) and National Science Foundation (DEB1556556) and to Smyth by the Robert E Moeller Research Fellow Award from the Lacawac Sanctuary Field Station.

References

Allen, J.G., N.B. Nelson and D.A. Siegel. 2017. Seasonal to multi-decadal trends in apparent optical properties in the Sargasso Sea. Deep Sea Research Part I: Oceanographic Research Papers 119: 58–67.

Austin, J. and S. Colman. 2008. A century of temperature variability in Lake Superior. Limnology and Oceanography 53: 2724–2730.

Baum, A., T. Rixen and J. Samiaji. 2007. Relevance of peat draining rivers in central Sumatra for the riverine input of dissolved organic carbon into the ocean. Estuarine, Coastal and Shelf Science 73: 563–570.

Beardall, J., C. Sobrino and S. Stojkovic. 2009. Interactions between the impacts of ultraviolet radiation, elevated CO_2, and nutrient limitation on marine primary producers. Photochemical & Photobiological Sciences 8: 1257–1265.

Behrenfeld, M.J., R.T. O'Malley, D.A. Siegel, C.R. McClain, J.L. Sarmiento, G.C. Feldman, A.J. Milligan, P.G. Falkowski, R.M. Letelier and E.S. Boss. 2006. Climate-driven trends in contemporary ocean productivity. Nature 444: 752–755.

Boehrer, B. and M. Schultze. 2008. Stratification of lakes. Reviews of Geophysics 46.

Bopp, L., L. Resplandy, J.C. Orr, S.C. Doney, J.P. Dunne, M. Gehlen, P. Halloran, C. Heinze, T. Ilyina, R. Séférian, J. Tjiputra and M. Vichi. 2013. Multiple stressors of ocean ecosystems in the 21st century: projections with CMIP5 models. Biogeosciences 10: 6225–6245.

Boyce, D.G., M.R. Lewis and B. Worm. 2010. Global phytoplankton decline over the past century. Nature 466: 591.

Boyce, F. 1974. Some aspects of great lakes physics of importance to biological and chemical processes. Journal of the Fisheries Research Board of Canada 31: 689–730.

Boyd, P.W. and S.C. Doney. 2002. Modelling regional responses by marine pelagic ecosystems to global climate change. Geophysical Research Letters 29: 53–51–53–54.

Boyd, P.W., R. Strzepek, F. Fu and D.A. Hutchins. 2010. Environmental control of open-ocean phytoplankton groups: Now and in the future. Limnology and Oceanography 55: 1353–1376.

Brainerd, K.E. and M.C. Gregg. 1995. Surface mixed and mixing layer depths. Deep Sea Research Part I: Oceanographic Research Papers 42: 1521–1543.

Bruce, L.C., M.A. Frassl, G.B. Arhonditsis, G. Gal, D.P. Hamilton, P.C. Hanson, A.L. Hetherington, J.M. Melack, J.S. Read, K. Rinke, A. Rigosi, D. Trolle, L. Winslow, R. Adrian, A.I. Ayala, S.A. Bocaniov, B. Boehrer, C. Boon, J.D. Brookes, T. Bueche, B.D. Busch, D. Copetti, A. Cortés, E. de Eyto, J.A. Elliott, N. Gallina, Y. Gilboa, N. Guyennon, L. Huang, O. Kerimoglu, J.D. Lenters, S. MacIntyre, V. Makler-Pick, C.G. McBride, S. Moreira, D. Özkundakci, M. Pilotti, F.J. Rueda, J.A. Rusak, N.R. Samal, M. Schmid, T. Shatwell, C. Snorthheim, F. Soulignac, G. Valerio, L. van der Linden, M. Vetter, B. Vinçon-Leite, J. Wang, M. Weber, C. Wickramaratne, R.I. Woolway, H. Yao and M.R. Hipsey. 2018. A multi-lake comparative analysis of the General Lake Model (GLM): Stress-testing across a global observatory network. Environmental Modelling & Software 102: 274–291.

Cabré, A., I. Marinov and S. Leung. 2015. Consistent global responses of marine ecosystems to future climate change across the IPCC AR5 earth system models. Climate Dynamics 45: 1253–1280.

Capotondi, A., A. Alexander Michael, A. Bond Nicholas, N. Curchitser Enrique and D. Scott James. 2012. Enhanced upper ocean stratification with climate change in the CMIP3 models. Journal of Geophysical Research: Oceans 117.

Church, M.J., M.W. Lomas and F. Muller-Karger. 2013. Sea change: Charting the course for biogeochemical ocean time-series research in a new millennium. Deep Sea Research Part II: Topical Studies in Oceanography 93: 2–15.

Cullen, J.J. and M.R. Lewis. 1988. The kinetics of algal photoadaptation in the context of vertical mixing. J. Plankton Res. 10: 1039–1063.

Dave, A.C. 2014. Correction to "Examining the global record of interannual variability in stratification and marine productivity in the low-and mid-latitude ocean". Journal of Geophysical Research: Oceans 119: 2121–2128.

Dave, A.C. and M.S. Lozier. 2013. Examining the global record of interannual variability in stratification and marine productivity in the low-latitude and mid-latitude ocean. Journal of Geophysical Research: Oceans 118: 3114–3127.

de Wit, H.A., S. Valinia, G.A. Weyhenmeyer, M.N. Futter, P. Kortelainen, K. Austnes, D.O. Hessen, A. Räike, H. Laudon and J. Vuorenmaa. 2016. Current browning of surface waters will be further promoted by wetter climate. Environmental Science & Technology Letters 3: 430–435.

Denman, K.L. and A.E. Gargett. 1983. Time and space scales of vertical mixing and advection of phytoplankton in the upper ocean. Limnol. Oceanogr. 28: 801–815.

Deppeler, S.L. and A.T. Davidson. 2017. Southern Ocean phytoplankton in a changing climate. Frontiers in Marine Science 4: 40.

Fee, E.J., R.E. Hecky, S.E.M. Kasian and D.R. Cruikshank. 1996. Effects of lake size, water clarity, and climatic variability on mixing depths in Canadian shield lakes. Limnol. Oceanogr. 41: 912–920.

Fichot, C.G., K. Kaiser, S.B. Hooker, R.M.W. Amon, M. Babin, S. Bélanger, S.A. Walker and R. Benner. 2013. Pan-Arctic distributions of continental runoff in the Arctic Ocean. Scientific Reports 3: 1053.

Fichot, C.G., S.E. Lohrenz and R. Benner. 2014. Pulsed, cross-shelf export of terrigenous dissolved organic carbon to the Gulf of Mexico. Journal of Geophysical Research: Oceans 119: 1176–1194.

Fink, G., M. Schmid, B. Wahl, T. Wolf and A. Wüest. 2014. Heat flux modifications related to climate-induced warming of large European lakes. Water Resources Research 50: 2072–2085.

Fu, W., J.T. Randerson and J.K. Moore. 2016. Climate change impacts on net primary production (NPP) and export production (EP) regulated by increasing stratification and phytoplankton community structure in the CMIP5 models. Biogeosciences 13: 5151–5170.

Gao, K., E.W. Helbling, D.P. Häder and D.A. Hutchins. 2012. Responses of marine primary producers to interactions between ocean acidification, solar radiation, and warming. Marine Ecology Progress Series 470: 167–189.

Gargett, A.E. and J. Marra. 2002. Effects of upper ocean physical processes - turbulence, advection, and air-sea interaction—on oceanic primary production. pp. 19–49 *In*: Robinson, A.R., J.J. McCarthy and B.J. Rothschild (eds.). The Sea: Biological-Physical Interactions in the Ocean. Wiley, New York.

González-Pola, C., J.M. Fernández-Díaz and A. Lavín. 2007. Vertical structure of the upper ocean from profiles fitted to physically consistent functional forms. Deep Sea Research Part I: Oceanographic Research Papers 54: 1985–2004.

Gorham, E. and F.M. Boyce. 1989. Influence of lake surface area and depth upon thermal stratification and the depth of the summer thermocline. Journal of Great Lakes Research 15: 233–245.

Häder, D.-P., C.E. Williamson, S.-A. Wangberg, M. Rautio, K.C. Rose, K. Gao, E.W. Helbling, R.P. Sinha and R. Worrest. 2015. Effects of UV radiation on aquatic ecosystems and interactions with other environmental factors. Photochemical & Photobiological Sciences 14: 108–126.

Heiskanen, J.J., I. Mammarella, A. Ojala, V. Stepanenko, K.-M. Erkkilä, H. Miettinen, H. Sandström, W. Eugster, M. Leppäranta, H. Järvinen, T. Vesala and A. Nordbo. 2015. Effects of water clarity on lake stratification and lake-atmosphere heat exchange. Journal of Geophysical Research: Atmospheres 120: 7412–7428.

Holte, J. and L. Talley. 2009. A new algorithm for finding mixed layer depths with applications to argo data and subantarctic mode water formation. Journal of Atmospheric and Oceanic Technology 26: 1920–1939.

King, J.R., B.J. Shuter and A.P. Zimmerman. 1997. The response of the thermal stratification of South Bay (Lake Huron) to climatic variability. Canadian Journal of Fisheries and Aquatic Sciences 54: 1873–1882.

Kirillin, G. and T. Shatwell. 2016. Generalized scaling of seasonal thermal stratification in lakes. Earth-Science Reviews 161: 179–190.

Kirk, J.T.O. 1994. Light and Photosynthesis in Aquatic Ecosystems. Cambridge University Press. Cambridge.

Koenings, J.P. and J.A. Edmundson. 2003. Secchi disk and photometer estimates of light regimes in Alaskan lakes: Effects of yellow color and turbidity. Limnology and Oceanography 36: 91–105.

Kraemer, B.M., O. Anneville, S. Chandra, M. Dix, E. Kuusisto, D.M. Livingstone, A. Rimmer, S.G. Schladow, E. Silow, L.M. Sitoki, R. Tamatamah, Y. Vadeboncoeur and P.B. McIntyre. 2015. Morphometry and average temperature affect lake stratification responses to climate change. Geophysical Research Letters 42: 4981–4988.

Leach, T.H., L.A. Winslow, F.W. Acker, J.A. Bloomfield, C.W. Boylen, P.A. Bukaveckas, D.F. Charles, R.A. Daniels, C.T. Driscoll, L.W. Eichler, J.L. Farrell, C.S. Funk, C.A. Goodrich, T.M. Michelena, S.A. Nierzwicki-Bauer, K.M. Roy, W.H. Shaw, J.W. Sutherland, M.W. Swinton, D.A. Winkler and K.C. Rose. 2018. Long-term dataset on aquatic responses to concurrent climate change and recovery from acidification. Scientific Data 5: 180059.

Letelier, R.M., A.E. White, R.R. Bidigare, B. Barone, M.J. Church and D.M. Karl. 2017. Light absorption by phytoplankton in the North Pacific Subtropical Gyre. Limnology and Oceanography 62: 1526–1540.

Lewis, W.M. 2011. Global primary production of lakes: 19th Baldi Memorial Lecture. Inland Waters 1: 1–28.

Lomas, M.W., N.R. Bates, R.J. Johnson, A.H. Knap, D.K. Steinberg and C.A. Carlson. 2013. Two decades and counting: 24-years of sustained open ocean biogeochemical measurements in the Sargasso Sea. Deep Sea Research Part II: Topical Studies in Oceanography 93: 16–32.

Lomas, M.W., D.K. Steinberg, T. Dickey, C.A. Carlson, N.B. Nelson, R.H. Condon and N.R. Bates. 2010. Increased ocean carbon export in the Sargasso Sea linked to climate variability is countered by its enhanced mesopelagic attenuation. Biogeosciences 7: 57–70.

Lozier, M.S., C. Dave Apurva, B. Palter Jaime, M. Gerber Lisa and T. Barber Richard. 2011. On the relationship between stratification and primary productivity in the North Atlantic. Geophysical Research Letters 38.

Luo, Y.W., H.W. Ducklow, M.A.M. Friedrichs, M.J. Church, D.M. Karl and S.C. Doney. 2012. Interannual variability of primary production and dissolved organic nitrogen storage in the North Pacific Subtropical Gyre. Journal of Geophysical Research: Biogeosciences 117.

MacIntyre, H.L., T.M. Kana and R.J. Geider. 2000. The effect of water motion on short-term rates of photosynthesis by marine phytoplankton. Trends Plant Sci. 5: 12–17.

Magnuson, J.J., S.R. Carpenter and E.H. Stanley. 2018a. North Temperate Lakes LTER: Physical Limnology of Primary Study Lakes 1981 - current.//lter.limnology.wisc.edu/dataset/north-temperate-lakes-lter-physical-limnology-primary-study-lakes-1981-current./lter.limnology.wisc.edu/dataset/north-temperate-lakes-lter-secchi-disk-depth-other-auxiliary-base-crew-sample-data-1981-curr.

Magnuson, J.J., S.R. Carpenter and E.H. Stanley. 2018b. North Temperate Lakes LTER: Secchi Disk Depth; Other Auxiliary Base Crew Sample Data 1981 - current.

Martinez, E., D. Antoine, F. D'Ortenzio and B. Gentili. 2009. Climate-Driven Basin-Scale Decadal Oscillations of Oceanic Phytoplankton. Science 326: 1253–1256.

Mason, L.A., C.M. Riseng, A.D. Gronewold, E.S. Rutherford, J. Wang, A. Clites, S.D.P. Smith and P.B. McIntyre. 2016. Fine-scale spatial variation in ice cover and surface temperature trends across the surface of the Laurentian Great Lakes. Climatic Change 138: 71–83.

Monteith, D.T., J.L. Stoddard, C.D. Evans, H.A. de Wit, M. Forsius, T. Høgåsen, A. Wilander, B.L. Skjelkvåle, D.S. Jeffries, J. Vuorenmaa, B. Keller, J. Kopácek and J. Vesely. 2007. Dissolved organic carbon trends resulting from changes in atmospheric deposition chemistry. Nature 450: 537.

Moore, J.K., W. Fu, F. Primeau, G.L. Britten, K. Lindsay, M. Long, S.C. Doney, N. Mahowald, F. Hoffman and J.T. Randerson. 2018. Sustained climate warming drives declining marine biological productivity. Science 359: 1139.

Moore, S., V. Gauci, C.D. Evans and S.E. Page. 2011. Fluvial organic carbon losses from a Bornean blackwater river. Biogeosciences 8: 901–909.

Morris, D.P., H. Zagarese, C.E. Williamson, E.G. Balseiro, B.R. Hargreaves, B. Modenutti, R. Moeller and C. Queimalinos. 1995. The attenuation of solar UV radiation in lakes and the role of dissolved organic carbon. Limnology and Oceanography 40: 1381–1391.

Neale, P.J., E.W. Helbling and H.E. Zagarese. 2003. Modulation of UV exposure and effects by vertical mixing and advection. pp. 107–134. *In*: Helbling, E.W. and H.E. Zagarese (eds.). UV Effects in Aquatic Organisms and Ecosystems. Royal Society of Chemistry, Cambridge, UK.

Neale, P.J., C. Sobrino and A. Gargett. 2012. Vertical mixing and the effects of solar radiation on photosystem II electron transport by phytoplankton in the Ross Sea polynya. Deep Sea Research Part I: Oceanographic Research Papers 63: 118–132.

Neori, A., O. Holmhansen, B.G. Mitchell and D.A. Kiefer. 1984. Photoadaptation in marine-phytoplankton - changes in spectral absorption and excitation of chlorophyll-a fluorescence. Plant Physiology 76: 518–524.

O'reilly, C.M., S. Sharma, D.K. Gray, S.E. Hampton, J.S. Read, R.J. Rowley, P. Schneider, J.D. Lenters, P.B. Mcintyre, B.M. Kraemer, G.A. Weyhenmeyer, D. Straile, B. Dong, R. Adrian, M.G. Allan, O. Anneville, L. Arvola, J. Austin, J.L. Bailey, J.S. Baron, J.D. Brookes, E.D. Eyto, M.T. Dokulil, D.P. Hamilton, K. Havens, A.L. Hetherington, S.N. Higgins, S. Hook, L.R. Izmest'eva, K.D. Joehnk, K. Kangur, P. Kasprzak, M. Kumagai, E. Kuusisto, G. Leshkevich, D.M. Livingstone, S. Macintyre, L. May, J.M. Melack, D.C. Mueller-Navarra, M. Naumenko, P. Noges, T. Noges, R.P. North, P.D. Plisnier, A. Rigosi, A. Rimmer, M. Rogora, L.G. Rudstam, J.A. Rusak, N. Salmaso, N.R. Samal, D.E. Schindler, S.G. Schladow, M. Schmid, S.R. Schmidt, E. Silow, M.E. Soylu, K. Teubner, P. Verburg, A. Voutilainen, A. Watkinson, C.E. Williamson and G. Zhang. 2015. Rapid and highly variable warming of lake surface waters around the globe. Geophysical Research Letters 42: 10: 773–710, 781.

Pilla, R.M., C.E. Williamson, J. Zhang, R.L. Smyth, J.D. Lenters, J.A. Bentrup and L.B. Knoll. 2018. Long-term trends in water temperature and thermal stratification in two small lakes resulting from browning-related decreases in water transparency. Journal of Geophysical Research-Biogeosciences 123(5): 1651–1665.

Polovina, J.J., E.A. Howell and M. Abecassis. 2008. Ocean's least productive waters are expanding. Geophysical Research Letters 35: L03618.

Poole, H.H. and W.R.G. Atkins. 1929. Photo-electric measurements of submarine illumination throughout the year. Journal of the Marine Biological Association of the United Kingdom 16: 297–324.

Read, J.S., D.P. Hamilton, A.R. Desai, K.C. Rose, S. MacIntyre, J.D. Lenters, R.L. Smyth, P.C. Hanson, J.J. Cole, P.A. Staehr, J.A. Rusak, D.C. Pierson, J.D. Brookes, A. Laas and H. Wu Chin. 2012. Lake-size dependency of wind shear and convection as controls on gas exchange. Geophysical Research Letters 39.

Read, J.S., D.P. Hamilton, I.D. Jones, K. Muraoka, L.A. Winslow, R. Kroiss, C.H. Wu and E. Gaiser. 2011. Derivation of lake mixing and stratification indices from high-resolution lake buoy data. Environmental Modelling & Software 26: 1325–1336.

Read, J.S. and K.C. Rose. 2013. Physical responses of small temperate lakes to variation in dissolved organic carbon concentrations. Limnology and Oceanography 58: 921–931.

Rhein, M., S.R. Rintoul, S. Aoki, E. Campos, D. Chambers, R.A. Feely, S. Gulev, G.C. Johnson, S.A. Josey, A. Kostianoy, C. Mauritzen, D. Roemmich, L.D. Talley and F. Wang. 2013. Observations: Ocean. pp. 255–315. *In*: Stocker, T.F., D. Qin, G.-K. Plattner, M. Tignor, S.K. Allen, J. Boschung, A. Nauels, Y. Xia, V. Bex and P.M. Midgley (eds.). Climate Change 2013: The Physical Science Basis. Contribution of Working Group I to the Fifth Assessment Report of the Intergovernmental Panel on Climate Change. Cambridge University Press, Cambridge, U.K.

Richardson, C.D., J.S. Melles, M.R. Pilla, L.A. Hetherington, B.L. Knoll, E.C. Williamson, M.B. Kraemer, R.J. Jackson, C.E. Long, K. Moore, G.L. Rudstam, A.J. Rusak, E.J. Saros, S. Sharma, E.K. Strock, C.K. Weathers and R.C. Wigdahl-Perry. 2017. Transparency, geomorphology and mixing regime explain variability in trends in lake temperature and stratification across Northeastern North America (1975–2014). Water 9,442.

Sallée, J.-B., J. Llort, A. Tagliabue and M. Lévy. 2015. Characterization of distinct bloom phenology regimes in the Southern Ocean. ICES Journal of Marine Science 72: 1985–1998.

Sarmiento, J.L., T.M.C. Hughes, R.J. Stouffer and S. Manabe. 1998. Simulated response of the ocean carbon cycle to anthropogenic climate warming. Nature 393: 245–249.

Sarmiento, J.L., R. Slater, R. Barber, L. Bopp, S.C. Doney, A.C. Hirst, J. Kleypas, R. Matear, U. Mikolajewicz, P. Monfray, V. Soldatov, S.A. Spall and R. Stouffer. 2004. Response of ocean ecosystems to climate warming. Global Biogeochemical Cycles 18.

Schmidtko, S., L. Stramma and M. Visbeck. 2017. Decline in global oceanic oxygen content during the past five decades. Nature 542: 335.

Seekell, D.A., J.F. Lapierre, J. Ask, A.K. Bergström, A. Deininger, P. Rodríguez and J. Karlsson. 2015. The influence of dissolved organic carbon on primary production in northern lakes. Limnology and Oceanography 60: 1276–1285.

Somavilla, R., C. González-Pola and J. Fernández-Diaz. 2017. The warmer the ocean surface, the shallower the mixed layer. How much of this is true? Journal of Geophysical Research: Oceans 122: 7698–7716.

Spencer, R.G.M., P.J. Mann, T. Dittmar, T.I. Eglinton, C. McIntyre, R.M. Holmes, N. Zimov and A. Stubbins. 2015. Detecting the signature of permafrost thaw in Arctic rivers. Geophysical Research Letters 42: 2830–2835.

Steinacher, M., F. Joos, T.L. Frölicher, L. Bopp, P. Cadule, V. Cocco, S.C. Doney, M. Gehlen, K. Lindsay, J.K. Moore, B. Schneider and J. Segschneider. 2010. Projected 21st century decrease in marine productivity: a multi-model analysis. Biogeosciences 7: 979–1005.

Talley, L.D., R.A. Feely, B.M. Sloyan, R. Wanninkhof, M.O. Baringer, J.L. Bullister, C.A. Carlson, S.C. Doney, R.A. Fine, E. Firing, N. Gruber, D.A. Hansell, M. Ishii, G.C. Johnson, K. Katsumata, R.M. Key, M. Kramp, C. Langdon, A.M. Macdonald, J.T. Mathis, E.L. McDonagh, S. Mecking, F.J. Millero, C.W. Mordy, T. Nakano, C.L. Sabine, W.M. Smethie, J.H. Swift, T. Tanhua, A.M. Thurnherr, M.J. Warner and J.Z. Zhang. 2016. Changes in ocean heat, carbon content, and ventilation: a review of the first decade of GO-SHIP global repeat hydrography. Annual Review of Marine Science 8: 185–215.

Verpoorter, C., T. Kutser, A.D. Seekill and L.J. Tranvik. 2014. A global inventory of lakes based on high-resolution satellite imagery. Geophysical Research Letters 41: 6396–6402.

Weyhenmeyer, G.A., R.A. Müller, M. Norman and L.J. Tranvik. 2016. Sensitivity of freshwaters to browning in response to future climate change. Climatic Change 134: 225–239.

Williamson, C.E., J.A. Brentrup, J. Zhang, W.H. Renwick, B.R. Hargreaves, L.B. Knoll, E.P. Overholt and K.C. Rose. 2014. Lakes as sensors in the landscape: Optical metrics as scalable sentinel responses to climate change. Limnology and Oceanography 59: 840–850.

Williamson, C.E., E.P. Overholt, R.M. Pilla, T.H. Leach, J.A. Brentrup, L.B. Knoll, E.M. Mette and R.E. Moeller. 2015. Ecological consequences of long-term browning in lakes. Scientific Reports 5: 18666.

Woolway, R.I., I.D. Jones, S.C. Maberly, J.R. French, D.M. Livingstone, D.T. Monteith, G.L. Simpson, S.J. Thackeray, M.R. Andersen, R.W. Battarbee, C.L. DeGasperi, C.D. Evans, E. de Eyto, H. Feuchtmayr, D.P. Hamilton, M. Kernan, J. Krokowski, A. Rimmer, K.C. Rose, J.A. Rusak, D.B. Ryves, D.R. Scott, E.M. Shilland, R.L. Smyth, P.A. Staehr, R. Thomas, S. Waldron and G.A. Weyhenmeyer. 2016. Diel Surface Temperature Range Scales with Lake Size. PLOS ONE 11: e0152466.

Wüest, A. and A. Lorke. 2003. Small-scale hydrodynamics in lakes. Annual Review of Fluid Mechanics 35: 373–412.

Yeates, P.S. and J. Imberger. 2003. Pseudo two-dimensional simulations of internal and boundary fluxes in stratified lakes and reservoirs. International Journal of River Basin Management 1: 297–319.

CHAPTER 7

Effects of Global Change on Aquatic Lower Trophic Levels of Coastal South West Atlantic Ocean Environments

Macarena S. Valiñas, Virginia E. Villafañe* and *E. Walter Helbling*

Introduction

Coastal areas can be defined as the interface zones between land and sea, with terrestrial environments deeply influencing marine ones and *vice versa* (Gattuso et al. 1998, Martinez et al. 2007). Although these areas do not have strict spatial boundaries, their seaward limit is often set to include the continental shelf, commonly defined as to the isobath of 200 m (Spalding et al. 2007). Approximately 60% of the human population lives now within 100 km of the coasts (Lotze et al. 2006, Halpern et al. 2008). Why are these areas so attractive for human settlement even since pre-historic times? One of the most probable answers is because of the favorable biophysical and climatic conditions together with the ease of communication and navigation that these sites frequently offer. In fact, many of the world's large cities are located in coastal areas, and a large portion of the economic activities are concentrated there (Costanza et al. 1997). Moreover, coastal areas harbor rich and valuable natural resources that provide important goods (e.g., food and raw materials) and services (e.g., disturbance regulation and nutrient cycling) for human beings. Due to their high primary productivity, these areas serve as feeding and spawning grounds and nurseries for many fish and invertebrates (including many of commercial interest) (Elliott and Dewailly 1995, Beck et al. 2001) therefore supporting high secondary production (ICES 2008).

Estación de Fotobiología Playa Unión and Consejo Nacional de Investigaciones Científicas y Técnicas (CONICET). Casilla de Correos N°15 (9103) Rawson, Chubut, Argentina.
* Corresponding author

Productivity of coastal areas is affected by several environmental factors, many of which are influenced by human activities. For instance eutrophication, a process typically triggered by the addition of nutrients, mainly land-derived (Nixon 1995, Cloern 2001, Valiela 2006) is considered as one of the most significant consequences of human alteration of coastal habitats (Valiela 2006). The growing human population (and thus the land use) along coastal areas has been the main source of increased nutrient inputs and of sewage and toxic elements coming into the water, changing the physical, chemical, and biological aquatic environment (UNEP 2006, Rabalais et al. 2009) (Fig. 7.1). As a consequence of the high nutrient loads into coastal waters there is an increase in primary production followed by changes in the flora and fauna community composition as seen in some studies (e.g., Duarte 1995, Rabalais et al. 2009). The anthropogenic influence has also resulted in global change, a process that has affected several variables and processes in coastal environments. For instance, global change involves changes in the precipitation patterns (i.e., it will either increase or reduce) which not only affects coastal salinity, but also the inputs of terrestrial-derived material, both in the form of particulate (organic and inorganic) and dissolved organic matter (DOM) (Häder et al. 2014b) that are then incorporated in the water column (Fig. 7.1). For example, for temperate latitudes of the South West (SW) Atlantic coasts (except for the southern tip of South America) an increase in precipitation (IPCC 2014) is expected which will incorporate higher amounts of terrestrial materials/nutrients into the water bodies that will lead to reduced penetration of solar radiation in coastal environments, as compared to that from pelagic areas (Häder et al. 2014a). However, in lower latitudes of the SW Atlantic coasts the precipitation will most probably be reduced (IPCC 2014) although coastal areas will still receive higher amounts of nutrients, just because of increased human activity. Another outcome of global change is the increase in atmospheric temperatures (0.4–0.8°C) with a concomitant warming

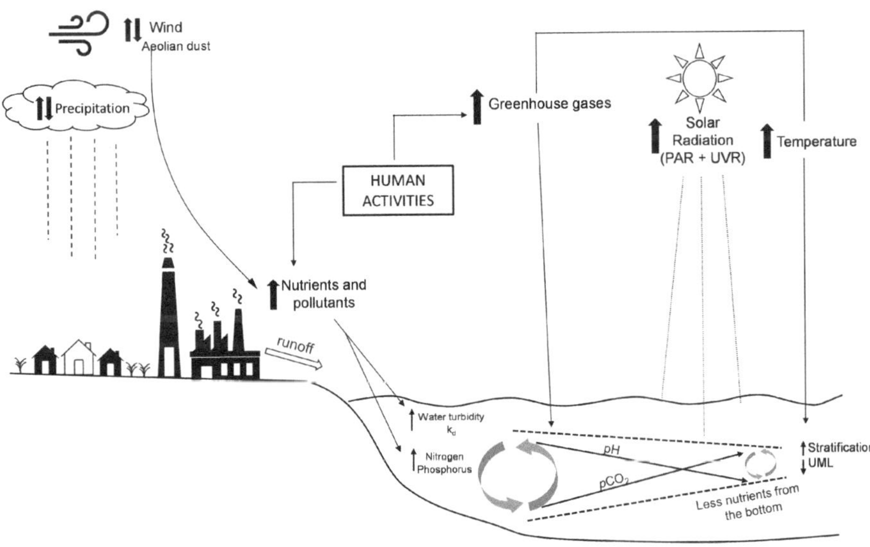

Fig. 7.1. Main consequences of global change related variables on coastal areas.

of the water bodies (IPCC 2013) which in turn increases stratification of the water column (Beardall et al. 2009, Häder et al. 2011). Such increased stratification results in that the organisms would be exposed to increased levels of solar radiation, including ultraviolet - UVR (280–400 nm) (Fig. 7.1).

The main cause of warming of the surface waters is the release of large amounts of greenhouse gases (mainly CO_2) into the atmosphere since the Pre-Industrial era (IPCC 2013) that trap some of the heat energy that would otherwise re-radiate to the space (Fig. 7.1). The rising of atmospheric CO_2 concentration also induces ocean acidification and alters the seawater chemistry (Turley and Gattuso 2012); the continuous uptake of atmospheric CO_2 is expected to substantially decrease the oceanic pH over the next few centuries, changing the saturation horizons of aragonite, calcite, and other minerals (Byrne 2011, Turley and Gattuso 2012). In addition, increasing CO_2 levels in the atmosphere contribute to deplete the ozone layer (Austin et al. 1992, UNEP 2012) leading to enhanced levels of UVR reaching the Earth's surface. However, due to the successful implementation of the Montreal Protocol and the subsequent amendments, it is expected that further ozone depletion will be limited and UV-B would slowly return to normal levels over the next decades (Ungar 2015). The recovery of the ozone layer is, however, slower than originally predicted as a result of the interaction between greenhouse gases and ozone (UNEP 2012). Overall, coastal areas are exposed to the simultaneous and combined effects of a variety of global change related variables. How organisms will respond to the joint impact of these variables is a rather complex issue, as they usually interact in a synergistic or antagonistic manner, thus enhancing or reducing individual effects (Crain et al. 2008, Gunderson et al. 2016).

The aim of this work is to review the status of knowledge on global change effects on organisms of lower trophic levels that inhabit rocky and sandy beaches, estuaries and bays, coastal lagoons, and coastal open waters of the SW Atlantic coast. The SW Atlantic extends south of the boundary of the Equatorial Counter Current (approximately at 8° N). In this review, the focus will be on research performed in the largest countries that have coasts on this part of the Atlantic Ocean: Brazil, Uruguay, and Argentina. The coastal area is influenced by two major rivers, Amazonas and La Plata, that discharge large amounts of freshwater and sediment into the Atlantic Ocean. Along the SW Atlantic coasts, there is a large diversity of habitats and environments: The Brazil coasts have a combination of freshwater, estuarine and marine ecosystems in the northern part, and sandy beaches, mangrove forests, rocky shores, lagoons and coral reefs to the south. Uruguay's coasts are dominated by sandy beaches whereas in Argentina there are mostly sandy beaches, with some rocky formations (Turra et al. 2016). In the following sections, the ecological effects of some variables affected by global change, i.e., nutrients, temperature, pH/pCO_2, and solar UVR in coastal areas of the SW Atlantic, focusing on aquatic lower trophic levels, i.e., autotrophs and invertebrates inhabiting both water column and intertidal zones will be discussed.

Ecological Effects of Increasing Temperature

The increase in water temperature as a result of global change has significant consequences on organisms and ecosystems, ranging from physiological effects

(Hochachka and Somero 2002) to changes in distributional boundaries of species and the associated replacement of cold-water species by warm-water ones. All these effects can ultimately affect the composition and structure of communities (Fields et al. 1993) as well as their interactions. Enhanced temperatures generally increase growth (Feng et al. 2008, Rosas-Navarro et al. 2016), photosynthesis (Feng et al. 2008, Hancke et al. 2008) and calcification rates (Xu et al. 2011, Rosas-Navarro et al. 2016) up to an optimum beyond which there is a decrease or death, evidencing the fact that species successfully develop within an optimal thermal window (Huertas et al. 2011). For instance, Sudatti et al. (2011) reported a positive relationship between biomass and temperature in the red macroalgae *Laurencia dendroidea* collected at Cabo Frio (Brazil), however, at high temperatures (35°C) the organisms died. In another study, Miranda et al. (2012) found that the optimal temperature for growth of the red macroalgae *Hydropuntia caudata* was 25ºC, while at 35ºC all the spores died. These results contrasts with those obtained by Yokoya and Oliveira (1993) and Macchiavello et al. (1998), who reported that the optimal growth temperature was 30ºC, hinting at the fact that different populations are locally adapted and have distinct tolerance limits to temperature (Miranda et al. 2012). Scherner et al. (2016) reported that increases in temperature did not affect the photosynthetic performance of most coralline macroalgal species tested, with the exception of *Halimeda cuneata* that showed high efficiency only at a modest increase (+ 1°C), probably related to its tropical nature (Bandeira-Pedrosa et al. 2004, Verbruggen et al. 2005). This might be related to the fact that, as compared with organisms living in the open ocean, those living in coastal areas seem to be relatively more resilient to temperature rises (as long as those occurs inside of the thermal window of each particular species) because they are already adapted to highly changing environments (Wilson 2008).

Enhanced water temperatures resulted in shifts in the abundance and/or distribution of species, including the worst case scenario of the disappearance or appearance of some of them (Whalter et al. 2002, Horta et al. 2012, Phillipart et al. 2003). Long-term studies performed in SW Atlantic coasts suggest that increased temperatures resulting from global warming, along with the overall impact of urbanization in recent decades (Oliveira and Qi 2003, de Faveri et al. 2010) are probably responsible for changes in macroalgae assemblages in this region. For instance, Horta et al. (2012) reported the introduction of exotic species, such as the red algae *Anotrichium yagii* and *Laurencia caduciramulosa* (Horta and Oliveira 2000, Cassano et al. 2008), as well as the expansion of economically important ones, such as *Hypnea* sp. (de Faveri et al. 2010). Also, Taouil and Yoneshigue-Valentin (2002) pointed to the disappearance of some representatives of the Dictyotales, known to produce important biotechnological substances. In another study, de Faveri et al. (2010) evaluated changes in macroalgae assemblages of Southern Brazil over a 30-years period and found significant differences in the composition of the flora. The authors suggested that sea surface warming could explain the absence of some brown macroalgal species that need temperatures below certain ranges (i.e., below 20ºC) to induce macrothallus formation (Orfanidis et al. 1996).

Long-term field observations also highlight the influence of warmer temperatures in conditioning the timing and composition of phytoplankton assemblages of the SW Atlantic coasts. In this sense, Guinder et al. (2010) observed the absence of a typical winter bloom and changes in phenology, together with the replacement of the dominant species, i.e., *Thalassiosira curviseriata*, and the appearance of different blooming ones, i.e., *Cyclotella* sp. and *T. minima*. In addition, shifts in the phytoplankton size structure towards small-sized diatoms, for example, small *Cyclotella* species and *Chaetoceros* spp. were noticed. Similarly, when analyzing summer phytoplankton patterns Guinder et al. (2013) and López Abbate et al. (2017) also observed changes from dominance of phytoflagellates and large diatoms towards smaller diatom species, such as *T. minima*.

Invertebrates inhabiting SW Atlantic coasts seem to benefit from increases in temperature, but only to a certain extent. For instance, increased temperature reduced the development time of larval stages to attain the first crab stage of *Pagurus criniticornis*; however, larvae were less tolerant when temperature exceeded 35°C (Blaszkowski and Moreira 1986). In another study, Ismael et al. (1997) found a decrease in larval tolerance to temperatures > 21°C in the burrowing crab *Chasmagnathus granulata* (= *Neohelice granulata*). Probably, this could explain its distribution limit, which has been found north up to Cabo Frio, Brazil (23 °S) (Boschi et al. 1992). Also changes in the distribution of species seems to be related to changes in water temperature, as seen in the drastic decrease in the populations of the yellow clam *Mesodesma mactroides* over the last 30 years (Odebrecht et al. 1995, Fiori et al. 2004, Ortega et al. 2012, 2016) with lower abundances during warm periods, coinciding with a poleward shift of the 20°C isotherm from 1985 to 2007. Changes in the size frequency distribution of the clam population were also reported, with fewer size classes and larger specimens being virtually absent during warm years (Ortega et al. 2016). On the contrary, warm-favoring species, such as the wedge clam *Donax hanleyanus* and the mole crab *Emerita brasiliensis* (Defeo 2003, Dadón 2005, Herrmann et al. 2009, Thompson and Sánchez De Bock 2009) benefited by increases in sea surface temperature in the last decades. In a recent study, Celentano and Defeo (2016) reported that the increase in abundance of *E. brasiliensis* was particularly higher in years when La Niña events occurred. Moreover, reproductive and recruitment periods were extended, being higher during warm years, and range size of classes were widely represented, which influenced the population structure in subsequent years. Thus, under a climate change scenario where an increasing frequency of extreme La Niña events is expected, this species might be a successful pioneer or colonizer, expanding its range polewards (Celentano and Defeo 2016, Poloczanska et al. 2013).

Ecological Effects of Ultraviolet Radiation

Solar radiation has a key role in aquatic ecosystems, as it constitutes the energy source for the entire food web via the photosynthetic process, which occurs mainly in the visible range of the solar electromagnetic spectrum (Photosynthetic Active Radiation, PAR, 400–700 nm). The depletion of the stratospheric ozone layer, caused by the release of ozone-depleting substances (chlorofluorocarbons (CFCs), halocarbons, carbon dioxide (CO_2), and methyl chloroform (MCF)) (Kerr and McElroy 1993) into the atmosphere (Ravishankara et al. 2009, García-Corral et al. 2016) resulted in an increase of solar UV-B radiation (280–315 nm) reaching the Earth's surface (Madronich et al. 1998) especially in polar regions and at mid-latitudes (McKenzie et al. 2011). In the water column, the penetration of solar radiation is further conditioned by several factors that include, among others, latitude, surface conditions and amount and type of particulate and dissolved materials (Kirk 1994). In particular, and for planktonic organisms, solar radiation is also conditioned by the depth of the upper mixed layer (UML) (Neale et al. 2003), i.e., the portion of the water column in which these organisms circulate. Either at normal or at increased levels, UVR, mainly UV-B, causes a number of deleterious effects on aquatic organisms (Helbling and Zagarese 2003). Among such effects, decreases in productivity and growth rates (Villafañe et al. 2003 and references therein), and damages to the photosynthetic pigments, such as chlorophyll and phycobiliproteins have been reported (Figueroa et al. 1997, Flores-Moya et al. 1999). These effects consequently reduce photosynthesis, i.e., causing photoinhibition (see Hanelt 1996, Villafañe et al. 2003) because of the weakening or destruction of oxidation sites and the reaction center of photosystem II (PSII) (Roleda et al. 2007). Moreover, UV-B energy induces photodamage in proteins, nucleic acids, DNA, lipids, and other compounds in biological tissues (Dunlap and Yamamoto 1995, Buma et al. 2003, Bischof and Steinhoff 2012) through, for instance, the formation of reactive oxygen species (ROS; Kieber et al. 2003). In zooplankton and benthic invertebrate species, studies reported negative effects of UVR on survival rates, reduced fertility and changes in the organisms' behavior, for example, motility and orientation and in feeding responses (Browman et al. 2003, Przeslawski et al. 2005, Hylander and Hansson 2010).

Unlike phytoplankton, by being fixed and restricted to their place of growth, benthic algae living in coastal intertidal areas are exposed to solar radiation (PAR and UVR) for extended periods, especially during low tide (Cabrera et al. 1995). Macroalgae from the SW Atlantic intertidal showed a variety of effects due to UVR exposure, for example, loss of chloroplast organisation (Schmidt et al. 2011, Simioni et al. 2014), changes in mitochondria (Schmidt et al. 2011, Pereira et al. 2014, de Oliveira et al. 2016), disrupted thylakoids (Bouzon et al. 2012, Pereira et al. 2014), increasing atrophy of the Golgi bodies (de Oliveira et al. 2016), inhibition of protein metabolism (Schmidt et al. 2011, 2012), delay of germination (Scariot et al. 2013), and changes in developmental patterns (de Oliveira et al. 2016). Even more, UVR also decreased the photosynthetic performance (Helbling et al. 2010, Schmidt et al. 2011, 2012, Simioni et al. 2014, Pereira et al. 2014), and consequently growth rates (Schmidt et al. 2011, 2012, Bouzon et al. 2012, Simioni et al. 2014, Pereira et al. 2014); in some cases tissue bleaching was also observed (Häder et al. 2001, Schmidt et al. 2012, Bouzon et al. 2012). The extent of such deleterious effects is highly variable, with an important component of species-specificity. However, species living under high radiation environments seem to be more resistant than those inhabiting low radiation environments, due to their previous light history (Helbling et al. 1992). For instance, in the red macroalga *Graciliaria birdiae* the northeastern Brazilian population is more resistant to the stress caused by UV-B than its southeastern counterparts (Ayres-Ostrock and Plastino 2014).

Many studies have addressed the effects of solar UVR on phytoplankton from temperate coastal areas from the SW Atlantic. In a coastal lagoon in Uruguay, Aubriot et al. (2004) found both inhibition and stimulation of phosphorus (P) uptakes due to UVR exposure that potentially would alter the P cycling. Conde et al. (2002) showed, for the same study area, that the inhibition of primary production was ca. 25%, with the contribution of UV-A to UV-B being close to a 2:1 ratio. In the southernmost part of Argentina (Beagle Channel) it was determined that UV-A was responsible for most of the inhibition of growth of phytoplankton communities; however, the reduction in photosynthesis was mostly due to UV-B (Hernando et al. 2006). In agreement with these findings, Hernando et al. (2005) working with isolated phytoplankton species from this area found significant oxidative stress induced by both, UV-A and UV-B. In a comparative study carried out between this sub-polar area with a tropical one in Brazil, Roy et al. (2006) determined important changes in the taxonomic composition of the communities due to UV-B exposure. In a northern Patagonian site, solar UVR caused significant photoinhibition (ca. 20%) with UV-A being responsible for the bulk of the inhibition (Villafañe et al. 2004b), with small cells (i.e., picoplankton, < 2 μm) being more resistant than larger ones (Helbling et al. 2001). However, these cells were more vulnerable to DNA damage as compared to large cells (Buma et al. 2001). This agrees with studies carried out throughout the annual cycle in the same study area that showed that cells characterizing the bloom (i.e., microplankton) were more sensitive to UVR (assessed as photoinhibition) than those from the pre-and post-bloom (pico/nanoplankton) (Villafañe et al. 2004a, Helbling et al. 2005). Winter blooms in the temperate Patagonian areas are frequently associated with calm weather conditions that promote stable and high stratification (Villafañe et al. 2004a, Bermejo et al. 2018). Thus, if such conditions are altered either by changes in wind patterns (Bermejo et al. 2018) or by global warming, the balance of UVR inhibition/utilization (Barbieri et al. 2002) would also change, with important consequences for the timing and intensity of the bloom, and so for the trophodynamics of the area.

Studies about the effects of UVR on benthic and planktonic invertebrates from coastal areas of the SW Atlantic are scarce when compared to those performed on primary producers. There are only two studies that addressed the effects of UVR as a single stressor on marine zooplankton in this region: Spinelli (2013) showed that UVR negatively affected feeding in adults of the copepod *Euterpina acutifrons* in the Patagonian region but not in their larvae. Hernández Moresino et al. (2011) reported that larvae of the crab *C. altimanus* inhabiting Patagonian coasts pre-exposed to UVR had a delay or absence of molting which was coupled to arrested body growth (Fig. 7.2). The authors found that larvae that received UVR remained as Zoea I stage during the whole experimental period, while 100% larvae Zoea I that received only PAR reached the Zoea II stage. Moreover, at the end of the experiment larvae were larger in the PAR as compared to the UVR treatment (Fig. 7.2). The extended larval period led to a longer planktonic stage as a result from a delay in molting; this might have important negative effects on this crab species as individuals would remain exposed to physical stress and predators for longer periods before reaching the juvenile stage, as seen in other studies (Morgan 1995).

Aquatic organisms have developed several mechanisms to cope with UVR that include: (i) protective strategies, through the presence of UV-absorbing compounds (UVACs), typically mycosporine-like amino acids (MAAs) or melanin and carotenoid pigments, (ii) repair of damage either photorepair or dark enzimatic repair (Roy 2000, Buma et al. 2003), and (iii) avoidance strategies (e.g., diel migration), by moving away from the radiation source (Roy 2000) or even hiding in shaded places, as seen in invertebrates living under rocks or within a macroalgae canopy. It has

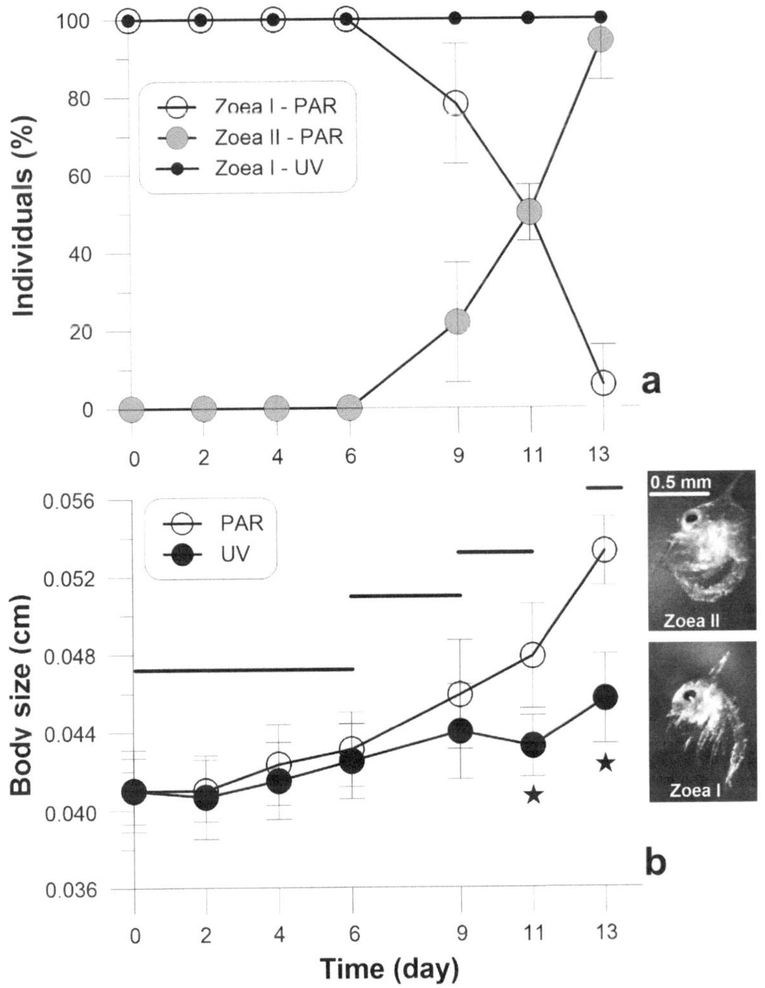

Fig. 7.2. Molt and growth of *Cyrtograpsus altimanus* crab larvae after exposure to UVR (black circles) and to PAR (white and gray circles). a) Proportion of Zoea I and II. b) Body size of larvae throughout the experimental period. The symbols represent the mean, and the vertical lines indicate the standard deviations, while the asterisks indicate significant differences among radiation treatments. The horizontal lines in panel b connect the PAR treatment samples that were not significantly different (Hernández Moresino et al. 2011, with permission of Elsevier).

been reported that several macroalgae species of the SW Atlantic deployed physical barriers to protect the inner tissues by increasing cell wall thickness (Schmidt et al. 2011, Pereira et al. 2014, Simioni et al. 2014, de Oliveira et al. 2016) and cytoplasm density (Bouzon et al. 2012) when exposed to UVR. Also, the presence of UVACs seem to be a rather common mechanism against UVR stress, as seen in macroalgae species inhabiting a wide latitudinal gradient of the Brazilian coast (Schmitz et al. 2018) as well as in several locations of the Patagonian coast. For instance, Richter et al. (2006) found diurnal changes in the UVACs concentration in the red algae *Corallina officinallis* that was higher around local noon. A similar pattern was reported for the red macroalgae *Ceramium* sp., with a high UVAC concentration during the day and decreasing in the evening, and being higher in samples exposed to UVR than in those receiving only PAR. However, the highest values of UVACs in the red algae *Porphyra columbina* were detected at night (Korbee Peinado et al. 2004). In natural phytoplankton communities from a tropical site in Brazil, high amounts of UVACs were found (Mohovic et al. 2006) although they were not enough to protect organisms against high UVR levels. However, in the Patagonian area, UVACs are virtually absent (Villafañe et al. 2004b, 2013) although they can be synthesized under enhanced (experimental) UVR levels (Helbling et al. 2008, Marcoval et al. 2008, Halac et al. 2014). Overall, natural phytoplankton communities from the SW Atlantic coasts mostly rely on other strategies to minimize the negative impact of UVR, such as xanthophyll pigments (Villafañe et al. 2008) and/or dissipation of excess energy in the form of heat (non-photochemical quenching, NPQ) (Villafañe et al. 2014, 2017).

Because MAAs can be only synthesized by organisms having the shikimate pathway, for example, phytoplankton, red algae and some cyanobacteria, the only way by which heterotrophs can acquire them is through their diet (Shick and Dunlap 2002). For instance, Helbling et al. (2002) reported that UVACs provided an effective protection against UV-B in the amphipod *Ampithoe valida* from Patagonian coasts, as higher survival rates were observed in individuals feeding on diets rich in these compounds. Nevertheless, in another study carried out with the same species, Valiñas and Helbling (2015) reported that amphipods had different sensitivity to UVR exposure (both UV-A and UV-B) depending on the sex: UVR decreased survival rates in females—even at higher proportions in those groups fed on poor UVACs diet, while survival rates of males were not affected by UVR independently of the type of diet. Also, Marcoval et al. (2017) reported that the bioaccumulation of UVACs from the diet protected early life stages of the shrimp *Pleoticus muelleri* from Argentinean coastal areas, by increasing their survival rates when organisms were exposed to UVR. Moreover, post-larvae of this species had a greater antioxidant activity when they were exposed to UVR and fed on a diet rich in UVACs. In another study, larvae of the crab *C. altimanus* that accumulated UVACs by direct (by feeding on phytoplankton rich in these compounds) and by indirect ways (UVACs that were transferred by mothers to the eggs) also showed higher survival than those without UVACs in their bodies, after UVR exposure (Hernández Moresino et al. 2011). However, the accumulation of photoprotective compounds does not always mean that they play a protective role against UVR, as seen in the isopod *Idotea baltica* in which the survival was negatively affected by UVR even in individuals with high amounts of UVACs in their bodies (Helbling et al. 2002).

Ecological Effects of Increased Nutrient Inputs

Human activities are responsible for coastal eutrophication, i.e., water enrichment with nutrients and organic matter (Nixon 1995), a phenomenon that has important consequences for coastal ecosystems. The discharges of domestic wastes into the rivers and the sea are one of the most important source of nutrients in coastal areas, particularly in developing countries where almost 100% of the waste passes untreated into rivers, streams, and the ocean because sewage treatment does not exist or is inadequate and/or inefficient (Marques et al. 2004). Nutrient inputs occur mostly in the form of nitrogen and phosphorus (Fong et al. 1998, Teichberg et al. 2010). The effects of such increments in nutrients are variable for organisms living in coastal areas, ranging from alteration of the biogeochemical cycles to changes in food webs and biodiversity (deYoung et al. 2008). One common symptom of eutrophication is the rapid increase in primary productivity through the growth of opportunistic algal species (i.e., Chlorophyta) causing algal blooms (e.g., Valiela et al. 1997, Raffaelli et al. 1998, de Faveri et al. 2010). The occurrence of these blooms can be beneficial, including the short-term nutrient sequestration (Howarth et al. 1996, Boyer et al. 2002) and the supply of more and better food particles of high nutritional quality for consumers (Valiela 2006). However, as nutrient input continues, huge densities of primary producers may negatively affect the ecosystem (Smith and Schindler 2009) as they consume oxygen at night, thus driving hypoxic events. During the day, the photosynthetic process can restore the oxygen concentration in the water; however, a succession of a few cloudy days may prevent this, resulting in anoxic events in sediments or near-bottom waters (D'Avanzo and Kremer 1994). Moreover, bacterial decomposition of algal biomass consumes oxygen which can cause extended and persistent anoxia (e.g., Duarte 1995, Valiela et al. 1997). These events of hypoxia and anoxia affect the survival of many organisms, such as fishes, shellfishes and other consumers, including algal grazers (D'Avanzo and Kremer 1994, Fox et al. 2009). Thus, coastal environments exposed to high nutrient loads are characterized by the dominance of fast-growing autotrophic species that leads to lower abundance and diversity of organisms, reorganizing the natural communities and ecosystem functions (Duarte 1995, Valiela et al. 1997, Osterling and Pihl 2001) and affecting its goods and services (McGlathery et al. 2007).

The SW Atlantic intertidal areas that are heavily impacted by nutrient inputs (e.g., due to discharge of effluents) are characterized by relatively low algae coverage, mainly green algae (e.g., *Enteromorpha* sp., *Ulva* sp.), cyanobacteria, tube-dwelling diatoms (e.g., *Berkeleya* sp.) and extensive areas of bare substrate (López Gappa et al. 1990, Martinetto et al. 2010, Martins et al. 2012, Becherucci et al. 2016b, 2016a, Fricke et al. 2016, 2017). For instance, Santi and Tavares (2009) reported that the diversity of macroalgae in the polluted areas of Guaranaba Bay (Brazil) was very low, with *Ulva* spp. having high colonization rates. The growth rate of *Ulva* was positively correlated with ambient supply of nutrients also in Patagonian coasts of Argentina (Teichberg et al. 2010). Similarly, high nutrient load due to increasing urbanization (Pagliosa and Rodrigues Barbosa 2006) led to species loss and community shifts in Brazilian coasts (Scherner et al. 2013, Portugal et al. 2017). Long-term studies reported a substantial loss of macroalgal richness and biodiversity along with changes in the assemblages in response to eutrophication in SW Atlantic coasts (Piriz et al. 2003, Horta et al. 2012). Also, significant decreases in calcareous algae coverage, along with a decrease in carbonates in the sediment were reported in urban areas from Brazil, probably due to phosphate inhibition of calcium carbonate crystal formation, and increased bioerosion (Hallock and Schlager 1986) together with a decrease in the acidity of the water, all caused by the excess of nutrients (Scherner et al. 2013). Because this loss of primary producers propagates through the food web affecting multiple levels, it could change the trophic structure of aquatic communities (Airoldi et al. 2008). Nevertheless, a recent study performed in Brazilian coasts showed a return of some species reported as 'lost' in previous studies; this increase in the number of species is thought to be a consequence of the efforts made by local governments to reduce pollution levels in the area (Oliveira and Qi 2003).

Several studies have also shown the influence of increased nutrient inputs on coastal phytoplankton of the SW Atlantic, either due to anthropogenic activity (domestic or industrial sewage) or increased riverine runoff, as well as aeolian transport due to changes in wind patterns. In coastal zones of Pernambuco in Brazil, Pereira et al. (2005) reported the dominance of some particular taxa (e.g., *Gyrosigma balticum* and *Bacillaria paxillifera*), low richness of species, diversity and equitability as indicators of eutrophication. In the hypereutrophic Recife harbor, Guenther et al. (2015) reported that phytoplankton production rates were seven times higher at present as compared to 20 years ago, and this was also tied to changes in size structure towards nanoplankton species. Also eutrophication was reported in the Guanabara Bay, through measurements of particulate and dissolved organic matter that were well above the average concentrations found in other less impacted coastal zones (Cotovicz Jr. et al. 2018). Studies carried out in Argentina also reported eutrophication effects on phytoplankton dynamics and structure. In this regard, López Abbate et al. (2015), in a two-year survey in two sites of the Bahía Blanca estuary with contrasting anthropogenic impact, found higher phytoplankton abundance dominated by smaller-sized, non-siliceous species in areas under severe eutrophication; also sewage pollution caused significant effects in the ecological stoichiometry, i.e., N and P excess with respect to Si, with potential consequences on the food web dynamics. Finally, in the Río de la Plata estuary, García and Bonel (2014) found signs of high eutrophication, i.e., with high picoplankton (chlorophytes and cyanobacteria) concentrations in the most polluted areas.

Studies on intertidal invertebrate communities of the SW Atlantic coasts showed a decrease in the abundance of different invertebrate taxa (e.g., molluscs, crustaceans and polychaetes) in areas impacted by continuous nutrient inputs (Defeo and de Alava 1995, Lercari and Defeo 1999, Breves-Ramos et al. 2005, Elias et al. 2006, Lozoya and Defeo 2006). It was found that highly nutrient-enriched areas are characterized by a low density and dominance of opportunistic species of polychaetes successfully colonizing the sediments (Elias et al. 2006, Bergamino et al. 2009), while non-opportunistic ones, e.g., other polychaetes (Elias et al. 2006), crustaceans (i.e., amphipods; Vallarino et al. 2002) and/or mollusks (Souza et al. 2013) were dominant at intermediate distances from the nutrient source. The low densities (or even absence) of filter-feeding species in areas with sewage inputs are attributed to the high proportion of suspended particles contained in the water, which could negatively affect the filtering mechanism, or to an insufficient oxygen supply characteristic during low tide (López Gappa et al. 1990). However, Martinetto et al. (2010) found higher abundances of small infaunal and epifaunal invertebrates in Patagonian sites affected by high nutrient inputs and suggested that the pronounced tidal flushing provides a continuous input of oxygen-rich water that limits anoxic or hypoxic events, and so it favors the development of benthic communities. Thus, the system remains permanently in an initial state of eutrophication where there is abundant high quality food for herbivores, without shifting towards the anoxia typical of advanced eutrophication (Martinetto et al. 2011).

Regarding zooplankton communities, studies performed in different areas of Argentinean coasts found the highest abundances of tintinnids in sites with high nutrient inputs (Barría de Cao et al. 2003, Garcia and Bonel 2014, Garcia and Barría de Cao 2018, but see López Abatte et al. 2016). Under high trophic conditions, i.e., with an increase in phytoplankton biomass due to high nutrient inputs, the relative importance of microzooplankton in relation to that of macrozooplankton is commonly observed, as smaller organisms can take advantage over larger ones. Kozlowsky-Suzuki and Bozelli (2002) working with individuals from a Brazilian coastal lagoon, experimentally demonstrated that the densities of the tintinid *Brachionus rotundiformis* increased while that of the copepod *Acartia tonsa* decreased in the nutrient-enriched treatments. Studies carried out with mesozooplankton also reported a wide variety of responses. For instance, Garcia and Bonel (2014) showed a decrease in mesozooplankton abundance in polluted areas of the Rio de la Plata estuary, as also seen in the Suape Bay (Brazil) (Pinto Silva et al. 2004). However, studies performed in the Bahia Blanca estuary showed that mesozooplankton is tolerant to sewage discharge and other types of stress (Fernández-Severini et al. 2009, Biancalana and Torres 2011, Dutto et al. 2012). Nevertheless, in some cases, differences in specific composition, evenness, and richness between impacted and non-impacted areas were observed, perhaps indicative of the modulating effect of the source of disturbance (Dutto et al. 2014). Also, no differences in zooplankton abundance (apart from that of cirripede larvae) between anthropogenically impacted and non-impacted areas of Paranaguá Bay (Brazil) were reported, probably as a consequence of the water circulation along the estuary, which may have diluted and dispersed the nutrients and pollutants from the bay to other areas of the estuary (Miyashita et al. 2012).

Increased nutrient effects on individuals′ fitness have been reported in zooplankton species inhabiting highly anthropogenic impacted areas of SW Atlantic coasts. For instance, Martinez et al. (2017) reported higher egg production rates in the copepod *A. tonsa* in such affected areas by anthropogenic activities; however, no differences were observed in egg hatching success. In another study, it was shown that the fecundity and survival of the copepod *Eurytemora americana* were negatively affected when organisms were grown in water collected near the sewage discharge (Berasategui et al. 2017). Probably, a high concentration of organic matter and nutrients which led to hypoxia and anoxia in near bottom water might explain the results obtained; however, other non-mutually exclusive causes, such as the high concentration of contaminants associated to the suspended sediments, and the high load of fine cohesive sediments (which reduces the feeding efficiency) cannot be ruled out.

Ecological Effects of Increased Acidification

The changes in the chemistry of the water, as a result of changes in pH/pCO_2, have biological, physiological, and evolutionary consequences for the marine biota and ecosystem processes to different degrees (Bibby et al. 2008). Carbonate ions are an important building block of structures, such as sea shells and coral skeletons; thus their decreases can especially affect organisms with calcareous structures (Riebesell et al. 2000). Regarding non-calcifying organisms, studies reported that some species may benefit while others may be harmed by the increased water acidity, evidencing that the responses to this stressor are highly species-specific (Fabry et al. 2008, Häder and Gao 2015). In coastal systems however, changes in pH result from several drivers/processes, such as impacts from watershed processes, nutrient inputs, and changes in ecosystem structure and metabolism (Duarte et al. 2013), as well as in the photosynthetic capacity (Villafañe et al. 2015). Therefore, the impacts of changes in pH in coastal systems are rather complex and difficult to predict.

Net calcification rates of different coastal organisms in response to lower pH (Ries et al. 2009, Hendriks et al. 2010) varies from strictly increasing to strictly decreasing, with several taxa showing the highest calcification rates at intermediate levels of pCO_2 (Ries et al. 2009). Primary producers also show assorted responses that go from species being positively affected from higher CO_2 levels (e.g., enhancing the photosynthetic performance (Johnson et al. 2012)) to being negatively affected (Riebesell 2004, Price et al. 2011) while others do not display any significant alterations (Israel and Hophy 2002, Riebesell 2004). Thus, there are no universal patterns, as responses of organisms to rising CO_2 concentrations will vary according to the organisms' sensitivities to decreased seawater pH, carbonate concentration, and carbonate saturation state. Nevertheless there is consensus that early life-stages are the most susceptible to lower pH, as it impairs fertilization (Byrne 2011), has negative effects on metamorphosis, growth and survivorship (Talmage and Gobler 2009) and produces abnormalities in larval morphologies (Dupont et al. 2008, Kurihara 2008).

We are not aware of any study that reports the effects of lower pH as a single stressor in phytoplankton species of the SW Atlantic coasts; however, some studies have been done with macroalgae and invertebrates. In a recent study performed with macroalgae collected from Brazilian coasts, Scherner et al. (2016) found that lower pH increased photosynthetic efficiency in a calcifying macroalga, but it did the opposite in a crustose coralline; in this latter the negative effects were only registered at the extreme acidity treatment, where the pH used was far lower than expected for the twenty-first century according to the IPCC projections (Diaz-Pulido et al. 2012). At lower pH treatments, in the calcifying macroalgae the photosynthetic performance was little affected, evidencing that some intertidal species are resilient to increased acidity, probably because they live in highly dynamic environments, and so they are acclimated to wide fluctuation in physical-chemical parameters. Contrary to what was expected, Malvé et al. (2016) in a survey covering 1,600 km of the SW Atlantic Ocean coast, reported that the mean shell length and relative shell weight of the gastropod *Trophon geversianus* was negatively correlated with pH. Laboratory experiments carried out with larval stages of the sea urchin *Lytechinus variegatus*, inhabiting the Brazilian coasts, showed that small decreases in the seawater's pH caused an inhibition of embryo-larval development (Passarelli et al. 2017). In this same species, and also in *Echinmetra lucunter* decreases in pH also affected the immune system and the coelomic fluid pH; however the effects were reversible if the exposure to low pH was for a short period of time (Figueiredo et al.

2016). Finally, in another recent study, Szalaj et al. (2017) found that low pH did not have an effect on the fertilization process in the mytilid *Perna perna*, but negatively affected their larval-development. Overall, all these scattered studies point out to the need of advancing in the research of this phenomenon in this region (Horta et al. 2012, Kerr et al. 2016).

Interactions Among Global Change Variables

Most studies based on single stressors allowed scientists to understand and to predict their effects on organisms, populations, and communities under a global change scenario. However, different environmental stressors, such as those discussed in this chapter (e.g., increased temperature, nutrients, UVR, low pH/high pCO_2) co-occur and interact in nature, complicating the predictability of biological responses. In a simple situation, the impact of the stressors may be additive so that the response can be predicted based on the sum of individual effects. However, other types of interactions involving synergistic or antagonistic responses (i.e., the response is greater or smaller than predicted from the sum of the individual effects) are more frequent (Crain et al. 2008, Gunderson et al. 2016). In a global change scenario, synergistic responses are of particular importance, as the future impacts on ecosystems predicted only on the basis of a single stressor effect might be underestimated.

Very few studies considering the combined effects of two or more variables have been performed with macroalgae and invertebrates of the SW Atlantic coasts. For instance, Gouvêa et al. (2017) reported that under low nutrient concentration, physiological performance and growth of *Laurencia catarinensis* inhabiting Brazilian coasts decreased with increasing temperatures, but increased when nutrient availability was higher and temperature was closer to that registered in summer. Comparatively, such studies considering interactions among variables are much more abundant for phytoplankton of the SW Atlantic. For example, in a comparative study carried out with coastal phytoplankton communities from temperate and tropical sites (Villafañe et al. 2017) it was found that UVR and nutrients acted antagonistically on the photosynthetic process at the temperate site as previously determined by Marcoval et al. (2008), whereas no differences were observed in the interactive and the sum of effects at a tropical site (Fig. 7.3). Such differences were attributed to differences in the taxonomic composition of the communities together with those in temperature among sites.

Other studies showed a decrease of 27% in the CO_2 sink capacity under high UVR levels in coastal phytoplankton from Patagonia when nutrients were added (regardless their origin, i.e., eolian or from hydric sources) (Cabrerizo et al. 2017). The interactive effects of temperature and UVR have also been studied with isolated phytoplankton species (i.e., the diatom *T. weissflogii*) such that the increased temperature counteracted the UVR inhibition an increase in RUBISCO activity and gene expression (Helbling et al. 2011). However, this beneficial effect is not universal, as temperature had little effect on pre-bloom communities but counteracted the photosynthetic inhibition during the bloom onset in Patagonian coastal waters (Villafañe et al. 2013) whereas during the winter bloom and during spring, temperature and UVR acted synergistically, thus

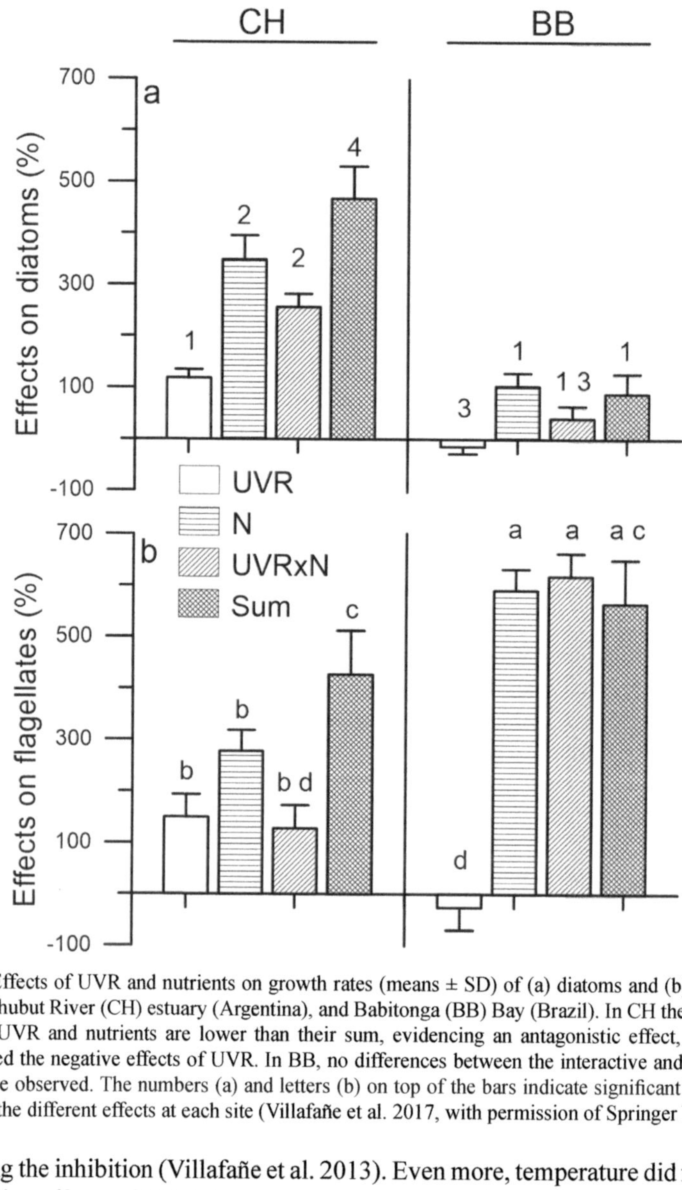

Fig. 7.3. Effects of UVR and nutrients on growth rates (means ± SD) of (a) diatoms and (b) flagellates from the Chubut River (CH) estuary (Argentina), and Babitonga (BB) Bay (Brazil). In CH the interactive effects of UVR and nutrients are lower than their sum, evidencing an antagonistic effect, as nutrient counteracted the negative effects of UVR. In BB, no differences between the interactive and the sum of effects were observed. The numbers (a) and letters (b) on top of the bars indicate significant differences among the different effects at each site (Villafañe et al. 2017, with permission of Springer Nature).

increasing the inhibition (Villafañe et al. 2013). Even more, temperature did not offset the negative effects of UV-B in mesocosms studies carried out with communities of the Beagle Channel, but it helped to shape its structure (Moreau et al. 2014, Hernando et al. 2018). The interactive effects of vertical mixing and UVR exposure have been addressed by Barbieri et al. (2002) who found that mixing enhanced UVR inhibition when 60% of the euphotic zone was mixed; however, this effect decreased and even reversed with increasing depth of the upper mixed layer. Thus an enhancement of photosynthesis was determined when 91% of the euphotic zone was mixed. Studies with spring phytoplankton communities of the Patagonian area also showed that simulated future conditions of higher nutrient inputs, UVR and lower pH shifted the community towards one dominated by large diatoms, with high growth rates, with little or no UVR inhibition of photosynthesis, thus increasing primary productivity (Villafañe et al. 2015). Similar increases in carbon incorporation were observed in a study conducted with summer communities in the same area under conditions of increased nutrients and lower pH (Durán et al. 2016). In another study evaluating the combined impacts of increased amounts of dissolved organic matter, nutrients and lower pH, it was found that the growth rates of phytoplankton significantly increased in these simulated future conditions, probably due to the nutrient addition and higher acidity (Villafañe et al. 2018). The highest growth rates were determined in those treatments in which DOM was acting as an absorber of solar radiation; however, cells in which DOM acted as a source of nutrients were photosynthetically less efficient probably due to a direct and negative effect of DOM, and/or due to mixotrophy of nanoflagellates.

Antagonistic effects between UVR and temperature have been reported for some crab larvae species from Patagonia (Hernández Moresino and Helbling 2010) with an increase in UVR tolerance (i.e., increased survival rates) with increasing temperature. Antagonistic effects of UVR and nutrient inputs were also observed in male amphipods from Patagonian coasts. Thus, individuals exposed to UVR had lower food consumption rates when fed on low quality macroalgae (without nutrient addition) than when fed on high-quality diets (with nutrient addition) (Fig. 7.4a). This suggests that nutrient addition partially counteracted the negative effects of UVR in these organisms (Fig. 7.4b) (Valiñas et al. 2014).

Conclusion

In the above sections we reviewed the state of knowledge of the effects of increased nutrient inputs, temperature, ultraviolet radiation, and low pH/high pCO_2 on organisms of the lower trophic levels of the SW Atlantic coastal areas. Based on the bibliography available and cited in this chapter, we calculated the proportion of studies dealing with the individual effects of nutrients, UVR, temperature and pH/pCO_2, as well as on multiple interactions on macroalgae, phytoplankton, zooplankton and benthos (Fig. 7.5). Though this figure may not be complete, it is representative for the total; thus, although the percentage could slightly change, the patterns would be maintained.

Nutrient inputs together with UVR are the stressors that have been studied most in SW Atlantic coastal areas (ca. 34 % and 30%, respectively), followed by temperature (ca. 20%) and pH/pCO_2 (ca. 6%) (Fig. 7.5). This general pattern slightly diverges from that reported by Harley et al. (2006), which showed temperature as the most studied single stressor worldwide, followed by UVR and nutrients. In agreement with the results, Harley et al. (2006) reports effects of decreased pH/increased pCO_2 on aquatic organisms as the least studied. The few papers dealing with decreased pH/increased pCO_2 is in part explained by the fact that the studies on these stressors are (compared with others) relatively recent (i.e., a few decades). In fact, the amount of studies on multiple stressors in SW Atlantic coasts surpasses those performed on pH/pCO_2 as a

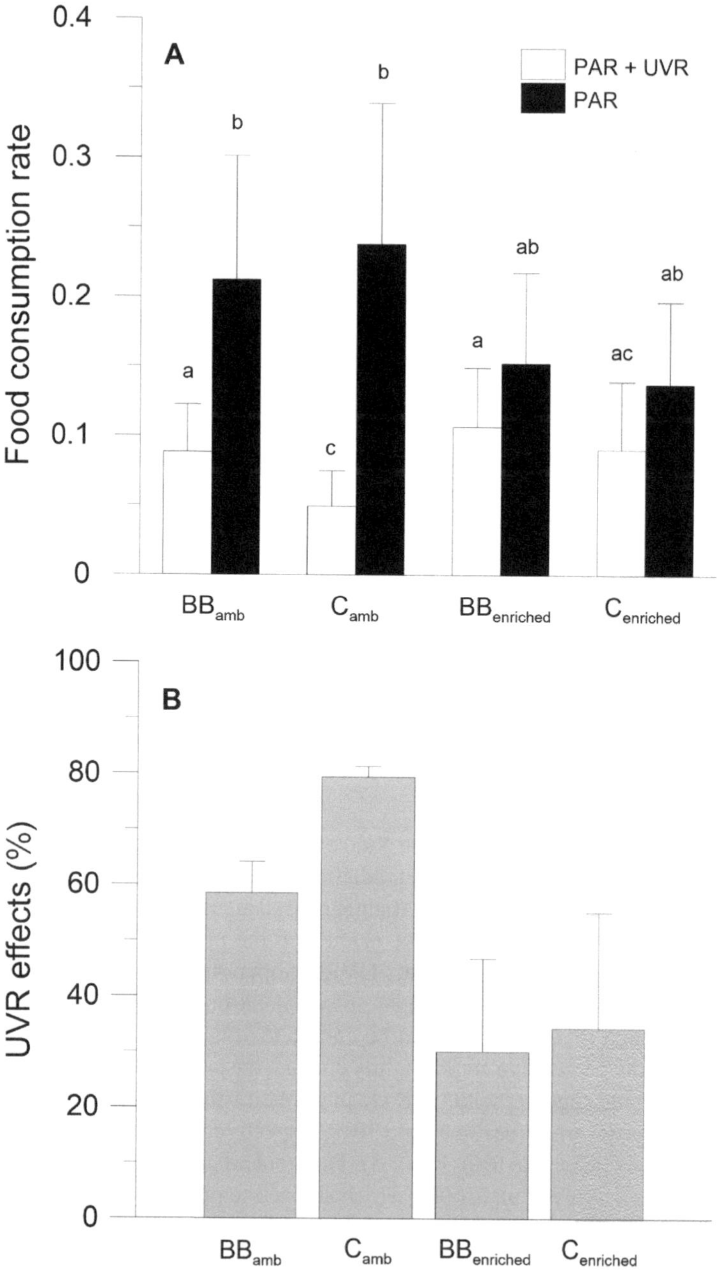

Fig. 7.4. (a) Daily mean food consumption rates (FCR, in (mg macroalgal tissue ingested (mg amphipod)$^{-1}$ day^{-1}) of males of the amphipod *Ampithoe valida* from Barrancas Blancas (BB) and Cangrejales (C) intertidal beaches exposed to PAR or PAR+UVR. (b) UVR effects (calculated as (P-PAB)/P)*100) on FCR. In both treatments were food was enriched with nutrients, the UVR effects decreased (antagonistic effect). The vertical lines indicate the standard deviation and the different letters indicate significant differences between nutrient and radiation treatments (after Valiñas et al. 2014).

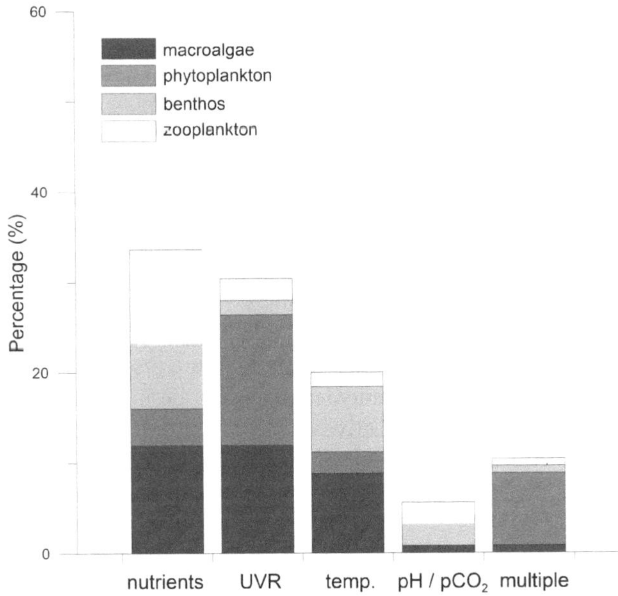

Fig. 7.5. Proportion (%) of studies on the single (increased nutrients, UVR, temperature and pH) and multiple effects on plankton and benthos from SW Atlantic coasts.

single factor, probably due to the growing interest in studying several global change stressors simultaneously, as it occurs in nature.

The proportion of studies of global change related variables on the different groups, i.e., macroalgae, phytoplankton, benthos, and zooplankton differs depending on the variable considered (Fig. 7.5). For instance, most studies on UVR effects have been performed on primary producers (i.e., macroalgae and phytoplankton) while less than 15% have been done on benthic and zooplanktonic organisms. Regarding the effects of increased temperature in SW Atlantic coasts, macroalgae and benthic organisms are among the most studied, while in both phytoplankton and zooplankton, temperature studies only account for 20% of the total. Nevertheless, studies on the effects of increased nutrients seems to be a bit more balanced; macroalgae (ca. 36%) have been the most studied group, followed by zooplankton (ca. 31%), benthos (ca. 21%), and phytoplankton (ca.12%). Most of the studies on the effects of increased nutrient inputs and temperature are descriptive, with only a few based on experiments. Thus, although the studies attribute the responses to eutrophication/increased temperature, it cannot be ruled out that other variables also contribute to the patterns observed. Evidently, more experimental studies are needed, that allow to manipulate the variable of interest and thus to corroborate that, for instance, changes in biomass, richness and biodiversity patterns can be effectively attributed to nutrient inputs.

Another important point is that considering the wide latitudinal range covered by SW Atlantic coasts, there is an imbalance in the amount of studies performed on the different groups, i.e., macroalgae, phytoplankton, benthos and zooplankton, depending on the stressor considered. For instance, increased temperature effects have been evaluated mainly on Brazilian coasts (ca. 52%), with macroalgae being the most studied group. In Argentina and Uruguay studies have been mainly done with benthic organisms (ca. 32 and 16%, respectively). On the contrary, UVR effects have been mostly evaluated by Argentinian researchers (ca. 64%), with phytoplankton being the most studied group. In Brazil and Uruguay the percentage of studies on UVR effects are significantly lower (ca. 25% and 11%) and have been mainly performed on macroalgae and phytoplankton (the latter only for Uruguay).

Efforts to manage and to conserve SW Atlantic coastal areas in the face of global change will require to improve the existing framework of knowledge in the region. This will be achieved, firstly, by performing more studies on the effects of global change stressors on particular organisms, in areas were the information is scarce (e.g., increase studies on UVR effects on phytoplankton in the sub-tropical and tropical areas of the SW Atlantic). Once this gap is filled, future research should address the global change impacts at the community level, by considering studies involving more than one trophic level and also interactive effects of multiple global change stressors. This will allow a better understanding of the functioning of aquatic ecosystems in SW Atlantic coasts, and their potential responses to a global change future scenario.

Acknowledgments

We would like to thank the Agencia Nacional de Promoción Científica y Tecnológica - ANPCyT (PICT 2013-0208 and PICT 2015-0462), Consejo Nacional de Investigaciones Científicas y Técnicas (CONICET) and Fundación Playa Unión for supporting this work. Also a special thanks to Cooperativa Eléctrica y de Servicios de Rawson for providing building' infrastructure for carrying out this work. This is Contribution N° 173 of Estación de Fotobiología Playa Unión.

References

Airoldi, L., D. Balata and M.W. Beck. 2008. The gray zone: Relationships between habitat loss and marine diversity and their applications in conservation. Journal of Experimental Marine Biology and Ecology 366: 8–15.

Aubriot, L., D. Conde, S. Bonilla and R. Sommaruga. 2004. Phosphate uptake behavior of natural phytoplankton during exposure to solar ultraviolet radiation in a shallow coastal lagoon. Marine Biology 144: 623–631.

Austin, J., N. Butchart and K.P. Shine. 1992. Possibility of an Arctic ozone hole in a double-CO_2 climate. Nature 360: 221–225.

Ayres-Ostrock, L.M. and E.M. Plastino. 2014. Effects of short-term exposure to ultraviolet-B radiation on photosynthesis and pigment content of red (wild types), greenish-brown, and green strains of *Gracilaria birdiae* (Gracilariales, Rhodophyta). Journal of Applied Phycology 26: 867–879.

Bandeira-Pedrosa, M.E., S.M.B. Pereira, Z.L. Bouzon and E.C. Oliveira. 2004. *Halimeda cuneata* (Bryopsidales, Chlorophyta), a new record for the Atlantic Ocean. Phycologia 43: 50–57.

Barbieri, E.S., V.E. Villafañe and E.W. Helbling. 2002. Experimental assessment of UV effects upon temperate marine phytoplankton when exposed to variable radiation regimes. Limnology and Oceanography 47: 1648–1655.

Barría de Cao, M.S., R. Pettigrosso, E. Parodi and R. Freije. 2003. Abundance and species composition of planktonic Ciliophora from the wastewater discharge zone in the Bahía Blanca Estuary, Argentina. Iheringia. Série Zoologia 93: 229–236.

Beardall, J., C. Sobrino and S. Stojkovic. 2009. Interactions between the impacts of ultraviolet radiation, elevated CO_2, and nutrient limitation on marine primary producers. Photochemical and Photobiological Sciences 8: 1257–1265.

Becherucci, M.E., E.N. Llanos, G.V. Garaffo and E.A. Vallarino. 2016a. Succession in an intertidal benthic community affected by untreated sewage effluent: A case of study in the SW Atlantic shore. Marine Pollution Bulletin 109: 95–103.

Becherucci, M.E., S. Lucerito, H.R. Benavides and E.A. Vallarino. 2016b. Assessing sewage impact in a South-West Atlantic rocky shore intertidal algal community Marine Pollution Bulletin 106: 388–394.

Beck, M.W., K.L. Heck, K.W. Able, D.L. Childers, D.B. Eggleston, B.M. Gillanders, B. Halpern, C.G. Hays, K. Hostino,, T.J. Minello, R.J. Orth, P. Sheridan and M.P. Weinstein. 2001. The role of nearshore ecosystems as fish and shellfish nurseries. BioScience 51: 633–641.

Berasategui, A.A., F. Biancalana, A. Fricke, M.D. Fernandez-Severini, R. Uibrig, M.S. Dutto, J. Marcovecchio, D. Calliari and M.S. Hoffmeyer. 2017. The impact of sewage effluents on the fecundity and survival of *Eurytemora americana* in a eutrophic estuary of Argentina. Estuarine, Coastal and Shelf Science. https://doi.org/10.1016/j.ecss.2017.1008.1034.

Bergamino, L., P. Muniz and O. Defeo. 2009. Effects of a freshwater canal discharge on polychaete assemblages inhabiting an exposed sandy beach in Uruguay. Ecological Indicators 9: 584–587.

Bermejo, P., E.W. Helbling, C. Durán-Romero, M.J. Cabrerizo and V.E. Villafañe. 2018. Abiotic control of phytoplankton blooms in temperate coastal marine ecosystems: A case study in the South Atlantic Ocean. Science of the Total Environment 612: 894–902.

Biancalana, F. and A.I. Torres. 2011. Variations of mesozooplankton composition in a eutrophicated semi-enclosed system (Encerrada Bay, Tierra del Fuego, Argentina). Brazilian Journal of Oceanography 59: 195–199.

Bibby, R., S. Widdicombe, H. Parry, J. Spicer and R. Pipe. 2008. Effects of ocean acidification on the immune response of the blue mussel *Mytilus edulis*. Aquatic Biology 2: 67–74.

Bischof, K. and F.S. Steinhoff. 2012. Impacts of stratospheric ozone depletion and solar UVB radiation on seaweeds. Seaweed Biology 219: 433–448.

Blaszkowski, C. and G.S. Moreira. 1986. Combined effects of temperature and salinity on the survival and duration of larval stages of *Pagurus criniticornis* (Dana) (Crustacea, Paguridae). Journal of Experimental Marine Biology and Ecology 103: 77–86.

Boschi, E.E., C.E. Fischbach and M.I. Iorio. 1992. Catalago ilustrado de los crustaceos estomatopodos y decapodos marinos de Argentina. Frente Maritimo 10: 7–94.

Bouzon, Z.L., F. Chow, C.S. Zitta, R.W. dos Santos, L.C.F. Ouriques,, M.R.d.L., Osorio, L.K.P., C. Gouveia, R.d.P. Martins, A. Latini, F. Ramlov, M. Maraschin and E.C. Schmidt. 2012. Effects of natural radiation, photosynthetically active radiation and artificial ultraviolet radiation-B on the chloroplast organization and metabolism of *Porphyra acanthophora* var. *brasiliensis* (Rhodophyta, Bangiales). Microscopy and Microanalysis 18: 1467–1479.

Boyer, E.W., C.L. Goodale, N.A. Jaworski and R.W. Howarth. 2002. Anthropogenic nitrogen sources and relationship to riverine nitrogen export in the north-eastern USA. Biogeochemistry 57/58: 137–169.

Breves-Ramos, A., H.P. Lavrado, A.O. Ribeiro Junqueira and S.H. Gonçalves da Silva. 2005. Succession in rocky intertidal benthic communities in areas with different pollution levels at Guanabara Bay (RJ-Brazil). Brazilian Archives of Biology and Technology 48: 951–965.

Browman, H.I., R.D. Vetter, C. Alonso Rodríguez, J.J. Cullen, F.R. Davis, E. Lynn and J.F. St-Pierre. 2003. Ultraviolet (280–400 nm) induced DNA damage in the eggs and larvae of *Calanus finmarchicus* G. (Copepoda) and atlantic cod (*Gadus morhua*). Photochemistry and Photobiology 77: 2–29.

Buma, A.G.J., P. Boelen and W.H. Jeffrey. 2003. UVR-induced DNA damage in aquatic organisms. pp. 291–327. *In*: Helbling, E.W. and H.E. Zagarese (eds.). UV Effects in Aquatic Organisms and Ecosystems. The Royal Society of Chemistry, Cambridge.

Buma, A.G.J., E.W. Helbling, M.K. de Boer and V.E. Villafañe. 2001. Patterns of DNA damage and photoinhibition in temperate South-Atlantic picophytoplankton exposed to solar ultraviolet radiation. Journal of Photochemistry and Photobiology, B: Biology 62: 9–18.

Byrne, M. 2011. Impact of ocean warming and ocean acidification on marine invertebrate life history stages: vulnerabilities and potential for persistence in a changing ocean. Oceanography and Marine Biology: An Annual Review 49: 1–42.

Cabrera, S., S. Bozzo and H. Fuenzalida. 1995. Variations in UV radiation in Chile. Journal of Photochemistry and Photobiology, B: Biology 28: 137–142.

Cabrerizo, M.J., P. Carrillo, V.E. Villafañe, J.M. Medina-Sáchez and E.W. Helbling. 2017. Increased nutrients from aeolian-dust and riverine origin decrease the CO_2-sink capacity of coastal South Atlantic waters under UVR exposure. Limnology and Oceanography. https://doi.org/10.1002/lno.10764.

Cassano, V., J.C. De-Pauloa, M.T. Fujii, B.A.P. da Gama and V.L. Teixeira. 2008. Sesquiterpenes from the introduced red seaweed *Laurencia caduciramulosa* (Rhodomelaceae, Ceramiales). Biochemical Systematics and Ecology 36: 223–226.

Celentano, E. and D. Defeo. 2016. Effects of climate on the mole crab *Emerita brasiliensis* on a dissipative beach in Uruguay. Marine Ecology Progress Series 552: 211–222.

Cloern, J.E. 2001. Our evolving conceptual model of the coastal eutrophication problem. Marine Ecology Progress Series 210: 223–253

Conde, D., L. Aubriot , S. Bonilla and R. Sommaruga. 2002. Marine intrusions in a coastal lagoon enhance the negative effect of solar UV radiation on phytoplankton photosynthetic rates. Marine Ecology Progress Series 240: 57–70.

Costanza, R., R. d'Arge, R. de Groot, S. Farber, M. Grasso, B. Hannon, K. Limburg, S. Naeem, R.V. O'Neill, J. Paruelo R.G. Raskin, P. Sutton and M. van den Belt. 1997. The value of the world's ecosystem services and natural capital. Nature 387: 253–260.

Cotovicz Jr., L.C., B.A. Knoppers, N. Brandini, D. Poirier, S.J. Costa Santos, R.C. Cordeiro and G. Abril. 2018. Predominance of phytoplankton-derived dissolved and particulate organic carbon in a highly eutrophic tropical coastal embayment (Guanabara Bay, Rio de Janeiro, Brazil). Biogeochemistry 137: 1–14.

Crain, C.M., K. Kroeker and B.S. Halpern. 2008. Interactive and cumulative effects of multiple human stressors in marine systems. Ecology Letters 11: 1304–1315.

D'Avanzo, C. and J.N. Kremer. 1994. Diel oxygen dynamics and anoxic events in a eutrophic estuary of Waquoit Bay, Massachusetts. Estuaries 17: 131–139.

Dadón, J.R. 2005. Changes in the intertidal community structure after a mass mortality event in sandy beaches of Argentina. Contributions to Zoology 74: 27–39.

de Faveri, C., F. Scherner, J. Farias, E.C. de Oliveira and P.A. Horta. 2010. Temporal changes in the seaweed flora in Southern Brazil and its potential causes. Pan-American Journal of Aquatic Sciences 5: 350–357.

de Oliveira, E.M., É.C. Schmidt, D.T. Pereira, Z.L. Bouzon and L.C. Ouriques. 2016. Effects of UV-B radiation on germlings of the red macroalga *Nemalion helminthoides* (Rhodophyta). Journal of Microscopy and Ultrastructure 4: 85–94.

Defeo, O. 2003. Marine invertebrate fisheries in sandy beaches: an overview. Journal of Coastal Research 35: 56–65.

Defeo, O. and A. de Alava.1995. Effects of human activities on long-term trends in sandy beach populations: the wedge clam *Donax hanleyanus* in Uruguay. Marine Ecology Progress Series 123: 73–82.

deYoung, B., M. Barange, G. Beaugrand, R. Harris, R.I. Perry, M. Scheffer and F. Werner. 2008. Regime shifts in marine ecosystems: detection, prediction and management. Trends in Ecology and Evolution 23: 402–409.

Diaz-Pulido, G., K.R.N. Anthony, D.I. Kline, S. Dove and O. Hoegh-Guldberg. 2012. Interactions between ocean acidification and warming on the mortality and dissolution of coralline algae. Journal of Phycology 48: 32–39.

Duarte, C. 1995. Submerged aquatic vegetation in relation to different nutrient regimes. OPHELIA 45: 87–112.

Duarte, C.M., I.E. Hendriks, T.S. Moore, Y.S. Olsen, A. Steckbauer, , L. Ramajo, J. Carstensen, J.A. Trotter and M. McCulloch. 2013. Is ocean acidification an open-ocean syndrome? Understanding anthropogenic impacts on seawater pH. Estuaries and Coasts 36: 221–236.

Dunlap, W.C. and Y. Yamamoto. 1995. Small-molecule antioxidants in marine organisms: antioxidant activity of mycosporine-glycine. Comparative Biochemistry and Physiology 112B: 105–114.

Dupont, S., J. Havenhand, L. Peck and M. Thorndyke. 2008. Near-future level of CO_2-driven ocean acidification radically affects larval survival and development in the brittlestar *Ophiothrix fragilis*. Marine Ecology Progress Series 373: 285–294.

Durán, C., J.M. Medina-Sánchez, G. Herrera and P. Carrillo. 2016. Changes in the phytoplankton-bacteria coupling triggered by joint action of UVR, nutrients, and warming in Mediterranean high-mountain lakes. Limnology and Oceanography 61: 413–429.

Dutto, M.S., G.A. Kopprio, M.S. Hoffmeyer, T.S. Alonso, M. Graeve and G. Kattner. 2014. Planktonic trophic interactions in a human-impacted estuary of Argentina: a fatty acid marker approach. Journal of Plankton Research 36: 776–787.

Dutto, M.S., M.C. López Abbate, F. Biancalana, A.A. Berasategui and M.S. Hoffmeyer. 2012. The impact of sewage on environmental quality and the mesozooplankton community in a highly eutrophic estuary in Argentina. ICES Journal of Marine Science 69: 399–409.

Elias, R., M.S. Rivero, J.R. Palacios and E.A. Vallarino. 2006. Sewage-induced disturbance on Polychaetes inhabiting intertidal mussel beds of *Brachidontes rodriguezii* off Mar del Plata (Southwestern Atlantic, Argentina). pp. 187–196. *In*: Sarda, R., G. San Martin, E. Lopez, D. Martin and D. George (eds.). Scientific Advances on Polychaete Research. Scientia Marina.

Elliott, M. and F. Dewailly. 1995. The structure and components of European estuarine fish assemblages. Netherland Journal of Aquatic Ecology 2-9: 397–417.

Fabry, V.J., B.A. Seibel, R.A. Feely, and J.C. Orr. 2008. Impacts of ocean acidification on marine fauna and ecosystem processes. ICES Journal of Marine Science. 65: 414–432.

Feng, Y., M.E. Warner, Y. Zhang, J. Sun, F.-X. Fu, J.M. Rose and D.A. Hutchins. 2008. Interactive effects of increased pCO_2, temperature and irradiance on the marine coccolithophore *Emiliania huxleyi* (Prymnesiophyceae). European Journal of Phycology 43: 87–98.

Fernández-Severini, M.D., S.E. Botte, M.S. Hoffmeyer and J.E. Marcovecchio. 2009. Spatial and temporal distribution of cadmium and copper in water and zooplankton in the Bahıa Blanca Estuary, Argentina. Estuarine Coastal and Shelf Science 85: 57–66.

Fields, P.A., J.B. Graham, R.J. Rosenblatt and G.N. Somero. 1993. Effects of expected global climate change on marine faunas. Trends in Ecology and Evolution 8: 361–367.

Figueiredo, D.A.L., P.C. Branco, D.A. dos Santo, A.K. Emerenciano, R.S. Iunes, J.C.S. Borges and J.R.M. Cunha da Silva. 2016. Ocean acidification affects parameters of immune response and extracellular pH in tropical sea urchins *Lytechinus variegatus* and *Echinometra luccunter*. Aquatic Toxicology 180: 84–94.

Figueroa, F.L., S. Salles, J. Aguilera, C. Jiménez, J. Mercado, B. Viñegla, A. Flores-Moya and M. Altamirano. 1997. Effects of solar radiation on photoinhibition and pigmentation in the red alga *Porphyra leucosticta*. Marine Ecology Progress Series 151: 81–90.

Fiori, S., V. Vidal-Martínez, R. Simá-Álvarez, R. Rodríguez-Canul, M.L. Aguirre-Macedo and O. Defeo. 2004. Field and laboratory observations of the mass mortality of the yellow clam *Mesodesma mactroides* in South America: The case of Isla del Jabalí, Argentina. Journal of Shellfish Research 23: 451–455.

Flores-Moya, A., D. Hanelt, F.L. Figueroa, M. Altamirano, B. Viñegla and S. Salles. 1999. Involvement of solar UV-B radiation in recovery of inhibited photosynthesis in the brown alga *Dictyota dichotoma* (Hudson) Lamouroux. Journal of Photochemistry and Photobiology, B: Biology 49: 129–135.

Fong, P., K.E. Boyer and J.B. Zedler. 1998. Developing an indicator of nutrient enrichment in coastal estuaries and lagoons using tissue nitrogen content of the opportunistic alga, *Enteromorpha intestinalis* (L. Link). Journal of Experimental Marine Biology and Ecology 231: 63–79.

Fox, S.E., M. Teichberg, Y.S. Olsen, L.E. Heffner and I. Valiela. 2009. Restructuring of benthic communities in eutrophic estuaries: lower abundance of prey leads to trophic shifts from omnivory to grazing. Marine Ecology Progress Series 380: 43–57.

Fricke, A., T.C. Kihara,, G.A. Kopprio, and M. Hoppenrath. 2017. Anthropogenically driven habitat formation by a tube dwelling diatom on the Northern Patagonian Atlantic coast. Ecological Indicators 77: 8–13.

Fricke, A., G.A. Kopprio, D. Alemany, M. Gastaldi, M. Narvarte, E.R. Parodi, L.J. Lara, F. Hidalgo, A. Martínez, E.A. Sar, O. Iribarne and P. Martinetto. 2016. Changes in coastal benthic algae succession trajectories and assemblages under contrasting nutrient and grazer loads. Estuaries and Coasts 39: 462–477.

García-Corral, L.S., J. Holding, P. Carrillo-de-Albornoz, A. Steckbauer, M. Pérez-Lorenzo, N. Navarro, P. Serret, C.M. Duarte and S. Agustí. 2016. Effects of UVB radiation on net community production in the upper global ocean. Global Ecology and Biogeography. https://doi.org/10.1111/geb.12513.

Garcia, M.D. and M.S. Barría de Cao. 2018. Anthropogenic pollution along the coast of a temperate estuary: effects on tintinnid assemblages. Hydrobiologia 809: 201–219.

Garcia, M.D. and N. Bonel. 2014. Environmental modulation of the plankton community composition and size-structure along the eutrophic intertidal coast of the Río de la Plata estuary, Argentina. Journal of Limnology 73: 562–573.

Gattuso, J.P., M. Frankignoulle and R. Wollast. 1998. Carbon and carbonate metabolism in coastal aquatic ecosystems. Annual Review of Ecology and Systematics 29: 405–434.

Gouvêa, L.P., N. Schubert, C.D. Leal Martins, M. Sissini, R. Ramlov, E.R. de Oliveira Rodrigues, E. Oliveira Bastos, V. Carvalho Freire, M. Maraschin, J.C. Simonassi, D.A. Varela, D. Franco, V. Cassano, A.L. Fonseca, J. Bonomi Barufi and P.A. Horta. 2017. Interactive effects of marine heatwaves and eutrophication on the ecophysiology of a widespread and ecologically important macroalga. Limnology and Oceanography 62: 2056–2075.

Guenther, M., M. Araújo, M. Flores-Monte, E. Gonzalez-Rodriguez and S. Neumann-Leitão. 2015. Eutrophication effects on phytoplankton size-fractioned biomass and production at a tropical estuary. Marine Pollution Bulletin 91: 537–547.

Guinder, V.A., C.A. Popovich, J.C.Molinero and J. Marcovecchio. 2013. Phytoplankton summer bloom dynamics in the Bahía Blanca Estuary in relation to changing environmental conditions. Continental Shelf Research 52: 150–158.

Guinder, V.A., C.A. Popovich, J.C. Molinero and G.M.E. Perillo. 2010. Long-term changes in phytoplankton phenology and community structure in the Bahía Blanca Estuary, Argentina. Marine Biology 157: 2703–2716.

Gunderson, A.R., E.J. Armstrong and J.H. Stillman. 2016. Multiple stressors in a changing world: The need for an improved perspective on physiological responses to the dynamic marine environment. Annual Review of Marine Science 8: 357–378.

Häder, D.-P. and K. Gao. 2015. Interactions of anthropogenic stress factors on marine phytoplankton. Frontiers in Environmental Science 3: 10.3389/fenvs.2015.00014.

Häder, D.-P., E.W. Helbling, C.E. Williamson and R.C. Worrest. 2011. Effects of UV radiation on aquatic ecosystems and interactions with climate change. Photochemical and Photobiological Sciences 10: 242–260.

Häder, D.-P., P.R. Richter, V.E. Villafañe and E.W. Helbling. 2014a. Influence of light history on the photosynthetic and motility responses of *Gymnodinium chlorophorum* exposed to UVR and different temperatures. Journal of Photochemistry and Biology B: Biology 138: 273–281.

Häder, D.P., M. Lebert and E.W. Helbling. 2001. Effects of solar radiation on the Patagonian macroalgae *Enteromorpha linza* (L.) J. Agardh - Chlorophyceae. Journal of Photochemistry and Photobiology, B: Biology 62: 43–54.

Häder, D.P., V.E. Villafañe and E.W. Helbling. 2014b. Productivity of aquatic primary producers under global climate change. Photochemical and Photobiological Sciences 13: 1370–1392.

Halac, S.R., V.E.Villafañe, R.J. Gonçalves and E.W. Helbling. 2014. Photochemical responses of three marine phytoplankton species exposed to ultraviolet radiation and increased temperature: Role of photoprotective mechanisms. Journal of Photochemistry and Photobiology B: Biology 141: 217–227.

Hallock, P. and W. Schlager. 1986. Nutrient excess and the demise of coral reefs and carbonate platforms. Palaios 1: 389–398.

Halpern, B.S., S.Walbridge, K.A. Selkoe, C.V. Kappel, F. Micheli, C. D'Agrosa, J.F. Bruno, K.S. Casey, C. Ebert, H.E. Fox, R. Fujita, D. Heinemann, H.S. Lenihan, E.M.P. Madin, M.T. Perry, E.R. Selig, M. Spalding, R. Steneck and R. Watson. 2008. A global map of human impact on marine ecosystems. Science 319: 948–952.

Hancke, K., T.B. Hancke, L.M. Olsen, G. Johnsen and R.N. Glud. 2008. Temperature effects on microalgal photosynthesis-light responses measured by O_2 production, pulse-amplitude-modulated fluorescence, and ^{14}C assimilation. Journal of Phycology 44: 501–514.

Hanelt, D. 1996. Photoinhibition of photosynthesis in marine macroalgae. Scientia Marina 60: 243–248.

Harley, C.D.G., A.R. Hughes, K.M. Hultgre, B.G. Miner, C.J.B. Sorte, C.S. Thornber, L.F. Rodriguez, L. Tomanek and S.L. Williams. 2006. The impacts of climate change in coastal marine systems. Ecology Letters 9: 228–241.

Helbling, E.W., E.S. Barbieri, M.A. Marcoval, R.J. Gonçalves and V.E. Villafañe. 2005. Impact of solar ultraviolet radiation on marine phytoplankton of Patagonia, Argentina. Photochemistry and Photobiology 81: 807–818.

Helbling, E.W., A.G.J. Buma, P. Boelen, H.J. van der Strate, M.V. Fiorda Giordanino and V.E. Villafañe. 2011. Increase in Rubisco activity and gene expression due to elevated temperature partially counteracts ultraviolet radiation–induced photoinhibition in the marine diatom *Thalassiosira weissflogii*. Limnology and Oceanography 56: 1330–1342.

Helbling, E.W., A.G.J. Buma, M.K. de Boer and V.E. Villafañe. 2001. *In situ* impact of solar ultraviolet radiation on photosynthesis and DNA in temperate marine phytoplankton. Marine Ecology Progress Series 211: 43–49.

Helbling, E.W., A.G.J. Buma, W. Van de Poll, M.V. Fernández Zenoff and V.E. Villafañe. 2008. UVR-induced photosynthetic inhibition dominates over DNA damage in marine dinoflagellates exposed to fluctuating solar radiation regimes. Journal of Experimental Marine Biology and Ecology 365: 96–102.

Helbling, E.W., C.F. Menchi and V.E. Villafañe. 2002. Bioaccumulation and role of UV-absorbing compounds in two marine crustacean species from Patagonia, Argentina. Photochemical and Photobiological Sciences 1: 820–825.

Helbling, E.W., D.E. Pérez, C.D. Medina, M.G. Lagunas and V.E. Villafañe. 2010. Phytoplankton distribution and photosynthesis dynamics in the Chubut River estuary (Patagonia, Argentina) throughout tidal cycles. Limnology and Oceanography 55: 55–65.

Helbling, E.W., V.E. Villafañe, M.E. Ferrario and O. Holm-Hansen. 1992. Impact of natural ultraviolet radiation on rates of photosynthesis and on specific marine phytoplankton species. Marine Ecology Progress Series 80: 89–100.

Helbling, E.W. and H.E. Zagarese. 2003 UV effects in aquatic organisms and ecosystems. The Royal Society of Chemistry, Cambridge.

Hendriks, I.E., C.M. Duarte and M. Álvarez. 2010. Vulnerability of marine biodiversity to ocean acidification: a meta-analysis. Estuarine, Coastal and Shelf Science 86: 157–164.

Hernández Moresino, R.D., R.J. Gonçalves and E.W. Helbling. 2011. Sublethal effects of ultraviolet radiation on crab larvae of *Cyrtograpsus altimanus*. Journal of Experimental Marine Biology and Ecology 407: 363–369.

Hernández Moresino, R.D. and E.W. Helbling. 2010. Combined effects of UVR and temperature on the survival of crab larvae (Zoea I) from Patagonia: The role of UV-absorbing compounds. Marine Drugs 8: 1681–1698.

Hernando, M., I. Schloss, S. Roy and G. Ferreyra. 2006. Photoacclimation to long-term ultraviolet radiation exposure of natural sub-Antarctic phytoplankton communities: Fixed surface incubations versus mixed mesocosms. Photochemistry and Photobiology 82: 923–935.

Hernando, M.P., G. Malanga and G.A. Ferreyra. 2005. Oxidative stress and antioxidant defenses generated by solar UV in a sub-Antarctic marine phytoflagellate. Scientia Marina 68: 287–295.

Hernando, M.P., G.F. Malanga, G.O. Almandoz, I.R. Schloss and G.A. Ferreyra. 2018. Responses of Sub-Antartic marine phytoplankton to ozone decrease and increased temperature. *In*: Hoffmeyer, M., S., M.E. Sabatini, F. Brandini, D. Calliari and N. Santinelli (eds.). Plankton Ecology of Atlantic South America. From the Subtropical to the Subantarctic Realm. Springer International Publishing, USA.

Herrmann, M., D. Carstensen, S. Fischer, J. Laudien, P.E. Penchaszadeh and W.E. Arntz. 2009. Population structure, growth, and production of the wedge clam *Donax hanleyanus* (Bivalvia: Donacidae) from northern Argentinean beaches. Journal of Shellfish Research 28: 511–526.

Hochachka, P. and G. Somero. 2002 Biochemical Adaptation: Mechanism and Process in Physiological Evolution. Oxford University Press, New York, USA.

Horta, P.A. and E.C. Oliveira. 2000. Morphology and reproduction of *Anotrichium yagii* (Ceramiales, Rhodophyta)—a new invader seaweed in the American Atlantic? Phycologia 39: 390–394.

Horta, P.A., T. Vieira-Pinto, C.D.L. Martins, M.N. Sissini, F. Ramlov, C. Lhullier, F. Scherner, P.F. Sanches, J.N. Farias, E. Bastos, J.L. Bouzon, P. Munoz, E. Valduga, N.P. Arantes, M.B. Batista, P. Riul, R.S. Almeida, E. Paes, A. Fonseca, E.P. Schenkel, L. Rorig, Z. Bouzon, J.B. Barufi, P. Colepicolo, N. Yokoya, M.S. Copertino and E.C. de Oliveira. 2012. Evaluation of impacts of climate change and local stressors on the biotechnological potential of marine macroalgae: a brief theoretical discussion of likely scenarios. Revista Brasilera de Farmacognosia 22: 768–774.

Howarth, R.W., G. Billen, D. Swaney, A. Townsend, N. Jaworski, K. Lajtha, J.A. Downing, R. Elmgren, N. Caraco, T. Jordan, F. Berendse, , J. Freney, V. Kudeyarov, P. Murdoch and Z. Zhao-Liang. 1996. Regional nitrogen budget and riverine N and P fluxes for the drainages to the Atlantic Ocean: natural and human influences. Biogeochemistry 35: 181e226.

Huertas, I.E., M. Rouco, V. López-Rodas and E. Costas. 2011. Warming will affect phytoplankton differently: Evidence through a mechanistic approach. Proceeding of the Royal Society of London B. 278: 3534–3543.

Hylander, S. and L.-A. Hansson. 2010. Vertical migration mitigates UV effects on zooplankton community composition. Journal of Plankton Research 32: 971–980.

ICES. 2008. ICES Science plan (2009–13). International Council for the Exploration of the Sea. 14.

IPCC. 2013 Climate Change 2013. The Physical Science Basis. Cambridge University Press, New York, USA.

IPCC. 2014 Climate change 2014: Impacts, Adaptation, and Vulnerability. Cambridge University Press Cambridge and New York

Ismael, D., K. Anger and G.S. Moreira. 1997. Influence of temperature on larval survival, development, and respiration in *Chasmagnathus granulata* (Crustacea, Decapoda). Helgoländer Meeresuntersuchungen 51: 463–475.

Israel, A. and M. Hophy. 2002. Growth, photosynthetic properties and Rubisco activities and amounts of marine macroalgae grown under current and elevated seawater CO_2 concentrations. Global Change Biology 8: 831–840.

Johnson, V.R., B.D. Russell, K.E. Fabricius, C. Brownlee and J.M. Hall-Spencer. 2012. Temperate and tropical brown macroalgae thrive, despite decalcification, along natural CO_2 grandients. Global Change Biology 18: 2792–2803.

Kerr, J.B. and C.T. McElroy. 1993. Evidence for large upward trends of ultraviolet-B radiation linked to ozone depletion. Science 262: 1032–1034.

Kerr, R., L.C. da Cunha, R.K. Kikuchi, P.A. Horta, R.G. Ito, M.N. Müller, I.B. Orselli, J.M. Lencina-Avila, M.R. de Orte, L. Sordo, B.R. Pinheiro, F.K. Bonou, N. Schubert, E. Bergstrom and M.S. Copertino. 2016. The Western South Atlantic ocean in a high-CO_2 world: Current measurement capabilities and perspectives. Environmental Management 57: 740–752.

Kieber, D.J., B.M. Peake and N.M. Scully. 2003. Reactive oxygen species in aquatic ecosystems. pp. 251–288. *In*: Helbling, E.W. and H.E. Zagarese (eds.). UV Effects in Aquatic Organisms and Ecosystems. RSC, Cambridge

Kirk, J.T.O. 1994. Optics of UV-B radiation in natural waters. Archiv für Hydrobiologie 43: 1–16.

Korbee Peinado, N., R.T. Abdala Díaz, F.L. Figueroa and E.W. Helbling. 2004. Ammonium and UV radiation stimulate the accumulation of mycosporine like amino acids in *Porphyra columbina* (Rhodophyta) from Patagonia. Argentina. Journal of Phycology 40: 248–259.

Kozlowsky-Suzuki, B. and R. Bozelli. 2002. Experimental evidence of the effect of nutrient enrichment on the zooplankton in a Brazilian coastal lagoon. Brazilian Journal of Biology 62: 835–846.

Kurihara, H. 2008. Effects of CO_2-driven ocean acidification on the early developmental stages of invertebrates. Marine Ecology Progress Series 373: 275–284.

Lercari, D. and D. Defeo. 1999. Effects of freshwater discharge in sandy beach populations: the mole crab *Emerita brasiliensis* in Uruguay. Estuarine Coastal and Shelf Science 49: 457–468.

López Abatte, M.C., M.S. Barría de Cao, R.E. Pettigrosso, V.A. Guinder, M.S. Dutto, A.A. Berasategui, C.J. Chazarreta and M.S. Hoffmeyer. 2016. Seasonal changes in microzooplankton feeding behavior under varying eutrophication level in the Bahía Blanca estuary (SW Atlantic Ocean). Journal of Experimental Marine Biology and Ecology 481: 25–33.

López Abbate, M.C., J.C. Molinero, V.A. Guinder, M.S. Dutto, M.S. Barría de Cao, L.A. Ruiz Etcheverry, R.E. Pettigrosso, M.C. Carcedo and M.S. Hoffmeyer. 2015. Microplankton dynamics under heavy anthropogenic pressure. The case of the Bahía Blanca Estuary, southwestern Atlantic Ocean. Marine Pollution Bulletin 95: 305–314.

López Abbate, M.C., J.C. Molinero, V.A. Guinder, G.M.E. Perillo, R.H. Freije, U. Sommer, C.V. Spetter and J.E. Marcovecchio. 2017. Time-varying environmental control of phytoplankton in a changing estuarine system. Science of the Total Environment 609: 1390–1400.

López Gappa, J.J., A. Tablado and N.H. Magaldi. 1990. Influence of sewage pollution on a rocky intertidal community dominated by the mytilid *Brachidontes rodriguezi*. Marine Ecology Progress Series 63: 163–175.

Lotze, H.K., H.S. Lenihan, B.j. Bourque, R.H. Bradbury, R.G. Cooke, M.C. Kay, S.M. Kidwell, M.X. Kirby, C.H. Peterson and J.B.C. Jackson. 2006. Depletion, degradation, and recovery potential of estuaries and coastal seas. Science 312: 1806–1809.

Lozoya, J.P. and O. Defeo. 2006. Effects of a freshwater canal discharge on an ovoviviparous isopod inhabiting an exposed sandy beach. Marine and Freshwater Research 57: 421–428.

Macchiavello, J., E.J. Paula and E.C. Oliveira. 1998. Growth rate responses of five commercial strains of *Gracilaria* (Rhodophyta, Gracilariales) to temperature and light. Journal of the World Aquaculture Society 29: 259–265.

Madronich, S., R.L. McKenzie, L.O. Björn and M.M. Caldwell. 1998. Changes in biologically active ultraviolet radiation reaching the Earth's surface. Journal of Photochemistry and Biology B: Biology 46: 5–19.

Malvé, M.E., S. Gordillo and M.M. Rivadeneira. 2016. Connecting pH with body size in the marine gastropod *Trophon geversianus* in a latitudinal gradient along the south-western Atlantic coast. Journal of the Marine Biological Association of the UK. https://doi.org/10.1017/S0025315416001557.

Marcoval, M.A., J. Pan, A.C. Díaz, L. Espino, N.S. Arzoz and J.L. Fenucci. 2017. Dietary photoprotective compounds ameliorate UV tolerance in shrimp (*Pleoticus muelleri*) through induction of antioxidant activity. Journal of the World Aquaculture Society. https://doi.org/10.1111/jwas.12482.

Marcoval, M.A., V.E. Villafañe and E.W. Helbling. 2008. Combined effects of solar ultraviolet radiation and nutrients addition on growth, biomass and taxonomic composition of coastal marine phytoplankton communities of Patagonia. Journal of Photochemistry and Photobiology, B: Biology 91: 157–166.

Marques, M., M.F. da Costa, O., M.M.I.d. and P.R.C Pinheiro. 2004. Water environments: anthropogenic pressures and ecosystem changes in the Atlantic drainage basins of Brazil. Ambio. 33: 68–77.

Martinetto, P., P. Daleo, M. Escapa, J. Alberti, J.P. Isacch, E. Fanjul, F. Botto, M.L. Piriz, G. Ponce, G. Casas and O. Iribarne. 2010. High abundance and diversity of consumers associated with eutrophic areas in a semi-desert macrotidal coastal ecosystem in Patagonia, Argentina. Estuarine, Coastal and Shelf Science 88: 357–364.

Martinetto, P., M. Teichberg, I. Valiela, D. Montemayor and O. Iribarne. 2011. Top-down and bottom-up regulation in a high nutrient–high herbivory coastal ecosystem. Marine Ecoloy Progress Series 432: 69–82.

Martinez, M., Rodríguez-Graña, L., Santos, L., Denicola, A. and Calliari, D. 2017. Oxidative damage and vital rates in the copepod *Acartia tonsa* in subtropical estuaries with contrasting anthropogenic impact. Journal of Experimental Marine Biology and Ecology 487: 79–85.

Martinez, M.L., A. Intralawan, G. Vázquez, O. Pérez-Maqueo, P. Sutton and R. Landgrave. 2007. The coasts of our world: Ecological, economic and social importance. Ecological Economics 63: 254–272.

Martins, C.D.L., N. Arantes, C. Faveri, M.B. Batista, E.C. Oliveira, P.R. Pagliosa, A.L. Fonseca, J.M.C. Nunes, F. Chow, S.B. Pereira and P.A. Horta. 2012. The impact of coastal urbanization on the structure of phytobenthic communities in southern Brazil. Marine Pollution Bulletin 64: 772–778.

McGlathery, K.J., K. Sundbäck and I.C. Anderson. 2007. Eutrophication in shallow coastal bays and lagoons: The role of plants in the coastal filter. Marine Ecology Progress Series 348: 1–18.

McKenzie, R., P.J. Aucamp, A. Bais, L.O. Björn, M. Ilyas and S. Madronich. 2011. Ozone depletion and climate change: impacts on UV radiation. Photochemical and Photobiological Sciences 10: 182–198.

Miranda, G.E.C., N.S. Yokoya and M.T. Fujii. 2012. Effects of temperature, salinity and irradiance on carposporeling development of *Hidropuntia caudata* (Gracilariales, Rhodophyta). Revista Brasilera de Farmacognosia 22: 818–824.

Miyashita, L.K., F.P. Brandini, J.E. Martinelli-Filho, L.F. Fernandes and R.M. Lopes. 2012. Comparison of zooplankton community structure between impacted and non-impacted areas of Paranaguá Bay Estuarine Complex, south Brazil. Journal of Natural History 46: 1557–1571.

Mohovic, B., S.M.F. Gianesella, I. Laurion and S. Roy. 2006. Ultraviolet B-photoprotection efficiency of mesocosm-enclosed natural phytoplankton communities from different latitudes: Rimouski (Canada) and Ubatuba (Brazil). Photochemistry and Photobiology 82: 952–961.

Moreau, S., B. Mostajir, G.O. Almandoz, S. Demers, M. Hernando, K. Lemarchand, M. Lionard, B. Mercier, S. Roy, I. Schloss, M. Thyssen and G.A. Ferreyra. 2014. Effects of enhanced temperature and ultraviolet B radiation on a natural plankton community of the Beagle Channel (Southern Argentina): A mesocosm study. Aquatic Microbial Ecology 72: 155–173.

Morgan, S.G. 1995. Life and death in the plankton: Larval mortality and adaptation. pp. 279–321. *In*: M.L. (ed.). Ecology of Marine Invertebrate Larvae. CRC Press, Boca Raton.

Neale, P.J., E.W. Helbling and H.E. Zagarese. 2003. Modulation of UVR exposure and effects by vertical mixing and advection. pp. 108–134. *In*: Helbling, E.W. and H.E. Zagarese (eds.). UV Effects in Aquatic Organisms and Ecosystems. Royal Society of Chemistry

Nixon, S.W. 1995. Coastal marine eutrophication: A definition, social causes, and future concerns. OPHELIA 41: 199–219.

Odebrecht, C., L. Rörig, V.T. Gracia and P.C. Araujo. 1995. Shellfish mortality and red tide event in southern Brazil. pp. 213–218. *In*: Lassus, P. (ed.). Harmful Marine Algal Blooms. Springer, New York.

Oliveira, E.C. and Y. Qi. 2003. Decadal changes in a polluted bay as seen from its seaweed flora: The case of Santos Bay in Brazil. Ambio. 32: 403–405.

Orfanidis, S., S. Haditonidis and I. Tsekos. 1996. Temperature requirements of *Scytosiphon lomentaria* (Scytosiphonales, Phaeophyta) from the Gulf of Thessaloniki, Greece, in relation to geographic distribution. Helgoländer Meeresuntersuchungen 50: 15–24.

Ortega, L., J.C. Castilla, M. Espino, C. Yamashiro and O. Defeo. 2012. Effects of fishing, market price, and climate on two South American clam species. Marine Ecology Progress Series 469: 71–85.

Ortega, L., E. Celentano, E. Delgado and O. Defeo. 2016. Climate change influences on abundance, individual size and body abnormalities in a sandy beach clam. Marine Ecology Progress Series 545: 203–213.

Osterling, M. and L. Pihl. 2001. Effects of filamentous green algal mats on benthic macrofaunal functional feeding groups. Journal of Experimental Marine Biology and Ecology 263: 159–183.

Pagliosa, P.R. and F.A. Rodrigues Barbosa. 2006. Assessing the environment–benthic fauna coupling in protected and urban areas of southern Brazil. Biological Conservation 129: 408–417.

Passarelli, M.C., A. Cesar, I. Riba and T.A. DelValls. 2017. Comparative evaluation of sea-urchin larval stage sensitivity to ocean acidification. Chemosphere 184: 224–234.

Pereira, D.T., E.C. Schmidt, Z.L. Bouzon and L.C. Ouriques. 2014. The effects of ultraviolet radiation-B response on the morphology, ultrastructure, and photosynthetic pigments of *Laurencia catarinensis* and *Palisada flagellifera* (Ceramiales, Rhodophyta): a comparative study. Journal of Applied Phycology 26: 2443–2452.

Pereira, L.C.C., J.A. Jimenez, M.L. Koening, F.F. Porto Neto, C. Medeiros and R.M. da Costa. 2005. Effect of coastline properties and wastewater on plankton composition and distribution in a stressed environment on the North Coast of Olinda-PE (Brazil). Brazilian Archives of Biology and Technology 48: 1013–1026.

Phillipart, C.J.M., H.M. van Aken, J.J. Beukema, O.G. Bos, G.C. Cadee and R. Dekker. 2003. Climate-related changes in recruitment of the bivalve *Macoma balthica*. Limnology and Oceanography 48: 2171–2185.

Pinto Silva, A., S. Neumann-Leitão, R. Schwamborn, L.M. de Oliveira and T. Almeida e Silva. 2004. Mesozooplankton of an impacted bay in North Eastern Brazil. Brazilian Archives of Biology and Technology 47: 485–493.

Piriz, M.L., M.C. Eyras and C.M. Rostagno. 2003. Changes in biomass and botanical composition of beach-cast seaweeds in a disturbed coastal area from Argentine Patagonia. Journal of Applied Phycology 15: 67–74.

Poloczanska, E.S., C.J. Brown, W.J. Sydeman, W. Kiessling, D.S. Schoeman, P.J. Moore, K. Brander, J.F. Bruno, L.B. Buckley, M.T. Burrows, C.M. Duarte, B.S. Halpern, J. Holding, C.V. Kappel, M.I. O'Connor, J.M. Pandolfi, C. Parmesan, F. Schwing, S.A. Thompson and A.J. Richardson. 2013. Global imprint of climate change on marine life. Nature Climate Change 3: 919.

Portugal, A.B., F.L. Carvalho, M.O. Soares, P.A. Horta and J.M. de Castro Nunes. 2017. Structure of macroalgal communities on tropical rocky shores inside and outside a marine protected area. Marine Environmental Research 130: 150–156.

Price, N.N., S.L. Hamilton, J.S. Tootell and J.E. Smith. 2011. Species-specific consequences of ocean acidification for the calcareous tropical green algae *Halimeda*. Marine Ecology Progress Series 440: 67–78.

Przeslawski, R., A.R. Davis and K. Benkendorff. 2005. Synergistic effects associated with climate change and the development of rocky shore molluscs. Global Change Biology 11: 515–522.

Rabalais, N.N., R.E. Turner, R.J. Díaz and D. Justic. 2009. Global change and eutrophication of coastal waters. ICES Journal of Marine Science 66: 1528–1537.

Raffaelli, D.G., J.A. Raven and L.J. Poole. 1998. Ecological impact of green macroalgal blooms. Oceanography and Marine Biology Annual Review 36: 97–125.

Ravishankara, A.R., J.S. Daniel and R.W. Portmann. 2009. Nitrous oxide (N_2O): The dominant ozone-depleting substance emitted in the 21st century. Science 326: 123–125.

Richter, P.R., R.J. Gonçalves, M.A. Marcoval, E.W. Helbling and D.-P. Häder. 2006. Diurnal changes in the composition of mycosporine-like amino acids (MAA) in *Corallina officinalis*. Trends in Photochemistry and Photobiology 11: 33–44.

Riebesell, U. 2004. Effects of CO_2 enrichment on marine phytoplankton. Journal of Oceanography 60: 719–729.

Riebesell, U., I. Zondervan, B. Rost, P.D. Tortell, R.E. Zeebe and F.M.M. Morel. 2000. Reduced calcification of marine plankton in response to increased atmospheric CO_2. Nature 407: 364–367.

Ries, J.B., A.L. Cohen and D.C. McCorkle. 2009. Marine calcifiers exhibit mixed responses to CO_2-induced ocean acidification. Geology 37: 1131–1134.

Roleda, M.Y., K. Zacher, A. Wulff, D. Hanelt and C. Wiencke. 2007. Photosynthetic performance, DNA damage and repair in gametes of the endemic Antarctic brown alga *Ascoseira mirabilis* exposed to ultraviolet radiation. Austral Ecology 32: 917–926.

Rosas-Navarro, A., G. Langer and P. Ziveri. 2016. Temperature affects the morphology and calcification of *Emiliania huxleyi* strains. Biogeosciences 13: 2913–2926.

Roy, S. 2000. Strategies for the minimization of UV-induced damage. pp. 177–205. *In*: De Mora, S.J., Demers, S. and Vernet, M. (eds.). The Effects of UV Radiation in the Marine Environment. Cambridge University Press, Cambridge.

Roy, S., B. Mohovic, S.M.F. Gianesella, I.R. Schloss, M.E. Ferrario and S. Demers. 2006. Effects of enhanced UV-B on pigment-based phytoplankton biomass and composition of mesocosm-enclosed natural marine communities from three latitudes. Photochemistry and Photobiology 82: 909–922.

Santi, L. and Tavares, M. 2009. Polychaete assemblage of an impacted estuary, Guanabara Bay, Rio de Janeiro, Brazil. Brazilian Journal of Oceanography 57: 287–303.

Scariot, L.A., Rover, T., Zitta, C.S., Horta, P.A., de Oliveira, E.C. and Bouzon, Z.L. 2013. Effects of UV-B radiation on *Gelidium floridanum* (Rhodophyta, Gelidiales): germination of tetraspores and early sporeling development. Journal of Applied Phycology 25: 537–544.

Scherner, F., P.A. Horta, E.C. Oliveira, J.C. Simonassi, J.M. Hall-Spencer, F. Chow, J.M.C. Nunes and S.M. Barreto Pereira. 2013. Coastal urbanization leads to remarkable seaweed species loss and community shifts along the SW Atlantic. Marine Pollution Bulletin 76: 106–115.

Scherner, F., C.M. Pereira, G. Duarte, P.A. Horta, C. Barreira e Castro, J. Bonomi Barufi and S.M. Barreto Pereira. 2016. Effects of ocean acidification and temperature increases on the photosynthesis of tropical reef calcified macroalgae. PLoS One 11: https://doi.org/10.1371/journal.pone.0154844.

Schmidt, É., R. dos Santos, C. de Faveri, P. Horta, R. de Paula Martins, A. Latini, F. Ramlov, M. Maraschin, and Z. Bouzon. 2012. Response of the agarophyte *Gelidium floridanum* after *in vitro* exposure to ultraviolet radiation B: changes in ultrastructure, pigments, and antioxidant systems. Journal of Applied Phycology 24: 1341–1352.

Schmidt, E.C., B. Pereira, C.L. Mansur Pontes, R. dos Santos, F. Scherner, P.A. Horta, R. de Paula Martins, A. Latini, M. Maraschin and Z.L. Bouzon. 2011. Alterations in architecture and metabolism induced by ultraviolet radiation-B in the carragenophyte *Chondracanthus teedei* (Rhodophyta, Gigartinales). Protoplasma 249: 353–367.

Schmitz, C., F. Ramlov, L.A. Ferreira de Lucena, V. Uarrota, M. Bernardes Batista, M.N. Sissini, I. Oliveira, B. Briani, C.D.L. Martins, J.M. de Castro Nunes, L. Rörig, P.A. Horta, F.L. Figueroa, N. Korbee, M. Maraschin and J.B. Barufi. 2018. UVR and PAR absorbing compounds of marine brown macroalgae along a latitudinal gradient of the Brazilian coast. Journal of Photochemistry and Photobiology, B: Biology 178: 165–174.

Shick, J.M. and W.C. Dunlap. 2002. Mycosporine-like amino acids and related gadusols: Biosynthesis, accumulation, and UV-protective functions in aquatic organisms. Annual Review of Physiology 64: 223–262.

Simioni, C., E.C. Schmidt, M.R. Felix, L.K. Polo, T. Rover, M. Kreusch, D.T. Pereira, F. Chow, F. Ramlov, M. Maraschin and Z.L. Bouzon. 2014. Effects of ultraviolet radiation (UVA+UVB) on young gametophytes of *Gelidium floridanum*: growth rate, photosynthetic pigments, carotenoids, photosynthetic performance, and ultrastructure. Photochemistry and Photobiology 90: 1050–1060.

Smith, V.H. and D.W. Schindler. 2009. Eutrophication science: Where do we go from here? Trends in Ecology and Evolution 24: 201–207.

Souza, F.M., K.M. Brauko, P.C. Lana, P. Muniz and M.G. Camargo. 2013. The effect of urban sewage on benthic macrofauna: A multiple spatial scale approach. Marine Pollution Bulletin 67: 234–240.

Spalding, M.D., H.E. Fox, G.R. Allen, N. Davidson, Z.A. Ferdaña, M. Finlayson, B.S. Halpern, M.A. Jorge, A. Lombana, D. Lourie Kirsten, S.A. Martin, E. McManus, J. Molnar, C.A. Recchia, and J. Robertson. 2007. Marine ecoregions of the World: A bioregionalization of coastal and shelf areas. BioScience 57: 573–583.

Spinelli, M.L. 2013 Ecología del mesozooplancton (Appendicularia y Copepoda) en aguas costeras Norpatagónicas (42°–46°S): ciclo anual y relaciones tróficas. PhD thesis. Universidad de Buenos Aires.

Sudatti, D.B., M.T. Fujii, S.V. Rodrigues, A. Turra and R.C. Pereira. 2011. Effects of abiotic factors on growth and chemical defenses in cultivated clones of *Laurencia dendroidea* J. Agardh (Ceramiales, Rhodophyta). Marine Biology 158: 1439–1446.

Szalaj, D., M.R. De Orte, T.A. Goulding, I.D. Medeiros, T.A. DelValls and A. Cesar. 2017. The effects of ocean acidification and a carbon dioxide capture and storage leak on the early life stages of the marine mussel *Perna perna* (Linneaus, 1758) and metal bioavailability. Environmental Science and Pollution Research 24: 765–781.

Talmage, S.C. and C.J. Gobler. 2009. The effects of elevated carbon dioxide concentrations on the metamorphosis, size, and survival of larval hard clams (*Mercenaria mercenaria*), bay scallops

(*Argopecten irradians*), and Eastern oysters (*Crassostrea virginica*). Journal of Limnology and Oceanography 54: 2072–2080.

Taouil, A. and Y. Yoneshigue-Valentin. 2002. Alterações na composição florística das algas da Praia de Boa Viagem (Niterói, RJ). Revista Brasilera de Botanica 25: 405–412.

Teichberg, M., S.E. Fox, I.S. Olsen, I. Valiela, P. Martinetto, O. Iribarne, E.Y. Muto, M.A.V. Petti, T.N. Corbisier, M. Soto-Jiménez, F. Páez-Osuna, P. Castro, H. Freitas, A. Zitelli, M. Cardinaletti and D. Tagliapietra. 2010. Eutrophication and macroalgal blooms in temperate and tropical coastal waters: Nutrient enrichment experiments with *Ulva* spp. Global Change Biology 16: 2624–2637.

Thompson, G.A. and M.F. Sánchez De Bock. 2009. Influence of beach morphodynamics on the bivalve *Donax hanleyanus* and *Mesodesma mactroides* populations in Argentina. Marine Ecology 30: 198–211.

Turley, C. and J.P. Gattuso. 2012. Future biological and ecosystem impacts of ocean acidification and their socioeconomic-policy implications. Current Opinion in Environmental Sustainability 4: 278–286.

Turra, A., A. Perez, F. Lucena-Frédou, M. Muelbert, A.R. Rey, L. Schejter, J. Calcagno, E. Marschoff and B. Ferreira. 2016. South Atlantic Ocean. First global integrated marine assesment. United Nations.

UNEP. 2006. Marine and coastal ecosystems and human well-being: A synthesis report based on the findings of the Millennium Ecosystem Assessment.

UNEP. 2012. Environmental effects of ozone depletion and its interactions with climate change: progress report, 2011. Photochemical & Photobiological Sciences 11: 13–27.

Ungar, S. 2015. Ozone depletion. Wiley

Valiela, I. 2006 Global Coastal Change. Willey-Blackwell.

Valiela, I., J. McClelland, J. Hauxwell, P.J. Behr, D. Hersh and K. Foreman. 1997. Macroalgal blooms in shallow estuaries: Controls and ecophysiological and ecosystem consequence. Limnology and Oceanography 42: 1105–1118.

Valiñas, M.S., P. Bermejo, L. Galbán, L. Laborda, D.P. Häder, V. Villafañe and E.W. Helbling. 2014. Combined impact of ultraviolet radiation and increased nutrients supply: A test of the potential anthropogenic impacts on the benthic amphipod *Amphitoe valida* from Patagonian waters (Argentina). Frontiers in Environmental Science - Environmental Toxicology 2: 1–10.

Valiñas, M.S. and E.W. Helbling. 2015. Sex-dependent effects of ultraviolet radiation on the marine amphipod *Ampithoe valida* (Ampithoidae). Journal of Photochemistry and Photobiology B: Biology 147: 75–82.

Vallarino, E.A., E.A. Vallarino, M.S. Rivero, M.C. Gravina and R. Elías. 2002. The community-level response to sewage impact in intertidal mytilid beds of the Southwestern Atlantic, and the use of the Shannon index to assess pollution. Revista de Biología Marina y Oceanografía 37: 25–33.

Verbruggen, H., O.D. Clerck, T. Schils, W.H.C.F. Kooistra and E. Coppejans. 2005. Evolution and phylogeography of *Halimeda* section Halimeda (Bryopsidales, Chlorophyta). Molecular Phylogenetics and Evolution 37: 789–803.

Villafañe, V.E., A.T. Banaszak, S.D. Guendulain-García, S.M. Strauch, S.R. Halac and E.W. Helbling. 2013. Influence of seasonal variables associated with climate change on photochemical diurnal cycles of marine phytoplankton from Patagonia (Argentina). Limnology and Oceanography 58: 203–214.

Villafañe, V.E., E.S. Barbieri and E.W. Helbling. 2004a. Annual patterns of ultraviolet radiation effects on temperate marine phytoplankton off Patagonia, Argentina. Journal of Plankton Research 26: 167–174.

Villafañe, V.E., M.J. Cabrerizo, G.S. Erzinger, P. Bermejo, S.M. Strauch, M.S. Valiñas and E.W. Helbling. 2017. Photosynthesis and growth of temperate and sub-tropical estuarine phytoplankton in a scenario of nutrient enrichment under solar ultraviolet radiation exposure. Estuaries and Coasts 40: 842–855.

Villafañe, V.E., G.S. Erzinger, S.M. Strauch and E.W. Helbling. 2014. Photochemical activity of PSII of tropical phytoplankton communities of Southern Brazil exposed to solar radiation and nutrient addition. Journal of Experimental Marine Biology and Ecology 459: 199–207.

Villafañe, V.E., P.J. Janknegt, M. de Graaff, R.J.W. Visser, W.H. van de Poll, A.G.J. Buma and E.W. Helbling. 2008. UVR-induced photoinhibition of summer marine phytoplankton communities from Patagonia. Marine Biology 154: 1021–1029.

Villafañe, V.E., M.A. Marcoval and E.W. Helbling. 2004b. Photosynthesis versus irradiance characteristics in phytoplankton assemblages off Patagonia (Argentina): Temporal variability and solar UVR effects. Marine Ecology Progress Series 284: 23–34.

Villafañe, V.E., J. Paczkowska, A. Andersson, C. Durán Romero, M.S. Valiñas and E.W. Helbling. 2018. Dual role of DOM in a scenario of global change on photosynthesis and structure of coastal phytoplankton from the South Atlantic Ocean. Science of the Total Environment. In press.

Villafañe, V.E., K. Sundbäck, F.L. Figueroa and E.W. Helbling. 2003. Photosynthesis in the aquatic environment as affected by UVR. pp. 357–397. *In*: Helbling, E.W. and H.E. Zagarese (eds.). UV effects in aquatic organisms and ecosystems. Royal Society of Chemistry.

Villafañe, V.E., M.S. Valiñas, M.J. Cabrerizo and E.W. Helbling. 2015. Physio-ecological responses of Patagonian coastal marine phytoplankton in a scenario of global change: Role of acidification, nutrients and solar UVR. Marine Chemistry 177: 411–420.

Whalter, G.R., E. Post, P. Convey, A. Menzel, C. Parmesan, T.J.C. Beebee, J.M. Fromentin, O. Hoegh-Guldberg and F. Bairlein. 2002. Ecological responses to recent climate change. Nature. 416: 389–395.

Wilson, J.G. 2008. Adaptations to life in estuaries. pp. 166–180. *In*: Safran, P. (ed.). Encyclopedia of Life Support Systems (EOLSS). Eolss Publishers, Oxford, UK

Xu, K., K. Gao, V.E. Villafaña and E.W. Helbling. 2011. Decreased calcification affects photosynthetic responses of *Emiliania huxleyi* exposed to UV radiation and elevated temperature. Biogeosciences Discuss 8: 857–884.

Yokoya, N.S. and E.C. Oliveira. 1993. Effects of temperature and salinity on spore germination and sporeling development in South American agarophytes (Rhodophyta). Japanese Journal of Phycology 41: 283–293.

CHAPTER 8

Effects of Climate Change on Corals

Donat-P. Häder

Introduction

Coral reefs are limited to the tropical and subtropical zones of the oceans and cover only 0.1–0.5% of the ocean floor. However, they harbour about 25% of all marine species and almost a third of the marine fish species. The main distribution areas are the Gulf of Mexico, the Red Sea, Maldives, the islands between Southeast Asia and Australia, including Indonesia and the Philippines, and the Great Barrier reef along the North East coast of Australia (Cairns 2001). Further habitats are scattered in the Pacific Ocean. Currently close to 800 coral species are known. The staghorn corals (genus *Acropora*) with some 370 species are the most important reef building corals throughout the world oceans (Wallace 1999). While most of the reef-building corals dwell close to the water surface, some scleractinian corals are found far below 40 m which are rarely visited by recreational SCUBA divers (Kahng and Maragos 2006). In addition to being of ecological primary importance, coral reefs generate large values from recreation, food production, and fisheries for human consumption. They also provide breakwaters protecting coastal areas from the open ocean; reportedly they reduce the wave energy by up to 97% and the wave height by up to 84% (Albright 2018). Losing the corals would affect 500 million people in tropical and subtropical coastal areas and result in a loss of $30 billion annually in goods and services.

Corals form colonies of polyps which excrete a common calcium carbonate skeleton. They grow heterotrophically by filtering organic matter from the water (Goreau et al. 1971). In addition, they harbour zooxanthellae, which are photosynthetic dinoflagellates (Rowan and Knowlton 1995) which give the corals their characteristic colors (Fig. 8.1). These symbionts contribute organic material resulting from photosynthetic carbon fixation to their hosts which helps in oligotrophic environments (Muscatine and Porter 1977). Reproduction can be asexual by fragmentation: pieces

Friedrich-Alexander University, Erlangen-Nürnberg, Neue Str. 9, 91096 Möhrendorf, Germany.
Email: donat@dphaeder.de

Fig. 8.1. Coral reef in the Red Sea near Sharm-el-Sheikh.

of the corals break off by the action of wind and waves or by fish or other animals or even by human activity, such as anchoring boats over a reef (Highsmith 1982). These fragments are displaced in the water and can regrow quickly after attaching to a different site. In addition, the polyps can form buds shedding small floating new polyps which can multiply by cloning before they settle down to form a new coral colony (Gateno and Rinkevich 2003, Gilmour 2004). In addition to asexual reproduction they multiply by sexual reproduction (Chornesky and Peters 1987, Szmant 1991). Sperms and eggs combine and after fertilization form young larvae which develop into swimming polyps; these can also grow to a polyp clone before settling and growing into a new colony. To increase chances of fertilization corals spawn synchronously: they release eggs and sperms after sunset on nights before the full moon in late summer (Harrison et al. 1984, Babcock et al. 1986). However, the success rate of sexual reproduction is still rather low: only one larva out of a million survives and forms a new colony (Albright 2018). Since most of the reproduction and growth depends on asexual mechanisms the genetic variability of corals is rather low and most organisms are members of the same clone (Mumby et al. 2011).

Coral reef communities depend on the interaction between organisms. Algae tend to overgrow corals. The sea urchin *Diadema antillarum* and fish graze on the algae (Hay 1984). Without these herbivores the corals would be overgrown. A reduction of the urchins would deem the corals to die. However, a delicate equilibrium has to be maintained. Other species pray on the corals: in the West Pacific, such as the crown-of-thorn starfish *Acanthaster planci*, the echinoderm *Culcita*, the gastropod *Drupella* and parrot fishes are a threat for corals (Done et al. 1996). *Acanthaster* is known to pray upon the corals themselves, feeding specifically on *Acropora* and other tabular corals (De'ath and Moran 1998). This species has been found to infest the Great Barrier Reef and deplete live corals on some reefs by 55–90% (Williams 1986). On the other hand damselfishes may protect corals because of their territorial behavior, which chases away puffer fishes and parrot fishes which pray on corals (Caley et al. 1997).

In recent years a number of catastrophic, climate change-related dying events have been observed in almost all coral-growing regions. Elevated temperatures induce the polyps to eject their zooxanthellae resulting in massive bleaching which eventually causes starvation due to the lack of food provided by their symbionts (Lesser 1997,

Jones et al. 1998). This phenomenon begins with an inhibition of CO_2 fixation in the dinoflagellates and subsequent oxidative stress. The first modern major global bleaching events were recorded in 1998 and 2010 (Berkelmans et al. 2004, Guest et al. 2012). Rising water temperatures were further elevated by massive El Niño events (Glynn et al. 2001). The following bleaching event (2014–17) was the longest and most extensive massive dying affecting 70% of the global coral reefs (Eakin et al. 2016). Two-thirds of the Great Barrier Reef were found to be killed or massively bleached (Hughes et al. 2017). Also this event coincided with a very strong El Niño phenomenon (Eghbert et al. 2017).

In addition to the tropical and subtropical, shallow-water and reef-building corals there is a second group of cold-water corals which are usually found in much deeper habitats around the world (Roberts 2009). Their ecology, biology, palaeontology, and geology are studied to a far smaller extent. Less than 300 publications can be found for the 20 years before 1996 but this number has more than doubled during the following 10 years.

The massive bleaching events and the numerous physiological reactions observed in coral reefs have been taken as a warning sign of an imminent disappearance of global coral reefs. In addition the rapid climate change does not leave room for substantial evolution and adaptation. However, recent evidence for responses to ocean acidification and geographical adaptation indicate that reefs are capable of greater temporal and spatial heterogeneity than currently assumed (Pandolfi et al. 2011). Improved understanding of responses to rapid climate change in the past, physiological reactions to interacting stress factors including temperature, acidification, and nutrients, is important to predict adaptation and acclimation and thus the future of global coral reefs.

Coral Reef Ecosystems and Services

Even though coral reefs cover a minute area of the global oceans they shelter about 25% of the marine organisms. They are characterized by a high diversity which rivals that of tropical rain forests (Connell 1978). Based on primary producers, such as phytoplankton and macroalgae, primary and secondary consumers dwell in the habitats which feed small fish which attract larger predators. The extensive calcium carbonate reef structure gives shelter and protection to many species including mollusks, worms, echinoderms and fish and provides a growth basis for sessile organisms, such as macroalgae. Being among the most productive and diverse ecosystems on earth, coral reefs are prone to dynamic fragility and structural complexity resulting in reduction of species diversity and loss of key functional groups of animals (Graham et al. 2006). Coral reefs can be subdivided into four basic types (Moberg and Folke 1999): (i) platform reefs are often found in the lagoons of atolls or barrier reefs, (ii) fringing reefs are located close to the shore separated by a shallow lagoon, (iii) barrier reefs are further offshore forming a wide, deep lagoon and (iv) atolls are often circular reefs with a central lagoon built on a volcano in the open ocean.

Coral reefs offer a plethora of goods and services including renewable resources, such as food for human consumption and inorganic reef material, such as coral blocks, rubble and sand for building. In addition they are exploited for raw materials for medicines, seaweeds for the production of agar, carrageenan and organic fertilizer as well as live fish and corals for the aquarium trade. The market for life organisms amounts to dozens of million dollars per year. In order to collect live fish, hundreds of tons of cyanide are distributed over coral reefs to stun reef dwelling fish. Clams and mother-of-pearl shells are collected for jewellery. For example, in 1978, 5,000 t of mother-of-pearl from the mollusk *Trochus niloticus* has been collected as curios (Craik et al. 1990). Other goods for the ornamental trade include the red coral (*Corallium rubrum*).

Overexploitation of corals, fish, and other members of the reef communities is a major threat for the ecosystem (Cesar 1996). Building material for the production of lime, mortar and cement are taken from the reefs. An estimated 20,000 m^3 of corals are mined every year in the Maldives to provide coral blocks, rubble and sand as the main construction material (Cesar 1996). Several reefs in Siberia, Saudi Arabia, USA, and Canada are supposed to contain mineral oil in porous limestone and are explored for exploitation (Birkeland 1997b).

Ecological services can be divided into physical structure, biotic, biogeochemical, information and social services and include production of seafood, recreation and protection of coastal regions (Moberg and Folke 1999). Coral reefs are among the most species-rich ecosystems. In addition, they provide shelter for spawning, nursing, breeding and feeding for numerous organisms and thus warrant the biological diversity and evolution for the future. About 60,000 plant and animal species have been described for coral reefs (Moberg and Folke 1999). Another major service of coral reefs is the protection of the coast from currents, waves and storms. Destruction of reefs leads to coast erosion which has to be countered by massive construction of artificial breakwaters (Weber 1993). Coral reefs contribute to recreation services: the fine white sand found in many tropical islands is the result of breaking down coral reef material.

Other services include export of organic material to pelagic food webs, nitrogen fixation and control of the Ca and CO_2 budgets. These services are compromised by anthropogenic activities resulting in loss of resilience and buffer capacity. Fish harvest from coral reef ecosystems amounts to about 10% of all marine fish caught for human consumption. Tens of millions of people in more than 100 countries use food from coral reefs as part of their staple diet. It has been calculated that 1 km^2 of a healthy reef could feed some 300 people with sufficient protein (Jennings and Polunin 1996). In addition to fish, crustaceans, mussels, sea cucumbers, and seaweeds are harvested (Birkeland 1997a).

Coral reefs are important for biogeochemical cycles. For example, they fix nitrogen based on cyanobacteria which is of importance for oligotrophic waters and is also beneficial for adjacent pelagic communities (Sorokin 1993). Near coastal reefs may receive nutrients by terrestrial run-off. Because of the high productivity coral reefs are important sinks for atmospheric carbon dioxide and thus in part mitigate the effects of global climate change (Gattuso et al. 1996). Of course, coral reefs play an important role in the global calcium cycle and are estimated to uptake more than half of the calcium which enters the sea (PAC 1978). In addition to the corals, calcified algae, worms and foraminifera bind calcium in the form of $CaCO_3$.

Recreation is another important service of coral reefs and can be calculated by income from tourism (Pendleton 1995). Driml calculated the value gained from tourism in the Great Barrier Reef World Heritage Area to be 682 million Australian dollars (Driml 1994) and the income earned by tourism in the Caribbean was estimated to amount to almost 9 billion US dollars in 1990, employing more than 350,000 people (Dixon et al. 1993). In addition, reefs provide jobs for fishermen: reefs damaged by overfishing and pollution in the Philippines resulted in the loss of 100,000 fishermens' jobs (McAllister 1988).

Effects of Increasing Temperatures

Corals are exposed to a plethora of environmental stress factors, but the most significant one is the rising temperature due to the anthropogenic global climate change. Even though the mean global temperature in the oceans has increased by 'only' 1°C, it has had dramatic effects on the biota (Fischetti 2013). Tropical sea surface temperatures have been reconstructed for the past four centuries from coral archives (Tierney et al. 2015). Analysis of the trends from data covering the Indian, western Pacific and western Atlantic Oceans indicate that these regions were cooling until modern warming started around 1830. Due to several volcanic eruptions in the early nineteenth century the Indo-Pacific ocean was characterized by exceptionally cool surface waters. Coral Sr:Ca ratios have been widely used to reconstruct past ocean temperatures, however, recent recordings indicated that the actual errors were underestimated and could not reliably reconstruct the amplitude or frequency of El Niño-Southern Oscillation events even when they were as large as ± 2°C (Alpert et al. 2016).

Excessive water temperatures are the main killer of coral reefs (Marshall et al. 2017). Since they are distributed in the tropics and subtropics they are already exposed to high temperatures near their permissive limits. The first major bleaching event in 1998 took the Maldives by surprise. After that a monitoring program has been created (Coral Bleaching Response Plan) which recorded 73% coral bleaching at the 71 sites under observation (Ibrahim et al. 2017). As indicated above, spawning in corals is synchronized to ensure that the gametes of both sexes have a high chance of meeting. A recent study looking into *Acropora* ssp. in 34 reefs in the Indian and Pacific Oceans to reveal which of the different environmental factors, such as photosynthetically available radiation, wind speed, water current speed, rainfall, sunset time and surface temperature, is the most important factor inducing spawning (Keith et al. 2016). The results indicate that a rapid increase in sea surface temperature is the main environmental clue to the initiation of spawning. These data are important to improve the fertilization efficiency of assisted evolution programs (see below). However, under higher temperatures (30 and 32°C) or elevated ammonium concentration larval development of *Diploria strigosa* was delayed or halted (Bassim and Sammarco 2003).

The evolutionary success of hermatypic (reef building) corals is due to the symbioses between the coral polyps and their dinoflagellate endosymbionts (zooxanthellae) (Muller-Parker et al. 2015). This is a true mutualistic relationship from which both partners benefit. While the algal partner provides organic material derived from its photosynthetic carbon fixation the polyp provides shelter and inorganic nutrients. Thus the stability of the symbiotic association could be used as a diagnostic indicator for the health of a coral reef under increasing environmental stressors.

Under excessive sea surface temperatures the symbiotic zooxanthellae are expelled from their host polyps, resulting in visible coral bleaching. The first indication for damage by heat stress on photosynthesis of the symbiotic dinoflagellates can be detected by pulse-amplitude-modulated (PAM) chlorophyll fluorescence and photorespirometry. When exposed to 34°C for 4 hours *Stylophora pistillata* showed a strong non-photochemical quenching (qN) of the fluorescence indicating a photoprotective dissipation of solar light energy in the form of heat (Jones et al. 1998). In addition, the oxygen production was significantly reduced and the optimal quantum yield (Fv/Fm) of photosystem II decreased. The quantum yield on the irradiated surfaces decreased more than on the shaded sides, indicating that excessive irradiation is an additional stress factor. The primary effect of heat damage in this coral has been proposed to be the reduced carbon fixation in the Calvin cycle. Extended exposure (7 hours) to elevated temperatures (30 and 32°C) induced expulsion rates of the zooxanthellae more than 1,000 times higher than in the controls kept at 27°C (Hoegh-Guldberg and Smith 1989). In two *Montastrea* species photosynthesis of the zooxanthellae was completely abolished by exposure of the corals to 30–36°C, while in *Siderastrea radians* and *Agarica agaricites* the decrease was less drastic (Warner et al. 1996).

Coral bleaching can be induced by short-term exposure (1–2 days) to temperatures 3–4°C above the normal summer ambient temperature or by long-term (several weeks) exposure at temperatures 1 or 2°C above the ambient summer temperatures (Jokiel and Coles 1990). Smaller temperature increases inhibit growth and reproduction. The mechanism of expelling zooxanthellae seems to be the release of host endoderm cells containing the symbionts as shown for the reef coral *Pocillopora damicornis* (Gates et al. 1992).

In addition to high temperatures, high solar irradiation causes bleaching of the photosynthetic pigments even when the algae are not expelled. This often occurs in shallow reef flats which may be even exposed to the air during low tides (Brown and Dunne 2016). This has been observed in colonies of *Goniastrea aspera* near Phuket, Thailand when low spring tides during the dry season coincide with high irradiances in the early afternoon. Solar UV radiation augments the bleaching induced by high irradiances and elevated temperatures (Jokiel and Coles 1990). This phenomenon can be observed in sensitive corals, such as *Pocillopora meandrina* to a depth of 20 m even though corals have a UV-absorbing pigment to protect them from short-wavelength solar radiation (Baird et al. 2009).

Effects of Ocean Acidification on Corals

Rapidly increasing atmospheric CO_2 concentrations due to fossil fuel burning, tropical deforestation and altered land usage is reflected by increasing CO_2 concentrations in oceanic surface waters (cf. Chapter 3, this volume) (Doney et al. 2009). Even though sea water is fairly well buffered, the pH has already decreased by 0.1 units with an increase of the proton concentration by 30% (Feely et al. 2004). Atmospheric CO_2

concentrations are predicted to reach 500 ppmv, a value which has not been recorded during the last 420,000 years when most marine organisms including corals evolved. According to the ICPP A1F1 scenario the pH will drop by 0.3–0.4 units by the end of the current century and the proton concentration increase by 100–150% (Houghton et al. 2001, Zeebe and Wolf-Gladrow 2001, Gattuso et al. 2015). These stress conditions together with a more than 2°C increase in water temperature will strongly affect coral reefs and possibly eliminate them (Hoegh-Guldberg et al. 2007).

Ocean acidification interferes with the calcification of many organisms, such as phytoplankton (cf. Chapter 5, this volume), macroalgae (cf. Chapter 9, this volume) and numerous zoological taxa, such as worms, molluscs and corals (cf. Chapter 15, this volume) which incorporate calcium carbonate to form exo- or endo-skeletons as a protection against predation and solar UV radiation (Andersson et al. 2008, Ries et al. 2009, Monteiro et al. 2016). In coral reefs ocean acidification has been found to reduce the calcification rate. In addition, it interferes with the relationship between the coral polyps and their symbiotic zooxanthellae compromising their productivity. An eight-week long mesocosm study using double, triple or quadruple CO_2 concentrations compared to current conditions was carried out using branching (*Acropora*) and massive (*Porites*) coral species (Anthony et al. 2008). The results confirmed that increased CO_2 concentrations contribute to bleaching, acting synergistically with warming by lowering the thermal threshold for bleaching. However, intermediate warm temperatures and slightly increased CO_2 concentrations resulted in a 30% increase in productivity in *Acropora*, while high CO_2 concentration stopped growth. Calcification in corals increases with increasing temperature up to an optimum and then decreases very fast (Kleypas et al. 2006). Therefore, the increasing calcification could mask the decrease due to ocean acidification. High nutrient concentrations counteract the effects of ocean acidification on calcification. Thus, evaluating the degree of calcification in the field is difficult because it depends on multiple stress factors including light, temperature, and nutrients. In addition, in reef flats the pH shows extreme diurnal variability because of antagonistic calcification and dissolution as well as photosynthesis and respiration.

Increasing atmospheric CO_2 concentrations resulting in decreased carbonate concentrations in the water also affect the growth of coral reefs. Low pH and low $CaCO_2$ saturation levels in the water result in microbioerosion and dissolution (DeCarlo et al. 2015). Corals possess carbonic anhydrases which play a major role in calcification; however, their capacity may be seriously challenged by ocean acidification (Zoccola et al. 2016). Also the feeding rate of the sclearactinian coral *Stylophora* was found to be reduced by ocean acidification (Houlbreque et al. 2015).

Corals show reduced skeletal growth under elevated CO_2 levels which also affects growth. Increased temperatures augment the negative effects of ocean acidification (Rodolfo-Metalpa et al. 2011). The effects of increased CO_2 concentrations were also measured in situ in shallow coastal sites where volcanic vents decrease the pH of the water (Hall-Spencer et al. 2008). One result of this study was the observation that many species resilient to naturally high CO_2 concentrations populated these areas, showing that these invasive species benefit from ocean acidification.

In contrast, under certain conditions corals can even calcify and grow faster as shown by a transplantation experiment along gradients of carbonate saturation state at Mediterranean CO_2 vents. Scleractinian corals can upregulate the pH at their calcification sites, so that the internal change is only half of that in the surrounding seawater as shown using the boron isotope systematic (McCulloch et al. 2012). Maintaining higher calcification rates require little addition energy costs.

The hard skeletons of stony corals are built from $CaCO_3$ using calcium and carbonate derived from water. However, the carbonate ion concentration in seawater is rather low, therefore, the corals need to use extra energy to increase the pH in isolated extracellular compartments to facilitate crystal growth (Cohen and Holcomb 2009). By this means bicarbonate ions are converted into carbonate ions which allows to produce $CaCO_3$ about 100 times faster than without the energy-dependent mechanism. The decreasing pH reduces the carbonate ion concentration which decreases the capability of corals for calcification or requires higher energy expenditure. Artificially feeding the corals heterotrophically or enhancing inorganic nutrient concentrations benefits corals at lower pH. However, under natural conditions ocean acidification is thought to decrease inorganic nutrient availability and phytoplankton concentrations in coral reef habitats so that the effect of ocean acidification may harm corals more than previously anticipated.

Other Environmental Factors Affecting Coral Reefs

Declining of coral reefs did not begin with climate change but centuries ago. Records going back a few thousand years including the trends in the major taxa of carnivores, herbivores, and reef-building species indicate that Atlantic reefs were degraded before those in the Red Sea and off Australia (Pandolfi et al. 2003). Large animals disappeared before small ones and reef-building species. Natural factors affecting coral reefs include hurricanes, volcanism, pests and predator outbreaks (Connell 1997).

Human impacts on coral reefs usually do not lead to a sustainable management. Coral mining, collecting organisms for the curio trade, destructive fishing using cyanide or dynamite, oil mining and uncontrolled tourism all affect coral reefs (Hawkins and Roberts 1994, Dulvy et al. 1995, Johannes and Riepen 1995). Increased pollution, increase of nutrients from terrestrial runoff and sediments are additional problems for coral reefs (Kuhlmann 1988). For example, polycyclic aromatic carbons are released from increased shipping and mineral oil exploration (Overmans et al. 2018). These detrimental effects are augmented by solar UV radiation due to phototoxicity: anthracene and phenanthrene were severely damaging to the larvae of the scleractinian coral *Acropora tenuis* in the presence of UV-A radiation which reduced the activity of superoxide dismutase.

Overfishing is another major problem (Pandolfi et al. 2005) and may lead to a loss in resilience (Bozec et al. 2016, Mora et al. 2016). Mass coral bleaching is often followed by the invasion of soft corals or cnidarians and especially macroalgae which result in a loss of diversity with diminished genetic diversity and loss of fish populations (Bouchon et al. 1992, Done 1992). The main reasons for the transition from corals to macroalgae are loss of grazers, increases in nutrient concentrations and inhibition of coral growth (Done 1992, Hughes 1994, Goreau et al. 1997). However, overfishing can be beneficial for a reef, for example, when the number of fish that predate on sea urchins diminish, the latter take over and graze on algae. In contrast, in Jamaica the urchin *Diadema* was found to be almost extinguished due to a pathogen, so that the algae had the upper hand supported by eutrophication (Goreau et al. 1997).

As indicated above, elevated temperatures often enhanced by a strong EL Niño are the main reason for coral bleaching. However, the delicate equilibrium between the corals and their symbiotic zooxanthellae can be disturbed by a number of natural and anthropogenic factors (Glynn and D'croz 1990, O'Connor et al. 2016). Other stress factors for coral bleaching are decreased salinity due to elevated runoff and urbanisation, distribution of heavy metals and other toxic substances as well as high solar UV radiation (Al-Naema et al. 2015, Bahr et al. 2015, Zhou et al. 2018).

Can the Corals be Saved?

The most recent global coral bleaching events have devastated more than 70% of the coral reefs (Albright 2018). Two-thirds of the Great Barrier Reef have been killed or severely bleached. It has been estimated that about half of the world's coral reefs have been lost during the last three decades and it has been estimated that only 10% will survive by mid-century (Walker 2017).

The ability of coral reefs to survive the predicted increases in water temperature resulting from global climate change will depend on their adaptation to elevated temperatures. A recent study of the widespread Indo-Pacific *Acropora millepora* has shown that under certain conditions adaptation is possible by acquiring different types of zooxanthellae (Berkelmans and Van Oppen 2006). In this study the dominant *Symbiodinium* type C was exchanged with the most thermally tolerant type D. Probably both types were already present in the host tissue but the more resistant one took over and provided an ecological advantage. Probably the only effective solution for the survival of corals is human assistance. One attempt are underwater nurseries: Researchers have installed underwater forests of plastic trees off the coast of the Florida Keys to grow corals (*Acropora cervicornis*) which will be later transplanted on to natural reefs (O'Donnell et al. 2017). In order to optimize the growth conditions several methods have been tested in a one-year study comparing the tree growth with that on concrete blocks. Colonies on trees grew up to three times faster than those reared on the concrete surfaces. During an extreme bleaching event colonies on the trees survived better than those on the blocks.

It is known that fragments of corals broken from a colony by storm or human interference can be displaced by the waves and start new colonies when finding a suitable support. This capability is exploited by coral nurseries around the world (Albright 2018). Some 90 species of the estimated 800 are currently farmed in nurseries. Tens of thousands of colonies are grown and transplanted onto degraded reefs by divers in the Caribbean and western Atlantic every year often funded by private donors, grants or official restoration projects (Lirman and Schopmeyer 2016). The Coral Reef Preservation and Rehabilitation Project has farmed and out-planted some 31,000 corals in Okinawa, Japan between 2012 and 2014 including 15 acroporid species (Omori et al. 2016). Later on the project concentrated on six species that were most tolerant to increased temperatures. More than 70% of the out-planted corals survived two years after transplantation. The growth rates were 3.4 cm for the first year and 5.6 cm in diameter for 2014, but even 19.9 cm for the six species planted in 2014. The methods were optimized by selecting successful species, choosing the size of the fragments and improving the processes of nursery farming. The price tag for producing, out-planting and monitoring a single colony was about US$20. Especially for slow-growing corals, such as the non branching *Echinopora lamellosa* and *Merulina scabricula* long times are required in the nurseries before the corals can be out-planted (dela Cruz et al. 2015). In addition, growth rates and survivorship are enhanced during the nursery phase. Small fragments with 1–3 cm^2 grow up to 50 times faster than colonies in the wild as shown with *Orbicella faveolata, Pseudodiploria clivosa* and *Porites lobata* (Forsman et al. 2015). When they are planted a few cm apart they can recombine and build a colony the size of a football field within a few months (Albright 2018). Twelve years ago this team succeeded in growing 600 corals per year, now they produce the same number in one afternoon and the projected numbers are 50,000 this year and 100,000 next year. The initial price tag of US$1,000 per colony has currently decreased to less than US$20 and the next goal to produce and plant one colony for US$2. This technique was also employed using two *Porites* species. However, recovery was significantly lower at a decreased pH (Hall et al. 2015). Most restoration projects are limited in size to a few ha. But the areas of reef degradation are on the order of hundred to thousand square kilometres. It has been estimated that it would require about US$300 billion to replant the full length of the Great Barrier Reef (2,300 km). However, this huge price tag may be worth it because it will create a large number of jobs, extensive fisheries and protect the coastline from devastating storms (Albright 2018).

Harvesting corals from the wild for aquarium trade has negative effects on coral reefs; therefore, it is preferable to produce cultured corals. The technique of growing corals from fragments is also employed to produce corals in aquaculture for aquarium hobbyists and public aquaria (Leal et al. 2017). One endeavour used *Duncanopsammia axifuga* waste fragments without polyps subsequently installed with one polyp to grow a clonal culture (Tagliafico et al. 2018). The genus *Sarcophyton* is well known to produce secondary metabolites with important biomedical applications. For this reason the corals were artificially reared using the asexual method of fragmentation (Santos 2015). The growth rates were improved by optimizing the quality and quantity of growth light.

Another approach of anthropogenically assisted coral survival is based on sexual reproduction. This endeavour takes advantage of the synchronized spawning of the gametes (Brady et al. 2009). In this method, eggs and sperms are collected underwater in tubes. Fertilization is brought about by bringing the gametes together. This procedure has the significant advantage that all eggs and sperm cells have a high chance of finding a sexual partner, while in nature the gametes are diluted in the vast ocean despite of the synchronized release. The resulting larvae are very small and run a high risk of being eaten in the wild. Another problem is that natural fertilization in the ocean can be inhibited by accidental oil spills (Negri and Heyward 2000). In *Acropora millepora* an inhibition by 25% was observed in the fertilization rate by exposure to very low concentrations (0.0721 mg/L) of total hydrocarbons. Also trace concentrations of heavy metals, such as copper, lead, zinc, cadmium, and nickel decreased the fertilization rate at concentrations of 15 to 40 µg/L (Reichelt-Brushett and Harrison 2005, Reichelt-Brushett and Michalek-Wagner 2005). Also reduced salinity decreases the success of fertilization and larval survival as shown in *Acropora millepora* and *Platygyra daedalea* (Scott et al. 2013) as well as by high inorganic nutrient concentrations (Lam et al. 2015).

An additional advantage of artificial fertilisation of corals is that gametes from different reefs can be combined, while in nature most fertilisation is from the same reef. For this reason many coral reefs are genetic clones. Cross fertilisation from different reefs broadens the genetic variation (Omori 2011) resulting in advantageous new features, such as enhanced temperature tolerance. The artificially reared larvae are kept under controlled conditions until they are large enough to be out-planted. Cross fertilization even works between species: in one-third of 42 species pairs from the genera *Acropora, Montipora,* and *Platygyra* isolated from Indo-Pacific reefs were cross fertilized with a success rate from 1–60% mainly in morphologically similar species (Willis et al. 1997). While in the wild the chance for a fertilized egg to grow into a new coral colony is about 0.0001%, artificial fertilization reaches almost 100% success rate (Albright 2018).

One goal of human-assisted evolution in corals is breeding clones with traits that allow them to survive the deteriorating growth conditions under climate change scenarios. Corals are subjected to sublethal stress conditions to adjust them to, for example, elevated temperatures. This epigenetic tuning may enable the corals to enhance their resilience and survival rates in the wild under higher temperatures. Another approach is selective breeding of corals and their symbionts for higher tolerance to environmental stress factors (Van Oppen and Gates 2006, van Oppen et al. 2015). However, this is a lengthy process since some corals may take up to a decade to fully mature and become sexually fertile.

If all these attempts fail there is another option. Eggs, sperm, larvae, and coral fragments are frozen in a bank and can be used for future attempts to regenerate coral reefs (Hagedorn and Carter 2016, Hagedorn et al. 2017). To date, 16 of the 800 coral species of the world have been banked and distributed in various countries.

The accumulated evidence indicates that recent effects of climate change, pollution and anthropogenic interference fasten the degradation. Reefs will not survive these stresses without protection. Genetic selection as well as aided evolution may help but these measures will only buy some time in view of the anthropogenically induced stresses which are bound to increase in the future.

References

Al-Naema, N., N. Deb, S. Saeed, J. Dupont and R. Ben-Hamadou. 2015. Acute orthogonal stress driven by temperature, salinity and light intensity on Qatari *Porites* photosynthesis and growth. Qscience Proceedings 17.

Albright, R. 2018. Can we save the corals? Scientific American 318: 38–45.

Alpert, A.E., A.L. Cohen, D.W. Oppo, T.M. DeCarlo, J.M. Gove and C.W. Young. 2016. Comparison of equatorial Pacific sea surface temperature variability and trends with Sr/Ca records from multiple corals. Paleoceanography 31: 252–265.

Andersson, A.J., F.T. Mackenzie and N.R. Bates. 2008. Life on the margin: Implications of ocean acidification on Mg-calcite, high latitude and cold-water marine calcifiers. Mar. Ecol. Prog. Ser. 373.

Anthony, K.R., D.I. Kline, G. Diaz-Pulido, S. Dove and O. Hoegh-Guldberg. 2008. Ocean acidification causes bleaching and productivity loss in coral reef builders. Proceedings of the National Academy of Sciences 105: 17442–17446.

Babcock, R., G. Bull, P.L. Harrison, A. Heyward, J. Oliver, C. Wallace and B. Willis. 1986. Synchronous spawnings of 105 scleractinian coral species on the Great Barrier Reef. Marine Biology 90: 379–394.
Bahr, K.D., P.L. Jokiel and K.S. Rodgers. 2015. The 2014 coral bleaching and freshwater flood events in Kāne'ohe Bay, Hawai'i. PeerJ 3: e1136.
Baird, A.H., R. Bhagooli, P.J. Ralph and S. Takahashi. 2009. Coral bleaching: the role of the host. Trends in Ecology & Evolution 24: 16–20.
Bassim, K. and P. Sammarco. 2003. Effects of temperature and ammonium on larval development and survivorship in a scleractinian coral (*Diploriastrigosa*). Marine Biology 142: 241–252.
Berkelmans, R., G. De'ath, S. Kininmonth and W.J. Skirving. 2004. A comparison of the 1998 and 2002 coral bleaching events on the Great Barrier Reef: spatial correlation, patterns, and predictions. Coral Reefs 23: 74–83.
Berkelmans, R. and M.J. Van Oppen. 2006. The role of zooxanthellae in the thermal tolerance of corals: a 'nugget of hope' for coral reefs in an era of climate change. Proceedings of the Royal Society of London B: Biological Sciences 273: 2305–2312.
Birkeland, C. 1997a. Life and death of coral reefs. Springer Science & Business Media. 1997b. Life and Death of Coral Reefs. Chapman and Hall, New York.
Bouchon, C., Y. Bouchon-Navaro and M. Louis. 1992. A first record of a *Sargassum* (Phaeophyta, algae) outbreak in a Caribbean coral reef ecosystem. Proceeding sof the 1st Gulf and Caribbean Fisheries Institute.
Bozec, Y.-M., S. O'Farrell, J.H. Bruggemann, B.E. Luckhurst and P.J. Mumby. 2016. Tradeoffs between fisheries harvest and the resilience of coral reefs. Proceedings of the National Academy of Sciences 113: 4536–4541.
Brady, A., J. Hilton and P. Vize. 2009. Coral spawn timing is a direct response to solar light cycles and is not an entrained circadian response. Coral Reefs 28: 677–680.
Brown, B.E. and R.P. Dunne. 2016. Coral Bleaching: the roles of sea temperature and solar radiation. Cheryl M. Woodley, Craig A. Downs, Andrew W. Bruckner, James W. Porter and Sylvia B. Galloway (eds): 266–283.
Cairns, S.D. 2001. Corals of the World. Science 292: 1492–1492.
Caley, M.J., M.H. Carr, M.A. Hixon, T.P. Hughes, G.P. Jones and B.A. Menge. 1997. Recruitment and the local dynamics of open marine populations. Annu. Rev. Ecol. Syst. 27: 477–500.
Cesar, H. 1996. Economic Analysis of Indonesian Coral Reefs. The World Bank.
Chornesky, E.A. and E.C. Peters. 1987. Sexual reproduction and colony growth in the scleractinian coral *Porites astreoides*. The Biological Bulletin 172: 161–177.
Cohen, A.L. and M. Holcomb. 2009. Why corals care about ocean acidification: uncovering the mechanism. Oceanography 22: 118–127.
Connell, J. 1997. Disturbance and recovery of coral assemblages. Coral Reefs 16: S101–S113.
Connell, J.H. 1978. Diversity in tropical rain forests and coral reefs. Science 199: 1302–1310.
Craik, W., R. Kenchington and G. Kelleher. 1990. Coral-reef management. pp. 453–467. *In*: Dubinsky, Z. (ed.). Ecosystems of the World: Coral Reefs. Elsevier, New York.
De'ath, G. and P. Moran. 1998. Factors affecting the behaviour of crown-of-thorns starfish (*Acanthaster planci* L.) on the Great Barrier Reef:: 2: Feeding preferences. Journal of Experimental Marine Biology and Ecology 220: 107–126.
DeCarlo, T.M., A.L. Cohen, H.C. Barkley, Q. Cobban, C. Young, K.E. Shamberger, R.E. Brainard and Y. Golbuu. 2015. Coral macrobioerosion is accelerated by ocean acidification and nutrients. Geology 43: 7–10.
dela Cruz, D.W., B. Rinkevich, E.D. Gomez and H.T. Yap. 2015. Assessing an abridged nursery phase for slow growing corals used in coral restoration. Ecological Engineering 84: 408–415.
Dixon, J.A., L. Fallon Scura and T. van't Hof. 1993. Meeting ecological and economic goals: marine parks in the Caribbean. Food and Agriculture Organization of the United Nations.
Done, T.J. 1992. Phase shifts in coral reef communities and their ecological significance. Hydrobiologia 247: 121–132.
Done, T.T., J.J. Ogden, W. Wiebe and B. Rosen. 1996. Biodiversity and ecosystem function of coral reefs. pp. 393–429. *In*: Global Biodiversity Assessment. John Wiley & Sons.
Doney, S.C., V.J. Fabry, R.A. Feely and J.A. Kleypas. 2009. Ocean acidification: The other CO_2 problem. Annual Reviews of Marine Sciences 1: 169–192.
Driml, S. 1994. Protection for profit: Economic and financial values of the Great Barrier Reef World Heritage Area and other protected areas. http://hdl.handle.net/11017/225.

Dulvy, N.K., D. Stanwell-Smith, W.R. Darwall and C.J. Horrill. 1995. Coral mining at Mafia Island, Tanzania: a management dilemma. Ambio, pp. 358–365.

Eakin, C., G. Liu, A. Gomez, J. De La Cour, S. Heron, W. Skirving, E. Geiger, K. Tirak and A. Strong. 2016. Global coral bleaching 2014–17: Status and an appeal for observations. Reef Encounter 31: 20–26.

Eghbert, E.A., O. Johan, C.E. Menkes, F. Niño, F. Birol, S. Ouillon and S. Andréfouët. 2017. Coral mortality induced by the 2015–16 El-Niño in Indonesia: the effect of rapid sea level fall. Biogeosciences 14: 817.

Feely, R.A., C.L. Sabine, K. Lee, W. Berelson, J. Kleypas, V.J. Fabry and F.J. Millero. 2004. Impact of anthropogenic CO_2 on the $CaCO_3$ system in the oceans. Science 305: 362–366.

Fischetti, M. 2013. Deep heat threatens marine life. Scientific American 308: 92.

Forsman, Z.H., C.A. Page, R.J. Toonen and D. Vaughan. 2015. Growing coral larger and faster: micro-colony-fusion as a strategy for accelerating coral cover. Peer J. 3: e1313.

Gateno, D. and B. Rinkevich. 2003. Coral polyp budding is probably promoted by a canalized ratio of two morphometric fields. Marine Biology 142: 971–973.

Gates, R.D., G. Baghdasarian and L. Muscatine. 1992. Temperature stress causes host cell detachment in symbiotic cnidarians: implications for coral bleaching. The Biological Bulletin 182: 324–332.

Gattuso, J.-P., A. Magnan, R. Billé, W. Cheung, E. Howes, F. Joos, D. Allemand, L. Bopp, S. Cooley and C. Eakin. 2015. Contrasting futures for ocean and society from different anthropogenic CO_2 emissions scenarios. Science 349: aac4722.

Gattuso, P.-P., M. Frankignoulle, S. Smith, J. Ware, R. Wollast, R.W. Buddemeier and H. Kayanne. 1996. Coral reefs and carbon dioxide. Science 271: 1298–1301.

Gilmour, J. 2004. Asexual budding in Fungiid corals. Coral Reefs 23: 595–595.

Glynn, P. and L. D'croz. 1990. Experimental evidence for high temperature stress as the cause of El Nino-coincident coral mortality. Coral Reefs 8: 181–191.

Glynn, P.W., J.L. Maté, A.C. Baker and M.O. Calderón. 2001. Coral bleaching and mortality in Panama and Ecuador during the 1997–98 El Niño–Southern Oscillation event: spatial/temporal patterns and comparisons with the 1982–83 event. Bulletin of Marine Science 69: 79–109.

Goreau, T., L. Daley, S. Ciappara, J. Brown, S. Bourke and K. Thacker. 1997. Community-based whole-watershed and coastal zone management in Jamaica. Proceedings of the Eighth International Coral Reef Symposium.

Goreau, T.F., N.I. Goreau and C. Yonge. 1971. Reef corals: autotrophs or heterotrophs? The Biological Bulletin 141: 247–260.

Graham, N.A., S.K. Wilson, S. Jennings, N.V. Polunin, J.P. Bijoux and J. Robinson. 2006. Dynamic fragility of oceanic coral reef ecosystems. Proceedings of the National Academy of Sciences 103: 8425–8429.

Guest, J.R., A.H. Baird, J.A. Maynard, E. Muttaqin, A.J. Edwards, S.J. Campbell, K. Yewdall, Y.A. Affendi and L.M. Chou. 2012. Contrasting patterns of coral bleaching susceptibility in 2010 suggest an adaptive response to thermal stress. PloS One 7: e33353.

Hagedorn, M. and V.L. Carter. 2016. Cryobiology: principles, species conservation and benefits for coral reefs. Reproduction, Fertility and Development 28: 1049–1060.

Hagedorn, M., V.L. Carter, E.M. Henley, M.J. Oppen, R. Hobbs and R.E. Spindler. 2017. Producing Coral Offspring with Cryopreserved Sperm: A Tool for Coral Reef Restoration. Scientific Reports 7: 14432.

Hall-Spencer, J.M., R. Rodolfo-Metalpa, S. Martin, E. Ransome, M. Fine, S.M. Turner, S.J. Rowley, D. Tedesco and M.-C. Buia. 2008. Volcanic carbon dioxide vents show ecosystem effects of ocean acidification. Nature 454: 96.

Hall, E.R., B.C. DeGroot and M. Fine. 2015. Lesion recovery of two scleractinian corals under low pH conditions: Implications for restoration efforts. Marine Pollution Bulletin 100: 321–326.

Harrison, P.L., R.C. Babcock, G.D. Bull, J.K. Oliver, C.C. Wallace and B.L. Willis. 1984. Mass spawning in tropical reef corals. Science 223: 1186–1189.

Hawkins, J.P. and C.M. Roberts. 1994. The growth of coastal tourism in the Red Sea: Present and future effects on coral reefs. Ambio 23: 503–508.

Hay, M.E. 1984. Patterns of fish and urchin grazing on Caribbean coral reefs: are previous results typical? Ecology 65: 446–454.

Highsmith, R.C. 1982. Reproduction by fragmentation in corals. Marine Ecology Progress Series, pp. 207–226.

Hoegh-Guldberg, O., P.J. Mumby, A.J. Hooten, R.S. Steneck, P. Greenfield, E. Gomez, C.D. Harvell, P.F. Sale, A.J. Edwards and K. Caldeira. 2007. Coral reefs under rapid climate change and ocean acidification. Science 318: 1737–1742.

Hoegh-Guldberg, O. and G.J. Smith. 1989. The effect of sudden changes in temperature, light and salinity on the population density and export of zooxanthellae from the reef corals *Stylophora pistillata* Esper and *Seriatopora hystrix* Dana. Journal of Experimental Marine Biology and Ecology 129: 279–303.

Houghton, J.T., Y. Ding, D.J. Griggs, M. Noguer, P.J. van der Linden, X.-. Dai, K. Maskell and C.-A. Johnson. 2001. Climate Change 2001: The Scientific Basis. Cambridge University Press. Cambridge, UK

Houlbrcque, F., S. Reynaud, C. Godinot, F. Oberhänsli, R. Rodolfo-Metalpa and C. Ferrier-Pagès. 2015. Ocean acidification reduces feeding rates in the scleractinian coral *Stylophora pistillata*. Limnology and Oceanography 60: 89–99.

Hughes, T.P. 1994. Catastrophes, phase shifts, and large-scale degradation of a Caribbean coral reef. Science 265: 1547–1551.

Hughes, T.P., J.T. Kerry, M. Álvarez-Noriega, J.G. Álvarez-Romero, K.D. Anderson, A.H. Baird, R.C. Babcock, M. Beger, D.R. Bellwood and R. Berkelmans. 2017. Global warming and recurrent mass bleaching of corals. Nature 543: 373.

Ibrahim, N., M. Mohamed, A. Basheer, H. Ismail, F. Nistharan, A. Schmidt, R. Naeem, A. Abdulla and G. Grimsditch. 2017. Status of Coral Bleaching in the Maldives in 2016. Centre, M. R., Malé, Maldives.

Jennings, S. and N.V. Polunin. 1996. Impacts of fishing on tropical reef ecosystems. Ambio: 44–49.

Johannes, R.E. and M. Riepen. 1995. Environmental economic and social implications of the live reef fish trade in Asia and the Western Pacific. Report to The Nature Conservancy and the Forum Fisheries Agency 83.

Jokiel, P. and S. Coles. 1990. Response of Hawaiian and other Indo-Pacific reef corals to elevated temperature. Coral Reefs 8: 155–162.

Jones, R.J., O. Hoegh-Guldberg, A.W. Larkum and U. Schreiber. 1998. Temperature-induced bleaching of corals begins with impairment of the CO_2 fixation mechanism in zooxanthellae. Plant, Cell & Environment 21: 1219–1230.

Kahng, S.E. and J.E. Maragos. 2006. The deepest, zooxanthellate scleractinian corals in the world? Coral Reefs 25: 254–254.

Keith, S.A., J.A. Maynard, A.J. Edwards, J.R. Guest, A.G. Bauman, R. Van Hooidonk, S.F. Heron, M.L. Berumen, J. Bouwmeester and S. Piromvaragorn. 2016. Coral mass spawning predicted by rapid seasonal rise in ocean temperature. Proc. R. Soc. B 283: 20160011.

Kleypas, J.A., R.A. Feely, V.J. Fabry, C. Langdon, C.L. Sabine and L.L. Robbins. 2006. Impacts of ocean acidification on coral reefs and other marine calcifiers: A guide for future research, Workshop held 18–20 April 2005, St. Petersburg, FL, sponsored by NSF, NOAA, and the U.S. Geological Survey: 88.

Kuhlmann, D. 1988. Sensitivity of Coral Reefs to Environmental Pollution.

Lam, E., A. Chui, C. Kwok, A. Ip, S. Chan, H. Leung, L. Yeung and P. Ang. 2015. High levels of inorganic nutrients affect fertilization kinetics, early development and settlement of the scleractinian coral *Platygyra acuta*. Coral Reefs 34: 837–848.

Leal, M.C., C. Ferrier-Pagès, D. Petersen and R. Osinga. 2017. Corals. Marine Ornamental Species Aquaculture: 406–436.

Lesser, M.P. 1997. Oxidative stress causes coral bleaching during exposure to elevated temperatures. Coral Reefs 16: 187–192.

Lirman, D. and S. Schopmeyer. 2016. Ecological solutions to reef degradation: optimizing coral reef restoration in the Caribbean and Western Atlantic. PeerJ 4: e2597.

Marshall, P., A.A. Abdulla, N. Ibrahim, R. Naeem and A. Basheer. 2017. Maldives Coral Bleaching.

McAllister, D.E. 1988. Environmental, economic and social costs of coral reef destruction in the Philippines. Galaxea 7: 161–178.

McCulloch, M., J. Falter, J. Trotter and P. Montagna. 2012. Coral resilience to ocean acidification and global warming through pH up-regulation. Nature Climate Change 2: 623–627.

Moberg, F. and C. Folke. 1999. Ecological goods and services of coral reef ecosystems. Ecological Economics 29: 215–233.

Monteiro, F.M., L.T. Bach, C. Brownlee, P. Bown, R.E. Rickaby, A.J. Poulton, T. Tyrrell, L. Beaufort, S. Dutkiewicz and S. Gibbs. 2016. Why marine phytoplankton calcify. Science Advances 2: e1501822.

Mora, C., N.A. Graham and M. Nyström. 2016. Ecological limitations to the resilience of coral reefs. Coral Reefs 35: 1271–1280.

Muller-Parker, G., C.F. D'elia and C.B. Cook. 2015. Interactions between corals and their symbiotic algae. pp. 99–116. *In*: Coral Reefs in the Anthropocene. Springer.

Mumby, P.J., I.A. Elliott, C.M. Eakin, W. Skirving, C.B. Paris, H.J. Edwards, S. Enríquez, R. Iglesias-Prieto, L.M. Cherubin and J.R. Stevens. 2011. Reserve design for uncertain responses of coral reefs to climate change. Ecology Letters 14: 132–140.

Muscatine, L. and J.W. Porter. 1977. Reef corals: Mutualistic symbioses adapted to nutrient-poor environments. Bioscience 27: 454–460.

Negri, A.P. and A.J. Heyward. 2000. Inhibition of fertilization and larval metamorphosis of the coral *Acropora millepora* (Ehrenberg, 1834) by petroleum products. Marine Pollution Bulletin 41: 420–427.

O'Connor, G., K. Cobb, H. Sayani, P. Grothe, A. Atwood, S. Stevenson, N. Hitt and J. Lynch-Stieglitz. 2016. The 2015/16 El Niño Event as Recorded in Central Tropical Pacific Corals: Temperature, Hydrology, and Ocean Circulation Influences. AGU Fall Meeting Abstracts.

O'Donnell, K.E., K.E. Lohr, E. Bartels and J.T. Patterson. 2017. Evaluation of staghorn coral (*Acropora cervicornis*, Lamarck 1816) production techniques in an ocean-based nursery with consideration of coral genotype. Journal of Experimental Marine Biology and Ecology 487: 53–58.

Omori, M. 2011. Degradation and restoration of coral reefs: Experience in Okinawa, Japan. Marine Biology Research 7: 3–12.

Omori, M., Y. Higa, C. Shinzato, Y. Zayasu, T. Nagata, R. Nakamura, A. Yokokura and S. Janadou. 2016. Development of active restoration methodologies for coral reefs using asexual reproduction in Okinawa, Japan. Proceedings of 13th International Coral Reef Symposium.

Overmans, S., M. Nordborg, R.D. Rua, D.L. Brinkman, A.P. Negri and S. Agustí. 2018. Phototoxic effects of PAH and UVA exposure on molecular responses and developmental success in coral larvae. Aquatic Toxicology.

PAC, S. 1978. Coral-reef area and the contributions of reefs to processes and resources of the world's oceans. Nature 273: 18.

Pandolfi, J.M., R.H. Bradbury, E. Sala, T.P. Hughes, K.A. Bjorndal, R.G. Cooke, D. McArdle, L. McClenachan, M.J. Newman and G. Paredes. 2003. Global trajectories of the long-term decline of coral reef ecosystems. Science 301: 955–958.

Pandolfi, J.M., S.R. Connolly, D.J. Marshall and A.L. Cohen. 2011. Projecting coral reef futures under global warming and ocean acidification. Science 333: 418–422.

Pandolfi, J.M., J.B. Jackson, N. Baron, R.H. Bradbury, H.M. Guzman, T.P. Hughes, C. Kappel, F. Micheli, J.C. Ogden and H.P. Possingham. 2005. Are US coral reefs on the slippery slope to slime? American Association for the Advancement of Science.

Pendleton, L.H. 1995. Valuing coral reef protection. Ocean & Coastal Management 26: 119–131.

Reichelt-Brushett, A.J. and P.L. Harrison. 2005. The effect of selected trace metals on the fertilization success of several scleractinian coral species. Coral Reefs 24: 524–534.

Reichelt-Brushett, A.J. and K. Michalek-Wagner. 2005. Effects of copper on the fertilization success of the soft coral *Lobophytum compactum*. Aquatic Toxicology 74: 280–284.

Ries, J.B., A.L. Cohen and D.C. McCorkle. 2009. Marine calcifiers exhibit mixed responses to CO_2-induced ocean acidification. Geology 37: 1131–1134.

Roberts, J.M. 2009. Cold-water Corals: The Biology and Geology of Deep-sea Coral Habitats. Cambridge University Press.

Rodolfo-Metalpa, R., F. Houlbrèque, É. Tambutté, F. Boisson, C. Baggini, F.P. Patti, R. Jeffree, M. Fine, A. Foggo and J. Gattuso. 2011. Coral and mollusc resistance to ocean acidification adversely affected by warming. Nature Climate Change 1: 308.

Rowan, R. and N. Knowlton. 1995. Intraspecific diversity and ecological zonation in coral-algal symbiosis. Proceedings of the National Academy of Sciences 92: 2850–2853.

Santos, S.L.M.d. 2015. Otimização da fragmentação e dos processos de fixação na reprodução assexuada do coral mole *Sarcophyton* sp. Maestrado, Universidade do Algarve.

Scott, A., P.L. Harrison and L.O. Brooks. 2013. Reduced salinity decreases the fertilization success and larval survival of two scleractinian coral species. Marine Environmental Research 92: 10–14.

1993. Coral Reef Ecology. Ecological Studies. Springer Verlag, Berlin.

Szmant, A.M. 1991. Sexual reproduction by the Caribbean reef corals *Montastrea annularis* and *M. cavernosa*. Marine Ecology Progress Series 74: 13–25.

Tagliafico, A., S. Rangel, B. Kelaher, S. Scheffers and L. Christidis. 2018. A new technique to increase polyp production in stony coral aquaculture using waste fragments without polyps. Aquaculture 484: 303–308.

Tierney, J.E., N.J. Abram, K.J. Anchukaitis, M.N. Evans, C. Giry, K.H. Kilbourne, C.P. Saenger, H.C. Wu and J. Zinke. 2015. Tropical sea surface temperatures for the past four centuries reconstructed from coral archives. Paleoceanography 30: 226–252.

Van Oppen, M.J. and R.D. Gates. 2006. Conservation genetics and the resilience of reef-building corals. Molecular Ecology 15: 3863–3883.

van Oppen, M.J., J.K. Oliver, H.M. Putnam and R.D. Gates. 2015. Building coral reef resilience through assisted evolution. Proceedings of the National Academy of Sciences 112: 2307–2313.

Walker, L. 2017. Impossible things: Science, denial and the Great Barrier Reef. Griffith REVIEW: 244.

Wallace, C. 1999. Staghorn corals of the world: a revision of the genus *Acropora*. CSIRO publishing.

Warner, M., W. Fitt and G. Schmidt. 1996. The effects of elevated temperature on the photosynthetic efficiency of zooxanthellae in hospite from four different species of reef coral: a novel approach. Plant, Cell & Environment 19: 291–299.

Weber, P. 1993. Reviving the coral reefs. pp. 42–60. *In*: Brown, L.R. (ed.). State of the World. W.W. Norton, New York.

Williams, D.M. 1986. Temporal variation in the structure of reef slope fish communities (central Great Barrier Reef): short-term effects of *Acanthaster planci* infestation. Marine Ecology Progress Series 28: 157–164.

Willis, B., R. Babcock, P.L. Harrison and C. Wallace. 1997. Experimental hybridization and breeding incompatibilities within the mating systems of mass spawning reef corals. Coral Reefs 16: S53–S65.

Zeebe, R.E. and D.A. Wolf-Gladrow. 2001. CO_2 in seawater: equilibrium, kinetics, isotopes. Gulf Professional Publishing.

Zhou, Z., X. Yu, J. Tang, Y. Wu, L. Wang and B. Huang. 2018. Systemic response of the stony coral *Pocillopora damicornis* against acute cadmium stress. Aquatic Toxicology 194: 132–139.

Zoccola, D., A. Innocenti, A. Bertucci, E. Tambutté, C.T. Supuran and S. Tambutté. 2016. Coral carbonic anhydrases: regulation by ocean acidification. Marine Drugs 14: 109.

CHAPTER 9

Responses of Calcifying Algae to Ocean Acidification

Kai Xu[1] and *Kunshan Gao*[2,*]

Introduction

The atmospheric CO_2 concentration increased from 280 ppmv at the beginning of industrial revolution to 400 ppmv today and it is predicted that CO_2 will increase to 800 ppmv by the end of this century (Feely et al. 2009, Monastersky 2013). About 25% of anthropogenic carbon emissions will be absorbed by the oceans, which will result in a decrease in surface seawater pH of 0.3 units from the current conditions by 2100. This process is commonly called 'ocean acidification (OA)' (Feely et al. 2009). The effects of future high CO_2 on seawater carbonate chemistry have gained a lot of attention especially on the ecological and biogeochemical effects on marine ecosystems and the carbon cycle. CO_2 is one of three major species of dissolved inorganic carbon (DIC) in the water. In coastal waters, the concentration of all the DIC species strongly depends on the respiration of organic carbon, tidal cycle and rain deposition, and the same holds for the pH (Koch et al. 2013, Shaw et al. 2013). However, in the open oceans, the carbonate chemistry is much more stable relative to that in coastal waters and variation of CO_2 is much larger than that of other DIC species (Gattuso et al. 2010). In addition, relative to the time scale of effects of increasing atmospheric CO_2 on seawater carbonate chemistry, the effects of tidal cycle and rain deposition are much shorter but stronger (see details below).

There is considerable evidence that future high atmospheric CO_2 will affect calcification by marine calcifying algae (Riebesell and Tortell 2011, Koch et al. 2013), but the prediction on ecological destiny of calcifying algae in future high CO_2 oceans is still largely uncertain. On the one hand, the acclimation to background DIC variations may result in that calcifying algae adapt to a moderate increase of CO_2. On the other

[1] Yindou Lu 43, Fisheries College, Jimei University, Xiamen, Fujian, 361021, China.
Email: kaixu@jmu.edu.cn

[2] Xiang-An Nanlu 4221, State Key Laboratory of Marine Environmental Science, Xiamen University, Xiamen, Fujian, 361102 China.

* Corresponding author: ksgao@jmu.edu.cn

hand, future high CO_2 may amplify the DIC variations (Shaw et al. 2013), that may have an even worse effect on benthic calcifying algae (Shaw et al. 2013, Lebrato et al. 2016). In addition, knowledge is limited on the mechanisms of calcification which complicates the understanding of effects of OA on calcification. Several review articles have recently summarized the interaction effects of OA with other environmental factors on calcifying algae (Riebesell and Tortell 2011, Koch et al. 2013, Beardall et al. 2014, Lebrato et al. 2016). The focus of this chapter is on the effects of ocean acidification on photosynthesis, calcification, and pH hemostasis of calcifying algae under a background of varying DIC concentrations in the environment.

Effects of OA on Calcification

Calcification

In the open oceans, the most ecologically important calcifying phytoplankton are coccolithophores (Rost and Riebesell 2004). In contrast, in coastal waters, the most common calcifying algae are benthic macroalgae, such as coralline red algae and calcareous green algae (Koch et al. 2013). All these algae play important roles in the marine carbon cycle due to their photosynthetic carbon fixation and calcification (Rost and Riebesell 2004, Fourqurean et al. 2012, Mazarrasa et al. 2015). The sites of $CaCO_3$ deposition of coccolithophores and the green alga *Halimeda* are in the intracellular space (Fig. 9.1A), while in coralline red algae deposition is extracellular (Fig. 9.1B). The common feature of these calcifying algae is that $CaCO_3$ precipitation is facilitated by high pH and high $CaCO_3$ saturation (Stanley 2008, Raven and Giordano 2009).

Unlike the DIC sources for photosynthesis, it is well accepted that HCO_3^- is the sole source for calcification in the coccolithophore *E. huxleyi* (Paasche 2002, Mackinder et al. 2010). Thus, *E. huxleyi* cells need to uptake and transport the HCO_3^- into the intracellular organelle, the coccolith vesicle (CV), and increase the pH of the CV by pumping out the proton to facilitate the deposition of $CaCO_3$ (Fig. 9.1A). Basically, photosynthesis provides the energy for the formation of CV and other calcification related organic components, and the deposition process of the $CaCO_3$ crystal. Previous studies assumed that photosynthesis may facilitate calcification by neutralizing the protons produced by calcification, and vice versa (Paasche 2002). However, knowledge is still limited on the mechanisms of cellular uptake and transport of HCO_3^- to facilitate calcification, and the detailed relationship between calcification and photosynthesis is still unclear (see details in following sections) (Brownlee et al. 2015). Hofmann et al. (2016) constructed a model which proposed the calcification mechanisms of coralline red algae; they suggested that the $CaCO_3$ deposition occurs at the cell wall where the pH of the cell surface is partly increased by photosynthetic HCO_3^- utilization and partly by a photosynthesis–independent proton pump. They also suggested that the inorganic carbon source for calcification is the CO_3^{2-} in seawater.

From a chemical point of view, obviously, ocean acidification will decrease the saturation of biologically important $CaCO_3$ skeletons of calcifying algae. That is the major reason that most calcifying algae tested so far show a decrease in calcification in response to OA (Riebesell and Tortell 2011, Koch et al. 2013). To date, only a few studies found positive effects of OA on calcification. Two studies on the

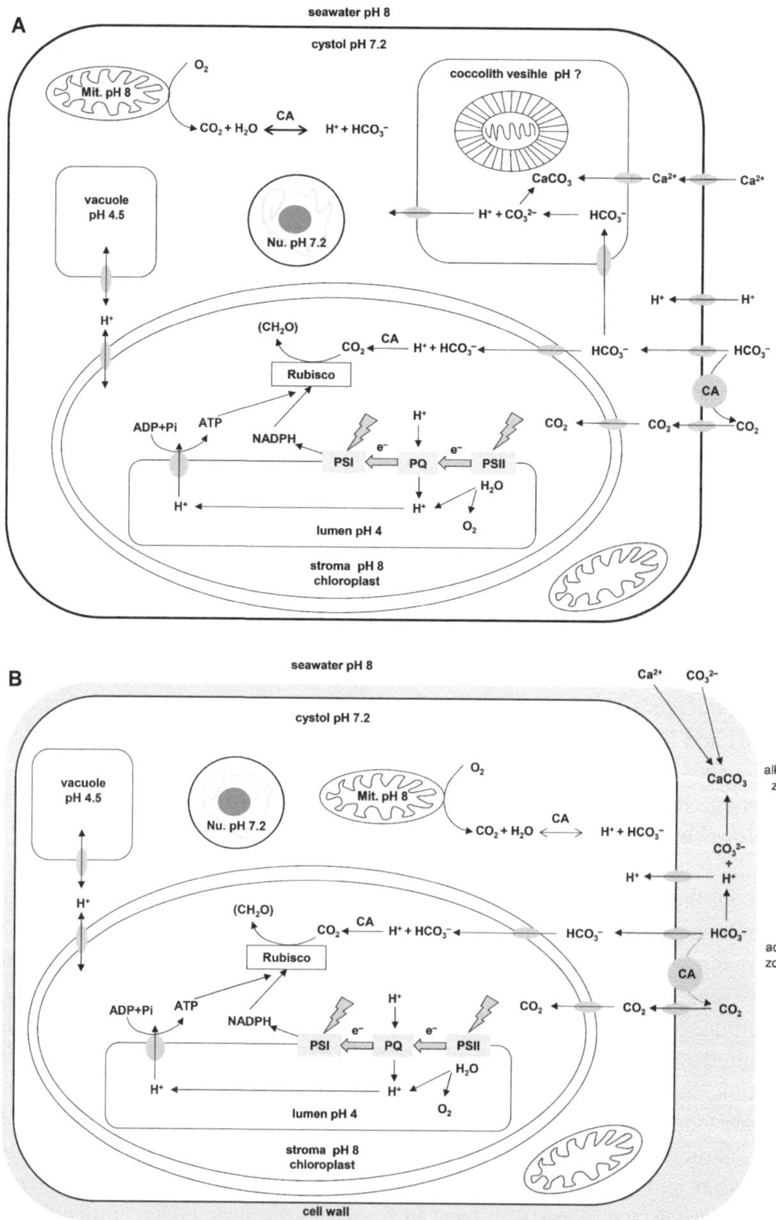

Fig. 9.1. The interaction between photosynthesis, calcification and pH hemostasis of two typical calcifying algae: coccolithophore (A) and coralline red algae (B). The major components of CCMs of these two algae are the same as in the model green alga *Chlamydomonas* (Wang et al. 2015). Both these algae absorb HCO_3^- and CO_2 from the surrounding seawater to fuel Rubisco. In the coccolithophore (A) the calcification occurs at the intracellular organelle coccolith vesicle which may share the same HCO_3^- transport system to move the HCO_3^- into the cytosol. The movement and conversion of DIC could help to adjust the pH hemostasis of the cytosol. In coralline red algae (B) the calcification happens at the cell wall where the pH is maintained by photosynthesis and photosynthesis–independent proton channels. Mit.: mitochondrion. Nu.: nucleus.

coccolithophore *E. huxleyi* appear to suggest that the calcification is stimulated by OA (Iglesias–Rodriguez et al. 2008, Shi et al. 2009). Beaufort et al. (2011) analyzed the relationship between coccolith mass and carbonate chemistry in the modern ocean and over the past 40,000 years, and found that the overall trend is that ocean acidification decreases calcification. However, these authors also reported that a heavily calcified *E. huxleyi* morphotype was found in modern high CO_2/low pH oceans. Smith et al. (2012) investigated the *E. huxleyi* population dynamics from summer to winter. The predominance of less calcified cells was shifted from lightly– to heavily–calcified cells along with the lowered pH and $CaCO_3$ saturation in the Bay of Biscay. Ries et al. (2009) also found that net calcification of coralline red algae and calcareous green algae increased under high levels of CO_2. Recently, Peach et al. (2017) reported that the calcification of six green *Halimeda* species was not significantly influenced by OA. Previous studies suggested that different effects of elevated CO_2 on calcification may attribute to different species or strains (Langer et al. 2006, Langer et al. 2009). Further studies are needed to explain the tolerance mechanisms of these calcifying algae to OA.

DIC Variations

The growth conditions of planktonic and benthic calcifying algae depend on the carbonate chemistry. In general, the daily and seasonal variations of pH and CO_2 in the open oceans are far less than in coastal waters (Dai et al. 2009, Takahashi et al. 2014). Assuming an atmospheric CO_2 increased to 800–1,000 ppmv by the end of this century, the pH would decrease by ~ 0.3 units and the seawater of the open oceans would still be $CaCO_3$–oversaturated. A previous study reported that the CO_2 in surface seawater in estuary areas could reach 7,000 ppmv due to aerobic respiration of organic carbon (Cai and Wang 1998, Zhai et al. 2005). Macroalgae inhabiting the intertidal zone periodically lose their seawater DIC sources and frequently suffer from low pH (down to 4.5) due to acid rain, a worldwide problem which is also a consequence of fossil fuel combustion (Vet et al. 2014). Some benthic and planktonic calcifying algae showed a tolerant ability to OA, which may be because they frequently experience DIC variations.

Coralline red algae are perennial calcifying algae frequently found in the intertidal zone, which means their perennial thallus could sustain a coverage of $CaCO_3$ in the intertidal zone under conditions with varying DIC concentrations and acid rain during periods of low tide. Gao et al. (2016) found that the coralline red algae could maintain their micro–environmental pH and photosynthesis by sacrificing their $CaCO_3$ skeleton, but the green alga *Ulva* could not survive at pH 3. Which means that acid rain could strongly decrease the macroalgae biodiversity in the intertidal zone. A recent study found that worldwide 24% of the benthic calcifiers currently live in $CaCO_3$–undersaturated seawater (Lebrato et al. 2016), but only a few studies analyzed the effects of large DIC variations on marine macroalgae (Gao et al. 2016, Li et al. 2017, Xu et al. 2017). In fact, a number of studies have suggested that some calcifiers have physiological mechanisms to compensate for undersaturation and to maintain their calcifying processes and internal pH (Lebrato et al. 2016). Thus, the highly fluctuating DIC and pH may underpin the responses of intertidal calcifying algae to OA.

However, increasing atmospheric CO_2 may amplify the DIC variations of future oceans (Shaw et al. 2013). It is predicted that future high atmospheric CO_2 will result in more benthic calcifiers (increase from the current 24% to 57% during the next 200 to 3000 years). Exposure to $CaCO_3$–undersaturated seawater (Lebrato et al. 2016), that will result in a situation where calcifiers are facing an even more serious challenge to maintain calcification in $CaCO_3$–undersaturated seawater. In addition, except for a few factors, most of the other factors in laboratory and mesocosm experiments during the past decades were identical to analyze the effects of target factors. But under natural conditions, the spatial–temporal scales of the changes of different biotic and abiotic factors are quite different (Levin 1992). The annual phytoplankton dynamics could be shaped by nutrient availability, temperature, light, and grazer pressure (Tilman et al. 1982, Calbet and Landry 2004, Boyd et al. 2010, Moore et al. 2013, Beardall et al. 2014). The interactive effects between OA and other environmental factors on calcification may mislead our conclusions especially concerning some factors closely related to each other (Boyd et al. 2010, Xu et al. 2014). Besides, the other environmental drivers may have even larger effects on the physiological responses of *E. huxleyi* than elevated CO_2 (Smith et al. 2012, Xu and Gao 2015, Feng et al. 2017). Overall, calcifying algae may be facing even worse effects of OA under a background of varying DIC. But it is still unclear whether the results of these studies could be used to predict the ecological destinies of the calcifying algae growing in a complex and changing environment.

OA Effects on Photosynthesis

CO_2 Concentrating Mechanisms

The major direct effects of OA on photosynthesis are that the CO_2 could be used as a substrate of Rubisco, the key enzyme of carbon assimilation. Like most algae, coccolithophores and calcifying macroalgae use CO_2 concentrating mechanisms (CCMs) to improve the CO_2 supply efficiency for photosynthetic carbon fixation (Giordano et al. 2005, Reinfelder 2011). The induction of CCMs could result in a higher intracellular DIC concentrations than in the environment (Raven et al. 2012). Previous studies suggested that photosynthesis of coccolithophores and most macroalgae is not saturated by DIC in current oceans (Rost and Riebesell 2004, Koch et al. 2013), which indicates that their photosynthesis is limited by the CCMs efficiency and DIC affinity. Thus, photosynthesis of these calcifying algae may benefit from the increasing CO_2 in the future oceans. To date, most of the relative studies focused on ecological and biogeochemical effects rather than on the mechanisms, although the latter could clarify how the CCMs of calcifying algae respond to OA.

A large number of studies investigated the responses of CCMs of algae to large CO_2 variations (from < 200 ppmv to $> 10,000$ ppmv) (Giordano et al. 2005, Baba and Shiraiwa 2012, Wang et al. 2015). Thereby two general principles were concluded: (i) The CCMs of some algae could quickly adjust to the changing environmental DIC on short–term time scales; for example, the daily adaptation of the green alga *Chlamydomonas* to natural CO_2 variation from ~ 400 ppmv to $> 10,000$ ppmv). (ii) The diversity of CCMs of different algal species indicates that the long–term adaptation to a specific DIC environment could irreversibly shape the CCMs. However, except for some model algae, the CCMs of most calcifying algae still lacks intensive studies.

Chlamydomonas, a model unicellular alga has three ways to assimilate and concentrate DIC in the cytosol (Wang et al. 2015). (i) The cells accumulate HCO_3^- directly through HCO_3^- transporters. Because the only substrate of Rubisco is CO_2, intracellular carbonic anhydrases (iCAs) convert the accumulated HCO_3^- to CO_2 at or near the Rubisco position. (ii) Extracellular carbonic anhydrases (eCAs) convert the seawater HCO_3^- to CO_2, which diffuses into the cytosol through membrane CO_2 channels. (iii) Diffusion of seawater CO_2 via CO_2 channels. Two kinds of channel proteins, rhesus protein and aquaporin, are supposed to take part in the diffusion of CO_2 across this membrane in photosynthetic organisms (Giordano et al. 2005, Wang et al. 2015). In addition, the CO_2 diffusion across the plasmalemma is under control of biological membrane systems (Endeward et al. 2014). Other eukaryotic algae mainly apply one or more of the above ways to uptake DIC (Giordano et al. 2005). These CCMs of *Chlamydomonas* are induced or depressed by different CO_2 concentrations which reflect the adaptation to the diverse natural habitats with greatly varying CO_2 concentrations (Baba and Shiraiwa 2012, Wang et al. 2015).

The CAs exist in all tested organisms, their most well–known function is to accelerate the inter–conversion of HCO_3^- and CO_2, this reaction will obviously change the proton concentration which is another function to maintain intracellular pH hemostasis (Krishnamurthy et al. 2008, Casey et al. 2010). Previous physiological and molecular studies confirmed that *E. huxleyi* contains extracellular CAs and HCO_3^- transporters (Fig. 9.1A), but their roles in photosynthetic carbon fixation are still under debate (Herfort et al. 2002, Bach et al. 2013, Stojkovic et al. 2013). To date, the DIC sources of the coccolithophore *E. huxleyi*, the ecologically most important calcifying phytoplankton, are still unclear because of the contradicting results (Paasche 2002, Holtz et al. 2015). Herfort et al. (2002) suggested that the calcifying *E. huxleyi* strain mainly absorbs HCO_3^- through transporters rather than via extracellular CAs, which indicates the presence of intracellular CAs. Bach et al. (2013) supported the idea that CO_2 in seawater is preferentially used in photosynthesis of the calcifying *E. huxleyi* strain, and the authors propose that cytosolic CA are absent. Obviously, even though the HCO_3^- is a less important DIC sources for photosynthesis, chloroplast stromal CA near the Rubisco is still needed to help the conversion of HCO_3^- to CO_2 (Wang et al. 2015). In addition, since the CA could greatly accelerate the inter–conversion between HCO_3^- and CO_2, if the CA does not exist in the cytoplasm, the uptake and cellular movement of inorganic carbon may disturb the pH hemostasis of the cytoplasm.

Stojkovic et al. (2013) also found that photosynthesis of a calcifying *E. huxleyi* strain prefers to use CO_2 in seawater, and a non–calcifying strain prefers to use HCO_3^-. This result indicates that the calcification, another DIC–requiring metabolic process, may contribute to photosynthetic utilization of inorganic carbon (Paasche 2002, Hofmann and Bischof 2014). The stop of calcification either by addition of an inhibitor or abruptly transferring the algae to Ca–free medium significantly increased photosynthesis of *E. huxleyi* (Sekino and Shiraiwa 1994, Herfort et al. 2002), but long–term (over 100 generations) treatment with lowered–Ca medium (0.1 mM) strongly inhibited photosynthesis (Xu et al. 2011, Xu and Gao 2012). Thus, the irreversible loss of calcification ability significantly shaped the CCMs of *E. huxleyi*. These results suggested that the DIC uptake systems could be switched to support photosynthesis and/or calcification. As a photosynthetic organism, *E. huxleyi* must preferentially maintain its photosynthesis which means calcification can be sacrificed if there exists a competition for DIC for both calcification and photosynthesis. In addition, calcification and photosynthesis may have a strong influence on the intracellular pH hemostasis and energy budget (Casey et al. 2010, Brownlee et al. 2015).

Very few studies focus on the CCM of calcifying macroalgae. Recently, a study reported the existence of HCO_3^- transporters and extracellular CAs in calcifying coralline macroalgae (Fig. 9.1B) which indicates that these algae also contain intracellular CAs to convert absorbed HCO_3^- into CO_2 to support photosynthesis (Hofmann et al. 2016). More studies are needed to test the results of the above study.

DIC Variations

For phytoplankton that live in open oceans, the daily and seasonal variations of pH and CO_2 are smaller than in those which live in coastal waters (Takahashi et al. 2014). Besides, many benthic macroalgae habit in intertidal zone which indicates that these algae experience a periodic emersed/submersed cycle. This cycle is controlled by the tidal cycle and is out of sync with the light/dark cycle. Under emersed conditions, the sole carbon source is atmospheric CO_2. Although macroalgae cannot directly absorb atmospheric CO_2 through stomata like higher plants, they can sustain a considerable photosynthetic rate which may be even higher than under submersed conditions (Gao et al. 1999, Ji and Tanaka 2002). However, when two coralline algae species were emersed in the air, their net photosynthetic and respiration rates were approximately zero (Guenther et al. 2014). To date, only a few studies analyzed the effects of emersed/submersed cycle on the responses of calcifying macroalgae to ocean acidification.

Another consequence of tidal cycles on benthic calcifiers is the seawater alkalization due to photosynthetic utilization of HCO_3^-. The rock pool green alga *Ulva* could increase the pH to above 10 (Björk et al. 2004). *Corallina*, a perennial calcifying red alga frequently found in rock pools, is naturally exposed to daily CO_2 variation between 70 and 1,000 ppmv. The respiration and gross primary production of *C. elongate* did not change by ocean acidification based on the prediction by IPCC (Egilsdottir et al. 2013). This study executed in the laboratory and the high CO_2 of seawater was achieved by bubbling the water with air with the target CO_2 concentration. The same method was also applied in another study, but the net photosynthesis rates of *C. sessilis* were significantly inhibited by high CO_2 (Gao and Zheng 2010). Such species– and strains–specific responses to OA have also been reported in coccolithophores (Langer et al. 2006, Langer et al. 2009). But for the different responses of the above two *Corallina* species, the major reason may be the fact that the DIC variation of *C. elongata* was larger than for *C. sessilis*. The former was collected from a rock pool, but the latter was collected from floating fish–farming rafts which were always sustained underwater. In addition, blooms of coccolithophore *Emiliania huxleyi* have been frequently observed in both open oceans and coastal areas (Tyrrell and Merico 2004). Thus, the planktonic and benthic calcifying algae are naturally exposed to variable CO_2 which may result in the fact that these calcifying algae are not sensitive to ocean acidification.

Previous results are generally based on laboratory studies. A disadvantage of laboratory studies is the use of old strains which were maintained at high biomass concentrations for years and even decades, with the result that the pH values of the medium were far larger than eight due to CCMs. The long–term high pH acclimation may have also already changed the physiological performance of algae. For example, *Gephyrocapsa oceanica* strain NIES–1318 lost its calcification ability after several years of laboratory incubation (Jin et al. 2013). A similar phenomenon has also been reported by Paasche (2002). Evolution could strongly influence the photosynthetic responses of coccolithophores to OA, as shown from the different effects of short– and long–term OA acclimation on particular organic carbon (POC) contents in *Gephyrocapsa oceanica* (Jin et al. 2013). On the other hand, the time scales of all the current experimental designs are much shorter than the natural processes of ocean acidification. In addition, the interaction effects between OA and other environmental factors may disturb the predictions based on experiments in laboratory or mesocosm studies (Boyd et al. 2010, Beardall et al. 2014).

OA Effects on Intracellular pH Hemostasis

The third direct physiological target of OA is the increasing proton concentration. Previous studies have suggested that in animal cells long–term cellular pH homeostasis is mainly achieved though the movements and metabolism of inorganic carbon (Roos and Boron 1981, Casey et al. 2010). Protons determine the pH of the solution and play a crucial role in all of the cellular biochemical reactions (Casey et al. 2010). Under normal physiological conditions, higher eukaryotes including mammals, plants, eukaryotic algae, and bacteria generally exhibit strict cytosolic pH homeostasis at a pH of around 7.2. Even the extracellular pH is far from neutral (Messerli et al. 2005, Casey et al. 2010, Taiz and Zeiger 2010, Martinez et al. 2012, Taylor et al. 2012). For example, phytoplankton live in the open oceans in which the typical seawater pH ranges between 7.9 and 8.2, but macroalgae live in coastal waters where the environmental pH ranges between 4.5 and 10 (Björk et al. 2004, Guenther et al. 2014, Takahashi et al. 2014, Vet et al. 2014). The acidophilic green alga *Chlamydomonas* sp. maintains an average cytosolic pH of 6.6 when grown in a medium with a pH between 2 and 7, while the neutrophilic relative *C. reinhardtii* maintains an average cytosolic pH of 7.1 in a pH 7 medium (Messerli et al. 2005). It is obvious that the cytosolic pH will influence the metabolism of a cell, but the mechanism controlling the maintenance of the cytosolic pH at such a conserved value under normal conditions needs to be explained. In addition, the relative complexity of highly compartmentalized eukaryotic cells, with different pH levels in the various compartments, makes it difficult to study the regulatory mechanisms of cytosolic pH. The opposite effects of photosynthesis and calcification on pH evoked the assumption that photosynthesis and calcification may benefit from each other by maintaining an intracellular pH (Paasche 2002).

Previous study found that the CCMs were applied to increase the intracellular DIC concentration to as high as 26 mmol/L in *E. huxleyi* cells (Sekino and Shiraiwa 1994), which suggests that the intracellular DIC pool could offers considerable buffer capacity to maintain intracellular pH hemostasis. Roos and Boron (1981) suggested that the long–term cellular pH buffering capacity depends on movement and production of HCO_3^-, an effective proton buffer which is the major chemical used to reduce the effects of changing the intracellular pH. In this way, the intracellular pH of calcifying algae is regulated by the proton transporters/channels and inorganic carbon transport and convert systems (for both CCMs and calcification) (Fig. 9.1). Taylor et al. (2011) showed that the coccolithophores *Coccolithus pelagicus* and *E. huxleyi* possess a proton channel which could rapidly remove protons from the cell during calcification and helps maintain a constant pH. Similarly, the higher expressed H^+–ATPase, heat shock proteins and loss of fermentation pathways that acidify the cytosol have played important roles in the adaptation of acidophilic green algae to extremely acidic environments (down to pH 1) (Hirooka et al. 2017). The characterization of proton channels in marine calcifying algae has resulted in the discovery of novel functions of these transport proteins in regulation of cellular pH, which could potentially help to understand the responses of calcifying algae to ocean acidification (Taylor et al. 2012).

Conclusion

It is generally accepted that calcifying algae are less calcified, but their photosynthesis may benefit from the increased CO_2 in the future high CO_2 oceans. However, some calcifying algae which inhabit coastal waters naturally experience large variations of DIC, which may introduce large uncertainties when evaluating the effects of OA on these algae. Thus, it is not surprising that a few studies found that calcification was not affected by OA. However, knowledge is still limited to understand the basic mechanisms which is necessary to explain the effects of OA on calcifying algae. Especially calcification, photosynthesis and pH hemostasis are linked together by the proton balance during cellular regulation of pH. Moreover, increasing atmospheric CO_2 may amplify the DIC variations of future oceans (Shaw et al. 2013), more calcifying algae will be facing the challenges of strongly varying $CaCO_3$ saturation states, CO_2 concentration and pH.

Acknowledgements

We thank Prof. Donat-P. Häder for his helpful comments and suggestions. This work was supported by the National Natural Science Foundation (41720104005, 41721005) and the Natural Science Foundation of Fujian Province (2016J01165).

References

Baba, M. and Y. Shiraiwa. 2012. High–CO_2 response mechanisms in microalgae. Advances in photosynthesis–fundamental aspects. M. M. Najafpour. Rijeka, Croatia, InTech: 299–320.

Bach, L.T., L.C. Mackinder, K.G. Schulz, G. Wheeler, D.C. Schroeder, C. Brownlee and U. Riebesell. 2013. Dissecting the impact of CO_2 and pH on the mechanisms of photosynthesis and calcification in the coccolithophore *Emiliania huxleyi*. New Phytol. 199: 121–134.

Beardall, J., S. Stojkovic and K. Gao. 2014. Interactive effects of nutrient supply and other environmental factors on the sensitivity of marine primary producers to ultraviolet radiation: Implications for the impacts of global change. Aquatic Biology 22: 6–23.

Beaufort, L., I. Probert, T. de Garidel–Thoron, E. M. Bendif, D. Ruiz–Pino, N. Metzl, C. Goyet, N. Buchet, P. Coupel, M. Grelaud, B. Rost, R.E.M. Rickaby and C. de Vargas. 2011. Sensitivity of coccolithophores to carbonate chemistry and ocean acidification. Nature 476(7358): 80–83.

Björk, M., L. Axelsson and S. Beer. 2004. Why is *Ulva intestinalis* the only macroalga inhabiting isolated rockpools along the Swedish Atlantic coast? Mar. Ecol. Prog. Ser. 284(1): 109–116.

Boyd, P.W., R. Strzepek, F.–X. Fu and D.A. Hutchins. 2010. Environmental control of open–ocean phytoplankton groups: Now and in the future. Limnology and Oceanography 55(3): 1353–1376.

Brownlee, C., G. Wheeler and A.R. Taylor. 2015. Coccolithophore biomineralization: New questions, new answers. Seminars in Cell & Developmental Biology 46: 11–16.

Cai, W. J. and Y. Wang. 1998. The chemistry, fluxes, and sources of carbon dioxide in the estuarine waters of the Satilla and Altamaha Rivers, Georgia. Limnology and Oceanography 43(4): 657–668.

Calbet, A. and M.R. Landry. 2004. Phytoplankton growth, microzooplankton grazing, and carbon cycling in marine systems. Limnology and Oceanography 49(1): 51–57.

Casey, J.R., S. Grinstein and J. Orlowski. 2010. Sensors and regulators of intracellular pH. Nat. Rev. Mol. Cell. Biol. 11(1): 50–61.

Dai, M., Z. Lu, W. Zhai, B. Chen, Z. Cao, K. Zhou, W.J. Cai and C.T.A. Chen. 2009. Diurnal variations of surface seawater pCO_2 in contrasting coastal environments. Limnology & Oceanography 54: 735–745.

Egilsdottir, H., F. Noisette, M.–L.N. Laure, J. Olafsson and S. Martin. 2013. Effects of pCO_2 on physiology and skeletal mineralogy in a tidal pool coralline alga *Corallina elongata*. Marine Biology 160(8): 2103–2112.

Endeward, V., S. Al–Samir, F. Itel and G. Gros. 2014. How does carbon dioxide permeate cell membranes? A discussion of concepts, results and methods. Front Physiol 4(382): 1–21.

Feely, R.A., S.C. Doney and S.R. Cooley. 2009. Ocean acidification: Present conditions and future changes in a high–CO_2 world. Oceangraphy 22: 36–47.

Feng, Y., M.Y. Roleda, E. Armstrong, P.W. Boyd and C.L. Hurd. 2017. Environmental controls on the growth, photosynthetic and calcification rates of a Southern Hemisphere strain of the coccolithophore *Emiliania huxleyi*. Limnology and Oceanography 62(2): 519–540.

Fourqurean, J.W., C.M. Duarte, H. Kennedy, N. Marbà, M. Holmer, M.A. Mateo, E.T. Apostolaki, G.A. Kendrick, D. Krause–Jensen, K.J. McGlathery and O. Serrano. 2012. Seagrass ecosystems as a globally significant carbon stock. Nature Geoscience 5(7): 505–509.

Gao, K., Y. Ji and Y. Aruga. 1999. Relationship of CO_2 concentrations to photosynthesis of intertidal macroalgae during emersion. Hydrobiologia 398: 355–359.

Gao, K. and Y. Zheng. 2010. Combined effects of ocean acidification and solar UV radiation on photosynthesis, growth, pigmentation and calcification of the coralline alga *Corallina sessilis* (Rhodophyta). Global Change Biology 16(8): 2388–2398.

Gao, S., Q. Sun, Y. Tao, X. Wang, W. Li, L. Huan, M. Wu and G. Wang. 2016. A decline in macro–algae species resulting in the overwhelming prevalence of *Corallina* species is caused by low–pH seawater induced by short–term acid rain. Journal of Experimental Marine Biology and Ecology 475: 144–153.

Gattuso, J.P., K. Gao, K. Lee, B. Rost and K.G. Schulz. 2010. Approaches and tools to manipulate the carbonate chemistry. Guide to best practices in ocean acidification research and data reporting. Luxembourg, Publications Office of the European Union: 41–52.

Giordano, M., J. Beardall and J.A. Raven. 2005. CO_2 concentrating mechanisms in algae: mechanisms, environmental modulation, and evolution. Annu. Rev. Plant Biol. 56(1): 99–131.

Guenther, R.J., P.T. Martone and C. Hurd. 2014. Physiological performance of intertidal coralline algae during a simulated tidal cycle. Journal of Phycology 50(2): 310–321.

Herfort, L., B. Thake and R.J. Roberts. 2002. Acquisition and use of bicarbonate by *Emiliania huxleyi*. New Phytologist 156(3): 427–436.

Hirooka, S., Y. Hirose, Y. Kanesaki, S. Higuchi, T. Fujiwara, R. Onuma, A. Era, R. Ohbayashi, A. Uzuka, H. Nozaki, H. Yoshikawa and S.Y. Miyagishima. 2017. Acidophilic green algal genome provides insights into adaptation to an acidic environment. Proc. Natl. Acad. Sci. U S A 114(39): E8304–E8313.

Hofmann, L.C. and K. Bischof. 2014. Ocean acidification effects on calcifying macroalgae. Aquatic Biology 22: 261–279.

Hofmann, L.C., K. Marguerite and D.B. Dirk. 2016. Biotic control of surface pH and evidence of light–induced H^+ pumping and Ca^{2+}-H^+ exchange in a tropical crustose coralline alga. Plos One 11(7): e0159057.

Holtz, L.–M., D. Wolf–Gladrow and S. Thoms. 2015. Numerical cell model investigating cellular carbon fluxes in *Emiliania huxleyi*. Journal of Theoretical Biology 364: 305–315.

Iglesias–Rodriguez, M.D., P.R. Halloran, R.E.M. Rickaby, I.R. Hall, E. Colmenero–Hidalgo, J.R. Gittins, D.R.H. Green, T. Tyrrell, S.J. Gibbs, P. Von Dassow, E. Rehm, E.V. Armbrust and K.P. Boessenkool. 2008. Phytoplankton calcification in a high–CO_2 world. Science 320(5874): 336–340.

Ji, Y. and J. Tanaka. 2002. Effect of desiccation on the photosynthesis of seaweeds from the intertidal zone in Honshu, Japan. Phycological Research 50(2): 145–153.

Jin, P., K. Gao and J. Beardall. 2013. Evolutionary responses of a coccolithophorid *Gephyrocapsa oceanica* to ocean acidification. Evolution 67(7): 1869–1878.

Koch, M., G. Bowes, C. Ross and X.H. Zhang. 2013. Climate change and ocean acidification effects on seagrasses and marine macroalgae. Glob. Chang. Biol. 19(1): 103–132.

Krishnamurthy, V.M., G.K. Kaufman, A.R. Urbach, I. Gitlin, K.L. Gudiksen, D.B. Weibel and G.M. Whitesides. 2008. Carbonic anhydrase as a model for biophysical and physical–organic studies of proteins and protein–ligand binding. Chemical Reviews 108(3): 946–1051.

Langer, G., M. Geisen, K.–H. Baumann, J. Kläs, U. Riebesell, S. Thoms and J.R. Young. 2006. Species–specific responses of calcifying algae to changing seawater carbonate chemistry. Geochemistry Geophysics Geosystems 7(9): Q09006.

Langer, G., G. Nehrke, I. Probert, J. Ly and P. Ziveri. 2009. Strain–specific responses of *Emiliania huxleyi* to changing seawater carbonate chemistry. Biogeosciences 6: 2637–2646.

Lebrato, M., A. Andersson, J. Ries, R. Aronson, M. Lamare, W. Koeve, A. Oschlies, M. Iglesias-Rodriguez, S. Thatje and M. Amsler.2016. Benthic marine calcifiers coexist with $CaCO_3$-undersaturated seawater worldwide. Global Biogeochemical Cycles 30(7): 1038–1053.

Levin, S.A. 1992. The Problem of Pattern and Scale in Ecology, Springer.

Li, Y.H., D. Wang, X.T. Xu, X.X. Gao, X. Sun and N.J. Xu. 2017. Physiological responses of a green algae (*Ulva prolifera*) exposed to simulated acid rain and decreased salinity. Photosynthetica 55: 623–629.

Mackinder, L., G. Wheeler, D. Schroeder, U. Riebesell and C. Brownlee. 2010. Molecular mechanisms underlying calcification in coccolithophores. Geomicrobiology Journal 27: 585–595.

Martinez, K.A., 2nd, R.D. Kitko, J.P. Mershon, H.E. Adcox, K.A. Malek, M.B. Berkmen and J.L. Slonczewski. 2012. Cytoplasmic pH response to acid stress in individual cells of *Escherichia coli* and *Bacillus subtilis* observed by fluorescence ratio imaging microscopy. Appl. Environ. Microbiol. 78(10): 3706–3714.

Mazarrasa, I., N. Marbà, C.E. Lovelock, O. Serrano, P.S. Lavery, J.W. Fourqurean, H. Kennedy, M.A. Mateo, D. Krause–Jensen, A.D.L. Steven and C.M. Duarte. 2015. Seagrass meadows as a globally significant carbonate reservoir. Biogeosciences 12(16): 4993–5003.

Messerli, M.A., L.A. Amaral–Zettler, E. Zettler, S.–K. Jung, P. J. Smith and M. L. Sogin. 2005. Life at acidic pH imposes an increased energetic cost for a eukaryotic acidophile. Journal of Experimental Biology 208(13): 2569–2579.

Monastersky, R. 2013. Global carbon dioxide levels near worrisome milestone. Nature 497(7447): 13.

Moore, C.M., M.M. Mills, K.R. Arrigo, I. Berman–Frank, L. Bopp, P.W. Boyd, E.D. Galbraith, R.J. Geider, C. Guieu, S.L. Jaccard, T.D. Jickells, J. La Roche, T.M. Lenton, N.M. Mahowald, E. Marañón, I. Marinov, J.K. Moore, T. Nakatsuka, A. Oschlies, M.A. Saito, T.F. Thingstad, A. Tsuda and O. Ulloa. 2013. Processes and patterns of oceanic nutrient limitation. Nature Geoscience 6(9): 701–710.

Paasche, E. 2002. A review of the coccolithophorid *Emiliania huxleyi* (Prymnesiophyceae), with particular reference to growth, coccolith formation, and calcification–photosynthesis interactions. Phycologia 40(6): 503–529.

Peach, K.E., M.S. Koch, P.L. Blackwelder and C. Manfrino. 2017. Calcification and photophysiology responses to elevated pCO_2 in six *Halimeda* species from contrasting irradiance environments on Little Cayman Island reefs. Journal of Experimental Marine Biology & Ecology 486: 114–126.

Raven, J.A. and M. Giordano. 2009. Biomineralization by photosynthetic organisms: Evidence of coevolution of the organisms and their environment? Geobiology 7(2): 140–154.

Raven, J.A., M. Giordano, J. Beardall and S.C. Maberly. 2012. Algal evolution in relation to atmospheric CO_2: Carboxylases, carbon–concentrating mechanisms and carbon oxidation cycles. Philosophical Transactions of the Royal Society B: Biological Sciences 367(1588): 493–507.

Reinfelder, J.R. 2011. Carbon concentrating mechanisms in eukaryotic marine phytoplankton. Annual Review of Marine Science 3(1): 291–315.

Riebesell, U. and P.D. Tortell. 2011. Effects of ocean acidification on pelagic organisms and ecosystems. Ocean acidification. J.P. Gattuso and L. Hansson. Oxyford, Oxyford University Press: 99–121.

Ries, J.B., A.L. Cohen and D.C. McCorkle. 2009. Marine calcifiers exhibit mixed responses to CO_2–induced ocean acidification. Geology 37(12): 1131–1134.

Roos, A. and W.F. Boron. 1981. Intracellular pH. Physiological Reviews 61: 296–434.

Rost, B. and U. Riebesell. 2004. Coccolithophores and the biological pump: responses to environmental changes. Coccolithophores: From molecular processes to global impact. Thierstein, H.R. and J.R. Young (eds.). Berlin, Germany, Springer: 76–99.

Sekino, K. and Y. Shiraiwa. 1994. Accumulation and utilization of dissolved inorganic carbon by a marine unicellular coccolithophorid, *Emiliania huxleyi*. Plant and Cell Physiology 35(3): 353–361.

Shaw, E.C., P. . Munday and B.I. McNeil. 2013. The role of CO_2 variability and exposure time for biological impacts of ocean acidification. Geophysical Research Letters 40(17): 4685–4688.

Shi, D., Y. Xu and F.M.M. Morel. 2009. Effects of the pH/pCO_2 control method on medium chemistry and phytoplankton growth. Biogeosciences 6(7): 1199–1207.

Smith, H.E.K., T. Tyrrell, A. Charalampopoulou, C. Dumousseaud, O.J. Legge, S. Birchenough, L. R. Pettit, R. Garley, S.E. Hartman and M.C. Hartman. 2012. Predominance of heavily calcified coccolithophores at low $CaCO_3$ saturation during winter in the Bay of Biscay. Proceedings of the National Academy of Sciences 109(23): 8845–8849.

Stanley, S.M. 2008. Effects of global seawater chemistry on biomineralization: Past, present, and future. Chem. Rev. 108(11): 4483–4498.

Stojkovic, S., J. Beardall and R. Matear. 2013. CO_2–concentrating mechanisms in three southern hemisphere strains of *Emiliania huxleyi*. J. Physiology 49(4): 670–679.

Taiz, L. and E. Zeiger. 2010. Plant Physiology. 5th. Sundeland, Massachusetts, Sinauer Associates.

Takahashi, T., S.C. Sutherland, D.W. Chipman, J.G. Goddard, T. Newberger and C. Sweeney. 2014. Climatological distributions of pH, pCO_2, total CO_2, alkalinity, and $CaCO_3$ saturation in the global surface ocean. Carbon Dioxide Information Analysis Center, Oak Ridge National Laboratory, U.S. Department of Energy, Oak Ridge, Tennessee.

Taylor, A.R., C. Brownlee and G.L. Wheeler. 2012. Proton channels in algae: Reasons to be excited. Trends in Plant Science 17(11): 675–684.

Taylor, A.R., A. Chrachri, G. Wheeler, H. Goddard and C. Brownlee. 2011. A voltage–gated H^+ channel underlying pH homeostasis in calcifying coccolithophores. PLoS Biol. 9(6): e1001085.

Tilman, D., S.S. Kilham and P. Kilham. 1982. Phytoplankton community ecology: the role of limiting nutrients. Annual Review of Ecology and Systematics 13: 348–372.

Tyrrell, T. and A. Merico. 2004. *Emiliania huxleyi*: Bloom observations and the conditions that induce them. Coccolithophores: From Molecular Processes to Global Impact. H.R. Thierstein and J.R. Young. Berlin, Germany, Springer.

Vet, R., R.S. Artz, S. Carou, M. Shaw, C.–U. Ro, W. Aas, A. Baker, V.C. Bowersox, F. Dentener and C. Galy–Lacaux. 2014. A global assessment of precipitation chemistry and deposition of sulfur, nitrogen, sea salt, base cations, organic acids, acidity and pH, and phosphorus. Atmospheric Environment 93: 3–100.

Wang, Y., D.J. Stessman and M.H. Spalding. 2015. The CO_2 concentrating mechanism and photosynthetic carbon assimilation in limiting CO_2: how *Chlamydomonas* works against the gradient. Plant Journal 82(3): 429–448.

Xu, K., H. Chen, W. Wang, Y. Xu, D. Ji, C. Chen and C. Xie. 2017. Responses of photosynthesis and CO_2 concentrating mechanisms of marine crop *Pyropia haitanensis* thalli to large pH variations at different time scales. Algal Research 28: 200–210.

Xu, K., F.–X. Fu and D.A. Hutchins. 2014. Comparative responses of two dominant Antarctic phytoplankton taxa to interactions between ocean acidification, warming, irradiance, and iron availability. Limnology and Oceanography 59(6): 1919–1931.

Xu, K. and K. Gao. 2012. Reduced calcification decreases photoprotective capability in the coccolithophorid *Emiliania huxleyi*. Plant and Cell Physiology 53(7): 1267–1274.

Xu, K. and K. Gao. 2015. Solar UV irradiances modulate effects of ocean acidification on the coccolithophorid *Emiliania huxleyi*. Photochemistry and Photobiology 91(1): 92–101.

Xu, K., K. Gao, V.E. Villafañe and E.W. Helbling. 2011. Photosynthetic responses of *Emiliania huxleyi* to UV radiation and elevated temperature: roles of calcified coccoliths. Biogeosciences 8: 1441–1452.

Zhai, W., M. Dai, W.–J. Cai, Y. Wang and Z. Wang. 2005. High partial pressure of CO_2 and its maintaining mechanism in a subtropical estuary: The Pearl River estuary, China. Marine Chemistry 93(1): 21–32.

CHAPTER 10

Effects of a Changing Climate on Freshwater and Marine Zooplankton

*Craig E. Williamson** and *Erin P. Overholt*

Introduction

Zooplankton are the dominant consumers in the pelagic ecosystems of inland waters and open oceans that collectively cover more than 70% of our planet. They are perhaps the most abundant multicellular organisms on the planet, with just the copepods alone estimated to be 1000 times as abundant as the incredibly diverse, abundant, and ubiquitous insects, and their total biomass exceeding total human biomass on our planet by 150 fold (Schminke 2007). The feeding habits of zooplankton range from bacterivory to herbivory and carnivory, with many species being omnivorous. By consuming phytoplankton and bacterioplankton, they can influence water quality, and mediate the transfer of energy and nutrients to higher trophic levels, providing the food source for the world's fisheries. Diel vertical migration of zooplankton can also mediate the transfer of energy and nutrients between the deep-water benthic and surface-water pelagic ecosystem compartments, thus influencing the transfer of photosynthetically fixed carbon with important implications for the fate of this carbon, greenhouse gas production, and thus feedbacks to climate change. Zooplankton feeding, migration, and consumption by higher trophic levels can influence the balance between the oceans being a sink vs. a source of greenhouse gases, thus providing a critical ecosystem service in mediating climate change. The estimated value of the ecosystem services provided globally by lakes, reservoirs, and oceans, of which zooplankton are a key component, has been estimated at over US$22 trillion per year, more than the value of the entire global gross national product (US$18 trillion per year) at the time (Costanza et al. 1997). Climate change threatens the ecosystem services provided

Department of Biology, Miami University, Oxford, OH 45056.
* Corresponding author: craig.williamson@miamioh.edu

by this critical trophic level in lakes and oceans. Furthermore, trophic amplification involving non-linear responses to temperature across multiple trophic levels may accentuate responses to temperature (Kirby and Beaugrand 2009). Concerns exist that the non-linear responses to climate change will push aquatic ecosystems beyond a critical tipping point, causing a switch to an alternative and potentially irreversible stable state (Richardson 2008) that could in turn jeopardize these critical ecosystem services.

Climate change has led to three fundamental types of changes in a wide variety of ectothermic organisms that have been observed so widely that they have been referred to as rules (Daufresne et al. 2009). Zooplankton have shown changes consistent with these rules that are also some of the strongest of any group of organisms. The first rule is that organisms are spreading to colder regions at higher latitudes and altitudes. A global meta-analysis reported that at over 1,000 km in a 40-year period, zooplankton range expansion is among the greatest of any organisms (Parmesan and Yohe 2003). The second rule is that the phenology of many organisms is changing (Walther et al. 2002). Both freshwater and marine zooplankton have exhibited numerous phenological changes with a wide variety of ecological consequences. The third rule is that body size decreases at warmer temperatures, and again, this smaller body size at warmer temperatures is a very common phenomenon in zooplankton. These three dimensions of response are discussed in more detail below in the context of responses available to freshwater vs. marine zooplankton in different types of habitats (Fig. 10.1).

Here we examine the multiple ways in which climate change is influencing zooplankton in both inland and oceanic waters. Our purpose is not to provide a comprehensive review, but rather to highlight several key ways in which both direct and indirect effects of climate change are influencing zooplankton. First, we provide a brief historical context, some brief notes on the climate context, and a perspective on the importance of climate oscillations and the potential for regime shifts. We then focus on some of the primary ways that climate change is altering zooplankton communities.

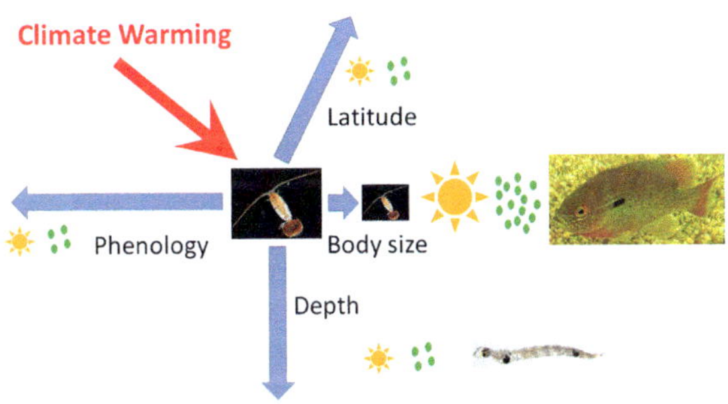

Fig. 10.1. Conceptual model showing the different dimensions in which zooplankton are responding to climate warming. Warmer conditions (large sun) can lead to smaller body size if zooplankton stay in place. The three primary dimensions of response include temporal changes in phenology, and spatial changes involving either migration deeper into the water column, or movement to higher latitudes. In all dimensions, light (small sun) and thus available phytoplankton for food (green dots) may be reduced. Predation regimes are also likely to change. For example, migration to cooler, deeper, darker waters may lead to a decrease in visual predation but an increase in exposure to tactile predators, themselves large zooplankton that migrate deeper to avoid visual predation. Photo credits: adult female *Aglaodiaptomus* carrying eggs by Robert Moeller, *Chaoborus* and *Lepomis* by Craig Williamson.

These include changes in their body size and metabolic rates, geographic distribution, vertical distribution in the water column, and phenology. We then conclude with a brief synthesis and identification of knowledge gaps.

A Brief Historical Context

Early studies of the effects of climate change on marine zooplankton have their origins in the analysis of long-term data sets associated with explaining fluctuations in commercial fisheries relative to climate oscillations. Analysis of changes in zooplankton of the North Sea from 1948–73 reported some of the first long-term trends that were attributed to climate (Colebrook 1978a, Colebrook 1978b). An analysis of long-term records in the California current reported lower zooplankton abundances in years with warmer surface waters during El Niño events (Bernal 1979, Chelton et al. 1982). An analysis of long-term data between 1960 and 1987 in the Bay of Biscay in the eastern North Atlantic revealed a lower abundance of summer populations of *Calanus* and higher abundance of autumn copepods in the genus *Eucheta* during warmer water periods (Southward et al. 1995). A closer analysis revealed an inverse relationship between two *Calanus* species in the eastern North Atlantic that was strongly related to the North Atlantic Oscillation (NAO). The generally more northerly-distributed *C. finmarchicus* was more abundant during negative phases of the NAO when surface ocean waters were cooler, while the more southerly-distributed *C. helgolandicus* was more abundant during the positive phases of the NAO and in warmer waters (Fromentin and Planque 1996). Wind stress and phytoplankton productivity were implicated in this response, but the relative importance of the direct vs. indirect effects of warmer temperatures remained enigmatic (Fromentin and Planque 1996). An El Niño event in 1992 that led to anomalously warm (28°C) surface waters in the tropical Pacific, was associated with a lower biomass of both total zooplankton and larger (> 1 mm) species (White et al. 1995).

One of the first clear signals that freshwater zooplankton were strongly influenced by climate change came from a long-term (1940–80) study in Lake Windermere in the legendary English Lake District where zooplankton biomass was found to be inversely correlated with June water temperatures (George and Harris 1985). It was later observed that this periodically lower zooplankton biomass was associated with the north vs. south position of the Gulf Stream in the western Atlantic, and the strength of thermal stratification in Lake Windermere (George and Taylor 1995). When the boundary of the Gulf Stream was more northerly located, early summer thermal stratification was stronger, the thermocline depth was shallower, and zooplankton biomass was lower.

Early modeling studies on the effects of climate change on zooplankton in lakes focused on changes in the diel vertical migration of *Daphnia* related to physical and biological changes in the vertical habitat gradient. Optimal depth distribution during the day and night were predicted to change in response to changes in thermal structure and vertical distribution of food (chlorophyll) and predators (fish; De Stasio et al. 1993, De Stasio et al. 1996). Temperature and food are important structural drivers of zooplankton diel vertical migration that are relatively constant in their vertical distribution day and night, while exposure to damaging ultraviolet (UV) radiation and visual predation are important dynamic drivers that vary greatly from day to night (Williamson et al. 2011). Biological responses to warming related to a doubling atmospheric carbon dioxide (CO_2) scenario were predicted to be much more variable than changes in thermal structure (De Stasio et al. 1996). Predicted physical changes in lakes included higher surface water temperatures, a shallower mixed layer depth, and longer and stronger periods of summer thermal stratification (De Stasio et al. 1996).

An early and comprehensive review of the direct and indirect effects of warmer temperatures on temperate freshwater zooplankton argued that mid-summer surface water temperatures greater than 25°C were already leading to reduced body sizes (Moore et al. 1996). These temperature-body size relationships were recognized as being more pronounced when tactile predators that select for smaller zooplankton prey were abundant, but less pronounced in the presence of visual predators that selectively feed on larger zooplankton (Fig. 10.2). Importantly, the effects of visual predation were seen to further reduce zooplankton body size beyond the temperature effects alone. The lower feeding rates of smaller zooplankton and the reduced availability of smaller zooplankton species as food for visually-feeding fishes led to the suggestion that future climate change scenarios may contribute to increases in harmful algal blooms (HABs) and reduce food availability for fish. The importance of cool-water refugia < 20°C also was recognized at that point in time in many lakes, especially for glacial relict species, such as the large calanoid copepods *Limnocalanus macrurus* and *Senecella calanoides* that have no resting stages to survive adversely warm surface waters and oxygen-depleted deeper waters. These species were predicted to be extinguished in lakes where warmer surface water temperatures combined with deeper anoxic waters to eliminate the cool-water refuge (Moore et al. 1996).

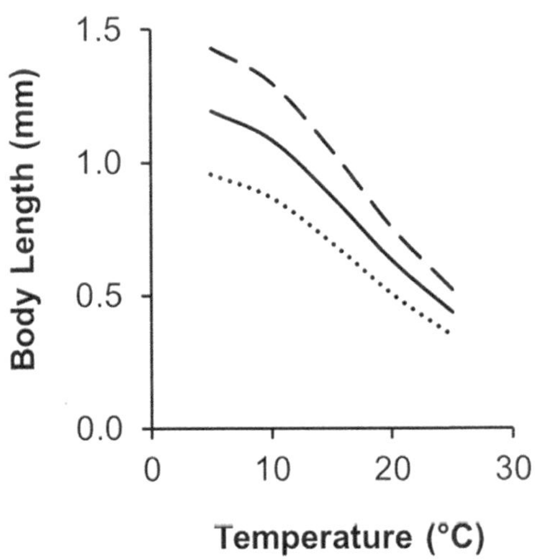

Fig. 10.2. Zooplankton body size decreases non-linearly with increasing temperature, with a more rapid decrease at warmer temperatures for freshwater cladocerans (solid line). Body size is also modified by predators. Visual predators feed selectively on larger prey, leading to further reductions in body size (dotted line) while tactile invertebrate predators prefer smaller prey, potentially offsetting the effects of warmer temperatures (dashed line). Modified from Moore et al. 1996, freshwater cladoceran relationship from Gillooly and Dodson 2000.

Climate Context: Not Just Temperature

Ocean temperatures have increased by over 1°C compared to the 20th century averages, with particularly strong warming in the last two years for which data are available (2015, 2016) (https://www.globalchange.gov/browse/indicators/indicator-sea-surface-temperatures). Surface water temperatures of lakes are warming even more rapidly, by up to 0.34°C per decade globally between 1985 and 2009 (O'Reilly et al. 2015), and up to 0.54°C per decade between 1975 and 2012 across some lake-rich regions, such as northeastern North America (Richardson et al. 2017). In many cases lake temperatures are actually warming more rapidly than regional air temperatures (Austin and Colman 2008; Schneider and Hook 2010, O'Reilly et al. 2015, Richardson et al. 2017, Weyhenmeyer et al. 2017).

In addition to warming, climate change is reducing water clarity through a variety of mechanisms ranging from increases in precipitation and runoff that flush dissolved organic matter (DOM) into inland and coastal waters, to the melting of permafrost, glaciers, snow, and ice. Following an extreme precipitation event in Lough Feagh, Ireland, *Daphnia* abundance was lower and calanoid copepod abundance higher than normal (de Eyto et al. 2016). Increases in the concentration and color of terrestrially-derived DOM of two-fold or more are inducing the browning of inland and coastal waters. Browning is largely driven by a combination of increases in precipitation and decreases in anthropogenic acidification (Evans et al. 2006, Monteith et al. 2007, Williamson et al. 2014, Strock et al. 2016). Browning alone can induce decreases in water clarity that lead to warmer surface water temperatures and stronger thermal stratification even in the absence of warmer air temperatures (Pilla et al. 2018). Decreasing water clarity also leads to a shallower compensation depth, the light depth below which respiration exceeds photosynthesis. These two factors combined accelerate oxygen depletion and the development of hypoxic or anoxic deep waters in inland and marine waters often referred to as 'dead zones' (Brothers et al. 2014, Williamson et al. 2015, Breitburg et al. 2018, Knoll et al. 2018). In coastal marine systems, the oxygen depletion appears to be driven by increases in nutrient pollution, but the potential contribution of browning and DOM to anoxia in these systems has yet to be examined.

Increases in DOM during browning are likely to mitigate damaging effects of UV radiation in surface waters through two primary mechanisms. DOM selectively absorbs the shorter wavelength, DNA-damaging UV-B radiation (Morris et al. 1995, Williamson et al. 1996). In addition, DOM increases surface water temperatures by absorbing sunlight, and these higher temperatures can increase the effectiveness of the enzyme photolyase which stimulates photoenzymatic repair of UV-B-induced DNA damage in zooplankton (Williamson et al. 2002; MacFadyen et al. 2004). These changes in UV in the surface waters can alter the vertical distribution of zooplankton (Williamson et al. 2011, Rose et al. 2012, Fischer et al. 2015, Leach et al. 2015, Urmy et al. 2016). Reduced UV radiation in surface waters can also compromise the valuable ecosystem service of solar disinfection of pathogens (Williamson et al. 2017), including important parasites of the keystone grazer *Daphnia* (Overholt et al. 2012).

The indirect effects of a warmer and wetter world range from changes in light availability, food supply, and predation regimes, to stronger thermal stratification, reductions in dissolved oxygen, and changes in pH related to not only ocean acidification, but to increases in DOM. Water transparency is a fundamental regulator of zooplankton vertical habitat gradients and thus their vertical distribution and abundance (Williamson et al. 2011). The vertical distribution and amplitude of diel vertical migration of zooplankton is highly correlated with water transparency (Dodson 1990, Wissel and Ramcharan 2003, Williamson et al. 2011). In eutrophic lakes and coastal and open oceans phytoplankton may be the primary attenuators of light, while in waters with minimal cultural eutrophication, DOM is the primary regulator of water transparency to both UV and photosynthetically active radiation (PAR) (Morris et al. 1995, Fee et al. 1996, Rose et al. 2009). In alpine and polar regions, melting of permafrost may contribute DOM and glaciers may contribute inorganic particulates that play a major role in controlling water transparency (Rose et al. 2014, Häder et al. 2015, Olson et al. 2017).

While inland and oceanic waters are experiencing many similarities in their responses to climate change, in some cases interactions of climate change with other types of environmental change lead to fundamentally different, if not even opposite trends in inland vs. oceanic ecosystems. For example, while elevated atmospheric CO_2 concentrations are leading to acidification with important implications for zooplankton in oceans (Hammill et al. 2018, Thor et al. 2018) and some large lakes (Weiss et al. 2018), long-term browning can lead to increases in pH of up to a full pH unit, accompanied by strong changes in zooplankton community structure (Williamson et al. 2015). This highlights the central importance of complex interactions between the direct and indirect effects of climate change.

Climate Oscillations and Regime Shifts

Regime shifts can be defined as abrupt, substantial, and persistent changes in natural communities or ecosystems (Beaugrand et al. 2015, Reid et al. 2016). These abrupt regime shifts remain one of the primary concerns regarding the effects of climate change on both humans and ecosystems (IPCC 2014). A recent comprehensive analysis documented a strong regime shift across the globe in the 1980s that included changes in the cryosphere, terrestrial ecosystems, and inland and oceanic waters (Reid et al. 2016). Some of the strongest evidence of regime shifts in marine zooplankton communities comes from studies of climate oscillations, such as the North Atlantic Oscillation (NAO), Pacific Decadal Oscillation (PDO) and El Niño Southern Oscillation (ENSO). Importantly, studies of major climate oscillations give us advanced insights into how long-term warming trends are likely to influence zooplankton, but in a much shorter time. Early studies on teleconnections demonstrated how climate oscillations led to pronounced changes in salmon fisheries (Mantua et al. 1997), as well as major regime shifts involving switches between sardine and anchovy dominance (Chavez et al. 2003). The non-linear biological responses to climate, such as these actually provide stronger signals of climate change than the climate metrics themselves (Hare and Mantua 2000). Climate-driven changes in zooplankton and other biota demonstrated distinct regime shifts in the Pacific Ocean in 1977 and 1989 (Hare and Mantua 2000), as well as in the North Atlantic in 1979 and 1988 (Weijerman et al. 2005). A more recent analysis across ocean basins provides evidence that zooplankton are some of the strongest responders and most effective sentinels of climate change for identifying the regime shifts in 1976–1978, 1988–1990, and 1997 (Beaugrand et al. 2015). Both freshwater and marine zooplankton can show stronger and clearer responses to subtle changes in climate than can the actual meteorological data (George and Taylor 1995, Taylor et al. 2002, Hays et al. 2005, Richardson 2008). Accumulating evidence suggests that these regime shifts are largely driven by human-caused temperature changes rather than other natural or global processes (Beaugrand et al. 2015, Reid et al. 2016).

These climate oscillations in the world's oceans also have a profound influence on zooplankton dynamics in inland waters. For example, in a small, shallow, eutrophic, polymictic lake in Germany, an analysis of 20 years of data from 1979–98 showed that the warmer years related to the NAO led to shorter periods of ice cover, and peak zooplankton abundances that were two weeks earlier than normal (Gerten and Adrian 2000). The timing of the seasonal peaks of the cladoceran *Bosmina* and the rotifer *Keratella* were significantly correlated with the NAO. But the timing of the peak abundance of the large cladoceran *Daphnia* and the onset of the clearwater phase, a period of high water transparency generated by its grazing, were significantly related only to spring water temperature and not the NAO. Further study revealed that in this same lake and in a larger, deeper lake in Germany, the abundance of *Daphnia* was significantly related to the timing of the NAO, but only during the months of April and May (Straile and Adrian 2000). Furthermore, the timing of the clearwater phase was closely synchronized with the NAO across 28 lakes in the region (Straile 2002), and the winter and spring abundance of *Daphnia* as well as the summer abundance of cyclopoids were similarly shown to be synchronized with the NAO across 18 lakes in this region over a 23-year period (Blenckner et al. 2007). Analysis of a 40-year database from 1962–2002 on Lake Washington found the timing of the peak abundances of several zooplankton species were inversely related to warmer spring water temperatures as well as to the PDO, and peak abundance of a rotifer was also inversely related to the ENSO index (Winder and Schindler 2004b). Such studies of climate oscillations demonstrate the strong sentinel nature of zooplankton dynamics and their importance in identifying regime shifts in response to climate change.

Three Dimensional Response Gradients and Indirect Effects

Here we suggest a general conceptual framework for identifying the major ways in which zooplankton respond to climate change. Marine and freshwater zooplankton at different latitudes have different spatial and temporal dimensions available for them to respond to warmer temperatures that may exceed their thermal optima. Marine zooplankton at higher latitudes can respond to climate warming in three dimensions. They can: (i) expand geographically to cooler waters at higher latitudes, (ii) migrate downwards in the water column to cooler waters at deeper depths, or (iii) alter their phenology such that their peak periods of growth and reproduction coincide with cooler seasonal temperatures (Fig. 10.1). Alternatively, if zooplankton stay in place, reduced body size is likely for several reasons including the effects of warmer temperatures, presence of predators, and light limitation or other mechanisms that can reduce food availability (Fig. 10.1). In equatorial regions zooplankton are more limited in their response dimensions due to the small seasonal variations in water temperature. Freshwater zooplankton are also physically constrained by the fact that they exist in what are effectively aquatic islands in the terrestrial landscape, so migration to higher latitudes has serious obstacles. Yet many freshwater zooplankton have resting eggs or other diapause stages that are highly resistant to extreme environmental conditions and may thus help overcome obstacles to geographic expansion.

Many other abiotic and biotic factors also change both with warming and as zooplankton respond along these three dimensions. This can lead to the potential for the indirect effects of climate change to be more important than temperature in altering zooplankton survival, growth, and reproduction. One of the most fundamental of these is a reduction in the incident solar radiation (Fig. 10.1) necessary for phytoplankton photosynthesis, which may in turn reduce food availability for zooplankton (Fig. 10.1). One example of the importance of reductions in food supply is in marine ecosystems where krill have declined by over an order of magnitude in recent decades, and this is thought to be due primarily to climate change-induced reductions in food supply rather than to thermal tolerances being exceeded (Hays et al. 2005). Food quality may also change at warmer temperatures. For example, phytoplankton community dominance may shift from diatoms that are of high food quality, to cyanobacteria that do better at warmer temperatures, but are of much poorer food quality (Paerl and Huisman 2008, Wagner and Adrian 2011, Ger et al. 2014, Ger et al. 2016, Rice and Stewart 2016). These shifts to cyanobacteria may also influence competitive interactions. For example, a combination of laboratory and field mesocosm experiments that manipulated temperature and cyanobacteria revealed that the invasive cladoceran *Daphnia lumholtzi* outcompeted the native *Daphnia pulicaria* at warmer temperatures, but only in the absence of cyanobacteria (Fey and Cottingham 2011).

By shifting their habitat to maintain more optimal temperatures, zooplankton may also alter their exposure to different types of predators or to damaging UV radiation. For example, shifting to deeper depths or earlier periods in the spring in response to warmer surface water temperatures may reduce light necessary for visual predation, yet increase exposure to tactile predators, which are generally large zooplankton that themselves seek to avoid visual predation (Fig. 10.1). UV exposure will also generally be lower as zooplankton shift to cooler temperatures earlier in the spring, at greater depths, or at higher latitudes.

Sorting out these multiple indirect and interactive effects of climate change is an ongoing challenge. One example is that warmer surface water temperatures are often associated with shallower mixing depths, making it very difficult to separate the effects of release from light limitation (due to a shallower mixing depth) from warmer temperatures. A clever set of *in situ* mesocosm experiments manipulated temperature and light limitation independently by controlling temperature with externally circulating water jackets and controlling mixing depths mechanically (Berger et al. 2007, Sebastian et al. 2012). The results of both sets of experiments demonstrated that light limitation resulting from deeper mixing depths was more important than temperature in determining the magnitude of peak abundance of *Daphnia* and their phytoplankton food (chlorophyll). Warmer temperatures stimulated earlier peaks in

Daphnia abundance in both experiments, but an earlier peak in food abundance in only one of the experiments (Sebastian et al. 2012). The absence of visual predators in these experiments makes it difficult to extrapolate these results to natural lake ecosystems. Whole-lake experiments that manipulated mixing and the thermocline to deeper depths, in fact suggested that declines in *Daphnia* and shifts to smaller cladocerans and copepods were most likely related to an increase in fish predation (Gauthier et al. 2014). Nevertheless, these experiments nicely demonstrate the potential importance of light limitation in determining zooplankton dynamics.

Acidification, upwelling, and nutrients can also accompany climate change and influence zooplankton dynamics. Increases in atmospheric CO_2 are acidifying many large lakes and oceans, with important consequences for zooplankton distribution and abundance (Hammill et al. 2018, Weiss et al. 2018). A combination of whole-lake and mesocosm experiments manipulating temperature and pH showed the potential for strong synergistic effects of these two variables on zooplankton, with potentially positive effects on specialists like *Daphnia catawba* that do better under acidic conditions and at warmer temperatures at this northerly latitude in Canada (Christensen et al. 2006). In the Antarctic and Southern ocean, zooplankton are responding to changes in not only warming, but freshening of the waters, a poleward shift in westerly winds, and consequent changes in current eddies (Constable et al. 2014). Coastal marine upwelling can also bring colder, nutrient-rich water to the surface, which stimulates large increases in zooplankton as surface waters become cooler (Tremblay et al. 2011).

Thermal Optima and Changes in Metabolism and Body Size

There is no single general pattern of change for zooplankton relative to warmer ocean water temperatures. This is not surprising given the variation in thermal tolerances among zooplankton species. Some studies show a positive effect of warmer temperatures on copepod abundance (Paul et al. 2016), while others show a negative effect with decreases in abundance at warmer temperatures (Garzke et al. 2015, Rice and Stewart 2016). Although knowledge on the thermal tolerances and metabolic responses of many groups of zooplankton is limited (Atkinson et al. 2012), warming is often implicated as an important regulator of the population and community dynamics of zooplankton. For example, different temperature tolerances of two species of *Calanus* appear to play a pivotal role in their major range expansions in the North Sea (Beaugrand et al. 2002, Beaugrand et al. 2015, Reid et al. 2016). Declines in primary productivity in the North Sea may also be driven by temperature and underlie declines in small but not large copepods (Capuzzo et al. 2018). A comparison of the differences in the population dynamics of three freshwater cyclopoid copepods to periods of warm (1992–1999) vs. cool summers (1980–91) revealed that the two more thermophilic species had longer periods of activity in the plankton with greater abundance during the warmer years, while the more eurythermal species showed little or no response to warming (Gerten and Adrian 2002).

The water temperature range being tested relative to the thermal optimum for a given species (cold or warm-water) will play a critical role in determining the response of zooplankton to warming temperatures (Klais et al. 2017). In lakes, smaller cladocerans, such as *Bosmina* may increase at warmer temperatures, while larger cladocerans, such as *Daphnia* decrease (Li et al. 2016). Yet in colder, more northerly lakes there is evidence for an increase in *Daphnia* during warmer years (Schindler et al. 2005, Fischer et al. 2011). Similarly, there has been over a three-fold increase in *Daphnia* in coldwater Lake Baikal in Siberia as surface water temperatures have warmed in recent decades (Hampton et al. 2008).

The thermal ecology is quite well understood for some zooplankton, such as *Daphnia*, (Moore et al. 1996, Gillooly 2000, Gillooly and Dodson 2000, Chen and Folt 2002, Lampert 2011, Straile et al. 2012, Wojtal-Frankiewicz 2012). An extensive examination of the population dynamics of *Daphnia* across latitudes in North America led the authors to suggest that temperature was more important than food resources or predators in regulating both their phenology and abundance across latitudes (Gillooly and Dodson 2000). The date of maximum abundance of *Daphnia* varied by over 100 days going from Florida to Alaska, with maximum abundance being observed consistently at temperatures of 18.5°C across all latitudes. Temperature phenology alone explained 57% of the variability in the timing of the maximum *Daphnia* population across 66 phenologies and 49 diverse lake sites (Straile et al. 2012). The authors suggested that this primacy of temperature was due to the primary growth period of *Daphnia* in the spring being a period of high food availability and low predator abundance. Collectively these studies point to temperature being important, but given the importance of mixing depth and other temperature-related variables, the relative importance of direct vs. indirect effects of temperature remains unclear.

One of the primary mechanisms through which warmer temperatures influence zooplankton is by altering both their body size and metabolic rates. A warming climate favors smaller body sizes in many taxa of organisms in aquatic ecosystems (Daufresne et al. 2009). One of the most widely known ecological rules that is relevant to the effects of climate change on zooplankton body size is Bergmann's rule, which, based on the concept of heat conservation, states that body size decreases at warmer temperatures (Bergmann 1847). Consistent with Bergmann's rule, an extensive examination of the population dynamics of freshwater cladocerans across over 1,100 lakes in the western hemisphere demonstrated that mean cladoceran body size is reduced by a factor of about two-fold as temperature increases from a peak body size at 6–8°C to over 25°C (Gillooly and Dodson 2000). A corresponding decrease in body size was observed at lower latitudes in both the northern and southern hemispheres. Part of this relationship is due to a decrease in body size within a genus or species, but most of the pattern is related to a dominance of smaller-bodied cladoceran species at warmer temperatures, with a transition from larger-bodied *Daphnia* to smaller-bodied genera like *Bosmina*. Lakes with more transparent waters often have colder waters. Thus zooplankton body sizes would be expected to be larger in more transparent lakes, a pattern that has been confirmed for cladocerans in 59 lakes in the northeastern United States (Stemberger and Miller 2003). The implication is that climate warming will lead to reduced body size in freshwater cladocerans. The lesser ability of smaller zooplankton to reduce phytoplankton biomass relative to available nutrients (Pace 1984) will increase chances of harmful algal blooms in inland waters. Smaller body size is similarly associated with warmer temperatures in marine zooplankton (Eisner et al. 2014, Garzke et al. 2015, Kelly et al. 2016). A ten year time series (2002–12) in the South Pacific included a five-year period of less intense thermal stratification with a colder, deeper mixed layer, and shallower oxygen minimum zone, conditions opposite many of the climate warming trends. During this period with a colder, deeper mixed layer, small copepods were replaced by large and medium copepods (Medellin-Mora et al. 2016).

Modeling and evidence from first principles have shown that metabolic rates of organisms across a wide variety of taxa, including zooplankton, increase exponentially with warming temperatures according to the Boltzmann-Arrhenius relationship (Gillooly et al. 2001, Gillooly et al. 2006). Mass-specific metabolic rate is also inversely related to body size as a function of the negative one-quarter- power (Gillooly et al. 2001, Gillooly et al. 2006). This points to climate warming having a dual effect on accelerating zooplankton mass-specific metabolic rates related to both warmer temperatures and smaller body sizes. Yet more important than the temperature dependence of metabolic rates alone, is the effect of temperature on metabolic balance (Alcaraz et al. 2014). At the ecosystem level, for example, the more rapid increase in respiration vs. photosynthesis in response to warmer temperatures could create a major imbalance that pushes ecosystems over thresholds or beyond tipping points to cause major regime shifts (Alcaraz et al. 2014). The importance of considering both sides of the metabolic balance has been nicely demonstrated with a case study in *Calanus glacialis*, a large calanoid that inhabits the Arctic (Alcaraz et al. 2014). The temperature dependence of ingestion rates was compared to the temperature dependence of respiration rates between 0°C and 10°C. While ingestion rates declined above a peak rate at 2°C, respiration rates continued to increase up to 6°C. This suggests that a critical tipping point exists at 6°C, above which *C. glacialis* cannot exist because respiration rates exceed ingestion. This pattern was supported by the natural distribution of this species which was rare or absent in waters above 6°C (Alcaraz et al. 2014).

Changes in Geographic Distribution and Diversity

While there have been some strong increases in the movement of marine zooplankton towards higher latitudes as expected with climate change (Parmesan and Yohe 2003), this pattern is not characteristic of freshwaters where zooplankton are more constrained in islands of water in the landscape. Marine zooplankton have shown a poleward shift of some warm-water species, and in some cases expansion of the latitudinal distribution of cold-water species (Richardson 2008, Hinder et al. 2014). Such poleward shifts have been observed in both the northern hemisphere (Mackas et al. 2007, Batten and Walne 2011, Feng et al. 2018) and the southern hemisphere (Mackey et al. 2012, Constable et al. 2014, Kelly et al. 2016). One of the most notable range expansions of zooplankton has been the 1,100 km extension of the range of *Calanus* in the North Atlantic Ocean over five decades (Beaugrand et al. 2002). This shift led to the replacement of the cold-water species *C. finmarchicus*, which exhibits its population peak in the spring, by the warmer-water species *C. helgolandicus*, which historically peaked in the autumn. These changes in zooplankton were accompanied by a strong decline in the recruitment of salmon (Beaugrand and Reid 2003) and cod (Beaugrand et al. 2003) in the North Sea. Shifts of the cold-water species *C. hyperboreus* to more southerly waters have also been observed, but this appears to be associated with cold-water currents that extend further southward (Richardson 2008). In contrast, mesozooplankton communities in the Southern Ocean showed little or no range shift in response to waters that warmed 0.74°C and a poleward shift of temperature isotherms of 500 km over 60 years (Tarling et al. 2018).

Latitudinal gradients in biodiversity have long been recognized, with generally greater diversity of many plants and animals in warmer climates in tropical regions than at higher latitudes (Pianka 1966). Though there are many potentially confounding factors, such as space-for-time analyses of changes in diversity with latitude are another approach that may inform us how zooplankton respond to climate warming. While marine copepods do not show any distinct pattern in diversity across latitudes in the southern hemisphere, in the northern hemisphere they do exhibit greater diversity at lower latitudes (Rombouts et al. 2009). Freshwater zooplankton communities similarly tend to be less diverse at both higher latitudes and higher elevations. In shallower alpine lakes exposure to high levels of UV radiation have been implicated in reducing zooplankton diversity to even a single species at higher elevations (Marinone et al. 2006). The biodiversity of foraminifera and some other marine zooplankton peaks at mid-latitudes. A previous study found that sea surface temperature remarkably explains almost 90% of the diversity of foraminiferans in the Atlantic Ocean (Rutherford et al. 1999). This suggests that all else being equal, climate warming is likely to increase diversity at higher latitudes but decrease it at lower latitudes.

Changes in Vertical Distribution

Simple shifts in the depth distribution and vertical migration of zooplankton in response to warmer temperatures are hard to separate from other drivers due to the presence of complex vertical habitats in lakes and oceans (Williamson et al. 2011). Changes in water transparency induce strong variations in vertical habitat gradients in light, temperature, oxygen, food, and the relative abundance of tactile vs. visual predators of zooplankton. Thus while surface temperatures of lakes and oceans are generally warmer, simple efforts to relate high temperature tolerance limits of zooplankton to their fitness will necessarily fall short. For example, an analysis of 45 years of data in Lake Baikal (1955–2000) indicated that rotifers, cladocerans, and juvenile copepods (but not adult copepods) have shifted to shallower depths during a period of warming surface temperatures (Hampton et al. 2014). While in this very cold Siberian lake, warmer surface waters could clearly provide a demographic advantage for the zooplankton, the authors note that the cause of this shallower depth distribution remains elusive.

Climate warming is melting glaciers and the resulting glacial flour is decreasing water transparency to PAR and UV radiation (Rose et al. 2014, Olson et al. 2017). A study looking at changes in *Daphnia* vs. calanoid copepods (*Boeckella*) along a 10-km-long section of a narrow Patagonian lake where glacial flour entered at one end, found an increase in the relative and absolute abundance of the copepods with increasing water transparency and distance from the glacial input (Hylander et al. 2011). Inorganic particles, such as silt, are known to inhibit feeding and decrease the abundance of *Daphnia* and other cladocerans, but higher turbidity also decreases fish predation (Kirk 1991, Jonsson et al. 2011). The implication is that the balance between the amount of turbidity from glacial flour and the abundance of planktivorous fish may determine zooplankton abundance and community structure as glaciers melt, and eventually disappear to leave waters of higher clarity but warmer temperatures.

Browning also decreases water transparency and strengthens the vertical habitat gradient. In one 27-year study in a clear-water lake where a doubling of DOM was observed, this long-term browning led to multiple changes in the vertical habitat gradient. There was more than a 10-fold decrease in hydrogen ion concentration, decreases in the 1% PAR depth from over 20 m to less than 15 m, decreases in the penetration of damaging UV radiation (1% of subsurface 320 nm irradiance) from over 10 m to less than 2 m, and substantial decreases in deep-water oxygen (Williamson et al. 2015, Pilla et al. 2018, Knoll et al. 2018). These changes in the vertical habitat gradients over this period were associated with major changes in the zooplankton (Williamson et al. 2015). The grazing crustaceans including *Daphnia* and calanoid copepods declined up to five-fold, while cyclopoid copepods and rotifers increased by up to 10-fold. The increase in hypolimnetic cyclopoids and decline of calanoids at all depths in this lake have led to a deeper depth distribution of copepods (Fig. 10.3). When browning and warmer temperatures were manipulated in mesocosm experiments in Sweden, rotifer populations peaked earlier in the year with warming, but these increases were offset by increases in predation by cyclopoid copepods (Zhang et al. 2015). Yet the mesocosm manipulations revealed little or no change in zooplankton related to the manipulation of browning alone. Clearly more research into the mechanisms that underlie long-term changes in lakes undergoing both warming and browning is needed to understand the relative importance of changes in water transparency and thermal stratification underlying the pronounced changes in zooplankton that have been observed.

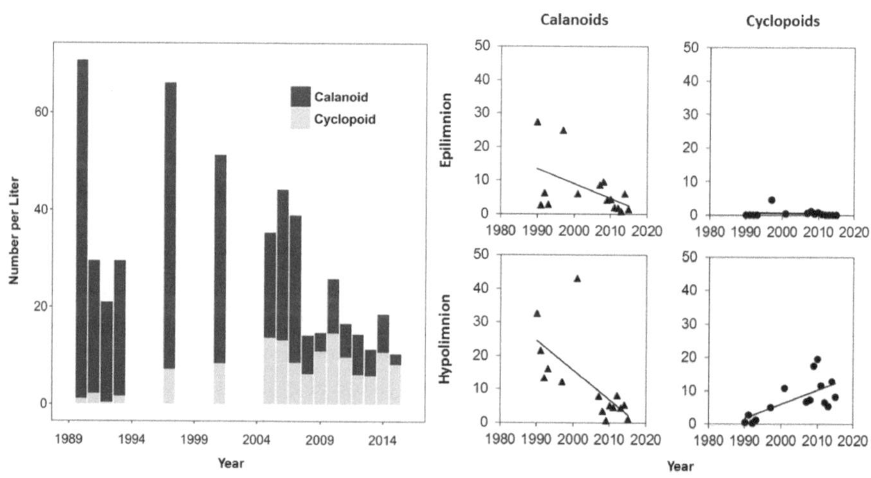

Fig. 10.3. Changes in summer copepod community structure in Lake Giles, northeastern Pennsylvania, over the past 27 yr. Browning in this lake has led to overall declines in copepod abundance, with a pronounced shift from calanoid dominance to more abundant cyclopoids (left). While the calanoids, primarily *Leptodiaptomus minutus*, have decreased in the epilimnion, the cyclopoid, *Cyclops scutifer*, a cold-water stenotherm rarely found in the epilimnion, has increased in the hypolimnion (right). Epilimnetic food availability for fish has consequently decreased substantially.

Decreases in deep-water oxygen in response to reductions in water transparency are likely to alter visual predation pressures on zooplankton by creating a vertical refuge. Planktivorous fish are generally less tolerant of low oxygen conditions than are zooplankton, with dissolved oxygen (DO) less than 2 mg L^{-1} often being lethal, while many zooplankton can survive DO concentrations down to less than 1 mg L^{-1} (Vanderploeg et al. 2009a, Vanderploeg et al. 2009b). Some zooplankton, such as *Mesocyclops* and *Chaoborus* can survive total anoxia for periods of many hours during vertical migration (Woodmansee and Grantham 1961; Williamson and Magnien 1982). In lakes, there is evidence that depths with hypoxic waters may thus be an effective refuge from visual predators, especially for larger zooplankton, such as *Daphnia* (Hanazato et al. 1989, Wright and Shapiro 1990, Tessier and Welser 1991), as well as tactile-feeding invertebrate predators of zooplankton, such as larvae of the phantom midge, *Chaoborus* (Wissel et al. 2003). The ability to use these hypoxic refugia will depend on the relative hypoxia tolerance of the particular zooplankton and fish species (Vanderploeg et al. 2009a). The presence of such a refuge from fish predation has been shown to be associated with up to a 10-fold increase in the predatory cladoceran *Bythotrephes* (Yan et al. 2001, Manca et al. 2007). Increases in the depth and duration of a deep-water refuge related to long-term warming water temperatures can explain increases in *Bythotrephes* better than changes in food resources or predator abundance (Manca et al. 2007, Manca and DeMott 2009). The consequence of these increases in *Bythotrephes* was a strong decline in *Daphnia*. This example highlights the importance of seasonal changes in not only water temperatures to key zooplankton species, but changes in critically important habitat gradients that determine vertical overlap of predator and prey and the availability of refugia. Periods of hypoxia can also compress *Daphnia* into narrow vertical strata in the water column where there is suboptimal habitat that can reduce their condition (RNA content) and body size (Goto et al. 2012), reproduction (Hanazato et al. 1989), and in some cases even increase fish predation as both fish and *Daphnia* become concentrated in the same strata (Vanderploeg et al. 2009b). When the hypoxia refuge hypothesis was tested in marine systems, there was no evidence for reduced vertical overlap between zooplankton and their fish predators (Sato et al. 2016).

The effects of DOM and anoxia on zooplankton in the context of climate change have also been observed in both synoptic studies and mesocosm experiments in marine and inland waters. Low oxygen conditions in the deep waters of the Baltic Sea associated with increasingly frequent climate change-induced stagnation events are contributing to declines in copepods (Moller et al. 2015). Mesocosm experiments where leaves were added in the autumn revealed lower water transparency, reduced dissolved oxygen, and higher phosphorus concentrations than controls the following spring, with a positive effect on cyclopoid copepods but a negative effect on calanoids and cladocerans (Fey et al. 2015). Evidence from both land-based mesocosms and lake sampling in May in southern Sweden found that warmer temperatures had stronger effects than browning on zooplankton recruitment and pelagic abundance and strongly favored cladocerans over copepods (Ekvall and Hansson 2012).

Changes in Phenology

Seasonal variation in incoming solar radiation to any given region on the Earth creates corresponding variations in underwater light and temperature with many consequences for zooplankton. Seasonal variations in temperature vary with latitude, with the greatest fluctuations at higher latitudes. In the world's oceans, the magnitude of seasonal water temperature ranges are on the order of 1–4°C in tropical and equatorial regions, 4–8°C at south temperate latitudes, and up to 8–12°C or greater in many regions at north temperate latitudes (Mackas et al. 2012). Phenological shifts during periods close to spring or autumn equinoxes will lead to the strongest changes in light environment because this is the period of the most rapid change in photoperiod and sun angle. During solstice periods, incident PAR changes at a rate of close to 1% per day, and incident DNA-damaging UV radiation by more than 2% per day (S. Madronich, personal communication).

Seasonal variations in water temperature in both inland and oceanic waters lead to alternating periods of optimal and suboptimal food resources and conditions for zooplankton reproduction, juvenile growth, dormancy, and predator avoidance. The phenological fine-tuning and synchrony of these life history events with optimal environmental conditions over a long-term evolutionary time scale is being disrupted by human-accelerated climate change. The seasonal timing of extreme climate-related events, such as heat waves, hurricanes, cyclones, or heavy precipitation, can thus be critically important to the magnitude of the effect on zooplankton (Huber et al. 2010). While uniform warming throughout the year may have little influence on the synchrony of consumers and their resources, disproportionally warm events during critical times in the life cycle may have severe consequences for zooplankton (Rusak et al. 2008, Wojtal-Frankiewicz 2012, Straile et al. 2015). One particularly critical event for zooplankton in this respect is the spring clearwater phase (Huber et al. 2010).

The most widely observed changes in zooplankton phenology related to climate warming are a shift to an earlier time period of critical events in the life cycle, such as periods of juvenile growth or peak population abundance (Mackas et al. 2012). In a few cases there is a shift to a later time period. The average phenological shift of 7.6 d earlier observed in marine zooplankton is much greater than what has been observed for terrestrial taxa, which range from an average of 2.3–6 days for a variety of herbs, trees, butterflies, amphibians, birds, and other taxa (Richardson 2008). This paper reported phenological shifts of 60 d earlier in warm vs. cool years in the peak biomass of *Neocalanus plumchrus* in the subarctic North Pacific. The release of larvae of many large marine invertebrates has moved to 27–47 d earlier over the past 4–5 decades (Richardson 2008). In contrast, holozooplankton, such as copepods that spend their entire lives in the pelagic zone have shifted their peak abundances to about 10 days earlier during this same time period (Richardson 2008). Analysis of changes in the phenology of *Daphnia* populations in Lake Windermere using ten different metrics spanning eight decades showed advances in timing of population events by between 3.7–6.7 d (Thackeray et al. 2012). Analysis of 24 years of data on a lake in Germany showed that peak abundance of *Daphnia* has advanced one month in response to warmer water temperatures in the spring (Adrian et al. 2006). In this same study zooplankton with more complex life cycles showed less pronounced and species-specific phenological responses.

Historically the greatest concern regarding climate change-induced changes in phenology has been that climate warming will lead to a mismatch between commercially important fish and their zooplankton prey, which could in turn decrease fishery production (Cushing 1990). From the perspective of the zooplankton, the potential for a trophic mismatch arises when temperature warming alters the synchrony between their periods of peak feeding, growth, and reproduction, with periods of peak food availability, or alternatively seasonal peaks in predator (e.g., fish) abundance (Fig. 10.1). For example, while marine holozooplankton have shown shifts in their peak abundances to 10 d earlier, peaks in their food resources, including diatoms and dinoflagellates, have shifted to 23–24 d earlier (Richardson 2008). In Lake Washington, phytoplankton were observed to shift their peaks to 19 d earlier over a 40-yr period while zooplankton shifted by 15 d earlier in the rotifer *Keratella*, but only 9 d earlier in the calanoid copepod *Leptodiaptomus*, and no significant shift in *Daphnia* (Winder and Schindler 2004b). Consequently, *Daphnia* exhibited strong declines in their abundance through this study period while *Keratella* did not (Winder and Schindler 2004a). A mismatch is also likely when consumer and resource dynamics are dependent upon different environmental cues. For example, a mismatch would be expected if phytoplankton growth rates are tied more to seasonal increases in light availability, while zooplankton growth rates are more temperature dependent (Wojtal-Frankiewicz 2012). In spite of numerous studies reporting mismatches between zooplankton and their prey, a meta-analysis of studies on the mismatch hypothesis in plankton argues that few studies have provided convincing evidence to support a climate change-induced mismatch between zooplankton and their food resources (Thackeray 2012). In some oceanic regions there is evidence that climate warming has actually increased the synchrony (match) between commercially important fish and their zooplankton food resources (Mackas et al. 2012). In spite of the difficulties in securing enough data for convincing evidence for individual cases of climate-induced mismatches involving zooplankton, the differences in the phenological responses among trophic levels from phytoplankton to zooplankton and fish (Thackeray et al. 2012), suggest mismatches are common, and may have serious consequences for food webs and hence the ecosystem services provided by zooplankton.

One of the best-studied systems is that of the trophic mismatch between juvenile cod and the timing of two prey species of copepods in the genus *Calanus* in the North Sea. During warmer years this mismatch is thought to contribute to the decline of cod (Beaugrand et al. 2003). In normal years, *C. finmarchicus* peak in abundance during the juvenile cod feeding period from March–August, but in warmer years *C. finmarchicus* is rare, and *C. helgolandicus*, which historically peaked in later fall months after the peak feeding period of juvenile cod, dominates the plankton. In some cases, warmer temperatures can increase the number of generations of copepods per year, which can help sustain or even increase abundance. For example, a study of the abundance of *C. finmarchicus* in the North Atlantic from 2001–11 found evidence that at warmer temperatures there may be a second generation, which in turn may enhance both bird and fish populations (Weydmann et al. 2018). In more recent years there is evidence for a second, earlier peak of *C. helgolandicus* in the North Sea that includes a greater abundance during the time of peak cod feeding (Fig. 10.4). In lakes, there is evidence that even when warmer temperatures lead to a second generation each year, declines in freshwater copepod populations may still occur (Winder et al. 2009).

Many zooplankton use diapause to survive periods of suboptimal temperature, food, or predators. In *C. finmarchicus*, climate warming may lead to substantial reductions in the duration of diapause, which in turn may alter the dynamics of lipid metabolism that could reduce reproduction and increase mortality (Wilson et al. 2016). In colder environments, warmer water temperatures may shorten development times and increase the ability of some large Arctic marine copepods, such as *C. hyperboreus* to successfully reach diapause (Feng et al. 2018). Changes in the timing of diapause may in turn lead to exposure to suboptimal conditions due to asynchrony with seasonal abundance of food supplies and predator avoidance, creating strong pressures for evolutionary adaptation (Miller et al. 2018).

The clearwater phase is a seasonal period of increased water transparency that in lakes is generated by increases in grazing by large cladocerans, such as *Daphnia* (Sommer et al. 2012). In lakes, since the seasonal timing of the peak in *Daphnia* populations is closely tied to water temperature phenology (Straile et al. 2012), water temperature can also regulate the seasonal timing of the clearwater phase. In some cases, spring precipitation can also play a key role in the interannual variation in the clearwater phase (Dröscher et al. 2009). There are potentially important feedbacks here as well because the intensity of a clearwater phase can regulate both surface and deep-water temperatures in more eutrophic lakes, and even regulate whether or not shallower lakes will be polymictic (Shatwell et al. 2016). Years with strong spring

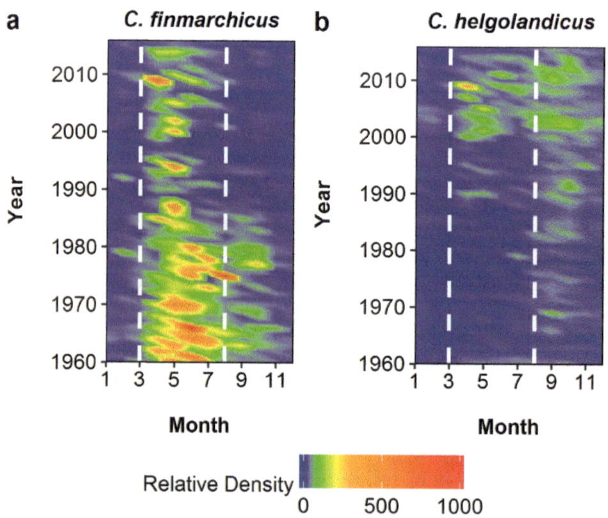

Fig. 10.4. Changes in abundance and phenology of the two major *Calanus* species in the North Sea. *C. finmarchicus* has shown substantial decreases in abundance over the past five decades. In contrast, *C. helgolandicus* has shown a modest increase in abundance as well as a phenological shift that increases its seasonal match with juvenile cod, which are most abundant in the North Sea between March and August (dashed vertical lines). *Calanus* species data for the North Sea were obtained courtesy of David Johns from the CPR Survey (www.cprsurvey.org DOI:10.7487/2018.100.1.1128).

phytoplankton blooms, little or no clearwater phase, and low water transparency, lead to longer and stronger thermal stratification, while years with weak spring blooms, a strong clearwater phase and higher water transparency lead to weaker thermal stratification and polymixis (Shatwell et al. 2016). Clearwater phase events have the potential to cause regime shifts in shallow lakes wherein rooted plants can become dominant in place of planktonic algae due to more light reaching the bottom of the lake (Scheffer et al. 2001). This demonstrates the critical feedbacks between the strength and magnitude of the population peaks of key zooplankton species like *Daphnia*, and the timing of warmer water temperatures and strength of thermal stratification. In marine systems the link between the clearwater phase and zooplankton grazers is less well defined, and if present, not always attributed to zooplankton grazing. This may be due to the abundance of copepods with their more diverse and omnivorous feeding habits and thus less distinct grazer-phytoplankton effects (Sommer et al. 2012).

One of the most pronounced phenological responses of aquatic ecosystems to climate change is the shorter duration of winter ice cover in temperate to polar regions. A study of 75 northern hemisphere lakes showed that ice duration has decreased on the order of 4.3 d per decade over the past 30 yr, accelerating from only 0.7 d per decade over the past 100 yr (Benson et al. 2012). The sensitivity of ice cover to warming air temperatures can vary greatly with latitude, for instance from 14 d per 1°C in Southern Sweden to 4 d per 1°C in Northern Sweden (Livingstone et al. 2010). The geographic extent of minimum summer ice cover in the Arctic Sea has similarly been showing strong trends of reduction on the order of 13% per decade (https://climate.nasa.gov/vital-signs/arctic-sea-ice/). Zooplankton respond to changes in ice cover. Freshwater zooplankton are less abundant under ice cover in the winter, though with no substantial overall differences in community structure (Hampton et al. 2017). The timing of ice cover can influence zooplankton abundance during the ice-free season as well. For example, the phenology of ice-cover was shown to be significantly related to several species of freshwater zooplankton in a suite of five north temperate lakes between 1982 and 2005 (Preston and Rusak 2010). Herbivorous cladoceran abundance in particular was higher in years with earlier ice-out in this study, though there was no strong relationship between omnivores or predators and ice-out date. Consistent with these field studies, experiments that exposed freshwater sediments to different temperatures and photoperiods showed that warmer temperatures accelerated hatching, and that the effect was stronger on cladocerans than on copepods or rotifers; photoperiod did not influence time to hatching (Jones and Gilbert 2016).

In marine systems, there is evidence that changes in ice thickness and extent of coverage can lead to a mismatch between zooplankton and their food resources. Marine copepods in the genus *Calanus* ascend from deep waters of the Arctic Sea to feed on the spring phytoplankton blooms. As they ascend they also develop increased concentrations of photoprotective compounds to reduce damage by UV radiation at the surface (Hylander et al. 2015). Their feeding on both ice algae and open-water algae is timed to take advantage of peaks in phytoplankton abundance and food quality (Soreide et al. 2010). The more rapid seasonal loss of Arctic Sea ice could jeopardise these important zooplankton grazers in two critical ways. First, it may generate a mismatch between the timing of reproduction and juvenile growth for the *Calanus*. In addition, UV exposure levels in these surface waters are predicted to increase by

10-fold with the disappearance of the Arctic Sea ice during the summer (Fountoulakis et al. 2014). The ability of these zooplankton to adapt to these changes is unknown.

Brief Synthesis and Knowledge Gap

The message that emerges here is that the effects of climate change on zooplankton are not just about warming, but about many indirect and interactive effects as well. Zooplankton are some of the strongest sentinels of climate change, with their responses often stronger than the changes that can be measured in the climate variables themselves. Zooplankton are responding to climate change with smaller body sizes, changes in their phenology, and changes in their vertical distribution in the water column as well as their geographic range. These patterns have far-reaching implications for both inland and marine waters. For example, smaller zooplankton create weaker links in aquatic food webs due to the lower feeding rates of smaller individuals and species. Their smaller size makes them less available as food for visually feeding fish, many of which are of great commercial importance and serve to feed a substantial portion of the human population on Earth. Large zooplankton such as *Daphnia* in freshwater systems have a greater ability to clear water of phytoplankton. The climate warming-induced smaller body sizes may thus lead to declines in water quality that could increase the likelihood of harmful algal blooms.

While there is evidence for a shift in marine food webs from copepods to gelatinous zooplankton (Purcell 2005, Purcell et al. 2007, Richardson 2008, Richardson et al. 2009), how widespread these trends are in the world's oceans as well as the role of climate change remain uncertain (Condon et al. 2012). This knowledge gap has important implications for both the global carbon cycle (Winder et al. 2017), as well as marine fisheries, tourism, and a variety of other critical ecosystem services (Purcell 2005, Purcell et al. 2007, Richardson 2008, Richardson et al. 2009). There is evidence from land-based mesocosm experiments that some jellyfish, such as appendicularians do better in warmer waters with a lower pH and that they may actually have the beneficial effect of accelerating the loss of fixed carbon from surface ocean waters (Winder et al. 2017). In lakes with low calcium concentrations there are also signs of a different type of jellification wherein zooplankton with high calcium requirements for their carapaces are being replaced by *Holopediium*, a cladoceran surrounded with a gelatinous sheath rather than a carapace (Jeziorski et al. 2014). These are just a few of the shifts in zooplankton community structure that demonstrate our inability to separate the relative importance of the direct effects of warming temperature from the multitude of interactive and indirect effects driven by both climate change and other environmental change. The indirect effects of climate and other types of anthropogenic environmental change may greatly augment, and in some cases be a greater threat than the direct effects of climate warming (Winder 2012, Domis et al. 2013, Pilla et al. 2018, Knoll et al. 2018).

The interactive effects of long-term trends of climate warming, climate oscillations, extreme climate events, trophic amplification, ocean acidification, and the development of dead zones and harmful algal blooms in inland and marine waters are all severe threats to aquatic ecosystems related to climate change. As the most abundant metazoans on Earth, zooplankton play a pivotal role in the future preservation of the water quality and pelagic food webs on which humans are intrinsically dependent for their survival. The strong sentinel responses of zooplankton to environmental change highlight their potential to play a key role in creating new insights into effective mitigation of these multiple threats to the critical ecosystem services provided by our world's lakes and oceans.

References

Adrian, R., S. Wilhelm and D. Gerten. 2006. Life-history traits of lake plankton species may govern their phenological response to climate warming. Global Change Biol. 12: 652–661.

Alcaraz, M., J. Felipe, U. Grote, E. Arashkevich and A. Nikishina. 2014. Life in a warming ocean: thermal thresholds and metabolic balance of arctic zooplankton. J. Plankton Res. 36: 3–10.

Atkinson, A., P. Ward, B.P.V. Hunt, E.A. Pakhomov and G.W. Hosie. 2012. An overview of southern ocean zooplankton data: abundance, biomass, feeding and functional relationships. CCAMLR Sci. 19: 171–218.

Austin, J. and S. Colman. 2008. A century of temperature variability in Lake Superior. Limnol. Oceanogr. 53: 2724–2730.

Batten, S.D. and A.W. Walne. 2011. Variability in northwards extension of warm water copepods in the NE Pacific. J. Plankton Res. 33: 1643–1653.

Beaugrand, G., K.M. Brander, J.A. Lindley, S. Souissi and P.C. Reid. 2003. Plankton effect on cod recruitment in the North Sea. Nature 426: 661–664.

Beaugrand, G., A. Conversi, S. Chiba, M. Edwards, S. Fonda-Umani, C. Greene, N. Mantua, S.A. Otto, P.C. Reid, M.M. Stachura, L. Stemmann and H. Sugisaki. 2015. Synchronous marine pelagic regime shifts in the Northern Hemisphere. Philos. Trans. R. Soc. B-Biol. Sci. 370. doi:10.1098/Rstb.2013.0272.

Beaugrand, G. and P.C. Reid. 2003. Long-term changes in phytoplankton, zooplankton and salmon related to climate. Global Change Biol. 9: 801–817.

Beaugrand, G., P.C. Reid, F. Ibanez, J. A. Lindley and M. Edwards. 2002. Reorganization of North Atlantic marine copepod biodiversity and climate. Science 296: 1692–1694.

Benson, B.J., J.J. Magnuson, O.P. Jensen, V.M. Card, G. Hodgkins, J. Korhonen, D.M. Livingstone, K.M. Stewart, G.A. Weyhenmeyer and N.G. Granin. 2012. Extreme events, trends, and variability in Northern Hemisphere lake-ice phenology (1855–2005). Clim. Change 112: 299–323.

Berger, S.A., S. Diehl, H. Stibor, G. Trommer, M. Ruhenstroth, A. Wild, A. Weigert, C.G. Jager and M. Striebel. 2007. Water temperature and mixing depth affect timing and magnitude of events during spring succession of the plankton. Oecologia 150: 643–654.

Bergmann, C. 1847. About the relationships between heat conservation and body size of animals. Goett. Stud. 3: 595–708.

Bernal, P.A. 1979. Large-scale biological events in the California Current. CalCOFI Report 20: 89–101.

Blenckner, T., R. Adrian, D.M. Livingstone, E. Jennings, G.A. Weyhenmeyer, D.G. George, T. Jankowski, M. Järvinen, C.N. Aonghusa, T. Nõges, D. Straile and K. Teubner. 2007. Large-scale climatic signatures in lakes across Europe: a meta-analysis. Global Change Biol. 13: 1314–1326.

Breitburg, D., L.A. Levin, A. Oschlies, M. Grégoire, F.P. Chavez, D.J. Conley, V. Garçon, D. Gilbert, D. Gutiérrez, K. Isensee, G.S. Jacinto, K.E. Limburg, I. Montes, S.W.A. Naqvi, G.C. Pitcher, N.N. Rabalais, M.R. Roman, K.A. Rose, B.A. Seibel, M. Telszewski, M. Yasuhara and J. Zhang. 2018. Declining oxygen in the global ocean and coastal waters. Science 359. Doi:10.1126/science.aam7240.

Brothers, S., J. Köhler, K. Attermeyer, H.P. Grossart, T. Mehner, N. Meyer, K. Scharnweber and S. Hilt. 2014. A feedback loop links brownification and anoxia in a temperate, shallow lake. Limnol. Oceanogr. 59: 1388–1398.

Capuzzo, E., C.P. Lynam, J. Barry, D. Stephens, R.M. Forster, N. Greenwood, A. McQuatters-Gollop, T. Silva, S.M. van Leeuwen and G.H. Engelhard. 2018. A decline in primary production in the North Sea over 25 years, associated with reductions in zooplankton abundance and fish stock recruitment. Global Change Biol. 24: e352–e364.

Chavez, F.P., J. Ryan, S.E. Lluch-Cota and M. Niquen. 2003. From anchovies to sardines and back: Multidecadal change in the Pacific Ocean. Science 299: 217–221.

Chelton, D.B., P.A. Bernal and J.A. Mcgowan. 1982. Large-scale interannual physical and biological interaction in the California Current. J. Mar. Res. 40: 1095–1125.

Chen, C.Y. and C.L. Folt. 2002. Ecophysiological responses to warming events by two sympatric zooplankton species. J. Plankton Res. 24: 579–589.

Christensen, M.R., M.D. Graham, R.D. Vinebrooke, D.L. Findlay, M.J. Paterson and M.A. Turner. 2006. Multiple anthropogenic stressors cause ecological surprises in boreal lakes. Global Change Biol. 12: 2316–2322.

Colebrook, J.M. 1978a. Changes in the zooplankton of the North Sea, 1948 to 1973. Rapp. P.-v Réun. Cons. Int. Explor. Mer. 172: 390–396.

Colebrook, J.M. 1978b. Continuous plankton records - zooplankton and environment, northeast Atlantic and North Sea, 1948–1975. Oceanol. Acta 1: 9–23.

Condon, R.H., W.M. Graham, C.M. Duarte, K.A. Pitt, C.H. Lucas, S.H.D. Haddock, K.R. Sutherland, K.L. Robinson, M.N. Dawson, M.B. Decker, C.E. Mills, J.E. Purcell, A. Malej, H. Mianzan, S.I. Uye, S. Gelcich and L.P. Madin. 2012. Questioning the rise of gelatinous zooplankton in the world's oceans. Bioscience 62: 160–169.

Constable, A.J., J. Melbourne-Thomas, S.P. Corney, K.R. Arrigo, C. Barbraud, D.K.A. Barnes, N.L. Bindoff, P.W. Boyd, A. Brandt, D.P. Costa, A.T. Davidson, H.W. Ducklow, L. Emmerson, M. Fukuchi, J. Gutt, M.A. Hindell, E.E. Hofmann, G.W. Hosie, T. Iida, S. Jacob, N.M. Johnston, S. Kawaguchi, N. Kokubun, P. Koubbi, M.A. Lea, A. Makhado, R.A. Massom, K. Meiners, M.P. Meredith, E.J. Murphy, S. Nicol, K. Reid, K. Richerson, M.J. Riddle, S.R. Rintoul, W.O. Smith, C. Southwell, J.S. Stark, M. Sumner, K.M. Swadling, K.T. Takahashi, P.N. Trathan, D.C. Welsford, H. Weimerskirch, K.J. Westwood, B.C. Wienecke, D. Wolf-Gladrow, S.W. Wright, J.C. Xavier and P. Ziegler. 2014. Climate change and Southern Ocean ecosystems I: how changes in physical habitats directly affect marine biota. Global Change Biol. 20: 3004–3025.

Costanza, R., R. dArge, R. deGroot, S. Farber, M. Grasso, B. Hannon, K. Limburg, S. Naeem, R.V., ONeill, J. Paruelo, R.G. Raskin, P. Sutton and M. vandenBelt. 1997. The value of the world's ecosystem services and natural capital. Nature 387: 253–260.

Cushing, D.H. 1990. Plankton production and year-class strength in fish populations: an update of the match/mismatch hypothesis. Adv. Mar. Biol. 26: 249–293.

Daufresne, M., K. Lengfellner and U. Sommer. 2009. Global warming benefits the small in aquatic ecosystems. Proc. Natl. Acad. Sci. USA 106: 12788–12793.

de Eyto, E., E. Jennings, E. Ryder, K. Sparber, M. Dillane, C. Dalton and R. Poole. 2016. Response of a humic lake ecosystem to an extreme precipitation event: physical, chemical, and biological implications. Inland Waters 6: 483–498.

De Stasio, B.T., Jr., D.K. Hill, J.M. Kleinhans, N.P. Nibbelink and J.J. Magnuson. 1996. Potential effects of global climate change on small north temperate lakes: Physics, fishes, and plankton. Limnol. Oceanogr. 41: 1136–1149.

De Stasio, B.T., Jr., N. Nibbelink and P. Olsen. 1993. Diel vertical migration and global climate change: A dynamic modeling approach to zooplankton behavior. Verh. Internat. Verein. Theor. Angew. Limnol. 25: 401–405.

Dodson, S.I. 1990. Predicting diel vertical migration of zooplankton. Limnol. Oceanogr. 35: 1195–1200.

Domis, L.N.D., J.J. Elser, A.S. Gsell, V.L.M. Huszar, B.W. Ibelings, E. Jeppesen, S. Kosten, W.M. Mooij, F. Roland, U. Sommer, E. Van Donk, M. Winder and M. Lurling. 2013. Plankton dynamics under different climatic conditions in space and time. Freshwat. Biol. 58: 463–482.

Dröscher, I., A. Patoine, K. Finlay and P.R. Leavitt. 2009. Climate control of the spring clear-water phase through the transfer of energy and mass to lakes. Limnol. Oceanogr. 54: 2469–2480.

Eisner, L.B., J.M. Napp, K.L. Mier, A.I. Pinchuk and A.G. Andrews. 2014. Climate-mediated changes in zooplankton community structure for the eastern Bering Sea. Deep-Sea Res. Pt. II 109: 157–171.

Ekvall, M.K. and L.A. Hansson. 2012. Differences in recruitment and life-history strategy alter zooplankton spring dynamics under climate-change conditions. PLOS One 7. doi:10.1371/journal.pone.0044614.

Evans, C.D., P.J. Chapman, J.M. Clark, D.T. Monteith and M.S. Cresser. 2006. Alternative explanations for rising dissolved organic carbon export from organic soils. Global Change Biol. 12: 2044–2053.

Fee, E.J., R.E. Hecky, S.E.M. Kasian and D.R. Cruikshank. 1996. Effects of lake size, water clarity, and climatic variability on mixing depths in Canadian Shield lakes. Limnol. Oceanogr. 41: 912–920.

Feng, Z., R. Ji, C. Ashjian, R. Campbell and J. Zhang. 2018. Biogeographic responses of the copepod *Calanus glacialis* to a changing Arctic marine environment. Global Change Biol. 24: e159–e170.

Fey, S.B. and K.L. Cottingham. 2011. Linking biotic interactions and climate change to the success of exotic *Daphnia lumholtzi*. Freshwat. Biol. 56: 2196–2209.

Fey, S.B., A.N. Mertens, L.J. Beversdorf, K.D. McMahon and K.L. Cottingham. 2015. Recognizing cross-ecosystem responses to changing temperatures: soil warming impacts pelagic food webs. Oikos 124: 1473–1481.

Fischer, J.M., M.H. Olson, N. Theodore, C.E. Williamson, K.C. Rose and J. Hwang. 2015. Diel vertical migration of copepods in mountain lakes: the changing role of ultraviolet radiation across a transparency gradient. Limnol. Oceanogr. 60: 252–262.

Fischer, J.M., M.H. Olson, C.E. Williamson, J.C. Everhart, P.J. Hogan, J.A. Mack, K.C. Rose, J.E. Saros, J.R. Stone and R.D. Vinebrooke. 2011. Implications of climate change for *Daphnia* in alpine lakes: predictions from long-term dynamics, spatial distribution, and a short-term experiment. Hydrobiologia 676: 263–277.

Fountoulakis, I., A.F. Bais, K. Tourpali, K. Fragkos and S. Misios. 2014. Projected changes in solar UV radiation in the Arctic and sub-Arctic Oceans: Effects from changes in reflectivity, ice transmittance, clouds, and ozone. J. Geophys. Res.: Atmos. 119: 8073–8090.

Fromentin, J.M. and B. Planque. 1996. *Calanus* and environment in the eastern North Atlantic. II. Influence of the North Atlantic Oscillation on *C. finmarchicus* and *C. helgolandicus*. Mar. Ecol. Prog. Ser. 134: 111–118.

Garzke, J., S.M.H. Ismar and U. Sommer. 2015. Climate change affects low trophic level marine consumers: warming decreases copepod size and abundance. Oecologia 177: 849–860.

Gauthier, J., Y. T. Prairie, and B.E. Beisner. 2014. Thermocline deepening and mixing alter zooplankton phenology, biomass and body size in a whole-lake experiment. Freshwat. Biol. 59: 998–1011.

George, D.G. and G.P. Harris. 1985. The effect of climate on long-term changes in the crustacean zooplankton biomass of Lake Windermere, UK. Nature 316: 536–539.

George, D.G. and A.H. Taylor. 1995. UK plankton and the Gulf Stream. Nature 378: 139.

Ger, K.A., L.A. Hansson and M. Lurling. 2014. Understanding cyanobacteria-zooplankton interactions in a more eutrophic world. Freshwat. Biol. 59: 1783–1798.

Ger, K.A., P. Urrutia-Cordero, P.C. Frost, L.A. Hansson, O. Sarnelle, A.E. Wilson and M. Lurling. 2016. The interaction between cyanobacteria and zooplankton in a more eutrophic world. Harmful Algae 54: 128–144.

Gerten, D. and R. Adrian. 2000. Climate-driven changes in spring plankton dynamics and the sensitivity of shallow polymictic lakes to the North Atlantic Oscillation. Limnol. Oceanogr. 45: 1058–1066.

Gerten, D. and R. Adrian. 2002. Species-specific changes in the phenology and peak abundance of freshwater copepods in response to warm summers. Freshwat. Biol. 47: 2163–2173.

Gillooly, J.F. 2000. Effect of body size and temperature on generation time in zooplankton. J. Plankton Res. 22: 241–251.

Gillooly, J.F., A.P. Allen, V.M. Savage, E.L. Charnov, G.B. West and J.H. Brown. 2006. Response to Clarke and Fraser: effects of temperature on metabolic rate. Funct. Ecol. 20: 400–404.

Gillooly, J.F., J.H. Brown, G.B. West, V.M. Savage and E.L. Charnov. 2001. Effects of size and temperature on metabolic rate. Science 293: 2248–2251.

Gillooly, J.F. and S.I. Dodson. 2000. Latitudinal patterns in the size distribution and seasonal dynamics of new world, freshwater cladocerans. Limnol. Oceanogr. 45: 22–30.

Goto, D., K. Lindelof, D.L. Fanslow, S.A. Ludsin, S.A. Pothoven, J.J. Roberts, H.A. Vanderploeg, A.E. Wilson and T.O. Hoeoek. 2012. Indirect consequences of hypolimnetic hypoxia on zooplankton growth in a large eutrophic lake. Aquat. Biol. 16: 217–227.

Häder, D.-P., C.E. Williamson, S.-Å. Wängberg, M. Rautio, K.C. Rose, K. Gao, E.W. Helbling, R.P. Sinha, and R. Worrest. 2015. Effects of UV radiation on aquatic ecosystems and interactions with other environmental factors. Photochem. Photobiol. Sci. 14: 108–126.

Hammill, E., E. Johnson, T.B. Atwood, J. Harianto, C. Hinchliffe, P. Calosi and M. Byrne. 2018. Ocean acidification alters zooplankton communities and increases top-down pressure of a cubozoan predator. Global Change Biol. 24: e128–e138.

Hampton, S.E., A.W.E. Galloway, S.M. Powers, T. Ozersky, K.H. Woo, R.D. Batt, S.G. Labou, C.M. O'Reilly, S. Sharma, N.R. Lottig, E.H. Stanley, R.L. North, J.D. Stockwell, R. Adrian, G.A. Weyhenmeyer, L. Arvola, H.M. Baulch, I. Bertani, L.L. Bowman, C.C. Carey, J. Catalan, W. Colom-Montero, L.M. Domine, M. Felip, I. Granados, C. Gries, H.-P. Grossart, J. Haberman, M. Haldna, B. Hayden, S.N. Higgins, J.C. Jolley, K.K. Kahilainen, E. Kaup, M.J. Kehoe, S. MacIntyre, A.W. Mackay, H.L. Mariash, R.M. McKay, B. Nixdorf, P. Nõges, T. Nõges, M. Palmer, D.C. Pierson, D.M. Post,

M.J. Pruett, M. Rautio, J.S. Read, S.L. Roberts, J. Rücker, S. Sadro, E.A. Silow, D.E. Smith, R.W. Sterner, G.E.A. Swann, M.A. Timofeyev, M. Toro, M.R. Twiss, R.J. Vogt, S.B. Watson, E.J. Whiteford, M.A. Xenopoulos and J. Grover. 2017. Ecology under lake ice. Ecol. Lett. 20: 98–111.

Hampton, S.E., D.K. Gray, L.R. Izmest'eva, M.V. Moore and T. Ozersky. 2014. The rise and fall of plankton: long-term changes in the vertical distribution of algae and grazers in Lake Baikal, Siberia. PLOS One 9. doi:10.1371/journal.pone.0088920.

Hampton, S.E., L.R. Izmest'eva, M.V. Moore, S.L. Katz and E.A. Silow. 2008. Sixty years of environmental change in the world's largest freshwater lake - Lake Baikal, Siberia. Global Change Biol. 14: 1947–1958.

Hanazato, T., M. Yasuno and M. Hosomi. 1989. Significance of a low oxygen layer for a *Daphnia* population in Lake Unoko, Japan. Hydrobiologia 185: 19–27.

Hare, S.R. and N.J. Mantua. 2000. Empirical evidence for North Pacific regime shifts in 1977 and 1989. Prog. Oceanogr. 47: 103–145.

Hays, G.C., A.J. Richardson and C. Robinson. 2005. Climate change and marine plankton. Trends Ecol. Evol. 20: 337–344.

Hinder, S.L., M.B. Gravenor, M. Edwards, C. Ostle, O.G. Bodger, P.L.M. Lee, A.W. Walne and G.C. Hays. 2014. Multi-decadal range changes vs. thermal adaptation for northeast Atlantic oceanic copepods in the face of climate change. Global Change Biol. 20: 140–146.

Huber, V., R. Adrian and D. Gerten. 2010. A matter of timing: Heat wave impact on crustacean zooplankton. Freshwat. Biol. 55: 1769–1779.

Hylander, S., T. Jephson, K. Lebret, J. von Einem, T. Fagerberg, E. Balseiro, B. Modenutti, M.S. Souza, C. Laspoumaderes, M. Jonsson, P. Ljungberg, A. Nicolle, P.A. Nilsson, L. Ranaker and L.A. Hansson. 2011. Climate-induced input of turbid glacial meltwater affects vertical distribution and community composition of phyto- and zooplankton. J. Plankton Res. 33: 1239–1248.

Hylander, S., T. Kiørboe, P. Snoeijs, R. Sommaruga and T.G. Nielsen. 2015. Concentrations of sunscreens and antioxidant pigments in Arctic *Calanus* spp. in relation to ice cover, ultraviolet radiation, and the phytoplankton spring bloom. Limnol. Oceanogr. 60: 2197–2206.

IPCC. 2014. Climate Change 2014: Impacts, Adaptation, and Vulnerability. Part A: Global and Sectoral Aspects. Contribution of Working Group II to the Fifth Assessment Report of the Intergovernmental Panel on Climate Change. Field, C.B., V.R. Barros, D.J. Dokken, K.J. Mach, M.D. Mastrandrea, T.E. Bilir, M. Chatterjee, K.L. Ebi, Y.O. Estrada, R.C. Genova, B. Girma, E.S. Kissel, A.N. Levy, S. MacCracken, P.R. Mastrandrea and L.L.White (eds.). Cambridge University Press, Cambridge, United Kingdom and New York, NY, USA, 1132 pp.

Jeziorski, A., A.J. Tanentzap, N.D. Yan, A.M. Paterson, M.E. Palmer, J.B. Korosi, J.A. Rusak, M.T. Arts, W. Keller, R. Ingram, A. Cairns and J.P. Smol. 2014. The jellification of north temperate lakes. Proc. R. Soc. London, Ser. B 282. doi:10.1098/rspb.2014.2449.

Jones, N.T. and B. Gilbert. 2016. Changing climate cues differentially alter zooplankton dormancy dynamics across latitudes. J. Anim. Ecol. 85: 559–569.

Jonsson, M., L. Ranaker, A. Nicolle, P. Ljungberg, T. Fagerberg, S. Hylander, T. Jephson, K. Lebret, J. von Einem, L.A. Hansson, P. A. Nilsson, E. Balseiro and B. Modenutti. 2011. Glacial clay affects foraging performance in a Patagonian fish and cladoceran. Hydrobiologia 663: 101–108.

Kelly, P., L. Clementson, C. Davies, S. Corney and K. Swadling. 2016. Zooplankton responses to increasing sea surface temperatures in the southeastern Australia global marine hotspot. Estuarine, Coastal Shelf Sci. 180: 242–257.

Kirby, R.R. and G. Beaugrand. 2009. Trophic amplification of climate warming. Proc. R. Soc. B-Biol. Sci. 276: 4095–4103.

Kirk, K.L. 1991. Inorganic particles alter competition in grazing plankton—the role of selective feeding. Ecology 72: 915–923.

Klais, R., S.A. Otto, M. Teder, M. Simm and H. Ojaveer. 2017. Winter-spring climate effects on small-sized copepods in the coastal Baltic Sea. ICES J. Mar. Sci. 74: 1855–1864.

Knoll, L.B., C.E. Williamson, R.M. Pilla, T.H. Leach, J.A. Brentrup and T.J. Fisher. 2018. Browning-related oxygen depletion in an oligotrophic lake. Inland Waters. doi.org/10.1080/20442041.2018.1452355.

Lampert, W. 2011. *Daphnia*: Development of a model organism in ecology and evolution. *In*: Kinne, O. (ed.). Excellence in Ecology, Book 21. International Ecology Institute, Oldendorf, Germany.

Leach, T.H., C.E. Williamson, N. Theodore, J.M. Fischer and M.H. Olson. 2015. The role of ultraviolet radiation in the diel vertical migration of zooplankton: an experimental test of the transparency-regulator hypothesis. J. Plankton Res. 37: 886–896.

Li, Y., P. Xie, D.D. Zhao, T.S. Zhu, L.G. Guo and J. Zhang. 2016. Eutrophication strengthens the response of zooplankton to temperature changes in a high-altitude lake. Ecology and Evolution 6: 6690–6701.

Livingstone, D.M., R. Adrian, T. Blenckner, G. George and G. A. Weyhenmeyer. 2010. Lake ice phenology, pp. 51–62. *In:* G. George (ed.). The Impact of Climate Change on European Lakes. Springer, New York, USA.

MacFadyen, E.J., C.E. Williamson, G. Grad, M. Lowery, W.H. Jeffrey and D.L. Mitchell. 2004. Molecular response to climate change: temperature dependence of UV-induced DNA damage and repair in the freshwater crustacean *Daphnia pulicaria*. Global Change Biol. 10: 408–416.

Mackas, D.L., S. Batten and M. Trudel. 2007. Effects on zooplankton of a warmer ocean: Recent evidence from the Northeast Pacific. Prog. Oceanogr. 75: 223–252.

Mackas, D.L., W. Greve, M. Edwards, S. Chiba, K. Tadokoro, D. Eloire, M.G. Mazzocchi, S. Batten, A.J. Richardson, C. Johnson, E. Head, A. Conversi and T. Peluso. 2012. Changing zooplankton seasonality in a changing ocean: Comparing time series of zooplankton phenology. Prog. Oceanogr. 97: 31–62.

Mackey, A.P., A. Atkinson, S.L. Hill, P. Ward, N.J. Cunningham, N.M. Johnston and E.J. Murphy. 2012. Antarctic macrozooplankton of the southwest Atlantic sector and Bellingshausen Sea: Baseline historical distributions (Discovery Investigations, 1928–1935) related to temperature and food, with projections for subsequent ocean warming. Deep-Sea Res. Pt. II 59: 130–146.

Manca, M.A. and W.R. DeMott. 2009. Response of the invertebrate predator *Bythotrephes* to a climate-linked increase in the duration of a refuge from fish predation. Limnol. Oceanogr. 54: 2506–2512.

Manca, M.M., M. Portogallo and M.E. Brown. 2007. Shifts in phenology of *Bythotrephes longimanus* and its modern success in Lake Maggiore as a result of changes in climate and trophy. J. Plankton Res. 29: 515–525.

Mantua, N.J., S.R. Hare, Y. Zhang, J. M. Wallace, and R. C. Francis. 1997. A Pacific interdecadal climate oscillation with impacts on salmon production. Bull. Amer. Meteorol. Soc. 78: 1069–1079.

Marinone, M.C., S.M. Marque, D.A. Suárez, M. del Carmen Diéguez, P. Pérez, P. De Los Rios, D. Soto and H.E. Zagarese. 2006. UV radiation as a potential driving force for zooplankton community structure in Patagonian lakes. Photochem. Photobiol. 82: 962–971.

Medellin-Mora, J., R. Escribano and W. Schneider. 2016. Community response of zooplankton to oceanographic changes (2002–12) in the central/southern upwelling system of Chile. Prog. Oceanogr. 142: 17–29.

Miller, D.D., Y. Ota, U.R. Sumaila, A.M. Cisneros-Montemayor and W.W.L. Cheung. 2018. Adaptation strategies to climate change in marine systems. Global Change Biol. 24. doi:10.1111/gcb.13829.

Moller, K.O., J.O. Schmidt, M.S. John, A. Temming, R. Diekmann, J. Peters, J. Floeter, A.F. Sell, J.P. Herrmann and C. Mollmann. 2015. Effects of climate-induced habitat changes on a key zooplankton species. J. Plankton Res. 37: 530–541.

Monteith, D.T., J.L. Stoddard, C.D. Evans, H.A. de Wit, M. Forsius, T. Høgåsen, A. Wilander, B.L. Skjelkvåle, D.S. Jeffries, J. Vuorenmaa, B. Keller, J. Kopácek and J. Vesely. 2007. Dissolved organic carbon trends resulting from changes in atmospheric deposition chemistry. Nature 450: 537–541.

Moore, M.V., C.L. Folt and R.S. Stemberger. 1996. Consequences of elevated temperatures for zooplankton assemblages in temperate lakes. Archiv. Hydrobbiol. 135: 289–319.

Morris, D.P., H. Zagarese, C.E. Williamson, E.G. Balseiro, B.R. Hargreaves, B. Modenutti, R. Moeller and C. Queimalinos. 1995. The attenuation of solar UV radiation in lakes and the role of dissolved organic carbon. Limnol. Oceanogr. 40: 1381–1391.

O'Reilly, C.M., S. Sharma, D.K. Gray, S.E. Hampton, J.S. Read, R.J. Rowley, P. Schneider, J.D. Lenters, P.B. McIntyre, B.M. Kraemer, G.A. Weyhenmeyer, D. Straile, B. Dong, R. Adrian, M.G. Allan, O. Anneville, L. Arvola, J. Austin, J.L. Bailey, J.S. Baron, J.D. Brookes, E. de Eyto, M.T. Dokulil, D.P. Hamilton, K. Havens, A.L. Hetherington, S.N. Higgins, S. Hook, L.R. Izmest'eva, K.D. Joehnk, K. Kangur, P. Kasprzak, M. Kumagai, E. Kuusisto, G. Leshkevich, D.M. Livingstone, S. MacIntyre, L. May, J.M. Melack, D.C. Mueller-Navarra, M. Naumenko, P. Noges, T. Noges, R.P. North, P.-D. Plisnier, A. Rigosi, A. Rimmer, M. Rogora, L.G. Rudstam, J.A. Rusak, N. Salmaso, N.R. Samal, D.E. Schindler, S.G. Schladow, M. Schmid, S.R. Schmidt, E. Silow, M.E. Soylu, K. Teubner, P. Verburg, A. Voutilainen, A. Watkinson, C.E. Williamson and G. Zhang. 2015. Rapid and highly variable warming of lake surface waters around the globe. Geophys. Res. Lett. 42: 10773–10781.

Olson, M.H., J.M. Fischer, C.E. Williamson, E.P. Overholt and N. Theodore. 2017. Landscape-scale regulators of water transparency in mountain lakes: implications of projected glacial loss. Can. J. Fish. Aquat. Sci. doi:10.1139/cjfas-2017-0215.

Overholt, E.P., S.H. Hall, C.E. Williamson, C.K. Meikle, M.A. Duffy and C.E. Cáceres. 2012. Solar radiation decreases parasitism in *Daphnia*. Ecol. Lett. 15: 47–54.

Pace, M.L. 1984. Zooplankton community structure, but not biomass, influences the phosphorus-chlorophyll a relationship. Can. J. Fish. Aquat. Sci. 41: 1089–1096.

Paerl, H.W. and J. Huisman. 2008. Blooms like it hot. Science 320: 57–58.

Parmesan, C. and G. Yohe. 2003. A globally coherent fingerprint of climate change impacts across natural systems. Nature 421: 37–42.

Paul, C., U. Sommer, J. Garzke, M. Moustaka-Gouni, A. Paul and B. Matthiessen. 2016. Effects of increased CO_2 concentration on nutrient limited coastal summer plankton depend on temperature. Limnol. Oceanogr. 61: 853–868.

Pianka, E.R. 1966. Latitudinal gradients in species diversity—a review of concepts. Am. Nat. 100: 33–46.

Pilla, R.M., C.E. Williamson, J. Zhang, R.L. Smyth, J.D. Lenters, J.A. Brentrup, L.B. Knoll and T.J. Fisher. 2018. Browning-related decreases in water transparency lead to long-term increases in surface water temperature and thermal stratification in two small lakes. J. Geophys. Res.: Biogeosci. Doi:10.1029/2017JG004321.

Preston, N.D. and J.A. Rusak. 2010. Homage to Hutchinson: does inter-annual climate variability affect zooplankton density and diversity? Hydrobiologia 653: 165–177.

Purcell, J.E. 2005. Climate effects on formation of jellyfish and ctenophore blooms: a review. J. Mar. Biol. Assoc. U.K. 85: 461–476.

Purcell, J.E., S. Uye and W.T. Lo. 2007. Anthropogenic causes of jellyfish blooms and their direct consequences for humans: a review. Mar. Ecol. Prog. Ser. 350: 153–174.

Reid, P.C., R.E. Hari, G. Beaugrand, D.M. Livingstone, C. Marty, D. Straile, J. Barichivich, E. Goberville, R. Adrian, Y. Aono, R. Brown, J. Foster, P. Groisman, P. Hélaouët, H.-H. Hsu, R. Kirby, J. Knight, A. Kraberg, J. Li, T.-T. Lo, R.B. Myneni, R.P. North, J.A. Pounds, T. Sparks, R. Stübi, Y. Tian, K.H. Wiltshire, D. Xiao and Z. Zhu. 2016. Global impacts of the 1980s regime shift. Global Change Biol. 22: 682–703.

Rice, E. and G. Stewart. 2016. Decadal changes in zooplankton abundance and phenology of Long Island Sound reflect interacting changes in temperature and community composition. Mar. Environ. Res. 120: 154–165.

Richardson, A.J. 2008. In hot water: zooplankton and climate change. ICES J. Mar. Sci. 65: 279–295.

Richardson, A.J., A. Bakun, G.C. Hays and M.J. Gibbons. 2009. The jellyfish joyride: causes, consequences and management responses to a more gelatinous future. Trends Ecol. Evol. 24: 312–322.

Richardson, D., S. Melles, R. Pilla, A. Hetherington, L. Knoll, C. Williamson, B. Kraemer, J. Jackson, E. Long, K. Moore, L. Rudstam, J. Rusak, J. Saros, S. Sharma, K. Strock, K. Weathers and C. Wigdahl-Perry. 2017. Transparency, geomorphology and mixing regime explain variability in trends in lake temperature and stratification across northeastern North America (1975–2014). Water 9. doi:10.3390/w9060442.

Rombouts, I., G. Beaugrand, F. Ibanez, S. Gasparini, S. Chiba and L. Legendre. 2009. Global latitudinal variations in marine copepod diversity and environmental factors. Proc. R. Soc. B-Biol. Sci. 276: 3053–3062.

Rose, K.C., D.P. Hamilton, C.E. Williamson, C.G. McBride, J.M. Fischer, M.H. Olson, J.E. Saros, M.G. Allan and N. Cabrol. 2014. Light attenuation characteristics of glacially-fed lakes. J. Geophys. Res.: Biogeosci. 119: 1446–1457.

Rose, K.C., C.E. Williamson, J M. Fischer, S.J. Connelly, M. Olson, A.J. Tucker and D.A. Noe. 2012. The role of UV and fish in regulating the vertical distribution of *Daphnia*. Limnol. Oceanogr. 57: 1867–1876.

Rose, K.C., C.E. Williamson, J.E. Saros, R. Sommaruga and J.M. Fischer. 2009. Differences in UV transparency and thermal structure between alpine and subalpine lakes: Implications for organisms. Photochem. Photobiol. Sci. 8: 1244–1256.

Rusak, J.A., N.D. Yan and K.M. Somers. 2008. Regional climatic drivers of synchronous zooplankton dynamics in north-temperate lakes. Can. J. Fish. Aquat. Sci. 65: 878–889.

Rutherford, S., S. D'Hondt and W. Prell. 1999. Environmental controls on the geographic distribution of zooplankton diversity. Nature 400: 749–753.

Sato, M., J.K. Horne, S.L. Parker-Stetter, T.E. Essington, J.E. Keister, P.E. Moriarty, L.B. Li and J. Newton. 2016. Impacts of moderate hypoxia on fish and zooplankton prey distributions in a coastal fjord. Mar. Ecol. Prog. Ser. 560: 57–72.

Scheffer, M., D. Straile, E.H. van Nes and H. Hosper. 2001. Climatic warming causes regime shifts in lake food webs. Limnol. Oceanogr. 46: 1780–1783.

Schindler, D.E., D.E. Rogers, M.D. Scheuerell and C.A. Abrey. 2005. Effects of changing climate on zooplankton and juvenile sockeye salmon growth in southwestern Alaska. Ecology 86: 198–209.

Schminke, H.K. 2007. Entomology for the copepodologist. J. Plankton Res. 29: i149–i162.

Schneider, P. and S.J. Hook. 2010. Space observations of inland water bodies show rapid surface warming since 1985. Geophys. Res. Lett. 37. doi:10.1029/2010GL045059.

Sebastian, P., H. Stibor, S. Berger and S. Diehl. 2012. Effects of water temperature and mixed layer depth on zooplankton body size. Mar. Biol. 159: 2431–2440.

Shatwell, T., R. Adrian and G. Kirillin. 2016. Planktonic events may cause polymictic-dimictic regime shifts in temperate lakes. Sci. Rep. 6: 24361.

Sommer, U., R. Adrian, L. De Senerpont Domis, J.J. Elser, U. Gaedke, B. Ibelings, E. Jeppesen, M. Lürling, J.C. Molinero, W.M. Mooij, E. van Donk and M. Winder. 2012. Beyond the Plankton Ecology Group (PEG) model: mechanisms driving plankton succession. Annu. Rev. Ecol. Evol. S. 43: 429–448.

Soreide, J.E., E. Leu, J. Berge, M. Graeve and S. Falk-Petersen. 2010. Timing of blooms, algal food quality and *Calanus glacialis* reproduction and growth in a changing Arctic. Global Change Biol. 16: 3154–3163.

Southward, A.J., S.J. Hawkins and M.T. Burrows. 1995. 70 years observations of changes in distribution and abundance of zooplankton and intertidal organisms in the western English Channel in relation to rising sea temperature. J. Therm. Biol. 20: 127–155.

Stemberger, R.S. and E.K. Miller. 2003. Cladoceran body length and Secchi disk transparency in northeastern US lakes. Can. J. Fish. Aquat. Sci. 60: 1477–1486.

Straile, D. 2002. North Atlantic Oscillation synchronizes food-web interactions in central European lakes. Proc. R. Soc. B-Biol. Sci. 269: 391–395.

Straile, D. and R. Adrian. 2000. The North Atlantic Oscillation and plankton dynamics in two European lakes - two variations on a general theme. Global Change Biol. 6: 663–670.

Straile, D., R. Adrian and D.E. Schindler. 2012. Uniform temperature dependency in the phenology of a keystone herbivore in lakes of the northern hemisphere. PLOS One 7: e45497.

Straile, D., O. Kerimoglu and F. Peeters. 2015. Trophic mismatch requires seasonal heterogeneity of warming. Ecology 96: 2794–2805.

Strock, K.E., J. E. Saros, S.J. Nelson, S.D. Birkel, J.S. Kahl and W.H. McDowell. 2016. Extreme weather years drive episodic changes in lake chemistry: implications for recovery from sulfate deposition and long-term trends in dissolved organic carbon. Biogeochemistry 127: 353–365.

Tarling, G.A., P. Ward and S.E. Thorpe. 2018. Spatial distributions of Southern Ocean mesozooplankton communities have been resilient to long-term surface warming. Global Change Biol. 24: 132–142.

Taylor, A.H., J.I. Allen and P.A. Clark. 2002. Extraction of a weak climate signal by an ecosystem. Nature 416: 629–632.

Tessier, A.J. and J. Welser. 1991. Cladoceran assemblages, seasonal succession and the importance of a hypolimnetic refuge. Freshwat. Biol. 25: 85–93.

Thackeray, S.J. 2012. Mismatch revisited: What is trophic mismatching from the perspective of the plankton? J. Plankton Res. 34: 1001–1010.

Thackeray, S.J., P.A. Henrys, I.D. Jones and H. Feuchtmayr. 2012. Eight decades of phenological change for a freshwater cladoceran: what are the consequences of our definition of seasonal timing? Freshwat. Biol. 57: 345–359.

Thor, P., A. Bailey, S. Dupont, P. Calosi, J.E. Søreide, P. De Wit, E. Guscelli, L. Loubet-Sartrou, I.M. Deichmann, M.M. Candee, C. Svensen, A.L. King and R.G.J. Bellerby. 2018. Contrasting physiological responses to future ocean acidification among Arctic copepod populations. Global Change Biol. 24. doi:10.1111/gcb.13870.

Tremblay, J.E., S. Belanger, D.G. Barber, M. Asplin, J. Martin, G. Darnis, L. Fortier, Y. Gratton, H. Link, P. Archambault, A. Sallon, C. Michel, W.J. Williams, B. Philippe and M. Gosselin. 2011. Climate forcing multiplies biological productivity in the coastal Arctic Ocean. Geophys. Res. Lett. 38. doi:10.1029/2011gl048825.

Urmy, S., C.E. Williamson, T.H. Leach, S.G. Schladow, E. Overholt and J.D. Warren. 2016. Vertical redistribution of zooplankton in an oligotrophic lake associated with reduction in ultraviolet radiation by wildfire smoke. Geophys. Res. Lett. 43: 3746–3753.

Vanderploeg, H.A., S.A. Ludsin, J.F. Cavaletto, T.O. Hook, S.A. Pothoven, S.B. Brandt, J.R. Liebig and G.A. Lang. 2009a. Hypoxic zones as habitat for zooplankton in Lake Erie: Refuges from predation or exclusion zones? J. Exp. Mar. Biol. Ecol. 381: S108–S120.

Vanderploeg, H.A., S.A. Ludsin, S.A. Ruberg, T.O. Hook, S.A. Pothoven, S.B. Brandt, G.A. Lang, J.R. Liebig and J.F. Cavaletto. 2009b. Hypoxia affects spatial distributions and overlap of pelagic fish, zooplankton, and phytoplankton in Lake Erie. J. Exp. Mar. Biol. Ecol. 381: S92–S107.

Wagner, C. and R. Adrian. 2011. Consequences of changes in thermal regime for plankton diversity and trait composition in a polymictic lake: a matter of temporal scale. Freshwat. Biol. 56: 1949–1961.

Walther, G.R., E. Post, P. Convey, A. Menzel, C. Parmesan, T.J.C. Beebee, J.M. Fromentin, O. Hoegh-Guldberg and F. Bairlein. 2002. Ecological responses to recent climate change. Nature 416: 389–395.

Weijerman, M., H. Lindeboom and A.F. Zuur. 2005. Regime shifts in marine ecosystems of the North Sea and Wadden Sea. Mar. Ecol. Prog. Ser. 298: 21–39.

Weiss, L.C., L. Pötter, A. Steiger, S. Kruppert, U. Frost and R. Tollrian. 2018. Rising pCO_2 in freshwater ecosystems has the potential to negatively affect predator-induced defenses in *Daphnia*. Curr. Biol. 28: 327–332.

Weydmann, A., W. Walczowski, J. Carstensen and S. Kwaśniewski. 2018. Warming of Subarctic waters accelerates development of a key marine zooplankton *Calanus finmarchicus*. Global Change Biol. 24: 172–183.

Weyhenmeyer, G.A., M. Mackay, J.D. Stockwell, W. Thiery, H.P. Grossart, P.B. Augusto-Silva, H.M. Baulch, E. de Eyto, J. Hejzlar, K. Kangur, G. Kirillin, D.C. Pierson, J.A. Rusak, S. Sadro and R.I. Woolway. 2017. Citizen science shows systematic changes in the temperature difference between air and inland waters with global warming. Sci. Rep. 7. doi:10.1038/Srep43890.

White, J.R., X.S. Zhang, L.A. Welling, M.R. Roman and H.G. Dam. 1995. Latitudinal gradients in zooplankton biomass in the tropical Pacific at 140-degrees-W during the JGOFS EqPac Study - Effects of El-Niño. Deep-Sea Res. Pt. II 42: 715–733.

Williamson, C.E., J.A. Brentrup, J. Zhang, W.H. Renwick, B.R. Hargreaves, L.B. Knoll, E. Overholt and K. Rose. 2014. Lakes as sensors in the landscape: Optical metrics as scalable sentinel responses to climate change. Limnol. Oceanogr. 59: 840–850.

Williamson, C.E., J.M. Fischer, S.M. Bollens, E.P. Overholt and J.K. Breckenridge. 2011. Toward a more comprehensive theory of zooplankton diel vertical migration: Integrating ultraviolet radiation and water transparency into the biotic paradigm. Limnol. Oceanogr. 56: 1603–1623.

Williamson, C.E., G. Grad, H.J. De Lange, S. Gilroy and D.M. Karapelou. 2002. Temperature-dependent ultraviolet responses in zooplankton: Implications of climate change. Limnol. Oceanogr. 47: 1844–1848.

Williamson, C.E., S. Madronich, A. Lal, R.E. Zepp, R.M. Lucas, E.P. Overholt, K.C. Rose, G. Schladow and J. Lee-Taylor. 2017. Climate change-induced increases in precipitation are reducing the potential for solar ultraviolet radiation to inactivate pathogens in surface waters. Sci. Rep. doi:10.1038/s41598-017-13392-2.

Williamson, C.E. and R.E. Magnien. 1982. Diel vertical migration in *Mesocyclops edax*: Implications for predation rate estimates. J. Plankton Res. 4: 329–339.

Williamson, C.E., E.P. Overholt, R.M. Pilla, T.H. Leach, J.A. Brentrup, L.B. Knoll, E.M. Mette and R.E. Moeller. 2015. Ecological consequences of long-term browning in lakes. Sci. Rep. doi:10.1038/srep18666.

Williamson, C.E., R.S. Stemberger, D.P. Morris, T.M. Frost and S.G. Paulsen. 1996. Ultraviolet radiation in North American lakes: attenuation estimates from DOC measurements and implications for plankton communities. Limnol. Oceanogr. 41: 1024–1034.

Wilson, R.J., N.S. Banas, M.R. Heath and D.C. Speirs. 2016. Projected impacts of 21st century climate change on diapause in *Calanus finmarchicus*. Global Change Biol. 22: 3332–3340.

Winder, M. 2012. Lake warming mimics fertilization. Nat. Clim. Change 2: 771–772.

Winder, M., J.M. Bouquet, J.R. Bermudez, S.A. Berger, T. Hansen, J. Brandes, A.F. Sazhin, J.C. Nejstgaard, U. Bamstedt, H.H. Jakobsen, J. Dutz, M.E. Frischer, C. Troedsson and E.M. Thompson. 2017. Increased appendicularian zooplankton alter carbon cycling under warmer more acidified ocean conditions. Limnol. Oceanogr. 62: 1541–1551.

Winder, M. and D.E. Schindler. 2004a. Climate change uncouples trophic interactions in an aquatic ecosystem. Ecology 85: 2100–2106.

Winder, M. and D.E. Schindler. 2004b. Climatic effects on the phenology of lake processes. Global Change Biol. 10: 1844–1856.

Winder, M., D.E. Schindler, T.E. Essington, A.H. Litt and W.T. Edmondson. 2009. Disrupted seasonal clockwork in the population dynamics of a freshwater copepod by climate warming. Limnol. Oceanogr. 54: 2493–2505.

Wissel, B. and C.W. Ramcharan. 2003. Plasticity of vertical distribution of crustacean zooplankton in lakes with varying levels of water colour. J. Plankton Res. 9: 1047–1057.

Wissel, B., N.D. Yan and C.W. Ramcharan. 2003. Predation and refugia: implications for *Chaoborus* abundance and species composition. Freshwat. Biol. 48: 1421–1431.

Wojtal-Frankiewicz, A. 2012. The effects of global warming on *Daphnia* spp. population dynamics: a review. Aquat. Ecol. 46: 37–53.

Woodmansee, R.A. and B.J. Grantham. 1961. Diel vertical migrations of two zooplankters (*Mesocyclops* and *Chaoborus*) in a Mississippi lake. Ecology 42: 619–628.

Wright, D. and J. Shapiro. 1990. Refuge availability: a key to understanding the summer disappearance of *Daphnia*. Freshwat. Biol. 24: 43–62.

Yan, N.D., A. Blukacz, W.G. Sprules, P.K. Kindy, D. Hackett, R.E. Girard and B.J. Clark. 2001. Changes in zooplankton and the phenology of the spiny water flea, *Bythotrephes*, following its invasion of Harp Lake, Ontario, Canada. Can. J. Fish. Aquat. Sci. 58: 2341–2350.

Zhang, H., M.K. Ekvall, J. Xu and L.-A. Hansson. 2015. Counteracting effects of recruitment and predation shape establishment of rotifer communities under climate change. Limnol. Oceanogr. 60: 1577–1587.

CHAPTER 11

UV-B Radiation and the Green Tide-forming Macroalga Ulva

Jihae Park,[1] Murray T. Brown,[2] Hojun Lee,[3] Christophe Vieira,[4] Lalit K. Pandey,[5] Eunmi Choi,[6] Stephen Depuydt,[1] Donat-P. Häder[7] and Taejun Han[1,3,]*

Introduction

For plants, sunlight not only provides a source of energy to drive primary production, but also information to guide photo-morphogenesis and reproduction (Kendrick and Kronenberg 1994). As well as light in the 400–700 nm waveband range (photosynthetic active radiation, PAR) solar radiation reaching the Earth's surface contains a small fraction of ultraviolet-B radiation (UV-B; 280–315 nm), which is harmful to the biota. UV-B quanta have high levels of energy and are effectively absorbed by important biological molecules, such as DNA, proteins, and lipids, which subsequently get destroyed. UV-B radiation directly alters the structure of DNA and indirectly damages nucleic acids (Mitchell and Karentz 1993). Proteins absorb UV-B because of their tryptophan, tyrosine, and phenylalanine contents (Yu and Bjorn 1997) and UV-B

[1] Lab of Plant Growth Analysis, Ghent University Global Campus, Songomunhwa-ro, 119, Yeonsu-gu, Incheon 21985, Republic of Korea.
[2] School of Marine Science & Engineering, Plymouth University, Plymouth, Devon, PL4 8AA, United Kingdom.
[3] Department of Marine Sciences, Incheon National University, 119, Academy-ro, Yeonsu-gu, Incheon 22012, Republic of Korea.
[4] Phycology Research Group and Center for Molecular Phylogenetics and Evolution, Ghent University, Krijgslaan 281 (S8), 9000, Gent, Belgium.
[5] Department of Plant Science, Faculty of Applied Sciences, MJP Rohilkhand University, Bareilly, 243006, India.
[6] Department of Chemistry, Incheon National University, 119, Academy-ro, Yeonsu-gu, Incheon 22012, Republic of Korea.
[7] Emeritus from Friedrich-Alexander University, Dept. Biology, Neue Str. 9, 91096, Möhrendorf, Germany.
* Corresponding author: hanalgae@hanmail.ne

radiation affects membranes by causing large reductions in the total lipid content (Kramer et al. 1991).

There is variability in the responses to UV-B radiation between different species or isolates of individual species, and even within the life cycle of a single species (Dring et al. 1996, Häder et al. 1996, Han 1996, Bischof et al. 1998, Hanelt 1998, Wiencke et al. 2000). The degree of UV-B sensitivity depends largely on the efficiency of constitutive and UV-induced protection and repair mechanisms, such as the light-driven repair of spore germination and reproduction (Pakker et al. 2000a,b, Han et al. 2003b, 2004), structural attenuation (Dring et al. 1996), UV absorption, and the synthesis of screening compounds (Karsten et al. 1998a,b, Sinha et al. 1998).

In seawater, UV-B radiation decreases exponentially with depth, although the rate of reduction depends on the productivity of a given area (Kirk 1994). In turbid coastal waters, UV-B radiation only penetrates to a few meters, while in clearer oceanic waters, this same reduction occurs at depths of greater than 30 m (Jerlov 1976). Since the water column does not screen UV radiation for marine organisms inhabiting shallow seas, the impacts of UV radiation on these ecosystems have been extensively studied, primarily focusing on the population growth rates, carbon assimilation, and nitrogen metabolism of microalgae (Häberlein and Häder 1992, Behrenfeld et al. 1993, Davidson et al. 1994, Lesser et al. 1994, Häder and Worrest 1997). While motility may provide protection for many microalgae by enabling them to avoid harmful UV radiation, exposure is unavoidable for sessile benthic macroalgae and they are, therefore, vulnerable to subsequent damage. Previous studies on the effects of UV-B on macroalgae have shown that radiation severely damages DNA, RNA, proteins, pigmentation, Rubisco, photosynthesis, growth, and reproduction (Franklin and Forster 1997, Häder 2001, Dring et al. 2001, Pang et al. 2001, Bischof et al. 2002b,c,d, Michler et al. 2002, Han et al. 2003b, 2004).

'Green tides' are a global phenomenon that occurs in coastal waters and estuaries as a result of anthropogenically-derived inputs of macronutrients, during which large blooms and accumulations of opportunistic green algae form. Amongst the most commonly occurring green algae species are those belonging to the macroalgal genus *Ulva* Linnaeus. 1753 (Valiela et al. 1997, Morand and Merceron 2005). *Ulva* spp. are cosmopolitan green algae found in fresh-, brackish-, and sea-waters. Presently, about 594 species of *Ulva* have been recorded, of which only 131 in Algaebase have been classified and of which 16 have a global distribution. Green tides are reported in many coastal waters around the world including China, Finland, France, Japan, New Zealand, South Korea, and the United States, and several species of *Ulva* have been identified as the causative organisms, including *U. linza*, *U. pertusa*, *U. procera*, and *U. prolifera*. Macroalgal blooms not only impact recreational activities and threaten commercial fisheries, but they also alter natural biological communities and ecosystem functions of the affected environments. One such large-scale algal bloom, which received worldwide media coverage, occurred off the coast of the 600 km^2 island of Qingdao in late June 2008, just prior to the start of the Beijing Olympics and threatened the sailing competition. This green tide lasted for two weeks and more than one million tons of *Ulva* were removed with the efforts of more than 10,000 people. The cost of managing the bloom in Qingdao was estimated to have exceeded US$100 million. The high costs of managing green tides and the devastating impacts they have on natural ecosystems, have provided the impetus for research efforts to monitor *Ulva* growth and develop effective methods to prevent formations of large-scale blooms.

The main driver of blooms is eutrophication that is mainly based on marine-level nitrogen discharge. Eutrophication is particularly important to address, as it is likely to continue and increase in the future; potentially becoming one of the greatest threats to our coastal and estuarine systems.

It is generally believed that the amount of UV radiation reaching the Earth's surface will continue to increase over the coming years due to a significant depletion of the ozone layer, and that this will likely lead to a relatively large increase in solar radiation in the UV-B range. Since marine macroalgae play key roles in coastal ecosystems, i.e., by providing food and shelter for a variety of consumers, it is important to understand the effects of UV-B on species, such as the intertidal *Ulva* spp. that cause green tides. This information is also necessary to predict future changes in coastal ecosystems due to enhanced UV-B radiation.

Hence, the purpose of this review is to present the current, collective understanding of the photobiological effects of UV-B radiation on the physiology of green tide-forming *Ulva* spp. and of the protection and recovery strategies that facilitate the successful survival of these species. This includes discussion on the effects on photosynthesis, morphology, growth, reproduction, spore liberation/spore germination, chloroplast, and cell movement, pigmentation, oxidative stress, antioxidant activity, UV-absorbing compounds and photoreactivation.

Effects of UV-B on Photosynthesis

Although UV-B accounts for only a small fraction of solar radiation, it has a considerable effect on photosynthesis. UV-B radiation reportedly causes significant photoinhibition, which is characterized by a decrease in quantum yield, reduced ability for photosynthetic O_2 evolution and fixation of photosynthetic CO_2. Excessive UV-B radiation causes significant thylakoid structural disruption and the inactivation of enzymes involved in CO_2 fixation and sugar production, as well as a decrease in primary and accessory photosynthetic pigment concentrations (Franklin and Forster 1997, Häder 2001, Sinha et al. 2001). Compared with the activities of internally bound photosynthesis components, thylakoid membranes of chloroplasts are more sensitive to UV-B. The expansion of the thylakoid membrane and the rupturing of the chloroplast double membrane can lead to changes in membrane permeability.

The inhibition of photosynthesis by UV-B exposure may be related to the direct effects on photosynthetic proteins and pigments, i.e., the degradation of D1 proteins in the reaction center or Photosystem II (PSII; Vass 1997), or may be indirectly related to the reduction in the catalytic activity of enzymes, such as Rubisco and the decrease in the expression of genes involved in photosynthesis (Jordan et al. 1992, Mackerness et al. 1999). The formation of reactive oxygen species (ROS) in chloroplasts may primarily destroy PSII by oxidative D1 degradation (Lesser 2006). The reaction center of PSII, one of the UV-B lesion sensitive sites, consists of a chlorophyll binding complex with two polypeptides: D1 and D2. These polypeptides degrade at high rates in response to UV radiation (Greenberg et al. 1989, Jansen et al. 1996). UV causes both photolysis of the PSII reaction centre subunits and the modification of the main steroid receptor Q_A.

Rubisco is one of the main targets of UV-B radiation in the photosynthesis of higher plants, phytoplankton, and benthic macroalgae (Strid et al. 1990, Jordan et al. 1992, Nogués and Baker 1995, Lesser et al. 1996, Allen et al. 1997, Bischof et al. 2000). The concomitant decrease in CO_2 assimilation may be the result of decreased activity and reduced enzyme levels. It has been reported that the reduction of RNA transcripts encoding Rubisco starts at a very early stage during UV exposure, even before further UV effects become apparent at a physiological level (Jordan et al. 1992, Mackerness et al. 1999).

There are significant differences in the degree of inhibition in different *Ulva* species. Measurements of chlorophyll *a* (Chl *a*) fluorescence can be used to investigate these degrees of inhibition on photosynthetic processes, such as light absorption, energy transfer, and photochemical reactions in PSII (Krause and Weis 1991) under different combinations of PAR and UVR (PAR, PAR + UV-A, and PAR + UV-A + UV-B), and which indicate that the UV-B component of radiation has an additive inhibitory effect on photosynthesis of *Ulva* spp. (Figueroa et al. 1997, Bischof et al. 2002a,b, Han et al. 2004).

UV-B directly destroys PSII in *Ulva*, and indirectly reduces photosynthetic pigments, which in turn hinders photosynthetic efficiency. After irradiation with different UV radiations, the decrease in maximum quantum yield, as measured by Fv/Fm, was accompanied by a gradual decrease in pigment content (Nogués and Baker 1995). UV-induced suppression of photosynthetic yield can lead to reductions in carbohydrate synthesis and growth. However, it has been shown that reducing CO_2 assimilation under UV-B exposure is not necessarily accompanied by a reduction in the maximum quantum efficiency of photosynthesis (Nogués and Baker 1995).

In contrast, dark respiration rates seem to be almost unaffected by UV exposure (Clendennen et al. 1996). This insensitivity may be explained in part by the structural resistance of mitochondria to UV, as observed in higher plants (Brandle et al. 1977).

Photoinhibition has been observed in macroalgae that were exposed to UV-B and subsequently recovered under dim light. Dynamic photoinhibition is a protective mechanism that is caused by an active down-regulation of photosynthesis and has often been described in the genus *Ulva* (Osmond 1994, Hanelt 1998).

Some differences have been found in the results when photosynthetic performance was measured by O_2 evolution or Chl *a* fluorescence. For example, in *U. rotundata*, the photoinhibition rate of P_{max} is always lower than that of Fv/Fm, which indicates that when the photochemical process is down-regulated, the production of net saturated O_2 begins to decrease (Figueroa et al. 2003). In contrast, net P_{max} is a parameter that is linked more to C assimilation and the Calvin cycle than to the processes related to photosynthetic efficiency or quantum yield at limited irradiances.

Flameling and Kromkamp (1998) proposed two possible explanations for the observed differences between O_2 evolution and Chl *a* fluorescence: (i) The production of net O_2 is affected by the consumption of O_2 or by the processes that affect linear electron transport, for example, cyclic electron transport around PSII, pseudocyclic electron transport in the Mehler reaction, Rubisco oxygenase activity, and light-dependent mitochondrial respiration; (ii) At saturating irradiances, photosynthesis turnover time may change, resulting in the effective quantum yield not matching the steady state O_2 yield. In addition to changes in the process of C assimilation, nitrate assimilation processes or nutrient restrictions can also affect the relationship between O_2 and fluorescence.

Exposure to solar UV-B can have many biological effects at the molecular, cellular, individual, and community levels. Species-specific adaptations to UV-B radiation have been reported, and even closely related species of the same genus, occupying different habitats, may have significantly different UV sensitivities. UV-resistant species are found in the intertidal zone, while more sensitive species are found in deeper waters. Seasonal variations in UV radiation also result in significant responses within and between species of marine macrobenthic communities throughout the year. Macroalgae can be classified by their ability to withstand solar UV radiation, and their extent of resistance is genetically determined, resulting in significant vertical stratification.

Possible differences in the adaptability of photosynthesis to UV radiation were tested in both the laboratory and in the field, in two *Ulva* spp.: *U. olivascens* from the intertidal zone and *U. rotundata* from the subtidal zone (Figueroa et al. 1997). *In situ*, UV-B radiation negatively affected photosynthesis and Rubisco in *U. rotundata*, while the UV doses required for 50% photoinhibition were lower in the subtidal species than in the intertidal species (Figueroa et al. 1997).

Although it has not been well studied in *Ulva* spp., some previous studies have examined UV adaptation at different life cycle stages. Typically, the early developmental stages of intertidal species have been found to be more sensitive to UV radiation than the adult stages (Major and Davison 1998, Coelho et al. 2000, Hoffman et al. 2003, Véliz et al. 2006). PSII activity in the filamentous stage of *Porphyra haitanesis* was impaired, unlike that in the adult blade stage, even when the solar radiation level decreased (Jiang and Gao 2008). It was reported that the higher UV sensitivity may be ascribed to a smaller amount of UV screening compounds.

A comparison of spore production, spore germination, and cell growth of *U. pertusa* after exposure to 1 W m^{-2} UV-B radiation for 40 min showed that sporulation rate was not affected whereas germination rate was reduced to about 92% of the control, and cell size increased by 21% (Han et al. 2003a, 2004).

A comparative study of UV tolerance between different thallus regions of *U. pertusa* showed UV-sensitive gradients (Han et al. 2003a). *Ulva* spp. have long been considered as simple and uniform algae with little functional differentiation within the thalli. When comparing sensitivities between different thallus regions, a seven-day treatment with PAR and PAR+UV-A resulted in a significant reduction in the effective quantum yield of the whole thallus with subsequent recovery towards the initial values, while PAR + UV-A + UV-B radiation resulted in greater photoinhibition and less recovery. Significant differences in UV-B susceptibility were observed, with the marginal region of the thallus being more sensitive and the basal portion being the most resistant (Han et al. 2003a). Different responses may be attributable to differences in the cell types that make up the different thallus parts, and this may eventually translate into functional differences, such as high productivity (i.e., photosynthesis and growth) and reproduction of the marginal parts and the regeneration of new cells in the basal parts (Burrows 1959, Moss and Marsland 1976). The division of tasks among different cell types within a thallus may have adaptive value for the success of opportunistic algae, such as pioneer species. One of the benefits of the opportunity-based survival strategy is high productivity and rapid growth, which perfectly matches the characteristics of the marginal thallus region of *U. pertusa*, allowing for rapid invasion into the primary substrata (Littler and Littler 1980). In addition, the UV resistance of the basal region and the possibility of regeneration of new cells may increase the chances of green algae surviving under adverse conditions (Fletcher 1976).

Effects of UV-B on Growth

In macroalgae, a decline in growth rate can be a result of UV stress. Relatively inconsistent changes in surface area and fresh weight have been recorded for *Ulva* spp., and it is, therefore, difficult to ascertain the effect of UV-B on these growth parameters.

The effects on growth rate based on fresh weight do not show much consistency between experiments performed by different researchers. No significant differences in fresh weight-based growth rate were detected for *U. pertusa*, between thalli treated with and without UV radiation (Han 1996), while the fresh weight of *U. expansa* increased more in UV-screened conditions than when exposed to full solar radiation conditions (Grobe and Murphy 1994). In the laboratory, artificial UV-B radiation of 2 W m^{-2} for various periods did not affect the fresh weight of *U. pertusa*. Furthermore, cell size was not affected by UV-B. Grobe and Murphy (1994) argued that the smaller-sized thalli grown under UV-B radiation were the result of an inhibition of cell division rather than of cell expansion. It is known that UV-B causes a delay in progression through the cell cycle, and an arrestment in the G1 and G2 phases by interfering with protein synthesis (Van't Hoff 1974).

Investigations into the possible mechanisms behind UV-B-induced growth reduction in *U. expansa* were made with two hypotheses: (i) the formation of cross-linking between cell wall components via the catalytic action of the peroxidase enzyme using hydrogen peroxide (H_2O_2) and (ii) a reduction in turgor pressure (Grobe and Murphy 1997). However, peroxidase enzymes and extracellular H_2O_2 did not appear to be involved in the reduction of the growth rate of *U. expansa* under UV-B exposure which did not affect cell turgor either. Thus, UV-B appears to inhibit the growth of *U. expansa* through several processes other than viscoelastic expansion (Grobe and Murphy 1997).

There are large species-specific differences in sensitivity to UV, and a spectrum of responses for coping with UV, even within a single species. It is noteworthy that many of the studies have focused on the establishment stage of algae, highlighting that those early stages are the most sensitive and are of paramount importance in determining the performance of the adult population in a given area.

When early germlings of the intertidal alga *U. lactuca* were exposed to ambient and elevated levels of UV-B for several weeks in outdoor tanks, no inhibition of growth was observed in terms of length and fresh weight compared with those of the controls (Kuhlenkamp and Lüning 1998). In contrast, Lee and Han (1998) reported that the surface area and cell number of eight-day-old *U. pertusa* germlings decreased significantly when exposed to UV-B at 4 W m^{-2} for 1 to 2 h. Although it is not possible to directly compare the results of the two different studies, there appears to be an age-related change in the ability of *Ulva* to withstand UV radiation. Changes in UV sensitivity in *U. pertusa* were partially confirmed by a direct comparison between UV-B responses in eight-day-old germlings and those in adult thalli (Lee and Han 1998). The cell surface area and cell number of the germlings decreased by 65–75% after 2 h of exposure to artificial UV-B at 4 W m^{-2}; however, only the number of cells of adult thalli decreased to 65% of that of controls after the same treatment for 2.5 h. Changes in susceptibility to strong light have already been observed between gametophytes and the 1–2-celled sporophytes of *Laminaria japonica* (Fei et al. 1989), although the underlying mechanisms are still unclear.

Further quantitative assessments are required to fully understand the impact of UV-B-induced growth inhibition on *Ulva* spp.

Effects of UV-B on Reproduction, Spore liberation, and Spore Germination

Ulva spp. are common, fast-growing macroalgae in the intertidal zone, and are considered to be the first colonizers of open substrates (Littler and Littler 1980, Beach et al. 1995). The worldwide presence of these green algae is attributed to their tolerance of a wide range of environmental conditions and their superior reproductive capacity (Smith 1947). *Ulva* sporulation involves the direct conversion of vegetative cells into reproductive divisions, resulting in zoospores or gametes (Føyn 1959). The environmental conditions required to initiate sporulation have previously been documented, and sporulation is known to be affected by light, temperature, nutrients, desiccation, water movement, tidal conditions, and the presence or absence of external symbiotic bacteria (Smith 1947, Christie and Evans 1962, Nørdby 1977, Provasoli and Pintner 1980, Stratmann et al. 1996). Since UV-B radiation is known to have various negative effects on macroalgae, it was surprising when the growth of *U. pertusa* was observed to be stimulated, rather than negatively affected, by UV-B exposure (Franklin and Forster 1997, Häder and Figueroa 1997, Han et al. 1998, Häder 2001). Considering that vegetative growth and reproduction represent antagonistic pathways (Lüning and Neushul 1978), and that the same genes are associated with the control of these two phenomena in *Ulva* (Løvlie and Bråten 1968), the enhanced growth and reduced reproduction after UV-B exposure may indicate that an appropriate level of UV-B quanta acts as a signal to switch the mechanism from growth to reproduction. It is, therefore, hypothesised that the UV-stimulated growth of *U. pertusa* was the result of a UV-induced delay in sporulation. In higher plants, it has been found that UV-B radiation can affect the timing and extent of reproduction by altering, rather than disrupting, regulatory mechanisms (Saile-Mark and Tevini 1997). The role of UV-B as a signal that affects the transcription rate of particular genes has been recognized in higher plants (Jenkins 2017).

Algal reproductive organs are poorly protected and are susceptible to the damaging effects of UV-B. Reductions in reproduction are expected to occur in *U. pertusa* inhabiting the intertidal zone when the algae are exposed to higher doses of UV-B radiation due to ebb tides, and this could cause potentially harmful consequences for population dynamics (Han et al. 2003b). In the laboratory, the incidence of sporulation in plants irradiated with UV-B was significantly lower, and the extent of reduction was greater in plants exposed to higher UV doses. When the results are extrapolated to events that may occur in nature, there are many difficulties owing to the differences in the energy distribution within the UV-B range emitted by the lamp and sunlight. However, expressing irradiance as biologically effective irradiance (BEI) allows treatments applied under laboratory conditions to be compared with the radiation experienced under natural conditions.

To generate a BEI of UV-B lamps and solar energy, the weighted spectral irradiance is defined as the product of the UV irradiance multiplied by the spectrum of plant DNA damage, normalized to 1 at 300 nm (Caldwell 1971). Considering that the vertical attenuation coefficient of UV-B (k = 300–320 nm) of KUV-B on the east coast of South Korea is 0.54 m^{-1} (Han et al. 2003b) and the total biologically effective UV-B dose of 50% inhibition of sporulation is 0.085 $Dose_{eff}$ kJ m^{-2}, the time required to achieve 50% inhibition below 1 m in Ahnin eastern coastal waters will be longer than 13 h. This analysis shows that the time required for 50% inhibition of sporulation in *U. pertusa* close to the surface of the water is 2.5 h and extends to > 13 h at depths below 1 m, which is too long to cause damage to the sporulation process. Therefore, it is expected that the inhibition of *Ulva* sporulation by UV-B radiation will not occur even in shallow water.

The spore liberation period of the UVR-treated alga *U. fasciata* was prolonged for 4 d. Spore liberation is an important event in the life cycle of macroalgae and the viability and development of spores are controlled by many environmental factors, such as temperature, light, photoperiod, and salinity, which may stimulate or inhibit these processes (Krupnova 1984, Makarov 1987, Voskoboinikov and Kamnev 1991). Any effects on spore liberation will eventually lead to the decline of the seaweed population.

The germination success of UV-B-irradiated *U. pertusa* spores was significantly lower than that of the unexposed controls, and the degree of reduction was related to the UV dose (Han et al. 2004). When exposed to December sunlight, the germination percentage of *U. pertusa* spores exposed to solar radiation for 1 h reached 100%, irrespective of irradiation treatment conditions. However, after 2 h of exposure to sunlight, complete inhibition of germination was observed in the PAR + UV-A + UV-B treatment compared with the 100% germination rate observed in the PAR and PAR + UV-A treatments (Han et al. 2004).

The establishment stage of macroalgae is obviously important and affects the performance of the adult population; however, most studies have evaluated the effect of UV-B radiation on the macroscopic stages of algae (Bañares et al. 2002).

Chloroplast and Cell Movement

Many marine algae show cell movement phenomena ranging from the transient motility of reproductive cells to chloroplast movement. The movement of macroalgal unicells has been shown to exhibit phototactic, and/or chemotactic responses (Jones and Babb 1968, Amsler and Neushul 1989).

Maintenance of the swimming behavior of reproductive cells can be an important determinant of algal dispersion and the reproductive strategies of macroalgae

(Amsler and Neushul 1991). To date, few studies have assessed the effects of UV-B on macroalgal cell motility, despite it having been well documented that movement and orientation responses in microalgae are impaired by UV-B radiation (Häder and Worrest 1991, 1997).

Park and Han (1998) noted that in *U. pertusa*, the motility of biflagellate cells was markedly reduced after short durations of UV-B exposure. After the addition of 10^{-4}–10^{-5} M DCMU, an inhibitor of electron transport and a specific and sensitive inhibitor of photosynthesis, the motility of *U. pertusa* gametes decreased to 50% of that of the control. When *U. pertusa* gametes were treated simultaneously with UV-B and DCMU, their motility was reduced by 90%, implying that the two agents attack the same photosynthetic site, and that the decrease in motility was due to the collapse of photosynthesis.

In some macroalgae, the chloroplasts move between a facial position with a low photon fluence rate (a high absorption area) and a profile position with a high photon fluence rate (a low absorption area). The facial position guarantees maximum energy reward because the highest percentage of chloroplasts within cells will be struck by light. Therefore, such chloroplast movement has been considered as an adaptive mechanism to ensure maximum light absorption by the chloroplast, as well as protecting the photosynthetic pigments from photo-degradation. For example, in the brown alga *Dictyota dichotoma* it has been suggested that movement of chloroplasts from facial to profile positions is a photoprotective mechanism synchronised with a gradual increase in the photon fluence rate according to changes in tidal levels (Hanelt and Nultsch 1990).

However, a study on the green alga *U. pertusa* found no significant difference in UV sensitivity with respect to the total chlorophyll content between different chloroplast configuration arrangements (Kong et al. 2002).

It has been reported that *U. pertusa* lost the rhythmical changes in chloroplasts after UV-B exposure, while the chloroplasts of control cells sustained such rhythmicity, with minimal and maximal changes occurring in the middle of the dark and light period, respectively (Kong et al. 2002). Previous studies have illustrated the mechanism of cell motility involving microtubules (MTs) and possibly actin filaments. The interaction between MTs and actin plays an important role in organizing the cytoskeleton and maintaining chloroplast movement (Melkonian et al. 1992). It is noteworthy that UV-B changes the MT structure as tubulin absorbs maximally at 280 nm because of its high content of amino acids with aromatic side chains (Zamansky et al. 1991). Thus, damage to MTs causes a cascade of effects on cell motility.

Plant movement is generally known to occur following a single reaction chain, starting with the recognition of the stimulus, continuing with a single transduction, resulting in the observed response. Photoreceptors participate in the first step of the reaction chain, and the driving force of the transduction process may be a change in ion transport across the cell membrane owing to the activation or inactivation of ion pumps, or a change in membrane permeability. UV-B has been observed to affect cell movement by damaging photoreceptor organelles and several components of the membrane channel (Hada et al. 1993, Sgarbossa et al. 1995).

Pigmentation

Photosynthetic pigments are important targets of UV-B radiation. The ambient level of UV-B radiation reduces the concentration of major photosynthetic pigments in phytoplankton and macroalgae with Chl *a* being more strongly affected than chlorophyll *b* (Chl *b*) (Teramura 1983, Strid et al. 1990).

The UV-induced decrease in the total chlorophyll concentration may reflect either a decrease in pigment synthesis through physical disturbances in chloroplast thylakoids or an increase in pigment destruction due to the absorption of high energy content. Considering that the amount of chlorophyll is related to the number of thylakoids in the chloroplast, it is noteworthy that *Chlorella* exposed to UV radiation showed swelling of the thylakoid membranes and destruction of the outer membranes (Ford et al. 1995). Regardless, it is known that UV quanta are absorbed by chlorophyll, indicating a direct impact of UV radiation on chlorophyll (Halldal 1967).

The destruction of carotenoids may also account for the loss of chlorophyll, as carotenoids protect photosynthetic components from photo-disruption. There is evidence that the xanthophyll cycle is also a target for UV-B radiation. Under field conditions, the de-epoxidation of violaxanthin to zeaxanthin in *U. lactuca* was inhibited following exposure to UV-B (Bischof et al. 2002a,b) and, as a consequence, the ability of the alga to efficiently respond to high radiation stress via thermal energy dissipation is diminished (Demmig-Adams 1990, Young and Frank 1996). This can increase the formation of ROS during photosynthesis, potentially leading to photooxidation of the photosynthetic system (Asada and Takahashi 1987).

In *U. pertusa* there is a significant correlation between total chlorophyll content and UV dosage and it seems that an overall change in chlorophyll concentration is a response to cumulative UV doses, even at a fixed UV-B irradiance (Han 1996). Intertidal macroalgae expectedly experience a wide variety of radiation conditions due to changes in tidal conditions, water clarity, and atmospheric transparency. If the total chlorophyll content and UV dosage is effectively correlated in *U. pertusa* in the natural environment, the destruction of chlorophyll by UV-B radiation would still occur, despite variations in UV radiation.

However adaptations, such as increased protection via carotenoids, have been shown to increase in *Ulva* spp. under exposure to UV-B (Beauchamp 2012). *Ulva* spp. in upper coastal areas showed a higher amount of chlorophyll than their lower coastal counterparts, suggesting an adaptability to protect themselves from the degradation of chlorophyll under high solar irradiance.

Oxidative Stress and Antioxidant Activity

Increased UV-B radiation (280–315 nm) can cause damage to aquatic plants, and lead to a reduction in growth and pigmentation, the inhibition of photosynthesis, and changes in morphology. One of the major causes of the biological effects of UV-B is the production of ROS via intracellular photo-sensitising chromophores, such as tryptophan, nicotinamide adenine dinucleotide, riboflavin, and porphyrin.

ROS are generated as by-products of the electron transport system during photosynthesis and mitochondrial respiration. ROS-mediated modifications of cellular components and associated cellular damage play important roles in metabolic disorders and cell death. Oxidative stress caused by increased ROS affects important biomolecules and causes changes in many biological processes. For example, excessive ROS can directly attack the polyunsaturated fatty acids in cell membranes, causing complicated processes of lipid peroxidation, and can oxidatively damage DNA and proteins. In plants, ROS cause the inactivation and degradation of Rubisco and other components of the Calvin cycle and interfere with photosynthesis.

However, seaweeds contain various antioxidant molecules and secondary metabolites that protect against oxidative damage (Choi et al. 2011, Park et al. 2016). Glutathione (GSH, L-γ-glutamyl-L-cysteinylglycine) is the main antioxidant compound in cells and is a reducing agent for many biochemical processes. Either directly or via glutathione peroxidase (GPx)-catalysed reactions, GSH is oxidised to glutathione disulfide (GSSG) and is secreted from cells, reducing the cellular glutathione content. The ratio of GSH/GSSG also decreases under oxidative stress conditions, unless GSSG is either reduced to regenerate GSH through a glutathione reductase (GR) system or secreted out of the cells. Therefore, the content of glutathione and the redox state are highly sensitive indicators for both the antioxidant capacity of and oxidative stress in cells (Choi et al. 2011). GSH is also involved in the ascorbic acid-glutathione pathway (Halliwell-Asada pathway), which plays an important role in protecting plant cells from ROS and requires GSH for the reduction of dehydroascorbic acid to ascorbic acid or H_2O_2 through an ascorbic acid peroxidase-catalysed reaction. In addition to its role as an antioxidant, GSH plays an important role in various processes, such as enzymatic reactions, transport, and protein and DNA synthesis. Plants also contain a variety of other antioxidant molecules, such as carotenoids and polyphenols, and several antioxidant enzymes, such as its various forms of superoxide dismutase (SOD) and catalase (CAT).

A comparative study of the oxidative stress and antioxidant effects of three types of marine algae exposed to UV-B was conducted using *U. pertusa* (a chlorophyte), *L. japonica* (a phaeophyte), and *Grateloupia lanceolata* (a rhodophyte) collected from the intertidal zone in Korea (Park et al. 2016). *Ulva pertusa* showed a strong adaptive antioxidant response to UV radiation, as indicated by a decrease in lipid peroxidation, a reduction in cell glutathione, and an increase in the ratio of reduced to oxidised glutathione, in line with a decrease in glutathione metabolism and ROS enzyme activity. This contrasted with the responses of two other seaweed species (*L. japonica* and *G. lanceolata*) that were significantly more sensitive to UV radiation, i.e., their exposure was correlated with a decrease in antioxidant enzyme activity, as indicated by an increase in lipid peroxidation and a decrease in glutathione content.

Depending on the degree of oxidative stress, antioxidant activity can be adjusted as an adaptive response. There were temporal changes in oxidative stress and antioxidant activity in *U. pertusa* when collected in July, September, and November (Choi et al. 2011). The level of lipid peroxidation, an oxidative stress marker, was lower in September and higher in November than in July. Glutathione content and redox state correlated well with lipid peroxidation, and the glutathione pool was in a more reduced state in September. The glutathione content was lower and more oxidised in November. The low lipid peroxidation in September was due to increased glutathione pools and increased ROS scavenging capacity supported by reduced GCL,

GPx, and SOD activities. The increased lipid peroxidation and oxidation and the depletion of glutathione in November suggested that ROS production was very high and overwhelmed the increased SOD and GR activity that occurred as an adaptive response to the increased oxidative stress. However, the decrease in catalase activity in September and November may be related to the control of the intrinsic temporal changes of the algal metabolism rather than antioxidant protection. It has also been observed that the antioxidant itself becomes a target for oxidative damage, resulting in activity loss. Thus, if the oxidative stress is too high, the cells suffer from oxidative damage, and if the level of stress is moderate, the cells will adapt to stress by an inducible increase in antioxidant function.

The strong UV-adaptive response via the changes in antioxidants in *U. pertusa* allows this species to efficiently recover from UV-induced oxidative stress, enabling it to occupy the shallow intertidal zone.

UV-absorbing Compounds

Several adaptive mechanisms to minimise UV-induced damage have been reported in macroalgae, including avoidance by the movement of cells and/or cell organelles, epidermal attenuation of UV radiation transmission, synthesis of screening pigments, and repair damage by photo-enzymatic activity.

UV-absorbing compounds act as sunscreens, preventing damaging UV radiation from reaching the DNA, proteins, and other sensitive molecules in cells (Franklin and Forster 1997, Häder 2001).

Various compounds have been identified from marine macroalgae, and some have been extensively studied, including such characteristic compounds as carotenoids, coumarins, phenolics and, in particular, mycosporine-like amino acids (MAAs) (Karsten et al. 1998a,b, Pérez-Rodríguez et al. 2001, 2003). From a survey of macroalgae from polar and temperate to tropical regions, the widespread presence of MAAs, especially in red algae (Karsten et al. 1998a,b, Hoyer et al. 2001, Shick and Dunlap 2002, Huovinen et al. 2004), has been noted. An MAA consists of a cyclohexenone ring linked to amino acid side groups. These compounds absorb wavelengths in the range of 310 to 365 nm across the UV-B and UV-A (315–400 nm) portions of the solar spectrum but transmit PAR (400–700 nm). Brown and green algae are reportedly deficient in MAAs and probably depend on other UV-protection compounds and mechanisms for UV defense (Shick and Dunlap 2002, Huovinen et al. 2004).

Seven MAAs have been isolated from marine macroalgae and have been identified as mycosporine-glycine, shinorine, porphyra-334, palythine, asterina-330, palythinol, usujirene, and palythene. The role of MAAs as UV protectants is inferred from the observation that their concentration correlates with the environmental UV fluence that an organism can experience. However, the association between the levels of UV-absorbing compounds and UV-B sensitivity is not always simple. For instance, since the main absorption band of MAAs is consistent with the UV-A spectrum, it is difficult to hypothesise that the compounds play a protective role in minimising UV-B damage. For example, methanol extracts of *U. pertusa* and *Kjellmaniella crassifolia* (phaeophyte) showed strong absorption characteristics at 300 nm and 340 nm, respectively, however, UV-B sensitivity was significantly different between the species, with the green alga being far less sensitive than the brown alga.

The biochemical detection and characterization of MAAs is well established in various macroalgae, but the physiological and ecological functions of the compounds are still relatively unknown. The possibility of MAAs as UV-protecting agents remains to be determined by continued and comprehensive studies of these compounds.

Ulva pertusa synthesizes a UV-B absorbing compound with a pronounced absorption peak at 294 nm in response to UV-B and the amount induced is proportional to the UV-B dose (Han and Han 2005). This is interesting as the induction of UV-absorbing compounds as a defense system has been reported in few green macroalgae (Hoyer et al. 2001) although they are present in some green microalgae, such as *Chlorella* and *Scenedesmus* (Xiong et al. 1999). In *Scenedesmus* spp. two UV-absorbing compounds with maximum absorptions at 292 nm and 302 nm were noted, but their chemical nature has not yet been elucidated.

The polychromatic spectrum for the induction of UV-B absorbing compounds in *U. pertusa* showed a prominent peak at 292 nm and a smaller peak at 311.5 nm (Han and Han 2005). *Ulva pertusa* appears to have UV-B-absorbing chromophores similar to those reported in higher plants, which reportedly have a maximum number of these chromophores between 290 and 310 nm (Herrlich et al. 1997). In higher plants, it has been suggested that the chemical entity of the chromophore is either a flavin or pterin (Ensminger and Schafer 1992), or possibly an aromatic residue of a protein (Kim et al. 1992).

The presence of UV-B photoreceptors has been suggested in several other green macroalgae. The action spectrum of the photoresponse in the rhizome of *Bryopsis plumosa*, a symbiotic alga, showed a maximum peak at 260 nm but another peak at 310 nm (Iseki and Wada 1995). The green macroalga *Prasiola stipitata* produces UV absorbing substances, and the action spectrum showed the maximum effectiveness at 300 nm (Gröniger and Häder 2002). Further studies are required to identify UV-B-absorbing chromophores in *U. pertusa*, because the response to UV-B may be a consequence of cellular effects, such as DNA damage and photooxidation products rather than UV-B itself (Portwich and Garcia-Pichel 2000).

Ulva pertusa shows a rhythmic induction of UV-B-absorbing compounds, even after 12 h of PAR + UV-A + UV-B followed by 12 h in the dark, with accumulation being confined to the light period (Han and Han 2005). The alga studied also appeared to accumulate UV-B exposure despite intermittent exposure to white light-dark periods, indicating the presence of a UV-B quantum counting system for at least 5 d. A similar pattern for MAA synthesis has been observed in some N_2-immobilised cyanobacteria (Sinha et al. 2001). In *Anabaena* spp., *Nostoc commune*, and *Scytonema* spp. an induction of shinorine was observed to occur during the light period. In an ecological context, the dependency of the accumulation of protection compounds on UV-B exposure makes it possible for *U. pertusa* to adapt to the continuous fluctuation of solar UV-B radiation reaching its natural habitat.

Under artificial UV-B irradiation, the accumulation of UV-B absorbing compounds in *U. pertusa* was accompanied by reduced growth (Han and Han 2005). Secondary product formation and vegetative growth are known to be antagonistic processes (Zenk et al. 1977). Additionally, it has been demonstrated that secondary metabolites that are not formed under optimal growth conditions begin to accumulate after a reduction in cell growth (Zenk et al. 1977). In *U. pertusa*, it appears that UV-B radiation causes a growth decline, thereby switching from primary metabolism to secondary metabolism, and inducing UV-B absorbing compounds (Han and Han 2005). In the same species, it is suggested that low levels of UV-B act as a signal for the transition from somatic growth to reproduction (Han et al. 2003).

However, UV-B-induced growth reduction can also be a by-product of the accelerated production of UV-B-absorbing compounds. To produce UV-absorbing compounds, considerable metabolic investment is required in terms of photon and nitrogen costs (Raven 1991). The accumulation of UV-B absorbing compounds by *U. pertusa* after UV-B exposure may also represent a mechanism for reducing growth, in order to reduce UV-B exposure, as a trade-off between growth and defense (Han and Han 2005).

Few studies have addressed the relationship between the concentration of UV-absorbing compounds and antioxidant activity in macroalgae. Therefore, the significant correlation between the concentration of UV-B-absorbing compounds and non-enzymatic antioxidant activity in *U. pertusa*, is notable (Han and Han 2005). This finding lends support to the hypothesis that these compounds serve as antioxidants for protection against UV-B-induced photo-oxidative stress. A similar concentration-dependent relationship between mycosporine-glycine and antioxidants has been reported for the ascidian *Lissoclinum patella* and the zoanthid *Palythoa tuberculosa* (Dunlap and Yamamoto 1995, Suh et al. 2003). Producing UV-B-absorbing compounds with dual functions (screening and antioxidant) may be a cost-saving strategy for *U. pertusa*. Since there is a spectral discrepancy between the absorption properties of UV-absorbing compounds and UV-B radiation, it is questionable whether these compounds offer complete protection against UV-B radiation. Also, the protective function of UV-absorbing compounds in many algae has been inferred from observations that the compounds accumulate in proportion to the degree of exposure to UV-B; however, the mere presence of UV-absorbing compounds does not guarantee protection against the harmful effects of UV radiation (Shick and Dunlap 2002). For example, in the dinoflagellate, *Prorocentrum micans*, the MAAs produced after UV exposure did not protect the photosynthetic mechanism, as cellular chlorophyll concentrations and Rubisco activity were significantly reduced despite their presence, whereas in *Gymnodinium sanguineum* the cells with higher concentrations of MAA were less sensitive to UV-B radiation (Lesser 1996, Neale et al. 1998). The results from a study on *U. pertusa*, which showed that the sensitivity of photosynthetic performance as measured by dark-adapted Chl *a* fluorescence, correlated inversely with the concentrations of UV-B-absorbing compound in the thallus imply that induction of UV-B-absorbing compounds under UV-B seemingly play an important role in protection UV-B radiation (Han and Han 2005). The induction of UV-B-absorbing compounds by UV-B radiation is a selective advantage of *U. pertusa*, which may initiate the defense mechanism in response to the harmful effects of the radiation. Further research is required to better understand the efficiency and mode of protection of these UV-B absorbing compounds (Lesser 1996, Neale et al. 1998).

Photo-reactivation

UV-B quanta are easily absorbed by biomolecules, such as nucleic acids (DNA, RNA, etc.), proteins, lipids, etc., which play important roles in the structure and function in cells of chloroxygenic organisms (Bischof et al. 2002a). UV damage to DNA involves structural changes, such as the formation of thymine dimer and 6–4 photoproducts (Pakker et al. 2000a, van De Poll et al. 2001).

On the contrary, the reversal of the harmful effects of UV radiation by subsequent illumination with longer wavelengths is almost universally observed in organisms (Jagger et al. 1964, Caldwell 1971). This phenomenon, known as photoreactivation, has been observed in some macroalgae with recovery from interference with DNA synthesis, nuclear division, and translocation, and from dimerization of DNA, loss of viability, delay in germination, and inhibition of spore production (Han and Kain 1992, 1993, Huovinen et al. 2000, Pakker et al. 2000a,b, Han et al. 2003b, 2004).

The mechanism of photoreactivation is relatively well described for DNA, the site of one of the most common UV lesions, and involves the restoration of UV-incurred damage in DNA by the photoreactivation (i.e., the absorption of light energy) using enzymes that monomerise dimers formed in UV-irradiated DNA (Karentz 1994). In general, DNA photoreactivating enzymes have UV and blue light requirements for optimum catalytic functions (Saito and Werbin 1970, O'Brien and Houghton 1982, Eker et al. 1990).

In contrast, Han and Kain (1992, 1993) reported that exposure to visible light enhanced the survival of UV-irradiated young laminarian sporophytes, but little is known about the functional role of photoreactivation. The blue light requirements for effective reactivation from UV-damage in young laminarian sporophytes corresponded to the report that the most effective reversal of UV damage is known to occur under near-UV and blue radiation (Jagger 1964, Caldwell 1971, Halldal and Taube 1972). It has also been reported that the degree of recovery increases with increasing photon irradiances (Han and Kain 1992, Han et al. 2003b, 2004).

DNA damage is known to result in decreased spore viability and to delay germination and the development of spores (Huovinen et al. 2000, Wiencke et al. 2000) at the initial stage of survival of macroalgae. The germination success of UV-B-irradiated spores of the intertidal green alga *U. pertusa* was significantly lower than that of the non-irradiated control spores, and the extent of the reduction was correlated with the UV dose (Han et al. 2004).

After exposure to moderate levels of UV-B radiation, subsequent exposure to visible light produced significantly higher rates of spore germination in *U. pertusa* under higher photon irradiances and blue light than it did under lower irradiances and red light (Han et al. 2004). The main targets of UV-B radiation in spores of *U. pertusa* are unknown, but it is noteworthy that the rate of spore germination has been shown to depend on photon irradiances during the post-illumination period.

Macroalgae living in the intertidal zone have a high degree of resistance to adverse environmental conditions (Davison and Pearson 1996). *Ulva* spp. with irradiance-dependent reactivation from UV-B damage may have a greater selective advantage than other species when competing for space in shallow water habitats that are frequently exposed to high intensity UV-B.

For UV-B irradiated *Ulva* spores, blue light was the most effective for reactivation of germination (Han et al. 2004). The lack of discernible germination in red light eliminates the involvement of photosynthetic pigments for the reactivation process. Similar reactivation by blue light has been reported for cyanobacteria (Saito and Werbin 1970, O'Brien and Houghton 1982, Eker et al. 1990). In *Ulva* spp., the phototaxis of zoospores is guided by blue light but not by green and red light (Callow and Callow 2000). The action spectrum of the photoreactivation of germination in *U. pertusa* spores irradiated with UV-B shows a large peak at 435 nm and a smaller but significant peak at 385 nm (Han et al. 2004). The improving effect of blue light on the germination of *U. pertusa* exposed to UV-B radiation seems to be related to the presence of a photoreactivation system that is similar to that found in other organisms. The action spectrum is similar to the DNA photoreactivation of cyanobacteria (van Baalen and O'Donnell 1972, Eker et al. 1990), with a major peak at 436 nm, and that of the higher plant, *Zea mays*, with a broad single peak at 385 nm (Ikenaga et al. 1974). The similarity between the action spectra suggests that the mechanism of photoreactivation may require DNA repair, and that the UV-B target involved in the *Ulva* spore germination may be DNA.

A variety of plant responses induced by blue/UV-A radiation have been reported, and there is now convincing evidence for the existence of multiple blue light/UV-A photoreceptors absorbing principally UV-A (315–400 nm) and blue (400–500 nm) wavelengths. Blue photoreceptors involved in the reactivation of spore germination in UV-B-irradiated *U. pertusa* may be one of these groups.

In addition to the matting properties that act as a selective UV-B filter for spores under the parent canopy, germination repair by blue/UV-A light after UV-B exposure probably explains the success of germination of *U. pertusa* spores in a shallow water environment that is influenced by strong solar UV radiation (Bischof et al. 2002d, Han et al. 2003b). If the early life stages are decisively important for the recruitment of the next generation, the presence of a light-driven repair system in the settled spores is a viable option for the successful continuation of *U. pertusa* in areas receiving intense solar UV-B radiation.

Mat-forming and Morphology

Ulva spp. often cause green tides. Thalli occur in a dense multi-layered mat floating on the water surface (Hernández et al. 1997, Vergara et al. 1997, 1998). The top layer is periodically exposed to high solar radiation, but the shaded layer receives reduced irradiance (Vergara et al. 1997, 1998). These so-called *Ulva* mats are frequently observed in soft bottom habitats in shallow coastal areas. A study on the influence of solar UV radiation on the *Ulva* canopy proposed canopy formation as an ecological strategy to shield the sub-canopy layers from harmful UV-B radiation by significant absorption of UV-B in the bleached top layer of the mat (Bischof et al. 2002a).

It has been reported that the morphological features of macroalgae affect their physiological responses to physical stress agents, such as high irradiance, desiccation, and wave motion (Davison and Pearson 1996). In higher plants, attenuation of UV radiation that reaches the mesophyll via the epidermis is one of the important factors determining the susceptibility of plants to UV radiation (Day et al. 1992). In algae, it is presumed that macroalgae with thin thalli are more susceptible to UV damage than those with thick thalli (Halldal 1964). However, there have been few direct comparisons of whether different macrophyte morphologies show different tolerances to UV radiation. In response to tests regarding the effects of artificial UV-B radiation, the brown macroalga *K. crassifolia* and the red macroalgae *Pachymeniopsis* spp. were seen to both possess thick thalli with several cell layers, and the green macroalga *U. pertusa* was seen to possess thin thalli with only two cell layers (Han et al. 1998). After 1 h of exposure to UV-B 2 W m^{-2}, the chlorophyll content was compared with that of the control specimens and in the brown alga was found to have decreased to 95.3%, in the green alga to 58.0%, and in the red algae to 80.6% of that of the control specimens. In addition, the fresh weights were reduced by 68.2% in *K. crassifolia* and by 21.4% in *U. pertusa*, whereas it was enhanced by 11.1% in the red algae *Pachymeniopsis* spp. (Han et al. 1998). These results may suggest that thallus form does not greatly affect the physiological response to UV radiation, at least in these three species.

Conclusion

Global climate change can also alter the exposure of ecosystems to UV-B radiation by affecting geochemical processes that influence ozone depletion. In addition, UV-B radiation may affect the cycling of C and inorganic nutrients, such as nitrogen, through changes in macroalgal communities. This could occur due to the effects of UV-B radiation on photosynthetic performance and the chemical properties of damaged and dead algal substances falling to the bottom. It is well known that seaweeds and seagrasses are the main vehicles for C assimilation of ~ 1 Pg C yr^{-1} (Chung et al. 2011) and may become C sinks for anthropogenic CO_2 emissions. The C assimilation capacity of *U. pertusa* is estimated at 9.49 kg ton^{-1} h^{-1}, i.e., an estimated 10.25 tons of C assimilation over 6 months, based on a daily photosynthetic period of 6 h (Han et al. unpublished). The effect of UV radiation on the C cycle may affect the cycling of metals and mineral nutrients (Zepp et al. 2007). It is noteworthy that the EC_{50} values of *U. pertusa* for four metals is lower than those of the standard Microtox and five other standard test organisms, and that *U. pertusa* can even be used as a model test organism to detect metal toxicity (Han and Choi 2005). The combined effects of UV radiation and metal toxicity can, therefore, have serious impacts on *Ulva* spp.

UV radiation can affect the nitrogen cycle in several ways, including changes in the decomposition of nitrogenous organics and effects on nitrogen fixation (Zepp et al. 2007). In addition, in aquatic environments UV interactions with inorganic nitrogen species, such as nitrate and nitrite are an important source of ROS, including the highly reactive hydroxyl radicals.

Alternatively, some adaptive mechanisms for UV-B radiation stress may depend on the nutritional status of seaweeds and the nutrient availability in seawater. For instance, nitrogen limitation can affect photosynthetic capacity and the content of protective compounds, thus increasing the sensitivity of seaweed to UV-B (Korbee et al. 2005). Increased levels of UV radiation and eutrophication can increase the energy investment needed for the synthesis of MAAs in macroalgae, thereby providing more protection under UV-B stress (Peinnado et al. 2004).

The release of ozone-depleting gases has presumably ceased since 1980, but a full recovery of the ozone layer is not expected until 2070 and, owing to its possible interaction with climate change, the effects of UV-B radiation are uncertain (Figueroa and Korbee 2010).

UV-B exists as a general abiotic factor, and the currently observed physiological patterns are the result of long-term adaptation and acclimation of individual species to various irradiations. Therefore, the adverse effects of UV-B exposure may consequently be offset to some extent, however, the permanent suppression of physiological processes may result. There is a possibility that the increased UV-B exposure caused by the thinning of the stratospheric ozone layer may disturb this balance, leading to further damage. It is still to be determined if, or to what extent, the adaptive and protective responses of macroalgae will be sufficient to re-establish this balance.

However, the presence and antioxidant capacity of UV-B-absorbing compounds in *U. pertusa* as well as its mat-forming properties and light-driven photo-repair give this species superior selective advantages over other macroalgae and enabling it to thrive in the presence of intense UV-B radiation (Fig. 11.1). *Ulva* spp. are evidently the fittest macroalgae of the intertidal zone where the macroalgal community is often exposed to high levels of solar radiation.

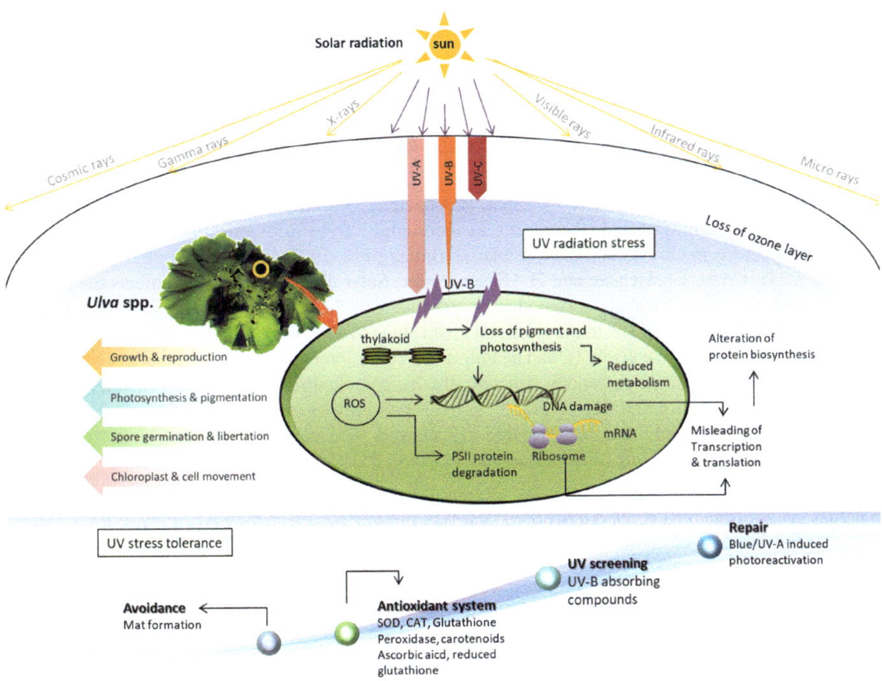

Fig. 11.1. A proposed model of the UV-B stree responses in *Ulva* spp. (for details, see text).

Acknowledgements

This work was supported by a Grant of Incheon National University Research (2012).

References

Allen, D.J. and I.F. McKee, P.K. Farage and N.R. Baker. 1997. Analysis of limitations to CO_2 assimilation on exposure of leaves of two *Brassica napus* cultivars to UV-B. Plant, Cell and Environment 20: 633–640.

Amsler, C.D. and M. Neushul. 1989. Chemotactic effects of nutrients on spores of the kelps *Macrocystis pyrifera* and *Pterygophora californica*. Marine Biology 102: 557–564.

Amsler, C.D. and M. Neushul. 1991. Photosynthetic physiology and chemical composition of spores of the kelps *Macrocystis pyrifera, Nereocystis luetkeana, Laminaria farlowii* and *Pterygophora californica* (Phaeophyceae). Journal of Phycology 27: 26–34.

Asada, K. and M. Takahashi. 1987. Production and scavenging of active oxygens in chloroplasts. pp. 227–287. *In*: Kyle, D.J., Osmond, C.B. and Arntzen, C.J. (eds.). In Photoinhibition. Elsevier, Amsterdam, Netherland.

Bañares, E., M. Altamirano, F.L. Figueroa and A. Flores-Moya. 2006. Influence of UV radiation on growth of sporelings of three non-geniculate coralline red algae from Southern Iberian Peninsula. Phycological Research 50: 23–30.

Beach, K.S., C.M. Smith, T. Michael and H.W. Shin. 1995. Photosynthesis in reproductive unicells of *Ulva fasciata* and *Enteromorpha flexuosa*: Implications for ecological success. Marine Ecology 125: 229–237.

Beauchamp, E. 2012. Effects of UV radiation and salinity on the intertidal macroalgae *Palmaria palmata* and *Ulva lactuca*; effects on photosynthetic performance, growth and pigments. The Plymouth Student Scientist 5: 3–22.

Behrenfeld, M., J. Hardy, H. Gucinski, A. Hanneman, H.L. II and A. Wones. 1993. Effects of ultraviolet-B radiation on primary production along latitudinal transects in the South Pacific ocean. Marine Environmental Research 35: 349–363.

Bischof, K., D. Hanelt and C. Wiencke. 1998. UV-radiation can affect depth-zonation of Antarctic macroalgae. Marine Biology 131: 597–605.

Bischof, K., D. Hanelt and C. Wiencke. 2000. Effects of ultraviolet radiation on photosynthesis and related enzyme reactions of marine macroalgae. Planta 211: 555–562.

Bischof, K., D. Hanelt and C. Wiencke. 2002a. UV radiation and Arctic marine macroalgae. *In*: Hessen D.O. (ed.). UV Radiation and Arctic Ecosystems. Ecological Studies (Analysis and Synthesis), Vol. 153. Springer, Berlin, Heidelberg.

Bischof, K., D. Hanelt, J. Aguilera, U. Karsten, B. Vögele, T. Sawall and C. Wiencke. 2002b. Seasonal variation in ecophysiological patterns in macroalgae from an Arctic fjord. I. Sensitivity of photosynthesis to ultraviolet radiation. Marine Biology 140: 1097–1106.

Bischof, K., G. Kräbs, C. Wiencke and D. Hanelt. 2002c. Solar ultraviolet radiation affects the activity of ribulose-1,5-biphosphate carboxylase-oxygenase and the composition of photosynthetic and xanthophyll cycle pigments in the intertidal green alga *Ulva lactuca* L. Planta 215: 502–509.

Bischof, K., G. Peralta, G. Kräbs, W.H. van de Poll, J.C. Pérez-Lloréns and A.M. Breeman. 2002d. Effects of solar UV-B radiation on canopy structure of *Ulva* communities from southern Spain. Journal of Experimental Botany 53: 2411–2421.

Brandle, J.R., W.F. Campbell, W.B. Sisson and M.M. Caldwell. 1977. Net photosynthesis, electron transport capacity, and ultrastructure of *Pisum sativum* L. exposed to ultraviolet-B radiation. Plant Physiology 60: 165–169.

Burrows, E.M. 1959. Growth form and environment in *Enteromorpha*. Botanical Journal of the Linnean Society 56: 204–206.

Caldwell, M.M. 1971. Solar UV irradiation and the growth and development of higher plants. pp. 131–177. *In*: Giese, A.C. (eds.). Photophysiology, Vol. 6, Academic Press, New York.

Callow, M.E. and J.A. Callow. 2000. Substratum location and zoospore behaviour in the fouling alga *Enteromorpha*. Biofouling 15: 49–56.

Choi, E.M., J.J. Park and T. Han. 2011. Temporal changes in oxidative stress and antioxidant activities in *Ulva pertusa* Kjellman. Toxicology and Environmental Health Science 3: 206–212.

Chung, I.K., J. Beardall, S. Mehta, D. Sahoo and S. Stojkovic. 2011. Using marine macroalgae for carbon sequestration: a critical appraisal. Journal of Applied Phycology 23: 877–886.

Christie, A.O. and L.V. Evans. 1962. Periodicity in the liberation of gametes and zoospores of *Enteromorpha intestinalis* Link. Nature 193: 193–194.

Clendennen, S.K., R.C. Zimmerman, D.A. Powers and R.S. Alberte. 1996. Photosynthetic response of the giant kelp *Macrocystis pyrifera* (Phaeophyceae) to ultraviolet radiation. Journal of Phycology 32: 614–620.

Coelho, S.M., J.W. Rijstenbil and M.T. Brown. 2000. Impacts of anthropogenic stresses on the early development stages of seaweeds. Journal of Aquatic Ecosystem Stress and Recovery 7: 317–333.

Davidson, A.T., D. Bramich, H.J. Marchant and A. McMinn. 1994. Effects of UV-B irradiation on growth and survival of Antarctic marine diatoms. Marine Biology 119: 507–515.

Davison, I.R. and G.A. Pearson. 1996. Stress tolerance in intertidal seaweeds. Journal of Phycology 32: 197–211.

Day, T.A., T.C. Vogelmann and E.H. DeLucia. 1992. Are some plant life forms more effective than others in screening out ultraviolet-B radiation? Oecologia 92: 513–519.

Demmig-Adams, B. 1990. Carotenoids and photoprotection in plants: A role for the xanthophyll zeaxanthin. Biochimica et Biophysica Acta (BBA) - Bioenergetics 1020: 1–24.

Dring, M.J., A. Wagner and K. Lüning. 2001. Contribution of the UV component of natural sunlight to photoinhibition of photosynthesis in six species of subtidal brown and red seaweeds. Plant Cell and Environment 24: 1153–1164.

Dring, M.J., V. Makarov, E. Schoschina, M. Lorenz and K. Lüning. 1996. Influence of ultraviolet-radiation on chlorophyll fluorescence and growth in different life-history stages of three species of *Laminaria* (Phaeophyta). Marine Biology 126: 183.

Dunlap, W.C. and Y. Yamamoto. 1995. Small-molecule antioxidants in marine organisms: antioxidant activity of mycosporine-glycine. Comparative Biochemistry and Physiology Part B: Biochemistry and Molecular Biology 112: 105–114.

Eker, A.P., P. Kooiman, J.K.C. Hessels and A. Yasui. 1990. DNA photoreactivating enzyme from the cyanobacterium *Anacystis nidulans*. Journal of Biological Chemistry 265: 8009–8015.

Ensminger, P.A. and E. Schäfer. 1992. Blue and ultraviolet-B light photoreceptors in parsley cells. Photochemistry and Photobiology 55: 437–447.

Fei, X.G., B. Jiang, M. Ding, Y. Wu, R. Huang and H. Li. 1989. Light demands of juvenile *Laminaria japonica*. Chinese Journal of Oceanology and Limnology 7: 1–9.

Figueroa, F.L., C. Nygard, N. Ekelund and I. Gomez. 2003. Photobiological characteristics and photosynthetic UV responses in two *Ulva* species (Chlorophyta) from southern Spain. Journal of Photochemistry and Photobiology B: Biology 72: 35–44.

Figueroa, F.L. and N. Korbee. 2010. Interactive effects of UV radiation and nutrients on ecophysiology: Vulnerability and adaptation to climate change. *In*: Seckbach, J., R. Einav and A. Israel (eds.). Seaweeds and their Role in Globally Changing Environments. Cellular Origin, Life in Extreme Habitats and Astrobiology, Vol 15. Springer, Dordrecht.

Figueroa, F.L., S. Salles, J. Aguilera, C. Jiménez, J. Mercado, B. Viñegla, A. Flores-Moya and M. Altamirano. 1997. Effects of solar radiation on photoinhibition and pigmentation in the red alga *Porphyra leucosticta*. Marine Ecology 151: 81–90.

Flameling, I.A. and J. Kromkamp. 1998. Light dependence of quantum yields for PSII charge separation and oxygen evolution in eukaryotic algae. Limnology and Oceanography 43: 284–297.

Fletcher, R.L. 1976. Post-germination attachment mechanisms in marine fouling algae. pp. 443–64. *In*: Sharpley, J.M. and A.M. Kaplan (eds.). Proceedings of the Third International Biodegradation Symposium. Applied Science Publishers, London.

Ford, T.W., A.M. Page and A.D. Stead. 1995. Effects of UV irradiation on the ultrastructure of the unicellular green alga *Chlorella*. The Phycologist 40: 19.

Føyn, B. 1959. Geschlechtskontrollierte Vererbung bei der marinen Grünalge *Ulva mutabilis*. Archiv für Protistenkunde 104: 236–253.

Franklin, L.A. and R.M. Forster. 1997. The changing irradiance environment: consequences for marine macrophyte physiology, productivity and ecology. European Journal of Phycology 32: 207–232.

Greenberg, B.M., V. Gaba, O. Canaani, S. Malkin, A.K. Mattoo, and M. Edelman. 1989. Separate photosensitizers mediate degradation of the 32-kDa photosystem II reaction center protein in the visible and UV-spectral regions. Proceedings of the National Academy of Sciences of the Unites States 86: 6617–6620.

Grobe, C.W. and T.M. Murphy. 1994. Inhibition of growth of *Ulva expansa* (Chlorophyta) by ultraviolet-B radiation, Journal of Plant Physiology 30: 783–790.

Grobe, C.W. and T.M. Murphy. 1997. Artificial ultraviolet radiation and cell expansion in the intertidal alga *Ulva expansa* (Setch.) S. and G. (Chlorophyta). Journal of Experimental and Marine Biological Ecology 217: 209–223.

Gröniger, A. and D,-P. Häder. 2002. Induction of the synthesis of an UV-absorbing substance in the green alga *Prasiola stipitata*. Journal of Photochemistry and Photobiology B: Biology 66: 54–59.

Häberlein, A. and D.-P. Häder. 1992. UV effects on photosynthetic oxygen production and chromoprotein composition in the freshwater flagellate *Cryptomonas* S2. Acta Protozoologica 31: 85–92.

Hada, M., M. Tada and T. Hashimoto. 1993. UV-B-induced absorbance changes in the yeast *Rhodotorula minuta*. Journal of Photochemistry and Photobiology B: Biology 17: 127–134.

Häder, D.-P. 1996. Effects of enhanced UV-B radiation on phytoplankton. Sientia Marina 60 (Suppl 1): 59–63.

Häder, D.P. 2001. Adaptation to UV stress in algae. *In:* Rai, L.C. and Gaur, J.P. (eds.) Algal Adaptation to Environmental Stresses. Springer, Berlin, Heidelberg.

Häder, D.-P. and F.L. Figueroa. 1997. Photoecophysiology of marine macroalgae. Photochemistry and Photobiology 66: 1–14.

Häder, D.-P., M. Lebert, and E.W. Helbling. 2001. Effects of solar radiation on the Patagonian macroalga *Enteromorpha linza* (L.) J. Agardh-Chlorophyceae. Journal of Photochemistry and Photobiology B: Biology 62: 43–54.

Häder, D.-P. and R.C. Worrest. 1991. Effects of enhanced solar ultraviolet radiation on aquatic ecosystems. Photochemistry and Photobiology 53: 717–725.

Häder, D.-P. and R.C. Worrest. Consequences of the effects of increased solar ultraviolet radiation on aquatic ecosystem pp. 11–30. *In:* D.-P. Häder. [eds.] 1997. The effects of ozone depletion on aquatic ecosystems, R.G. Landes, Austin, Texas, USA.

Halldal, P. 1967. Ultraviolet action spectra in algology. Photochemistry and Photobiology 6: 445–460.

Halldal, P. and Ö. Taube. 1972. Ultraviolet action and photoreactivation in algae. pp. 163–188. *In*: Giese, A.C. (eds.). Photophysiology, Vol. 7, Academic Press, New York.

Han, T. 1996. Effect of ultraviolet-B radiation on *Ulva pertusa* Kjellman (Chlorophyta) I. Growth and pigment content. Algae 11: 155–159.

Han, T. and G.W. Choi. 2005. A novel marine algal toxicity bioassay based on sporulation inhibition in the green macroalga *Ulva pertusa* (Chlorophyta). Aquatic Toxicology 75: 202–212.

Han, T., H. Chung and S.H. Kang. 1998. UV photobiology of marine macroalgae. Korean Journal of Polar Research 9: 37–46.

Han, T., Y.S. Han, J.M. Kain and D.-P. Häder. 2003a. Thalllus differentiation of photosynthesis, growth, reproduction, and UV-sensitivity in the green alga *Ulva pertusa* (Chlorophyceae). Journal of Phycology 39: 712–721.

Han, T., Y.S. Han, K.Y. Kim, J.H. Kim, H.W. Shin, J.M. Kain, J.A. Callow and M.E. Callow. 2003b. Influences of light and UV-B on growth and sporulation of the green alga *Ulva pertusa* Kjellman. Journal of Experimental Marine Biology and Ecology 290: 115–131.

Han, T., J.A. Kong, Y.S. Han, S.H. Kang and D.-P. Häder. 2004. UV-A/blue light–Induced reactivation of spore germination in UV-B irradiated *Ulva pertusa* (chlorophyta). Journal of Phycology 40: 315–322.

Han, T. and J.M. Kain. 1992. Blue light sensitivity of UV-irradiated young sporophytes of *Laminaria hyperborea*. Journal of Experimental Marine Biology and Ecology 158: 219–230.

Han, T. and J.M. Kain. 1993. Blue light photoreactivation of ultraviolet irradiated young sporophytes of *Alaria esculenta* and *Laminaria saccharina* (Phaeophyta). Journal of Phycology 29: 79–81.

Han, Y. and T. Han. 2005. UV-B induction of UV-B protection in *Ulva pertusa* (Chlorophyta). Journal of Phycology 41: 523–530.

Hanelt, D. 1998. Capability of dynamic photoinhibition in Arctic macroalgae is related to their depth distribution. Marine Biology 131: 361–369.

Hanelt, D. and W. Nultsch. 1990. Daily changes of the phaeoplast arrangement in the brown alga *Dictyota dichotoma* as studied in field experiments. Marine Ecology 61: 273–279.

Hernández, I., G. Peralta, J.L. Pérez-Lloréns, J.J. Vergara and F.X. Niell. 1997. Biomass and dynamics of growth of *Ulva* species in Palmones river estuary. Journal of Phycology 33: 764–772.

Herrlich, P., C. Blattner, A. Knebel, K. Bender and H.J. Rahmsdorf. 1997. Nuclear and non-nuclear targets of genotoxic agents in the induction of gene expression: shared principles in yeast, rodents, man and plants. Biological Chemistry 378: 1217–1229.

Hoffman, J.R., L.J. Hansen and T. Klinger. 2003. Interactions between UV radiation and temperature limit inferences from single-factor experiments. Journal of Phycology 39: 268–272.

Hoyer, K., U. Karsten, T. Sawall and C. Wiencke. 2001. Photoprotective substances in Antarctic macroalgae and their variation with respect to depth distribution, different tissues and developmental stages. Marine Ecological Progress Series 211: 117–29.

Huovinen, P.S., A.O.J. Oikari, M.R. Soimasuo and G.N. Cherr. 2000. Impact of UV radiation on the early development of the giant kelp (*Macrocystis pyrifera*) gametophytes. Photochemistry and Photobiology 72: 308–313.

Huovinen, P.S., I. Gómez, F.L. Figueroa, N. Ulloa and V. Morales. 2004. Ultraviolet-absorbing mycosporine-like amino acids in red macroalgae from Chile. Botanica Marina 47: 21–29.

Ikenaga, M. and S. Kondo, T. Fujii. 1974. Action spectrum for enzymatic photoreactivation in maize. Photochemistry and Photobiology 19: 109–113.

Iseki, M. and S. Wada. 1995. Action spectrum in the ultraviolet region for phototropism of *Bryopsis* rhizoids. Plant and Cell Physiology 36: 1033–1040.

Jagger, J., W.C. Wise and R.S. Stafford. 1964. Delay in growth and division induced by near ultraviolet radiation in *Escherichia coli* B and its role in photoprotection and liquid holding recovery. Photochemistry and Photobiology 3: 11–24.

Jansen, M.A.K., V. Gaba, B.M. Greenberg, A.K. Mattoo and M. Edelman. 1996. Low threshold levels of ultraviolet-B in a background of photosynthetically active radiation trigger rapid degradation of the D2 protein of photosystem-II. The Plant Journal 9: 693–699.

Jenkins, G.I. 2017. Photomorphogenic responses to ultraviolet-B light. Plant, Cell & Environment 40: 2544–2557.

Jerlov, N.G. 1976. Marine Optics. Elsevier, Amsterdam.

Jiang, H. and K. Gao. 2008. Effects of UV Radiation on the photosynthesis of conchocelis of *Porphyra haitanensis* (Bangiales, Rhodophyta). Phycologia 47: 241–248.

Jones, W.E. and M.S. Babb. 1968. The motile period of swarmers of *Enteromorpha intestinalis* (L.) link. British Phycological Bulletin 3: 525–528.

Jordan, B.R., J. He, W.S. Chow and J.M. Anderson. 1992. Changes in mRNA levels and polypeptide subunits of ribulose 1,5-bisphosphate carboxylase in response to supplementary ultraviolet-B radiation. Plant, Cell and Environment 15: 91–98.

Karentz, D. 1994. Ultraviolet tolerance mechanisms in Antarctic marine organisms. *In*: Weiler, C.S. and P.A. Penhale (eds.). Ultraviolet Radiation in Antarctica: Measurements and Biological Effects. Antarctic Research Series. Vol. 62. American Geophysical Union, Washington, DC, pp. 93–110.

Karsten, U., T. Sawall, D. Hanelt, K. Bischof, F.L. Figueroa, A. Flores-Moya and C. Wiencke. 1998a. An Inventory of UV-absorbing mycosporine-like amino acids in macroalgae from polar to warm-temperate regions. Botanica Marina 41: 443–453.

Karsten, U., T. Sawall, K. Lüning and C. Wiencke. 1998b. A survey of the distribution of UV-absorbing substances in tropical macroalgae. Phycological Research 46: 271–278.

Kendrick, R.E. and G.H.M. Kronenberg. 1994. Photomorphogenesis in Plants, Ed 2. (eds.). Kluwer Academic, Dordrecht, Netherlands.

Kim, S.T., Y.F. Li and A. Sancar. 1992. The third chromophore of DNA photolyase: Trp-277 of *Escherichia coli* DNA photolyase repairs thymine dimers by direct electron transfer. Proceedings of the National Academy of Sciences of the Unites States 89: 900–904.

Kirk, J.T.O. 1994. Light and photosynthesis in aquatic ecosystems, 2nd ed. Cambridge University Press.

Kong, J.A., Y.S. Han and T. Han. 2002. Rhythmic phenomena in the green alga *Ulva pertusa* Kjellman. Algae 17: 259–265.

Korbee, N., F.L. Figueroa and J. Aguilera. 2005. Effect of light quality on the accumulation of photosynthetic pigments, proteins and mycosporine-like amino acids in the red alga *Porphyra leucosticta* (Bangiales, Rhodophyta). Journal of Photochemistry and Photobiology B: Biology 80: 71–78.

Kramer, G.F., H.A. Norman, D.T. Krizek and R.M. Mirecki. 1991. Influence of UV-B radiation on polyamines, lipid peroxidation and membrane lipids in cucumber. Phytochemistry 30: 2101–2108.

Krause, G.H. and E. Weis. 1991. Chlorophyll fluorescence and photosynthesis: the basics. Annual Review of Plant Physiology and Plant Molecular Biology 42: 313–349.

Krupnova, T.N. 1984. Reproductive strategies of the Japanese kelp (*Laminaria japonica* Aresch.), an object of the mariculture. Ph. D. Thesis. Pacific Fisheries Research Centre, Vladivostok.

Kuhlenkamp, R. and K. Lüning. 1998. Are natural surface radiation levels of PAR and UV-light detrimental to early germling stages of intertidal macroalgae? The Phycologist 49: 38.

Lee, M.S. and T. Han. 1998. Comparative sensitivity of UV-irradiated germlings and adult thalli of *Ulva pertusa* to light. The Phycologist 49: 30.

Lesser, M.P. 1996. Acclimation of phytoplankton to UV-B radiation: Oxidative stress and photoinhibition of photosynthesis are not prevented by UV-absorbing compounds in the dino-flagellate *Prorocentrum micans*. Marine Ecology Progress Series 132: 287–297.

Lesser, M.P. 2006. Oxidative stress in marine environments: Biochemistry and physiological ecology. Annual Review of Physiology 68: 253–278.

Lesser, M.P., J.J. Cullen and P.J. Neale. 1994. Carbon uptake in a marine diatom during acute exposure to ultraviolet B radiation: relative importance of damage and repair. Journal of Phycology 30: 183–192.

Littler, M.M. and D.S. Littler. 1980. The evolution of thallus form and survival strategies in benthic marine macroalgae: field and laboratory tests of a functional form model. The American Naturalist 116: 25–44.

Løvlie, A. and T. Bråten. 1968. On the division of cytoplasm and chloroplast in the multicellular green alga *Ulva mutabilis* Føyn. Experimental Cell Research 51: 211–220.

Lüning, K. and M. Neushul. 1978. Light and temperature demands for growth and reproduction of laminarian gametophytes in southern and central California. Marine Biology 45: 297–309.

Mackerness, S.A.H., S.L. Surplus, P. Blake, C.F. John, V. Buchanan-Wollaston, B.R. Jordan and B. Thomas. 1999. Ultraviolet-B-induced stress and changes in gene expression in *Arabidopsis thaliana*: role of signalling pathways controlled by jasmonic acid, ethylene and reactive oxygen species. Plant, Cell and Environment 22: 1413–1423.

Major, K.M. and I.R. Davison. 1998. Influence of temperature and light in growth and photosynthetic physiology of *Fucus evanescens* (Phaeophyta) embryos. European Journal of Phycology 33: 129–138.

Makarov, V.N. 1987. The behaviour of zoospores and the early ontogenic stages in the kelp *Laminaria saccharina* (L.) Lamour. in the White and the Barents Seas. Ph. D. Thesis. Botanical Institute of Russian Academy of Sciences, Leningrad.

Melkonian, M., P.L. Beech, C. Katsaros and D. Schulze. 1992. Centrin-Mediated Cell Motility in Algae. Pp. 179–221. *In:* Melkonian, M. (eds.). Algal Cell Motility. Current Phycology. Springer, Boston, MA.

Michler, T., J. Aguilera, D. Hanelt, K. Bischof and C. Wiencke. 2002. Long-term effects of ultraviolet radiation on growth and photosynthetic performance of polar and cold-temperate macroalgae. Marine Biology 140: 1117–1127.

Mitchell, D.L. and D. Karentz. 1993. The induction and repair of DNA photodamage in the environment. *In:* Young, A.R., J. Moan, L.O. Björn and W. Nultsch (eds.). Environmental UV Photobiology. Springer, Boston, MA

Morand, P. and M. Merceron. 2005. Macroalgal population and sustainability. Coastal Research 21: 1009–1020.

Moss, B. and A. Marsland. 1976. Regeneration of *Enteromorpha*. British Phycological Journal 11: 309–313.

Neale, P.J., A.T. Banaszak and C.R. Jarriel. 1998. Ultraviolet sun-screens in *Gymnodinium sanguineum* (Dinophyceae): mycosporine-like amino acids protect against inhibition of photosynthesis. Journal of Phycology 34: 928–938.

Nogués, S. and N.R. Baker. 1995. Evaluation of the role of damage to photosystem II in the inhibition of CO_2 assimilation in pea leaves on exposure to UV-B radiation. Plant, Cell and Environment 18: 781–787.

Nørdby, Ø. 1977. Optimal conditions for meiotic spore formation in *Ulva mutabilis* Føyn. Botanica Marina 20: 19–28.

O'Brien, P.A. and J.A. Houghton. 1982. Photoreactivation and excision repair of UV induced pyrimidine dimers in the unicellular cyanobacterium *Gloeocapsa alpicola* (*Synechocystis* PCC6308) Photochemistry and Photobiology 35: 359–364.

Osmond, C.B. 1994. What is photoinhibltion? Some insights from comparisons of shade and sun plants. pp. 1–24. *In:* Baker, N.R. and J.R. Bowyer (eds.). Photoinhibition of Photosynthesis, from the Molecular Mechanisms to the Field. BIOS Scientific Publ, Oxford.

Pakker, H., C.A.C. Beekman and A.M. Breeman. 2000a. Efficient photoreactivation of UVBR-induced DNA damage in the sub-littoral macroalga *Rhodymenia pseudopalmata* (Rhodophyta). European Journal of Phycology 35: 109–114.

Pakker, H., R.S.T. Martins, P. Boelen, A.G.J. Buma, O. Nikaido and A.M. Breeman. 2000b. Effects of temperature on the photoreactivation of ultraviolet-B-induced DNA damage in *Palmaria palmata* (Rhodophyta). Journal of Phycology 36: 334–341.

Pang, S., I. Gómez and K. Lüning. 2001. The red macroalga *Delesseria sanguinea* as a UVB-sensitive model organism: selective growth reduction by UVB in outdoor experiments and rapid recording of growth rate during and after UV pulses. European Journal of Phycology 36: 207–216.

Park, B.J. and T. Han. 1998. Effect of UV-B radiation on gamete motility of *Ulva pertusa*. The Phycologist 49: 40.

Park, J.J., T. Han and E.M. Choi. 2016. Differences in the oxidative stress and antioxidant responses of three marine macroalgal species upon UV exposure. Toxicology and Environmental Health Science 8: 101–107.

Peinado, N.K., R.T.A. Díaz, F.L. Figueroa and E.W. Helbling. 2004. Ammonium and UV radiation stimulate the accumulation of mycosporine-like amino acids in *Porphyra columbina* (Rhodophyta) from Patagonia, Argentina. Journal of Phycology 40: 248–259.

Pérez-Rodríguez, E., J. Aguilera and F.L. Figueroa. 2003. Tissular localization of coumarins in the green alga *Dasycladus vermicularis* (Scopoli) Krasser: a photoprotective role? Journal of Experimental Botany 54: 1093–1100.

Pérez-Rodríguez, E., J. Aguilera, I. Gómez and F. Figueroa. 2001. Excretion of coumarins by the Mediterranean green alga *Dasycladus vermicularis* in response to environmental stress. Marine Biology 139: 633–639.

Portwich, A. and F. Garcia-Pichel. 2000. A novel prokaryotic UVB photoreceptor in the cyanobacterium *Chlorogloeopsis* PCC 6912. Photochemistry and Photobiology 71: 493–498.

Provasoli, L. and I.J. Pintner. 1980. Bacteria induced polymorphism in an axenic laboratory strain of *Ulva lactuca* (Chlorophyceae). Journal of Phycology 16: 196–201.

Raven, J.A. 1991. Responses of aquatic photosynthetic organisms to increased solar UV-B. Journal of Photochemistry and Photobiology B: Biology 9: 239–244.

Saile-Mark, M. and M. Tevini. 1997. Effects of solar UV-B radiation on growth, flowering and yield of central and southern European bush cultivars (*Phaseolus vulgaris* L.). Plant Ecology 128: 114–125.

Saito, N. and H. Werbin. 1970. Purification of a blue-green algal deoxyribonucleic acid photoreactivating enzyme. An enzyme requiring light as a physical cofactor to perform its catalytic function. Biochemistry 9: 2610–2620.

Sgarbossa, A., S. Lucia, F. Lenci. D. Gioffré, F. Ghetti and G. Checcucci. 1995. Effects of UV-B irradiation on motility and photoresponsiveness of the coloured ciliate *Blepharisma japonicum*. Journal of Photochemistry and Photobiology B: Biology 27: 243–249.

Shick, J.M. and W.C. Dunlap. 2002. Mycosporine-like amino acids and related gadusols: biosynthesis, accumulation, and UV-protective functions in aquatic organisms. Annual Review of Physiology 64: 223–262.

Sinha R.,M. Dautz and D.-P. Häder. 2001. A simple and efficient method for the quantitative analysis of thymine dimers in cyanobacteria, phytoplankton and macroalgae. Acta Protozoologica 40: 187–195.

Sinha, R.P., M. Klisch, A. Gröniger and D.-P. Häder. 1998. Ultraviolet-absorbing/screening substances in cyanobacteria, phytoplankton and macroalgae. Journal of Photochemistry and Photobiology B: Biology 47: 83–94.

Smith, G.M. 1947. On the reproduction of some pacific coast species of *Ulva*. American Journal of Botany 34: 80–87.

Stratmann, J., G. Paputsoglu and W. Oertel. 1996. Differentiation of *Ulva mutabilis* (Chlorophyta) gametangia and gamete release are controlled by extracellular inhibitor. Journal of Phycology 32: 1009–1021.

Strid, A., W.S. Chow and J.M. Anderson. 1990. Effects of supplementary ultraviolet-B radiation on photosynthesis in *Pisum sativum*. Biochimica et Biophysica Acta (BBA) - Bioenergetics 1020: 260–268.

Suh, H.J., H.W. Lee and J. Jung. 2003. Mycosporine glycine protects biological systems against photodynamic damage by quenching singlet oxygen with a high efficiency. Photochemistry and Photobiology 78: 109–113.

Teramura, A.H. 1983. Effects of ultraviolet-B radiation on the growth and yield of crop plants. Physiologia Plantarum 58: 415–427.

Valiela, J., J. McClelland and P.J. Hauxwell, P.J. Behr, D. Hersh and K. Foreman. 1997. Macroalgal blooms in shallow estuaries: controls and ecophysiological and ecosystem consequences. Limnology and Oceanography 42: 1105–1118.

van Baalen, C. and R. O'Donnell. 1972. Action spectra for ultraviolet killing and photoreactivation in the blue-green alga *Agmenellum quadruplicatum*. Photochemistry and Photobiology 15: 269–274.

van De Poll, W.H., A. Effert, A.G.J. Buma and A.M. Breeman. 2001. Effects of UV-B-induced DNA damage and photoinhibition on growth of temperate marine red macrophytes: habitat-related differences in UV-B tolerance. Journal of Phycology 37: 30–38.

Van't Hoff, J. 1974. Regulation of cell division in higher plants. Brookhaven Symposium in Biology 96: 143–150.

Vass, I. 1997. Adverse effects of UVB light on the structure and function of the photosynthetic apparatus. pp. 931–949. *In*: Pessarakali, M. (eds.). Handbook of Photosynthesis, Marcel Dekker, New York.

Véliz, K., M. Edding, F. Tala and I. Gómez. 2006. Effects of ultraviolet radiation on different life cycle stages of the South Pacific kelps, *Lessonia nigrescens* and *Lessonia trabeculata* (Laminariales, Phaeophyceae). Marine Biology 149: 1015–1024.

Vergara, J.J., J.L. Pérez-Lloréns, G. Peralta, I. Hernández and F.X. Niell. 1997. Seasonal variation of photosynthetic performance and light attenuation in *Ulva* canopies from palmones river estuary. Journal of Phycology 33: 773–779.

Vergara, J.J., M. Sebastián, J.L. Pérez-Lloréns and I. Hernández. 1998. Photoacclimation of *Ulva rigida* and *U. rotundata* (Chlorophyta) arranged in canopies. Marine Ecology 165: 283–292.

Voskoboinikov, G.M. and A.N. Kamnev. 1991. Morphofunctional changes of the chloroplasts during the seaweed ontogenesis. Nauka Publishing House, Moscow.

Wiencke, C., I. Gómez, H. Pakker, A. Flores-Moya, M. Altamirano, D. Hanelt, K. Bischof and F.L. Figueroa. 2000. Impact of UV-radiation on viability, photosynthetic characteristics and DNA of brown algal zoospores: implications for depth zonation. Marine Ecology 197: 217–229.

Xiong, F., J. Kopecky and L. Nedbal. 1999. The occurrence of UV-B absorbing mycosporine-like amino acids in freshwater and terrestrial microalgae (Chlorophyta). Aquatic Botany 63:37–49.

Young, A.J. and H.A. Frank. 1996. Energy transfer reactions involving carotenoids: Quenching of chlorophyll fluorescence. Journal of Photochemistry and Photobiology B: Biology 36: 3–15.

Yu, S.G. and L.O. Björn. 1997. Effects of UVB radiation on light-dependent and light-independent protein phosphorylation in thylakoid proteins. Journal of Photochemistry and Photobiology B: Biology 37: 212–218.

Zamansky, G.B., B.A. Perrino and I.N. Chou. 1991. Disruption of cytoplasmic microtubules by ultraviolet radiation. Experimental Cell Research 195: 269–273.

Zenk, M.H., H. El-Shagi, H. Arens, J. Stöckigt, E.W. Weiler and B. Deus. 1977. Formation of the indole alkaloids serpentine and ajmalicine in cell suspension cultures of *Catharanthus roseus*. pp. 27–43. *In*: Barz, W., E. Reinhard and M.H. Zenk (eds.). Plant Tissue Culture and its Bio-Technological Application. Proceedings in Life Sciences. Springer, Berlin, Heidelberg.

Zepp, R.G., D.J. Erickson III, N.D. Paul and B. Sulzberger. 2007. Interactive effects of solar UV radiation and climate change on biogeochemical cycling. Photochemical and Photobiological Sciences 10: 261.

CHAPTER 12

Mid-Latitude Macroalgae

Donat-P. Häder

Introduction

Aquatic ecosystems contribute about half of the global primary biomass production and incorporate about the same amount of atmospheric carbon per year as all terrestrial ecosystems combined (Falkowski 2013). More than 99% of the open water surface is represented by marine ecosystems which consequently have the largest share in productivity. While phytoplankton is responsible for most of the productivity, macroalgae are mainly restricted to the coastal ocean and the continental shelves (Falkowski 2012). However, they play major roles as the basis of the intricate food webs and provide shelter for larval and adult stages of many animal species. In addition, they are of large economic importance because several hundred thousand tons of seaweeds are harvested for the production of food, biofuel, biofertilizers and technological utilization (Chen et al. 2015, Yun et al. 2015, Montingelli et al. 2016).

While phytoplankton undergo active or passive vertical migration in the water column to adapt to the environmental growth and stress factors, most macroalgae and seagrasses are benthic and attached to the substratum at their growth site (Häder and Figueroa 1997, Figueroa et al. 2002, Canal-Vergés et al. 2014). Because of this lack of mobility they have to cope with the changing photo-climate caused by tidal movement and changes in water transparency which are not synchronized with the diel solar rhythm. Thus, macroalgae in the upper littoral can be exposed to excessive visible and UV solar radiation during daytime or covered by turbid water which limits photosynthesis and biomass production (Migné et al. 2015, Kendrick et al. 2016). The coastal habitats are heavily exposed to global climate change stress factors so that the algae have to cope with thermal and desiccation stress (Ji et al. 2016). In addition, coastal ecosystems are affected by eutrophication and the impact of pesticides and other pollutants (Lyons et al. 2014, Gubelit et al. 2016). Ocean warming and ocean acidification are additional stress factors of climate change (Gao et al. 2012b, Koch et al. 2013).

Friedrich-Alexander University, Erlangen-Nürnberg, Neue Str. 9, 91096 Möhrendorf, Germany.
Email: donat@dphaeder.de

Distribution of Macroalgae in the Littoral

The littoral zone extends from the high water mark, which is rarely covered by water, to the permanently submerged regions of the coastal zone to the edge of the continental shelf. In marine biology it is often subdivided into the supralittoral zone (also called spray or supratidal zone) which extends from the high tide mark and receives water only from spray. The eulittoral (also called intertidal) zone extends from the spring high tide line to the spring low tide line. This is followed by the sublittoral zone which runs dry only under extreme conditions, such as strong offshore winds during spring low tide (Mitra and Zaman 2016). Other definitions use the terms upper, mid and lower littoral (Miller and Perry 2016). Since most freshwater ecosystems rarely show any tidal action the littoral zone is not subdivided but often neighbored by more or less extended wetland areas (Keddy 2010).

The supralittoral zone is often occupied by lichens and some animals, such as periwinkles, Isopoda and Neritidae which can cope with temperature extremes, high solar visible and UV radiation as well as changing salinity. Only a few macroalgae are adapted to these habitats, such as the terrestrial green *Prasiola calophylla* (Hartmann et al. 2016). The upper eulittoral of rocky shores is often inhabited by Chlorophytes, such as *Ulva* and *Enteromorpha* (Rico et al. 2014). These species have to tolerate occasional desiccation, variable temperatures as well as exposure to excessive solar radiation (Xu et al. 2016). The bulk of the marine macroalgae is found in the eulittoral which rarely falls dry and thus provides well-buffered temperatures. But due to the variable thickness of the water column above them they are exposed to changing irradiances due to the tidal action (Puente et al. 2017). Species in this habitats include the large kelps found in the lower littoral (Sakanishi et al. 2017). Many Rhodophytes, such as *Rhodymenia*, *Gracillaria* or *Kallymenia* are restricted to the sublittoral where they are protected from desiccation, temperature extremes and high solar irradiances (Gersun et al. 2016). Some very light-sensitive Rhodophyte species are even restricted to crevices and caves with very low irradiances, such as *Peyssonnelia* (Monteiro 2013). Some coralline red algae are adapted to extremely low irradiances at depths of up to 200 m which are characterized by the general term maerl. Due to the very low irradiances in their subtidal habitats they have growth rates of approximately 1 mm per year (Wilson et al. 2004).

Measurement of Photosynthetic Parameters in Macroalgae

Oxygen Production

Photosynthetic yield can be determined using several physiological parameters. Most of them are restricted to the laboratory, but some have been adapted for field work. The increase in biomass due to photosynthetic productivity can be used, but this is very time consuming and not practical in the field. One option is to determine photosynthetic carbon incorporation using C^{14}. For this method algal thalli are enclosed in a container with some radioactive carbon (such as $C^{14}O_2$) added. After exposure to light the excessive radioactive carbon is removed and the remaining radioactivity of the sample is determined in a counter (Bidwell 1958). Use of this method in the field is problematic because of potential radioactive contamination.

Measurement of oxygen production is more commonly used both in the laboratory and in the field. For this purpose Clark electrodes are utilized which indicate the oxygen concentration in the form of an easily measured electrical signal (Häder and Schäfer 1994a, Häder and Schäfer 1994b, Häder et al. 1997a). A submersible instrument with a chamber which holds algal thalli under a quartz window has been developed (Fig. 12.1). The seawater in the chamber is agitated by a magnetic stirrer. The electric signal from the Clark electrode inserted into the chamber is amplified in an electronic circuit and then sent to a computer on board a research vessel or on the beach where it is stored in memory. The oxygen development or consumption is calculated from the change in the oxygen concentration over time. However, it is important to make sure that the oxygen concentration does not exceed saturation.

Fig. 12.1. Instrument to determine oxygen production of algal thalli *in situ*.

As expected, the photosynthetic oxygen production of macroalgae depends on the available solar irradiation. Thalli of the Chlorophyte *Ulva lactuca* were exposed at different depths of the water column in a sealed box which was suspended in open water off the island of Helgoland, German Bight (Fig. 12.2a). In darkness respiration consumed oxygen at a steady rate (Häder and Schäfer 1994b). Under solar radiation at the water surface oxygen was produced photosynthetically and the rate decreased as the box with the thallus was lowered into the water column. Although the illuminance decreased significantly at a depth of even 1 m, oxygen production exceeded respiration and only at 5 m water depth and below there was a net oxygen consumption.

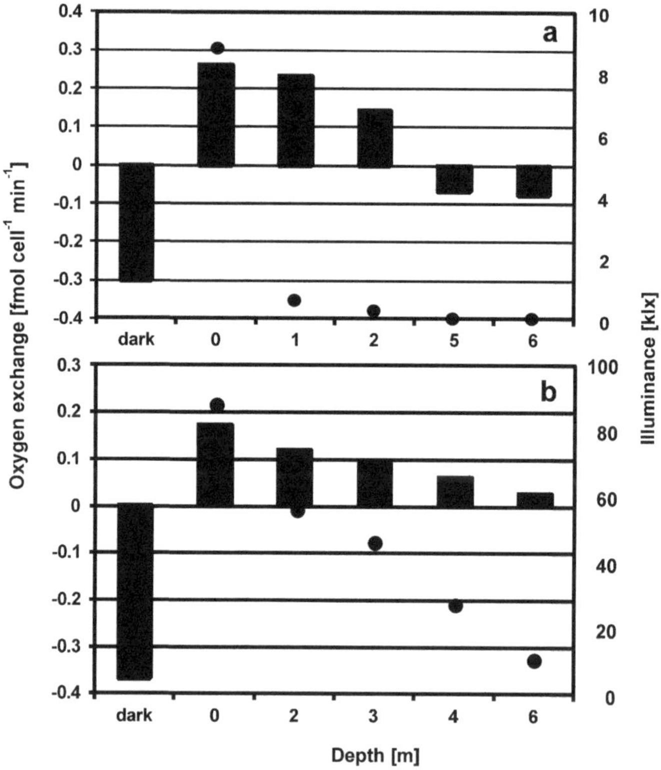

Fig. 12.2. Photosynthetic oxygen production (solid bars) of *Ulva lactuca* (a) in comparison to the illuminance of solar radiation (circles) measured on a cloudy day in the open sea northwest of the island of Helgoland, German Bight and of *Laminaria saccharina* (b) measured in the Northwest harbor of Helgoland on a clear day. Redrawn after (Häder and Schäfer 1994b).

Similar values were found for the Chlorophyte *Enteromorpha intestinalis* measured in the open sea east of the island of Hiddensee, Baltic Sea (Häder and Schäfer 1994b). In contrast, the Phaeophyte *Laminaria saccharina* showed positive oxygen production down to a depth of 6 m; but the illuminance was much higher during a clear day (Fig. 12.2b).

In the experiments described above, the thalli had been exposed at each depth for a short period (about 2 min) starting from the lowest level. Therefore, they had no time to adapt to the increasing illuminance by photoinhibition. A different situation occurs when macroalgae are exposed to solar radiation for an extended period of time (Fig. 12.3). Thalli of the Phaeophyte *Pilayella litoralis* were exposed to solar radiation at 60.9 klx on a clear day. In the dark control there was oxygen consumption due to respiration. After the onset of light oxygen production prevailed for up to 10 min and then dropped rapidly and oxygen exchange became negative after about 30 min of exposure. This decline of photosynthetic oxygen production is due to photoinhibition of photosystem II while respiration is less affected.

Under high solar irradiation in Mediterranean coasts photoinhibition can be even more pronounced. The Rhodophyte *Corallina elongata* was harvested from under an

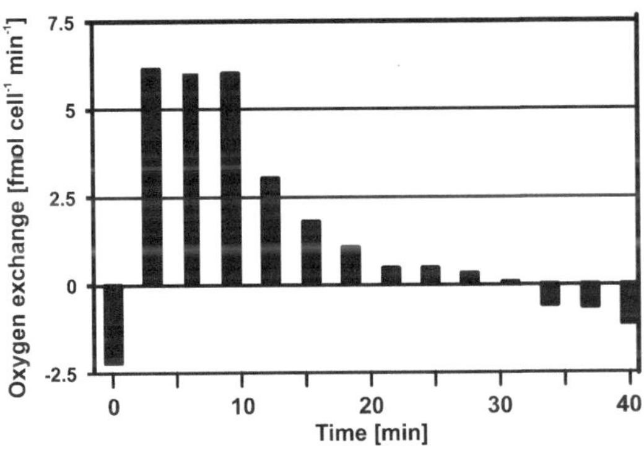

Fig. 12.3. Photosynthetic oxygen exchange of the brown alga *Pilayella litoralis* exposed to solar radiation on the surface of the water column in the Harbor of Kloster (Hiddensee, Baltic Sea).

overhanging rock on the island of Sardinia and exposed to unfiltered solar radiation of about 107 klx, which is about 10 times higher than at the growth site (Fig. 12.4). Under these conditions the oxygen production dropped to about one-third and after 10 min exposure the net oxygen exchange was negative indicating that photosynthetic oxygen production had ceased while respiration continued. This and other examples show that many algae undergo photoinhibition when exposed to unfiltered solar radiation at the surface. This often occurs when low tide coincides with high solar radiation on a clear day around noon time. Thus main productivity of upper littoral algae is mainly restricted to the morning hours and in the late afternoon after recovery.

The phenomenon of photoinhibition is due to a decrease in photosynthetic activity induced by excessive visible and/or UV radiation which damages the D1 protein in the reaction center of photosystem II (PS II) resulting in a reduced electron transport (Keren et al. 1997). Excessive irradiance cannot be utilized effectively by chlorophyll

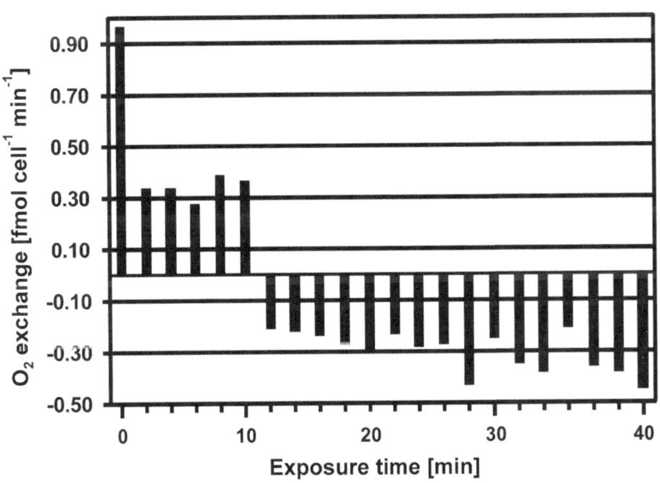

Fig. 12.4. Photosynthetic oxygen exchange of *Corallina elongata* during exposure to unfiltered solar radiation of 107 klx measured on 12 October 1993 at the water surface in Baia del Conta, Sardinia, Italy.

so that it undergoes a transition to the excited triplet state (Guidi et al. 2017). Usually this activation energy can be quenched by the action of β-carotene and the xanthophyll cycle which converts the excessive energy in the form of heat (Christa et al. 2017, Khoroshyy et al. 2017); but when their capacity is exceeded, the energy of this electronic state can be transferred to ground-state oxygen (triplet) which is thereby excited to the singlet state ($^1O_2$) (Lüttge et al. 2005, Telfer 2014). In this form it can damage proteins and other important biomolecules in its vicinity (Kale et al. 2017). Photoinhibition is a reversible process and can be relieved by *de novo* synthesis of the D1 protein after the light stress subsides.

In addition to damaging effects on the PS II reaction center there are a number of other targets for deleterious solar UV radiation. In addition to the photosynthetic chlorophylls, all algal groups use a number of accessory pigments which form the antenna complex. These pigments increase the absorption, harvest solar radiation and funnel the energy to the reaction centers (De Martino et al. 2000). Rhodophyta use an array of phycobilins—a feature they share with cyanobacteria and Cryptophyta—to harvest solar energy in the green wavelength band which is not absorbed by the chlorophylls (Gantt et al. 2003). These linear tetrapyrrole pigments are characterised by a high sensitivity and are easily damaged by UV radiation (Sinha et al. 2001a).

The filamentous Rhodophyte *Ceramium* contains phycoerythrin (absorption at 495, 540 and 565 nm) and phycocyanin (absorption at 615 nm) as accessory pigments. Figure 12.5 shows the concentration of phycoerythrin over a period of 36 h under unfiltered solar radiation and under PAR only (Sinha et al. 2003). The concentration is low during the late afternoon and evening but starts to rise around midnight and reaches a maximum at 5 am. Subsequently, it decreases again towards the evening. The same behavior was found for phycocyanin and for other Rhodophyta (*Callithamnion* and *Corallina*). This indicates that the accessory pigments are increasingly destroyed during daytime and resynthesized over night. However, it is interesting that not only unfiltered solar radiation affects the pigments but even excessive PAR destroys the phycobilins.

Pulse Amplitude Modulation Fluorometry

In addition to oxygen evolution and carbon assimilation, pulse amplitude modulation (PAM) fluorometry is an invaluable tool to determine photosynthetic parameters under controlled laboratory conditions as well as under field conditions with changing stress parameters. Recent developments over the last few decades have resulted in a multitude of portable instruments which can be used for measurements of photosynthetic processes in algae above and even under water (Häder et al. 2001c, Howe et al. 2017) (Fig. 12.6). These instruments have the advantage of measuring the parameters of interest easily, non-invasively and rapidly. However, it has to be kept in mind that these instruments exclusively use fluorescence from photosystem II and do not give information on net photosynthesis (including respiration) nor on the downstream processes, such as the biochemical pathways of carbon assimilation. Therefore, it is an error to use PAM parameters as a proxy for growth and biomass production (Logan et al. 2014).

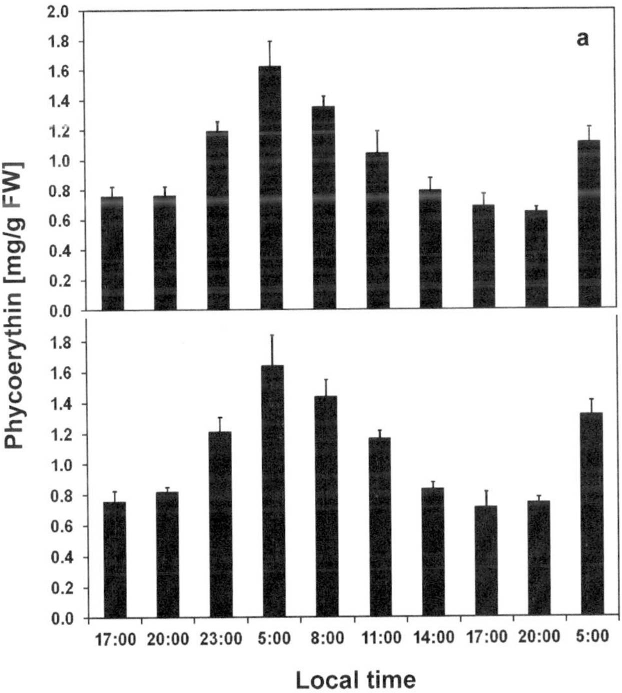

Fig. 12.5. Phycoerythrin concentration in *Ceramium* under unfiltered solar radiation (a) and under photosynthetic active radiation (b). Redrawn after Sinha et al. 2003.

Fig. 12.6. Diving-PAM-II instrument. Courtesy Walz, Effeltrich, Germany.

When a photon is absorbed by the photosynthetic apparatus its energy can be lost as heat, result in fluorescence or it can be used in a photochemical reaction (Häder and Tevini 1987); only the latter pathway results in photosynthetic carbon assimilation. Thus, ignoring energy quenching by heat production, fluorescence competes with photochemical activity for the energy from an excited photosynthetic pigment. Based on this idea Schreiber et al. (1986) developed the PAM method which allows to calculate the quantum yield of PSII from the emitted fluorescence.

When being dark-adapted, the primary electron acceptor (Q_A) of PS II is in the open state and is ready to accept electrons from the reaction center. Under strong irradiation the reaction center chlorophyll (P680) is ionized and the acceptor reduced (closed state) (Schreiber and Bilger 1993). In order to maintain the reaction center basically open, low light signals are used for the measuring light; under these conditions only little energy is lost as fluorescence which is detected by the PAM instrument in the form of Fo (Hanelt 2017) (Fig. 12.7). The measuring light is modulated and the detector is synchronized to this frequency and phase (lock-in amplifier) so that the much higher actinic light can be ignored (Bicanic et al. 1989). When a plant or alga receives a moderate level of actinic or solar radiation, photosystem II undergoes a transition to the light-acclimated state and the fluorescence increases to Ft (temporary fluorescence measured in the light-acclimated state). Subsequently a very short oversaturating light pulse is given from a halogen lamp inside the PAM instrument via a glass fiber optic cable which oxidizes (closes) all reaction centers, so that during the light flash no more energy can be funneled into the photochemical reaction and a maximal fluorescence is monitored (Fm). Under these conditions the amount of excitation which is dissipated as heat can be determined from the difference between the absorbed light energy and the energy dissipated in the form of fluorescence. The difference between Fm and Fo is called variable fluorescence (Fv). The ratio Fv/Fm indicates the optimal quantum yield (Y) of PS II in the dark-acclimated state and indicates which percentage of the absorbed light energy is funneled into photosynthesis. It is independent of the amount of absorbed energy and

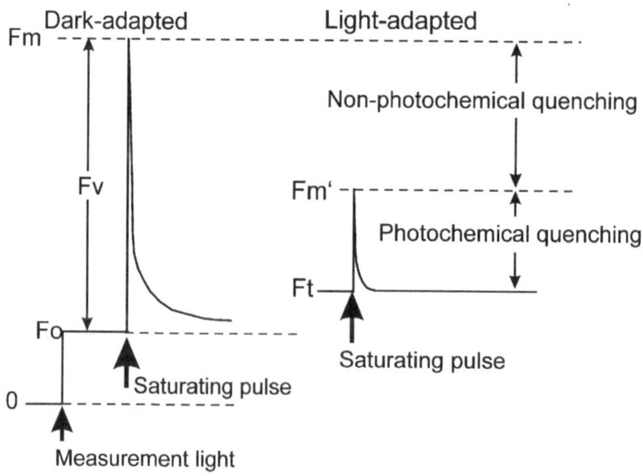

Fig. 12.7. Fluorescence levels of plant materials in the dark- and light-acclimated state in response to saturating light pulses.

the concentration of photosynthetic pigments (Schreiber et al. 1994). The maximal fluorescence in the light-acclimated state is called Fm' and the ratio Fm'-Ft/Fm' is called effective quantum yield which has a linear relationship to the CO_2 fixation (Genty et al. 1989).

From these fluorescence signals the fraction of absorbed light which is funnelled into photosynthesis can be determined as photochemical quenching (qP) as well as the fraction which is dissipated as heat (non-photochemical quenching, qN) according to the equations

qN = (Fm - Fm')/(Fm - Fo') and

qP = (Fm' - Ft)/(Fm' - Fo')

qN ranges from 0 to 1 and is often inversely related to qP (Büchel and Wilhelm 1993). An increase of qN can indicate an increase in the zeaxanthin content in PS II (Adams III and Demmig-Adams 1992), an aggregation of the light-harvesting complexes (Ruban et al. 1993) or an increase in the fraction of inactive PS II centers (Guenther and Melis 1990, Öquist and Chow 1992, Critchley and Russell 1994) all of which protect the active reaction centers from damage by excessive irradiation.

Application of PAM Fluorometry to Measure Photosynthesis in Macroalgae on Site

PAM fluorometers have been used above and in the water to determine the effects of environmental stress parameters on photosynthesis at mid-latitudes in, for example, Greece (Häder et al. 1996b, Häder et al. 1996d, Häder et al. 1997a, Häder et al. 1997b), Sardinia (Herrmann et al. 1995), Gran Canaria (Häder et al. 2000b, Häder et al. 2001d, Häder et al. 2001c), Southern Spain (Häder et al. 1996c, Häder and Figueroa 1997, Häder et al. 1998, Jiménez et al. 1998, Häder et al. 1999a, Häder et al. 1999b) and Patagonia, Argentina (Häder et al. 2000a, Häder et al. 2001a, Häder et al. 2001b, Häder et al. 2003a, Häder et al. 2003b, Häder et al. 2004, Helbling et al. 2010).

The values of the various fluorescence parameters change with the actinic irradiance. Thalli of the Rhodophyte *Corallina mediterranea* were adapted to an irradiance of 23 W m^{-2} for 10 min. Subsequently they were exposed to exponentially increasing white light irradiances for 6.5 min each after which the fluorescence parameters were determined (Fig. 12.8). Up to about 10 W m^{-2} most values were fairly stable. With increasing irradiances qP and Fm' as well as the quantum yield steadily dropped, while qN sharply rose at irradiances above 30 W m^{-2}. Ft and Fo' stayed basically at the same level.

Thalli of the Chlorophyte *Ulva laetevirens* have been exposed to unfiltered solar radiation at the water surface during a sunny summer day on the island of Sardinia, Italy (Herrmann et al. 1995). During exposure Fo first rose slightly and then gradually decreased. In contrast, Fm steadily fell from an initial value of 0.53 to about 0.1 within 70 min (Fig. 12.9). Consequently, the quantum yield dropped fast from the initial value close to 0.7 to almost 0. These dramatic kinetic changes are partially due to the excessive visible radiation. In addition, the strong UV-A and UV-B solar radiation

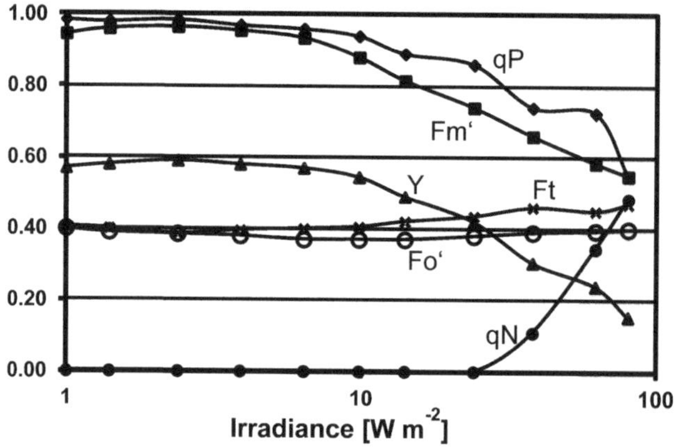

Fig. 12.8. Fluorescence parameters measured in *Corallina mediterranea* at increasing irradiances.

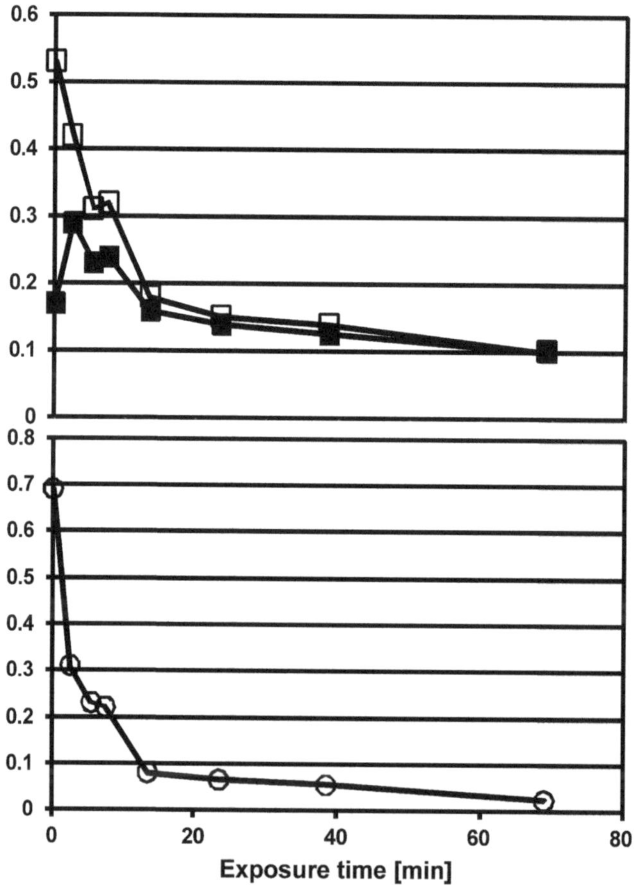

Fig. 12.9. Chlorophyll fluorescence parameters Fo (open squares), Fm (closed squares) and Fv/Fm (open circles) of *Ulva laetevirens* before and during exposure to unfiltered solar radiation. Redrawn after (Herrmann et al. 1995).

augmented the sharp declines as shown by parallel experiments under a WG 320 nm filter, which cuts off UV-B and under a GG 400 filter, which removes all solar UV radiation.

The calcium incrusted Chlorophyte *Halimeda tuna* is a common macroalga in the eulittoral on rocky shores at mid-latitudes, such as the Mediterranean (Vroom et al. 2003). Thalli were harvested from a rock pool in Saronikos Gulf, Corinth, Greece and dark-adapted for 30 min and then subsequently exposed to solar radiation for 2.5 h under different filters which transmit all solar radiation or radiation deprived of UV-B and/or UV-A (Häder et al. 1996d) (Fig. 12.10). The yield dropped considerably in all treatments, but significantly more under the unfiltered solar radiation. During recovery the yield increased but even after 1.5 h in the shade the initial values were not reached, especially after radiation including UV-A or total UV.

Depending on their adaptation to different habitats marine algae can show different sensitivity to solar visible and UV radiation. The Phaeophyta *Padina pavonica* (previously *P. pavonia*) grows on rocky shores at various depths from surface habitats in rock pools down to several meters in the Atlantic Ocean and the Mediterranean Sea as well as in the Pacific and Indian Ocean (Chappuis et al. 2014). *Padina* thalli were harvested from a rock pool near the surface and from 7 m depth and exposed to unfiltered solar radiation (WG 295 nm) and to visible radiation deprived of UV (Fig. 12.11) (Häder et al. 1996a). The deep water alga showed a significantly higher photoinhibition of the quantum yield than the thallus retrieved from the water surface and did not reach the initial value during a 2-h recovery in darkness. Unfiltered solar

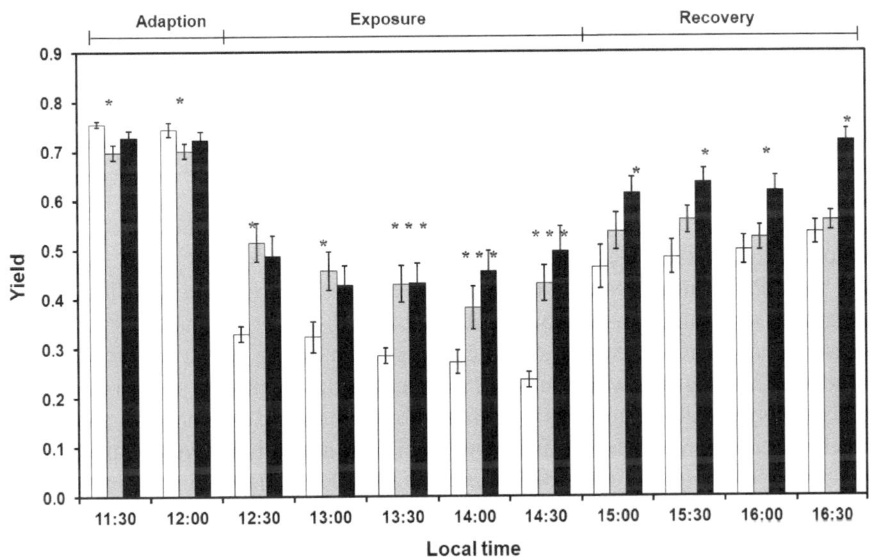

Fig. 12.10. Photosynthetic quantum yield of *Halimeda tuna* during dark adaptation, exposure to solar radiation and recovery in the shade. The thalli were exposed under different cut-off filters: WG 295 (open bars), WG 335 (gray bars) and GG 400 (black bar). Each data point is the average of at least eight measurements and asterisks indicate statistically significant deviations from the WG 295 nm values. Redrawn after Häder et al. 1996d.

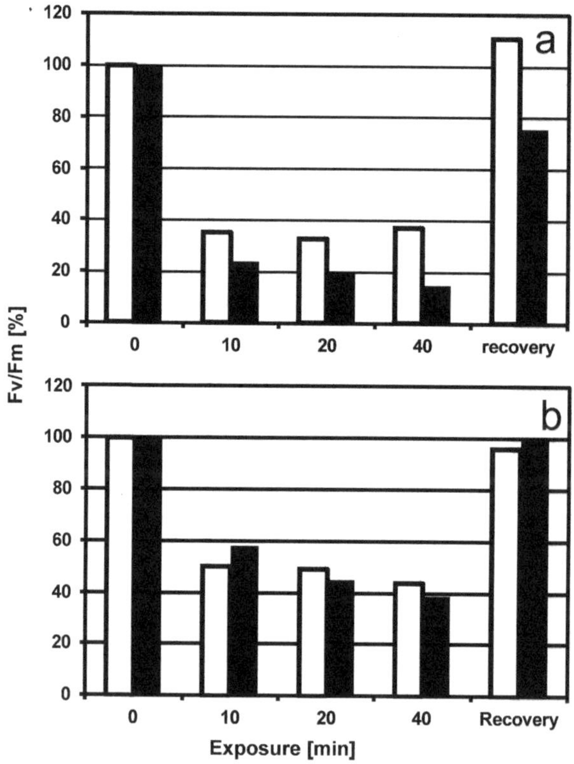

Fig. 12.11. Effective quantum yield of *Padina pavonica* exposed to solar radiation under a 295 nm cut-off filter: (a) and a GG 400 cutoff filter nm (b) in percentage of the initial values (time 0). Open bars indicate values for a sample retrieved from a rock pool and black bars values for a sample collected at 7 m depth. The absolute value of Fv/Fm at 0 min in the sample from the rock pool was 0.66 and that from 7 m was 0.72. Recovery values were measured after 2 h in darkness. Redrawn after Häder et al. 1996a.

radiation had a stronger effect than visible radiation even though the UV component has only 10% irradiance compared to the visible wavelength band.

Most PAM measurements on algae were made by exposing the thalli at the surface, on a research vessel or on the beach. In order to approach natural conditions, measurements were made using specimen retrieved by diving. However, even then the algae were manipulated which may have influenced the results. In order to determine true photosynthetic values it is preferably to do the measurements on algae still attached at their growth site. This is now possible using a diving PAM which allows non-invasive monitoring of plants *in situ* at depth. Using such an instrument a diver measured the photosynthetic quantum yield in several red, brown and green algae at 15 m depth on the coast of Gran Canaria at regular time intervals (Häder et al. 2000b, Häder et al. 2001d, Häder et al. 2001c). The Phaeophyte *Dictyota dichotoma* showed a photosynthetic quantum yield of close to 0.8 in the morning (Fig. 12.12). By noon it decreased to below 0.7 and then slowly recovered up to the evening. This result is remarkable since the irradiances of PAR reaching the algae was less than 25% of those measured at the surface and of UV-B less than 20%. Similar results were found

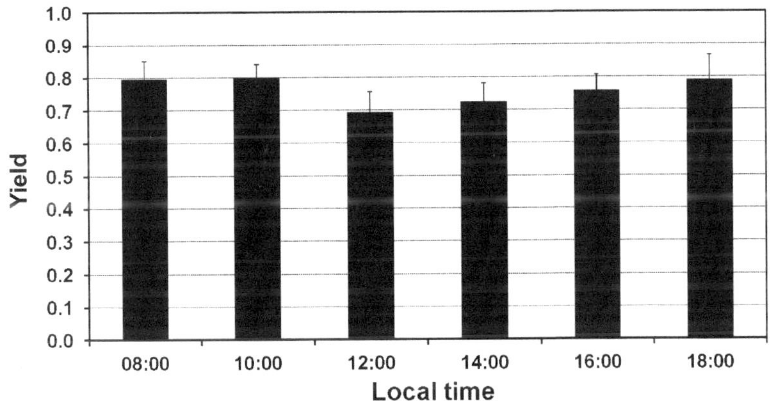

Fig. 12.12. Photosynthetic quantum yield of *Dictyota dichotoma* measured on a clear day on site at 15 m depth on the coast of Gran Canaria using a diving PAM. Redrawn after Häder et al. 2001d.

for *Sargassum, Cystoseira, Valonia, Caulerpa, Codium, Asparagopsis, Gelidium, Halopithys* and *Lobosphora* (Häder et al. 2000b, Häder et al. 2001d, Häder et al. 2001c).

Recovery after photoinhibition is thought to be due to resynthesis of UV-damaged D1 protein in PS II. In order to prove this hypothesis three Patagonian macroalgae (the Chlorophyte *Ulva*, the Rhodophyte *Porphyra* and the Phaeophyte *Dictyota*) were exposed to unfiltered solar radiation, and the decrease in photosynthetic quantum yield as well as the subsequent recovery in darkness was measured (Häder et al. 2002). In parallel experiments streptomycin and chloramphenicol were added which are known inhibitors of protein synthesis (Vázquez 1979). Both substances retarded the recovery considerably. The xanthophyll cycle is thought to ameliorate the inhibition by excessive light (Goss and Jakob 2010). Therefore, we applied dithiothreitol, a known inhibitor of the violaxanthin de-epoxidase (Winter and Königer 1989). This drug also increased photoinhibition and retarded recovery even in red algal species which at the time of the experiment were thought not to possess the xanthophyll cycle. However, recent research indicated that several Rhodophyta contain several xanthophylls and that xanthophyll cycle-related pigments were found in four of 12 analysed orders (Schubert et al. 2006).

Mitigating Strategies to Decrease Damage by Solar UV

Aquatic biomass producers require solar radiation in order to utilise its energy for carbon assimilation. Simultaneously they cannot avoid being exposed to the deleterious UV radiation. In order to decrease the level of damage they have developed a number of mitigating strategies. Free moving phytoplankton are capable of active or passive vertical migrations to escape excessive solar UV radiation (Richter et al. 2007, Halac et al. 2014). Alternatively, some organisms form layers and crusts to be protected by self shading (Pathak et al. 2015). Due to their sessile growth macroalgae cannot undergo vertical migrations, but, as indicated above, they can choose their growth depth within the littoral. Some very UV-sensitive red algae grow epiphytically on the bases of kelp algae being protected from direct radiation by the shading blades (Reed and Foster 1984). One mechanism to protect from photodamage of the photosynthetic apparatus is the movement of chloroplasts inside the cell. In strong light the phaeoplasts of the brown alga *D. dichotoma* move to positions at cell walls parallel to the light beam (Hanelt and Nultsch 1990). Another strategy is to use efficient mechanisms based on enzymatic repair. The biosynthesis of these proteins is augmented by higher temperatures so that one of the effects of climate change is an increase in these mechanisms which are used to repair the photosynthetic apparatus and damaged DNA (Pakker et al. 2000, Häder et al. 2011). In addition to direct damage of biomolecules solar UV may produce reactive oxygen species (ROS) which in turn destroy cellular components (He and Häder 2002, Häder et al. 2015). Mitigating these effects algae use quenchers of ROS to minimize the effects of solar UV radiation (Mallick and Mohn 2000). Detoxification of ROS can be accomplished by the involvement of organic quenchers, such as carotenoids or ascorbate (Vincent and Neale 2000, He and Häder 2002) or quenching enzymes, such as superoxide dismutase (Sun et al. 2014). Also these mitigating mechanisms against solar UV damage are augmented by increased temperatures due to climate change. Another mitigating effect due to climate change results from the increased DOM concentrations in coastal waters which specifically absorb in the UV region (Oliver et al. 2014)

Many phytoplankton (including cyanobacteria) and macroalgae are capable of synthesizing UV-absorbing pigments which protect them from damaging solar radiation (Sinha et al. 2006, Sinha et al. 2007). Scytonemin is restricted to some cyanobacteria (Sinha et al. 1999, Singh et al. 2010). In addition to attenuating solar UV, scytonemin has been shown to have the capability of scavenging oxygen radicals (Matsui et al. 2012). Cyanobacteria, eukaryotic phytoplankton and macroalgae use mycosporine-like amino acids (MAAs) as absorbers for UV radiation (Sinha et al. 2001b).

MAAs are water-soluble UV-absorbing compounds which have a cyclohexenone or cyclohexenimine chromophore linked with the nitrogen atom of an amino acid or its imino alcohol (Sinha and Häder 2003b). They have absorption maxima in the range from 310 to 360 nm. Figure 12.13 shows the molecular structure of some of the more than 20 MAAs isolated so far (Sinha et al. 2007). They originate from the first part of the Shikimate pathway starting with erythrose-4-phosphate which reacts with phosphoenol pyruvate to 3-dehydroquinate which is a precursor for the synthesis of fungal mycosporines and MAAs via gadusol. The first MAA is mycosporine-glycine from which other secondary MAAs are synthesized. It should be mentioned in passing that animals do not possess the Shikimate pathway and therefore cannot produce MAAs. Rather they take them up with their diet and incorporate them into their cells (Carroll and Shick 1996). The absorption spectra of some MAAs are shown in Fig. 12.14.

The concentration of UV-absorbing compounds changes with the exposure to solar radiation. *Corallina officinalis* grows on rocky shores and has two different morphological forms in the mid-intertidal and lower intertidal (Häder et al. 2003a). Samples of the low intertidal form were collected during daytime at 1-h intervals (Richter et al. 2006). Since the sites were accessible only for 2 to 5 h during low tide, samples from six different days were used and dried for further evaluation. For the night hours samples were collected at 19 h and kept in darkness over night and samples

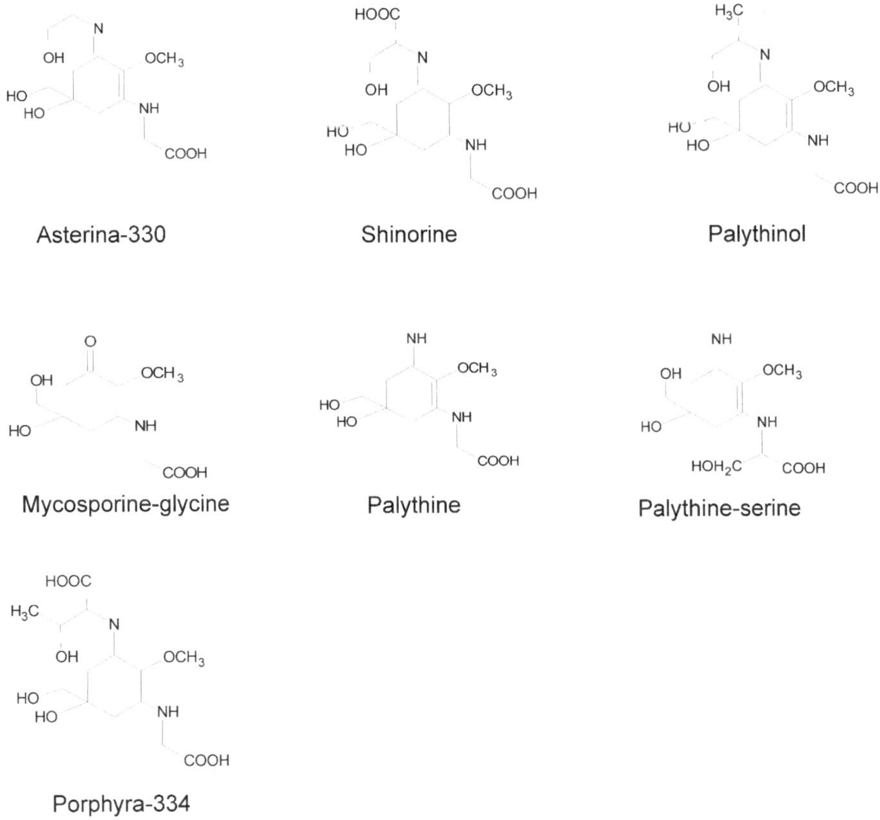

Fig. 12.13. Molecular structures of some of the more than 20 known MAAs. Redrawn after (Sinha and Häder 2003a).

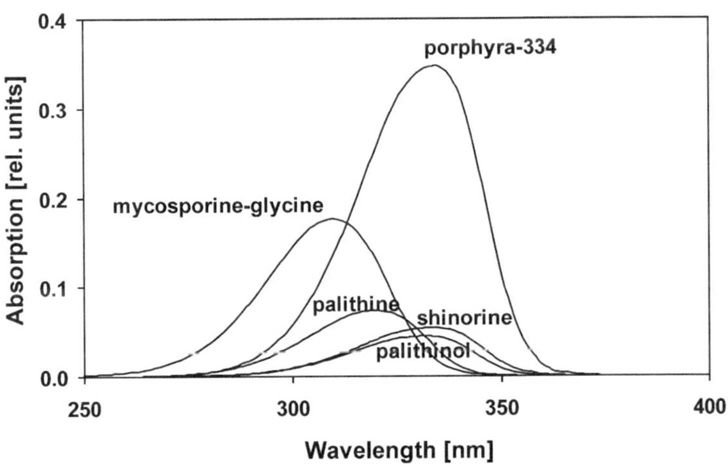

Fig. 12.14. Absorption spectra of some MAAs. Redrawn after (Sinha and Häder 2003a).

were harvested every hour. All samples were extracted and the MAAs separated by HPLC. The main MAAs in this alga are shinorine and palythine (Richter et al. 2006). Both MAAs show clear diurnal changes in their concentration with a main peak around noon (Fig. 12.15).

Four Rhodophyte species (*Ceramium*, *Corallina*, *Callithamnion* and *Porphyra*) were exposed to full solar radiation and to photosynthetic active radiation (PAR), respectively, during short-term (46 h) experiments (Helbling et al. 2004). Algae were exposed to solar radiation under two treatments (PAR only: 400–700 nm, and PAR+ UVR: 280–700 nm). Samples were taken at regular time intervals, absorption spectra measured and the concentrations of UV-absorbing compounds, carotenoids and photosynthetic pigments determined. While the photosynthetic pigments did not show large changes over the day, the UV-absorbing pigments showed considerable changes which were species specific. Figure 12.16 shows the concentration of UV-absorbing

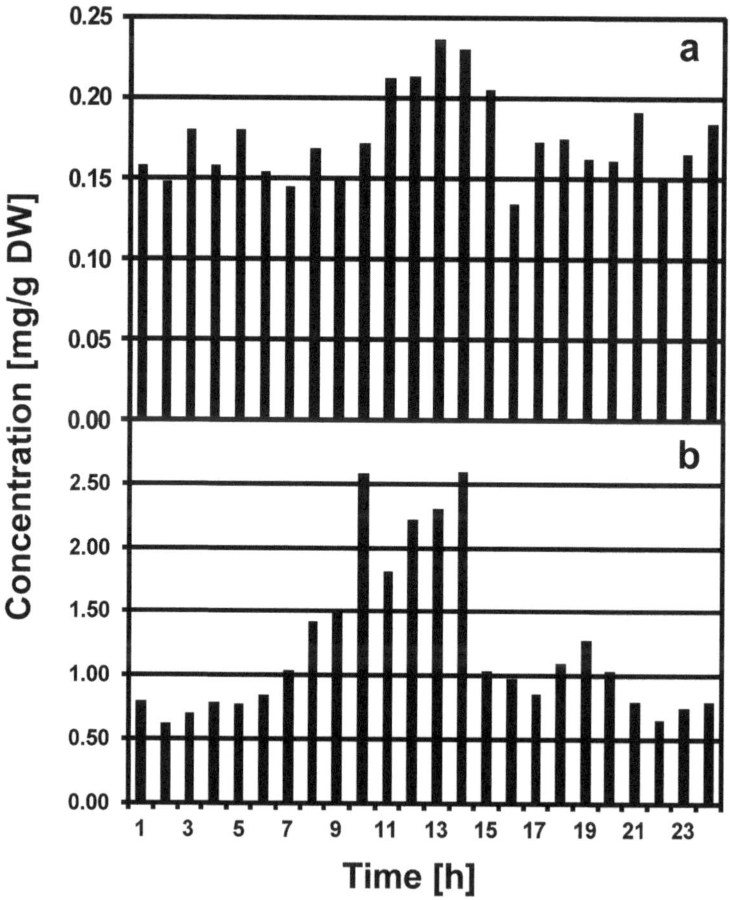

Fig. 12.15. Diurnal changes in the concentration (mg/g DW of shinorine (a) and palythine (b) in *Corallina officinalis* harvested from the low intertidal. Differences between midday samples and morning or evening samples were statistically significant for shinorine ($p < 0.05$) and palythine ($p < 0.01$). Redrawn after Richter et al. 2006.

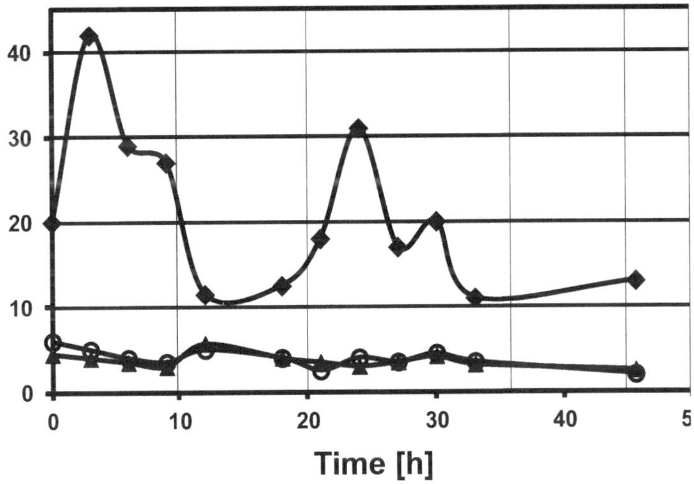

Fig. 12.16. Mean absorption (OD/fresh weight in g) at 330 nm (UV-absorbing pigments), 470 nm (carotenoids) and 665 nm (chlorophyll *a*) measured at regular time intervals during the 48-h experiment exposed to unfiltered solar radiation. Redrawn after Helbling et al. 2004.

compounds, carotenoids, and photosynthetic pigments in *Ceramium* spec. during the experiment. In the samples exposed to unfiltered solar radiation the absorption at 330 nm strongly increased during daytime and fell to a lower level at night, while in the PAR-only exposure there was no clear correlation indicating that the biosynthesis of the UV-absorbing pigments was induced by solar UV radiation (Helbling et al. 2004).

The hypothesis that biosynthesis of UV-absorbing pigments is induced by the exposure to UV radiation was supported by measuring a polychromatic action spectrum in the Chlorophyte *Prasiola stipitata* using a large number of cut-off filters (Gröniger and Häder 2002). In addition to the photosynthetic pigments, the absorption spectrum of this alga shows a pronounced peak at 324 nm which was also found in the HPLC analysis of the extracted pigments. The molecular nature of the UV-absorbing pigment is still unknown. This pigment was also found in the terrestrial *P. crispa* in Antarctica, the absorption of which far exceeded that of chlorophyll (Post and Larkum 1993). Recently, a new MAA was identified in *Prasiola calophylla* as prasiolin (*N*-[5,6 hydroxy-5(hydroxymethyl)-2-methoxy-3-oxo-1-cycohexen-1-yl] glutamic acid) (Hartmann et al. 2016).

The action spectrum shows a clear induction by UV-B radiation (Fig. 12.17) (Gröniger and Häder 2002). The photoreceptor for the UV-induced induction is not known. In higher plants the molecular nature and its biochemical reactions of the photoreceptor (UVR8) for UV-induced photomorphogenesis has been identified (Rizzini et al. 2011). This protein has also been identified in the Chlorophytes *Chlorella* and *Volvox* (Kianianmomeni 2014, Fernández et al. 2016) as well as in the liverwort *Marchantia* and the moss *Physcomitrella*. Therefore it could be speculated that the UV-induced biosynthesis of the UV-absorbing pigment in *Prasiola* could be controlled by the same photoreceptor.

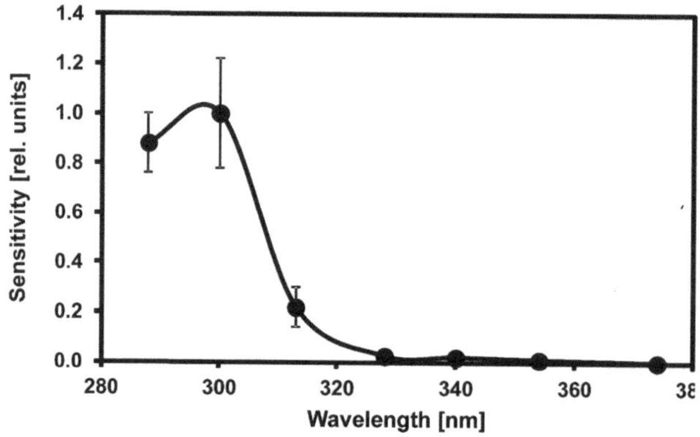

Fig. 12.17. Polychromatic action spectrum for the induction of an unidentified UV-absorbing pigment in *Prasiola stipitata*. Redrawn after (Gröniger and Häder 2002).

Effect of Ocean Acidification on Macroalgae

The atmospheric CO_2 concentration has increased from about 280 ppm before the Industrial Revolution to about 400 ppm due to fossil fuel consumption and changes in land use (Temme et al. 2014). Part of this anthropogenically released CO_2 is taken up by the oceans resulting in ocean acidification. If the emissions continue under the 'business as usual' scenario an increase in the atmospheric CO_2 to 800–1000 ppm is expected by the end of the century (Hein and Sand-Jensen 1997) corresponding to a pH reduction in the oceans by 0.3–0.4 (Caldeira and Wickett 2003, Feely et al. 2004, Gao et al. 2012b). Usually CO_2 is a limiting growth factor for plants, so that an increase in productivity could be expected due to ocean acidification. However, most species have an enzymatic CO_2-concentrating mechanism so that productivity is not limited by the current dissolved inorganic carbon (DIC) (Cornwall et al. 2015). Likewise, diatom-dominated phytoplankton communities showed no increased productivity under augmented CO_2 concentrations (Tortell et al. 2000) while others seem to benefit from increased CO_2 concentrations (Hutchins et al. 2007, Riebesell and Tortell 2011).

On the other hand ocean acidification interferes with calcification. Many phytoplankton and animals have an exoskeleton made from calcium carbonate the synthesis of which is hampered by a lower pH. Also macroalgae have incrustations of calcium carbonate in their thalli, such as the red Corallinaceae *Jania* and *Corallina* (Pope 1963), the Chlorophyte *Acetabularia* (Ricci et al. 2016) or the Phaeophyte *Padina* (Vuayaraghavan and Sokhi 1986). This widespread ocean acidification decreases productivity in calcifying macroalgae (Gao et al. 2012a). One of the reasons for this result is that the calcium incrustations filter out some of the deleterious solar UV radiation (Beardall et al. 2014, Häder et al. 2015). The combined effects of ocean acidification and solar UV radiation on photosynthesis, growth, pigmentation and calcification have been studied in the Rhodophyte *Corallina sessilis* (Gao and Zheng 2010). In some Coralline algae the effect of ocean acidification is augmented by increased temperatures (Vásquez-Elizondo and Enríquez 2016). Similar effects were found in some Arctic Rhodophytes and Phaeophytes (Gordillo et al. 2016). In contrast, increased CO_2 concentrations in the water may favor some non-calcareous benthic algae (Johnson et al. 2014). Because of the contrasting effects of ocean acidification, changes in the species composition and biomass productivity are expected.

Effect of Increasing Temperatures on Macroalgae

Increasing atmospheric CO_2 concentrations result in rising temperatures. Also the surface water temperatures of the oceans have been found to rise accordingly (Fischetti 2013). Coral bleaching has been found to be strongly associated with elevated temperatures (Hughes et al. 2003). The Ligurian Sea is one of the coldest areas of the Mediterranean. Temperatures have been recorded systematically starting in the 1950s (Parravicini et al. 2015). The records show a cooling phase from 1958 to 1980, a period of rapid warming up to 1990 and a slower warming up to 2010. During the warming periods alien species increased while the number of native cold-water species decreased. In the tropics, many algae dwell close to their thermal limits and may not survive climate change-induced higher temperatures (Koch et al. 2013). The link between temperature and growth, distribution and abundance of macroalgae is fairly well investigated (Harley et al. 2006). In contrast, the interaction between temperature and other stress factors is not well understood. In a mesocosms study the combined effects of elevated CO_2 and increased temperature have been studied on macroalgal assemblages from intertidal rock pools (Olabarria et al. 2013). Photosynthetic efficiency, productivity, respiration and biomass production were measured in the native *Cystoseira* and the non-indigenous *Sargassum*. High temperature (20°C) and increased CO_2 significantly reduced the biomass, decreased productivity and affected respiration rates. Different species within the assemblages responded differently to theses stress factors which changed the species composition. *Sargassum* seems to be more tolerant to climate change and may benefit from its lower sensitivity. Climate change has also been found to affect growth rates and distribution of seagrasses (Short and Neckles 1999). South-Eastern Australian temperate coasts are a hotspot for marine biodiversity but have experienced ocean warming considerably well above global average rates (Wernberg et al. 2013). One result of this is a marked decline of the giant kelp *Macrocystis* linked with a poleward range extension of the key herbivore sea urchin. Recent research has indicated that the synergy with other stress factors, such as ocean acidification, overharvesting and reduced water quality will reduce the resilience of temperate communities of important habitat-dominating macroalgae to storms and diseases. Also the benthic macroalgae of the Great Barrier Reef including Rhodophyta, Phaeophyta, and Chlorophyta have been found to be vulnerable to climate change (Diaz-Pulido et al. 2007). Temperatures are predicted to increase by 1–3° C by the end of the century. Ocean temperature controls biogeographic distribution of seaweeds and seasonality of reef algae which are also impacted by predators. Moderate ocean warming may increase metabolism and productivity and also change seasonality of growth and reproduction (Beardall et al. 1998). Even though many macroalgal species have a wide thermal window for survival of 8–35°C they may not survive permanently at elevated temperatures (Pakker et al. 1995). Increasing ocean temperatures result in thermal expansion and melting of glaciers and ice sheets result in sea level rise at a rate of 1–2 mm per year. This may affect shallow coastal habitats, especially platform reefs and may exceed depth limits for survival of shallow-water algal turfs and upright macroalgae.

Conclusion

Mid-latitude marine macroalgae are exposed to a number of environmental stress factors in their habitat mainly on temperate rocky shores. Anthropogenically induced global climate change alters many factors which changes growth and productivity, habitat selection as well as species composition. Even though many macroalgae are adapted to wide thermal windows they may be endangered in extreme habitats, such as cold-water algae or reef algae for which the water temperatures exceed their limits. In most cases increasing CO_2 concentrations in the water does not augment algal growth since most species possess carbon dioxide-accumulating mechanisms (Raven et al. 2011). Exceptions to the rule are some freshwater and marine supralittoral, littoral, and sublittoral Florideae (Raven et al. 2005) which benefit from increasing CO_2 concentrations in the water column.

Ocean acidification resulting from increasing CO_2 concentrations has different consequences for algal photosynthesis, productivity, and growth which depends on the species as well as interacting other stress factors, such as nutrient availability and temperature and are not yet fully understood.

References

Adams III, W.W. and B. Demmig-Adams. 1992. Operation of the xanthophyll cycle in higher plants in response to diurnal changes in incident sunlight. Planta 186: 390–398.

Beardall, J., S. Beer and J.A. Raven. 1998. Biodiversity of marine plants in an era of climate change: some predictions based on physiological performance. Botanica Marina 41: 113–123.

Beardall, J., S. Stojkovic and K. Gao. 2014. Interactive effects of nutrient supply and other environmental factors on the sensitivity of marine primary producers to ultraviolet radiation: Implications for the impacts of global change. Aquatic Biology 22: 5–23.

Bicanic, D., F. Harren, J. Reuss, E. Woltering, J. Snel, L. Voesenek, B. Zuidberg, H. Jalink, F. Bijnen and C. Blom. 1989. Trace detection in agriculture and biology. pp. 213–245. *In*: Hess, P. (ed.). Photoacoustic, Photothermal and Photochemical Processes in Gases. Springer, Berlin, Heidelberg.

Bidwell, R.G.S. 1958. Photosynthesis and metabolism of marine algae: II A survey of rates and products of photosynthesis in $C^{14}O_2$. Canadian Journal of Botany 36: 337–349.

Büchel, C. and C. Wilhelm. 1993. *In vivo* analysis of slow chlorophyll fluorescence induction kinetics in algae: progress, problems and perspectives. Photochemistry and Photobiology 58: 137–148.

Caldeira, K. and M.E. Wickett. 2003. Oceanography: anthropogenic carbon and ocean pH. Nature 425: 365–365.

Canal-Vergés, P., M. Potthoff, F.T. Hansen, N. Holmboe, E.K. Rasmussen and M.R. Flindt. 2014. Eelgrass re-establishment in shallow estuaries is affected by drifting macroalgae–Evaluated by agent-based modeling. Ecological Modelling 272: 116–128.

Carroll, A.K. and J.M. Shick. 1996. Dietary accumulation of UV-absorbing mycosporine-like amino acids (MAAs) by the green sea urchin (*Strongylocentrotus droebachiensis*). Marine Biology 124: 561–569.

Chappuis, E., M. Terradas, M.E. Cefalì, S. Mariani and E. Ballesteros. 2014. Vertical zonation is the main distribution pattern of littoral assemblages on rocky shores at a regional scale. Estuarine, Coastal and Shelf Science 147: 113–122.

Chen, H., D. Zhou, G. Luo, S. Zhang and J. Chen. 2015. Macroalgae for biofuels production: progress and perspectives. Renewable and Sustainable Energy Reviews 47: 427–437.

Christa, G., S. Cruz, P. Jahns, J. Vries, P. Cartaxana, A.C. Esteves, J. Serôdio and S.B. Gould. 2017. Photoprotection in a monophyletic branch of chlorophyte algae is independent of energy-dependent quenching (qE). New Phytologist 214: 1132–1144.

Cornwall, C.E., A.T. Revill and C.L. Hurd. 2015. High prevalence of diffusive uptake of CO_2 by macroalgae in a temperate subtidal ecosystem. Photosynthesis Research 124: 181–190.

Critchley, C. and A.W. Russell. 1994. Photoinhibition of photosynthesis *in vivo*: The role of protein turnover in photosystem II. Physiologia Plantarum 92: 188–196.

De Martino, A., D. Douady, M. Quinet-Szely, B. Rousseau, F. Crepineau, K. Apt and L. Caron. 2000. The light-harvesting antenna of brown algae. European Journal of Biochemistry 267: 5540–5549.

Diaz-Pulido, G., L.J. McCook, A.W. Larkum, H.K. Lotze, J.A. Raven, B. Schaffelke, J.E. Smith and R.S. Steneck. 2007. Vulnerability of Macroalgae of the Great Barrier Reef to Climate Change.

Falkowski, P. 2012. Ocean science: The power of plankton. Nature 483: S17–S20.

Falkowski, P. 2013. Primary productivity in the sea. Springer Science & Business Media.

Feely, R.A., C.L. Sabine, K. Lee, W. Berelson, J. Kleypas, V.J. Fabry and F.J. Millero. 2004. Impact of anthropogenic CO_2 on the $CaCO_3$ system in the oceans. Science 305: 362–366.

Fernández, M.B., V. Tossi, L. Lamattina and R. Cassia. 2016. A comprehensive phylogeny reveals functional conservation of the UV-B photoreceptor UVR8 from green algae to higher plants. Frontiers in Plant Science 7.

Figueroa, F.L., C. Jiménez, B. Viñegla, E. Pérez-Rodríguez, J. Aguilera, A. Flores-Moya, M. Altamirano, M. Lebert and D.-P. Häder. 2002. Effects of solar UV radiation on photosynthesis of the marine angiosperm *Posidonia oceanica* from southern Spain. Marine Ecology Progress Series 230: 59–70.

Fischetti, M. 2013. Deep heat threatens marine life. Scientific American 308: 92.

Gantt, E., B. Grabowski and F.X. Cunningham Jr. 2003. Antenna systems of red algae: Phycobilisomes with photosystem II and chlorophyll complexes with photosystem I. pp. 307–322. *In*: Green, B.R. and W.W. Parson (eds.). Light-Harvesting Antennas in Photosynthesis. Advances in Photosynthesis and Respiration. Springer, Dordrecht.

Gao, K., E.W. Helbling, D.-P. Häder and D.A. Hutchins. 2012a. Responses of marine primary producers to interactions between ocean acidification, solar radiation, and warming. Marine Ecology Progress Series 470: 167–189

Gao, K., J. Xu, G. Gao, Y. Li, D.A. Hutchins, B. Huang, Y. Zheng, P. Jin, X. Cai, D.-P. Häder, W. Li, K. Xu, N. Liu and U. Riebesell. 2012b. Rising carbon dioxide and increasing light exposure act synergistically to reduce marine primary productivity. Nature Climate Change 2: 519–523.

Gao, K. and Y. Zheng. 2010. Combined effects of ocean acidification and solar UV radiation on photosynthesis, growth, pigmentation and calcification of the coralline alga *Corallina sessilis* (Rhodophyta). Global Change Biology 16: 2388–2398.

Genty, B., J.M. Briantais and N.R. Baker. 1989. The relationship between the quantum yield of photosynthetic electron transport and quenching of chlorophyll fluorescence. Biochimica and Biophysica Acta 990: 87–92.

Gersun, L., R. Anderson, J. Hart, G. Maneveldt and J. Bolton. 2016. Sublittoral seaweed communities on natural and artificial substrata in a high-latitude coral community in South Africa. African Journal of Marine Science 38: 303–316.

Gordillo, F.J., R. Carmona, B. Viñegla, C. Wiencke and C. Jiménez. 2016. Effects of simultaneous increase in temperature and ocean acidification on biochemical composition and photosynthetic performance of common macroalgae from Kongsfjorden (Svalbard). Polar Biology 39: 1993–2007.

Goss, R. and T. Jakob. 2010. Regulation and function of xanthophyll cycle-dependent photoprotection in algae. Photosynthesis Research 106: 103–122.

Gröniger, A. and D.-P. Häder. 2002. Induction of the synthesis of an UV-absorbing substance in the green alga *Prasiola stipitata*. Journal of Photochemistry and Photobiology B: Biology 66: 54–59.

Gubelit, Y., Y. Polyak, G. Dembska, G. Pazikowska-Sapota, L. Zegarowski, D. Kochura, D. Krivorotov, E. Podgornaya, O. Burova and C. Maazouzi. 2016. Nutrient and metal pollution of the eastern Gulf of Finland coastline: Sediments, macroalgae, microbiota. Science of the Total Environment 550: 806–819.

Guenther, J.E. and A. Melis. 1990. The physiological significance of photosystem II heterogeneity in chloroplasts. Photosynthesis Research 23: 105–109.

Guidi, L., M. Tattini and M. Landi. 2017. How does chloroplast protect chlorophyll against excessive light? pp. Jacob-Lopes, E., L. Queiroz Zepka and M. Isabel Queiroz (eds.). Chlorophyll. InTech.

Häder, D.-P. and F.L. Figueroa. 1997. Photoecophysiology of marine macroalgae. Photochemistry and Photobiology 66: 1–14.

Häder, D.-P., A. Gröniger, C. Hallier, M. Lebert, F.L. Figueroa and C. Jiménez. 1999a. Photoinhibition by visible and ultraviolet radiation in the red macroalga *Porphyra umbilicalis* grown in the laboratory. Plant Ecology 145: 351–358.

Häder, D.-P., E.W. Helbling, C.E. Williamson and R.C. Worrest. 2011. Effects of UV radiation on aquatic ecosystems and interactions with climate change. Photochememical and Photobiological Sciences 10: 242–260.

Häder, D.-P., H. Herrmann and R. Santas. 1996a. Effects of solar radiation and solar radiation deprived of UV-B and total UV on photosynthetic oxygen production and pulse amplitude modulated fluorescence in the brown alga *Padina pavonia*. FEMS Microbiology Ecology 19: 53–61.

Häder, D.-P., H. Herrmann, J. Schäfer and R. Santas. 1996b. Photosynthetic fluorescence induction and oxygen production in corallinacean algae measured on site. Botanica Acta 109: 285–291.

Häder, D.-P., H. Herrmann, J. Schäfer and R. Santas. 1997a. Photosynthetic fluorescence induction and oxygen production in two Mediterranean *Cladophora* species measured on site. Aquatic Botany 56: 253–264.

Häder, D.-P., M. Lebert, F.L. Figueroa, C. Jiménez, B. Viñegla and E. Perez-Rodriguez. 1998. Photoinhibition in Mediterranean macroalgae by solar radiation measured on site by PAM fluorescence. Aquatic Botany 61: 225–236.

Häder, D.-P., M. Lebert and E.W. Helbling. 2000a. Photosynthetic performance of the chlorophyte *Ulva rigida* measured in Patagonia on site. Recent Research in Developments in Photochemistry & Photobiology 4: 259–269.

Häder, D.-P., M. Lebert and E.W. Helbling. 2001a. Effects of solar radiation on the Patagonian macroalga *Enteromorpha linza* (L.) J. Agardh - Chlorophyceae. Journal of Photochemistry and Photobiology B: Biology 62: 43–54.

Häder, D.-P., M. Lebert and E.W. Helbling. 2001b. Photosynthetic performance of marine macroalgae measured in Patagonia on site. Trends in Photochemistry and Photobiology 8: 145–152.

Häder, D.-P., M. Lebert and E.W. Helbling. 2003a. Effects of solar radiation on the Patagonian rhodophyte *Corallina officinalis* (L.). Photosynthesis Research 78: 119–132.

Häder, D.-P., M. Lebert and E.W. Helbling. 2003b. *In situ* effects of solar radiation on photosynthesis in the Patagonian rhodophyte, *Porphyra columbina* Montagne. Recent Research in Developments in Biochemistry 4: 931–944.

Häder, D.-P., M. Lebert and E.W. Helbling. 2004. Variable fluorescence parameters in the filamentous Patagonian rhodophytes, *Callithamnion gaudichaudii* and *Ceramium* sp. under solar radiation. Journal of Photochemistry and Photobiology B: Biology 73: 87–99.

Häder, D.-P., M. Lebert, C. Jiménez, S. Salles, J. Aguilera, A. Flores-Moya, J. Mercado, B. Viñegla and F.L. Figueroa. 1999b. Pulse amplitude modulated fluorescence in the green macrophytes, *Codium adherens, Enteromorpha muscoides, Ulva gigantea* and *Ulva rigida*, from the Atlantic coast of Southern Spain. Environmental and Experimental Botany 41: 247–255.

Häder, D.-P., M. Lebert, J. Mercado, J. Aguilera, S. Salles, A. Flores-Moya, C. Jiménez and F.L. Figueroa. 1996c. Photosynthetic oxygen production and PAM fluorescence in the brown alga *Padina pavonica* (Linnaeus) Lamouroux measured in the field under solar radiation. Marine Biology 127: 61–66.

Häder, D.-P., M. Lebert, R.P. Sinha, E.S. Barbieri and E.W. Helbling. 2002. Role of protective and repair mechanisms in the inhibition of photosynthesis in marine macroalgae. Photochemical & Photobiological Sciences 1: 809–814.

Häder, D.-P., M. Porst, H. Herrmann, J. Schäfer and R. Santas. 1996d. Photoinhibition in the Mediterranean green alga *Halimeda tuna* Ellis et Soland measured *in situ*. Photochemistry and Photobiology 64: 428–434.

Häder, D.-P., M. Porst, H. Herrmann, J. Schäfer and R. Santas. 1997b. Photosynthesis of the Mediterranean green alga *Caulerpa prolifera* measured in the field under solar irradiation. Journal of Photochemistry and Photobiology B: Biology 37: 66–73.

Häder, D.-P., M. Porst and M. Lebert. 2000b. On site photosynthetic performance of Atlantic green algae. Journal of Photochemistry and Photobiology B: Biology 57: 159–168.

Häder, D.-P., M. Porst and M. Lebert. 2001c. Photoinhibition in common Atlantic macroalgae measured on site in Gran Canaria. Helgoland Marine Research 55: 67–76.

Häder, D.-P., M. Porst and M. Lebert. 2001d. Photosynthetic performance of the Atlantic brown macroalgae, *Cystoseira abies-marina, Dictyota dichotoma* and *Sargassum vulgare*, measured in Gran Canaria on site. Environmental and Experimental Botany 45: 21–32.

Häder, D.-P. and J. Schäfer. 1994a. *In situ* measurement of photosynthetic oxygen production in the water column. pp. 259–268. *In*: Wiersma, G.B. and J.A. Santolucito (eds.). Environmental Monitoring Assessment 32. Kluwer Academic Publishers, NL.

Häder, D.-P. and J. Schäfer. 1994b. Photosynthetic oxygen production in macroalgae and phytoplankton under solar irradiation. Journal of Plant Physiology 144: 293–299.

Häder, D.-P. and M. Tevini. 1987. General Photobiology. Pergamon. Oxford, UK.

Häder, D.-P., C.E. Williamson, S.-A. Wangberg, M. Rautio, K.C. Rose, K. Gao, E.W. Helbling, R.P. Sinha and R. Worrest. 2015. Effects of UV radiation on aquatic ecosystems and interactions with other environmental factors. Photochemical & Photobiological Sciences 14: 108–126.

Halac, S., V. Villafañe, R. Gonçalves and E. Helbling. 2014. Photochemical responses of three marine phytoplankton species exposed to ultraviolet radiation and increased temperature: Role of photoprotective mechanisms. Journal of Photochemistry and Photobiology B: Biology 141: 217–227.

Hanelt, D. 2017. Photosynthesis assessed by chlorophyll fluorescence. pp. 169–198. *In*: Häder, D.-P. and G.S. Erzinger (eds.). Bioassays. Advanced Methods and Applications. Elsevier, Atlanta, GA.

Hanelt, D. and W. Nultsch. 1990. Daily changes of the phaeoplasts arrangement in the brown alga *Dictyota dichotoma* as studied in field experiments. Marine Ecology Progress Series 61: 273–279.

Harley, C.D., A. Randall Hughes, K.M. Hultgren, B.G. Miner, C.J. Sorte, C.S. Thornber, L.F. Rodriguez, L. Tomanek and S.L. Williams. 2006. The impacts of climate change in coastal marine systems. Ecology Letters 9: 228–241.

Hartmann, A., A. Holzinger, M. Ganzera and U. Karsten. 2016. Prasiolin, a new UV-sunscreen compound in the terrestrial green macroalga *Prasiola calophylla* (Carmichael ex Greville) Kützing (Trebouxiophyceae, Chlorophyta). Planta 243: 161–169.

He, Y.Y. and D.-P. Häder. 2002. UV-B-induced formation of reactive oxygen species and oxidative damage of the cyanobacterium *Anabaena* sp.: Protective effects of ascorbic acid and *N*-acetyl-L-cysteine. Journal of Photochemistry and Photobiology B: Biology 66: 115–124.

Hein, M. and K. Sand-Jensen. 1997. CO_2 increases oceanic primary production. Nature 388: 526–527.

Helbling, E.W., E.S. Barbieri, R.P. Sinha, V.E. Villafañe and D.-P. Häder. 2004. Dynamics of potentially protective compounds in Rhodophyta species from Patagonia (Argentina) exposed to solar radiation. Journal of Photochemistry and Photobiology B: Biology 75: 63–71.

Helbling, E.W., V.E. Villafañe and D.-P. Häder. 2010. Ultraviolet radiation effects on macroalgae from Patagonia, Argentina. pp. 199–214. *In*: Israel, A.R., R. Einav and J. Seckbach (eds.). Seaweeds and their Role in Globally Changing Environments. Springer, Heidelberg.

Herrmann, H., F. Ghetti, R. Scheuerlein and D.-P. Häder. 1995. Photosynthetic oxygen and fluorescence measurements in *Ulva laetevirens* affected by solar irradiation. Journal of Plant Physiology 145: 221–227.

Howe, P.L., A.J. Reichelt-Brushett, M.W. Clark and C.R. Seery. 2017. Toxicity estimates for diuron and atrazine for the tropical marine cnidarian *Exaiptasia pallida* and in-hospite *Symbiodinium* spp. using PAM chlorophyll-*a* fluorometry. Journal of Photochemistry and Photobiology B: Biology 171: 125–132.

Hughes, T.P., A.H. Baird, D.R. Bellwood, M. Card, S.R. Connolly, C. Folke, R. Grosberg, O. Hoegh-Guldberg, J.B.C. Jackson, J. Kleypas, J.M. Lough, P. Marshall, M. Nyström, S.R. Palumbi, J.M. Pandolfi, B. Rosen and J. Roughgarden. 2003. Climate change, human impacts, and the resilience of coral reefs. Science 301: 929–933.

Hutchins, D., F.-X. Fu, Y. Zhang, M.E. Warner, Y. Feng, K. Portune, P.W. Bernhardt and M.R. Mulholland. 2007. CO_2 control of *Trichodesmium* N_2 fixation, photosynthesis, growth rates, and elemental ratios: implications for past, present, and future ocean biogeochemistry. Limnol Oceanogr 52: 1293–1304.

Ji, Y., Z. Xu, D. Zou and K. Gao. 2016. Ecophysiological responses of marine macroalgae to climate change factors. Journal of Applied Phycology 28: 2953–2967.

Jiménez, C., F.L. Figueroa, S. Salles, J. Aguilera, J. Mercado, B. Vinegla, A. Flores-Moya, M. Lebert and D.-P. Häder. 1998. Effects of solar radiation on photosynthesis and photoinhibition in red macrophytes from an intertidal system of Southern Spain. Botanica Marina 41: 329–338.

Johnson, M.D., N.N. Price and J.E. Smith. 2014. Contrasting effects of ocean acidification on tropical fleshy and calcareous algae. PeerJ 2: e411.

Kale, R., A.E. Hebert, L.K. Frankel, L. Sallans, T.M. Bricker and P. Pospíšil. 2017. Amino acid oxidation of the D1 and D2 proteins by oxygen radicals during photoinhibition of Photosystem II. Proceedings of the National Academy of Sciences 114: 2988–2993.

Keddy, P.A. 2010. Wetland ecology: Principles and conservation. Cambridge University Press.

Kendrick, G., M. Vanderklift, D. Bearham, J. Mclaughlin, J. Greenwood, C. Säwström, B. Laverock, L. Chovrelat, A. Zavala-Perez, L. De Wever, M. Trapon, M. Grol, E. Guilbault, D. Oades, P. McCarthy, K. George, T. Sampi, D. George, C. Sampi, Z. Edgar, K. Dougal and A. Howard. 2016. Benthic primary productivity: production and herbivory of seagrasses, macroalgae and microalgae. Western Australian Marine Science Institution.

Keren, N., A. Berg, P.J. Van Kan, H. Levanon and I. Ohad. 1997. Mechanism of photosystem II photoinactivation and D1 protein degradation at low light: The role of back electron flow. Proceedings of the National Academy of Sciences 94: 1579–1584.

Khoroshyy, P., D. Bína, Z. Gardian, R. Litvín, J. Alster and J. Pšenčík. 2018. Quenching of chlorophyll triplet states by carotenoids in algal light-harvesting complexes related to fucoxanthin-chlorophyll protein. Photosynthesis Research 135: 213–225.

Kianianmomeni, A. 2014. More light behind gene expression. Trends in Plant Science 19: 488–490.

Koch, M., G. Bowes, C. Ross and X.H. Zhang. 2013. Climate change and ocean acidification effects on seagrasses and marine macroalgae. Global Change Biology 19: 103–132.

Logan, B.A., B. Demmig-Adams, W.W. Adams III and W. Bilger. 2014. Context, quantification, and measurement guide for non-photochemical quenching of chlorophyll fluorescence. pp. 187–201. *In*: Demmig-Adams, B., G. Garab, W. Adams III and Govindjee (eds.). Non-Photochemical Quenching and Energy Dissipation in Plants, Algae and Cyanobacteria. Advances in Photosynthesis and Respiration (Including Bioenergy and Related Processes). Springer, Dordrecht.

Lüttge, U., M. Kluge and G. Bauer. 2005. Botanik. 5. vollst. überarb. Auflage. Wiley-VCH, Weinheim.

Lyons, D.A., C. Arvanitidis, A.J. Blight, E. Chatzinikolaou, T. Guy-Haim, J. Kotta, H. Orav-Kotta, A.M. Queirós, G. Rilov and P.J. Somerfield. 2014. Macroalgal blooms alter community structure and primary productivity in marine ecosystems. Global Change Biology 20: 2712–2724.

Mallick, N. and F.H. Mohn. 2000. Reactive oxygen species: response of algal cells. Journal of Plant Physiology 157: 183–193.

Matsui, K., E. Nazifi, Y. Hirai, N. Wada, S. Matsugo and T. Sakamoto. 2012. The cyanobacterial UV-absorbing pigment scytonemin displays radical scavenging activity. Journal of General and Applied Microbiology 58: 137–144.

Migné, A., G. Delebecq, D. Davoult, N. Spilmont, D. Menu and F. Gévaert. 2015. Photosynthetic activity and productivity of intertidal macroalgae: *in situ* measurements, from thallus to community scale. Aquatic Botany 123: 6–12.

Miller, W.R. and E.S. Perry. 2016. The coastal marine Tardigrada of the Americas. Zootaxa 4126: 375–396.

Mitra, A. and S. Zaman. 2016. Basics of Marine and Estuarine Ecology. Springer. New Delhi, Heidelberg, New York, Dordrecht, London.

Monteiro, P. 2013. Atlantic Area Eunis Habitats. Adding new habitat types from European Atlantic coast to the EUNIS Habitat Classification. CCMAR-Universidade do Algarve.

Montingelli, M.E., K.Y. Benyounis, B. Quilty, J. Stokes and A.G. Olabi. 2016. Optimisation of biogas production from the macroalgae *Laminaria* sp. at different periods of harvesting in Ireland. Applied Energy 177: 671–682.

Olabarria, C., F. Arenas, R.M. Viejo, I. Gestoso, F. Vaz-Pinto, M. Incera, M. Rubal, E. Cacabelos, P. Veiga and C. Sobrino. 2013. Response of macroalgal assemblages from rockpools to climate change: Effects of persistent increase in temperature and CO_2. Oikos 122: 1065–1079.

Oliver, A., I. Giesbrecht, S. Tank, B. Hunt and K. Lertzman. 2014. Dom Export from Coastal Temperate Bog Forest Watersheds to Marine Ecosystems: Improving Understanding of Watershed Processes and Terrestrial-Marine Linkages on the Central Coast of British Columbia. AGU Fall Meeting Abstracts.

Öquist, G. and W.S. Chow. 1992. On the relationship between the quantum yield of photosystem II electron transport, as determined by chlorophyll fluorescence and the quantum yield of CO_2-dependent O_2 evolution. Photosynthesis Research 33: 51–62.

Pakker, H., A.M. Breeman, W.F. Prud'homme van Reine and C. Hock. 1995. A comparative study of temperature responses of Caribbean seaweeds from different biogeographic groups. Journal of Phycology 31: 499–507.

Pakker, H., R.S.T. Martins, P. Boelen, A.G.J. Buma, O. Nikaido and A.M. Breeman. 2000. Effects of temperature on the photoreactivation of ultraviolet-B-induced DNA damage in *Palmaria palmata* (Rhodophyta). Journal of Phycology 36: 334–341.

Parravicini, V., L. Mangialajo, L. Mousseau, A. Peirano, C. Morri, M. Montefalcone, P. Francour, M. Kulbicki and C.N. Bianchi. 2015. Climate change and warm-water species at the north-western boundary of the Mediterranean Sea. Marine Ecology 36: 897–909.

Pathak, J., R. Richa, A.S. Sonker, V.K. Kannaujiya and R.P. Sinha. 2015. Isolation and partial purification of scytonemin and mycosporine-like amino acids from biological crusts. Journal of Chemical and Pharmaceutical Research 7: 362–371.

Pope, B.M. 1963. Calcification in red coralline algae. Department of Biological Sciences, Stanford University.

Post, A. and A. Larkum. 1993. UV-absorbing pigments, photosynthesis and UV exposure in Antarctica: comparison of terrestrial and marine algae. Aquatic Botany 45: 231–243.

Puente, A., X. Guinda, J.A. Juanes, E. Ramos, B. Echavarri-Erasun, F. Camino, S. Degraer, F. Kerckhof, N. Bojanić and M. Rousou. 2017. The role of physical variables in biodiversity patterns of intertidal macroalgae along European coasts. Journal of the Marine Biological Association of the United Kingdom 97: 549–560.

Raven, J.A., L.A. Ball, J. Beardall, M. Giordano and S.C. Maberly. 2005. Algae lacking carbon-concentrating mechanisms. Canadian Journal of Botany 83: 879–890.

Raven, J.A., M. Giordano, J. Beardall and S.C. Maberly. 2011. Algal and aquatic plant carbon concentrating mechanisms in relation to environmental change. Photosynthesis Research 109: 281–296.

Reed, D.C. and M.S. Foster. 1984. The effects of canopy shadings on algal recruitment and growth in a giant kelp forest. Ecology 65: 937–948.

Ricci, S., F. Antonelli, C.S. Perasso, D. Poggi and E. Casoli. 2016. Bioerosion of submerged lapideous artefacts: Role of endolithic rhizoids of *Acetabularia acetabulum* (Dasycladales, Chlorophyta). International Biodeterioration & Biodegradation 107: 10–16.

Richter, P.R., R.J. Gonçalves, A. Marcoval, E.W. Helbling and D.-P. Häder. 2006. Diurnal changes in the composition of mycosporine-like amino acids (MAA) in *Corallina officinalis*. Trends in Photochemistry and Photobiology 11: 33–44.

Richter, P.R., D.-P. Häder, R.J. Gonçalves, M.A. Marcoval, V.E. Villafañe and E.W. Helbling. 2007. Vertical migration and motility responses in three marine phytoplankton species exposed to solar radiation. Photochemistry and Photobiology 83: 810–817.

Rico, A., P. Lanas and J.L. Gappa. 2014. Colonization of *Ulothrix flacca*, *Urospora penicilliformis* and *Blidingia minima* (Chlorophyta) in Comodoro Rivadavia harbor (Chubut, Argentina). Revista del Museo Argentino de Ciencias Naturales nueva serie 5: 93–97.

Riebesell, U. and P.D. Tortell. 2011. Effects of ocean acidification on pelagic organisms and ecosystems. pp. 99–116. *In*: Gattuso, J.P. and L. Hansson (eds.). Ocean Acidification. Oxford University Press, Oxford, UK.

Rizzini, L., J.-J. Favory, C. Cloix, D. Faggionato, A. O'Hara, E. Kaiserli, R. Baumeister, E. Schäfer, F. Nagy and G.I. Jenkins. 2011. Perception of UV-B by the *Arabidopsis* UVR8 protein. Science 332: 103–106.

Ruban, A.V., A.J. Young and P. Horton. 1993. Induction of non-photochemical energy dissipation and absorbance changes in leaves. Evidence for changes in the state of the light-harvesting system of Photosystem II in vivo. Plant Physiology 102: 741–750.

Sakanishi, Y., H. Kasai and J. Tanaka. 2017. Trade-off relationship between productivity and thallus toughness in Laminariales (Phaeophyceae). Phycological Research 65: 103–110.

Schreiber, U. and W. Bilger. 1993. Progress in chlorophyll fluorescence research: Major developments during the past years in retrospect. pp. 151–153. *In*: Lüttge, U. and H. Ziegler (eds.). Progress Botany. Springer Verlag, Berlin.

Schreiber, U., W. Bilger and C. Neubauer. 1994. Chlorophyll fluorescence as a nonintrusive indicator for rapid assessment of in vivo photosynthesis. pp. 49–70. *In*: Schulze, E.D. and M.M. Caldwell (eds.). Ecophysiology of Photosynthesis. Ecological Studies. Springer Verlag, Berlin.

Schreiber, U., U. Schliwa and W. Bilger. 1986. Continuous recording of photochemical and non-photochemical chlorophyll fluorescence quenching with a new type of modulation fluorometer. Photosynthesis Research 10: 51–62.

Schubert, N., E. García-Mendoza and I. Pacheco-Ruiz. 2006. Carotenoid composition of marine red algae. Journal of Phycology 42: 1208–1216.

Short, F.T. and H.A. Neckles. 1999. The effects of global climate change on seagrasses. Aquatic Botany 63: 169–196.

Singh, S.P., S. Kumari, R.P. Rastogi, K.L. Singh, Richa and R.P. Sinha. 2010. Photoprotective and biotechnological potentials of cyanobacterial sheath pigment, scytonemin. African Journal of Biotechnology 9: 580–588.

Sinha, R.P., E.S. Barbieri, M. Lebert, E.W. Helbling and D.-P. Häder. 2003. Effects of solar radiation on phycobiliproteins of marine red algae. Trends in Photochemistry and Photobiology 10: 149–157.

Sinha, R.P. and D.-P. Häder. 2003a. Biochemistry of mycosporine-like amino acids (MAAs) synthesis: Role in photoprotection. Recent Research in Developments in Biochemistry 4: 971–983.

Sinha, R.P. and D.-P. Häder. 2003b. Biochemistry of phycobilisome disassembly by ultraviolet-B radiation in cyanobacteria. Recent Research in Developments in Biochemistry 4: 945–955.

Sinha, R.P., G. Keshari, S. Kumari, S.P. Singh, R.P. Rastogi and K.L. Singh. 2006. Screening of mycosporine-like amino acids (MAAs) in cyanobacteria. Modern Journal of Life Sciences 5: 1–6.

Sinha, R.P., M. Klisch, A. Gröniger and D.-P. Häder. 2001a. Responses of aquatic algae and cyanobacteria to solar UV-B. pp. 221–236. *In*: Rozema, J., Y. Manetas and L.O. Björn (eds.). Plant Ecology. Special Issue: Responses of Plants to UV-B Radiation. Kluwer Academic Publishers, Dordrecht, Boston, London.

Sinha, R.P., M. Klisch, E.W. Helbling and D.-P. Häder. 2001b. Induction of mycosporine-like amino acids (MAAs) in cyanobacteria by solar ultraviolet-B radiation. Journal of Photochemistry and Photobiology B: Biology 60: 129–135.

Sinha, R.P., M. Klisch, A. Vaishampayan and D.-P. Häder. 1999. Biochemical and spectroscopic characterization of the cyanobacterium *Lyngbya* sp. inhabiting mango (*Mangifera indica*) trees: presence of an ultraviolet-absorbing pigment, scytonemin. Acta Protozoologica 38: 291–298.

Sinha, R.P., S.P. Singh and D.-P. Häder. 2007. Database on mycosporines and mycosporine-like amino acids (MAAs) in fungi, cyanobacteria, macroalgae, phytoplankton and animals. Journal of Photochemistry and Photobiology B: Biology 89: 29–35.

Sun, X., Y. Zhong, Z. Huang and Y. Yang. 2014. Selenium accumulation in unicellular green alga *Chlorella vulgaris* and its effects on antioxidant enzymes and content of photosynthetic pigments. PLoS One 9: e112270.

Telfer, A. 2014. Singlet oxygen production by PSII under light stress: mechanism, detection and the protective role of β-carotene. Plant and Cell Physiology 55: 1216–1223.

Temme, A., W. Cornwell, H. Cornelissen and R. Aerts. 2014. From the low past to the high future: Plant growth across CO_2 levels. EGU General Assembly Conference Abstracts.

Tortell, P.D., G.H. Rau and F.M. Morel. 2000. Inorganic carbon acquisition in coastal Pacific phytoplankton communities. Limnology and Oceanography 45: 1485–1500.

Vásquez-Elizondo, R.M. and S. Enríquez. 2016. Coralline algal physiology is more adversely affected by elevated temperature than reduced pH. Scientific Reports 6: 19030.

Vázquez, D. 1979. Protein synthesis and translation inhibitors. pp. 1–14. *In*: Vázquez, D. (ed.). Inhibitors of Protein Biosynthesis. Springer, Berlin, Heidelberg.

Vincent, W.F. and P.J. Neale. 2000. Mechanisms of UV damage to aquatic organisms. pp. 149–176. *In*: de Mora, S.J., S. Demers and M. Vernet (eds.). The Effects of UV Radiation on Marine Ecosystems. Cambridge Univ. Press, Cambridge.

Vroom, P.S., C.M. Smith, J.A. Coyer, L.J. Walters, C.L. Hunter, K.S. Beach and J.E. Smith. 2003. Field biology of *Halimeda tuna* (Bryopsidales, Chlorophyta) across a depth gradient: comparative growth, survivorship, recruitment, and reproduction. Hydrobiologia 501: 149–166.

Vuayaraghavan, M.R. and G. Sokhi. 1986. Phaeophyceae—An ultrastructural and histochemical overview. Proc. Indian Natn. Sci. Acad. 4: 529–546.

Wernberg, T., D.A. Smale, F. Tuya, M.S. Thomsen, T.J. Langlois, T. De Bettignies, S. Bennett and C.S. Rousseaux. 2013. An extreme climatic event alters marine ecosystem structure in a global biodiversity hotspot. Nature Climate Change 3: 78.

Wilson, S., C. Blake, J.A. Berges and C.A. Maggs. 2004. Environmental tolerances of free-living coralline algae (maerl): implications for European marine conservation. Biological Conservation 120: 279–289.

Winter, K. and M. Königer. 1989. Dithiothreitol, an inhibitor of violaxanthin de-epoxidation, increases the susceptibility of leaves of *Nerium oleander* L. to photoinhibition of photosynthesis. Planta 180: 24–31.

Xu, D., X. Zhang, Y. Wang, X. Fan, Y. Miao, N. Ye and Z. Zhuang. 2016. Responses of photosynthesis and nitrogen assimilation in the green-tide macroalga *Ulva prolifera* to desiccation. Marine Biology 163: 9.

Yun, E.J., I.-G. Choi and K.H. Kim. 2015. Red macroalgae as a sustainable resource for bio-based products. Trends in Biotechnology 33: 247–249.

CHAPTER 13

Polar Macroalgae

Donat-P. Häder

Introduction

The Arctic and Antarctic Oceans are similar with respect to temperature and UV irradiation. Both regions are characterized by high changes in solar irradiance, temperature and salinity throughout the year. In addition, they are covered by ice in the winter. However they differ in many respects (Zacher et al. 2009). The Arctic Ocean is almost completely enclosed by continental land masses, while the Antarctic Ocean is open to the South Atlantic and South Pacific Ocean. The Antarctic circumpolar current flows clockwise around Antarctica and isolates the waters from the adjacent temperate waters. In contrast, there is an influx of warm Atlantic water into the Arctic Ocean through the Fram Strait. Cold waters are thought to have existed in the Antarctic Ocean for about 14 million years, while the Arctic became cold only two million years ago and this difference has influenced the evolution of the algal communities. The Antarctic macroalgae flora consists of about 33% endemic species while there are only few endemic species in the Arctic. Nutrient content in the Arctic is depleted during the summer while they are replete year round in the Antarctic Ocean.

Stratospheric Ozone Depletion and Increased Solar UV-B

Polar regions are the areas on our planet most affected by climate change including solar UV radiation (Thompson et al. 2011). About 90% of the atmospheric ozone is concentrated in the stratosphere. Anthropogenic emissions of ozone-depleting gases, such as chlorofluorocarbons (CFCs) have depleted the total ozone concentration since the 1970s by about 10% (Solomon 1999). Since ozone is the major absorber of solar UV-B radiation (280–315 nm) stratospheric ozone depletion results in increased short-wavelength radiation on the earth's surface (Williamson et al. 2014). A more dramatic depletion occurs in Antarctica each Austral spring, when about half of the total column ozone is catalytically destroyed by ClO, forming the Antarctic ozone hole

Friedrich-Alexander University, Erlangen-Nürnberg, Neue Str. 9, 91096 Möhrendorf, Germany.
Email: donat@dphaeder.de

(Fig. 13.1). A video showing the development of the ozone hole from 1979 to 2017 can be seen at https://ozonewatch.gsfc.nasa.gov/ozone_maps/movies/OZONE_D1979-10%25P1Y_G%5e1920X1080.IOMPS_PNPP_V21_MMERRA2_LSH.mp4.

Ozone depletion is due to heterogeneous chlorine chemistry that occurs at cold temperatures in polar stratospheric clouds and is particularly strong in the Antarctic, but also in the Arctic where it leads to smaller isolated holes which travel on circumpolar paths and touch Northern Europe, North America and Asia during spring time (Waibel et al. 1999). Ozone loss at both Northern and Southern mid latitudes is less dramatic and no ozone loss has been recorded in the tropics. In addition to anthropogenic release of ozone-depleting substances, natural volcanic halogen emissions can result in ozone depletion as observed in 1992–93 after the Mt. Pinatubo eruption (Hofmann et al. 1994). The tremendous reduction in the emission of ozone-depleting substances after the ratification of the Montreal Protocol and its subsequent Amendments signifies a spectacular scientific success story (DeSombre 2000). Stratospheric ozone depletion has levelled off during the beginning of this century and is expected to recover to pre-1970s levels by about 2065 (McKenzie et al. 2011). The sluggish recovery is due to the long lifetimes of ozone-depleting substances in the stratosphere in the order of decades up to a century (Douglass et al. 2008). For this reason, the Antarctic ozone hole only gradually decreases and the elevated UV-B levels in the Austral spring persists for some time in the future (Abbasi and Abbasi 2017).

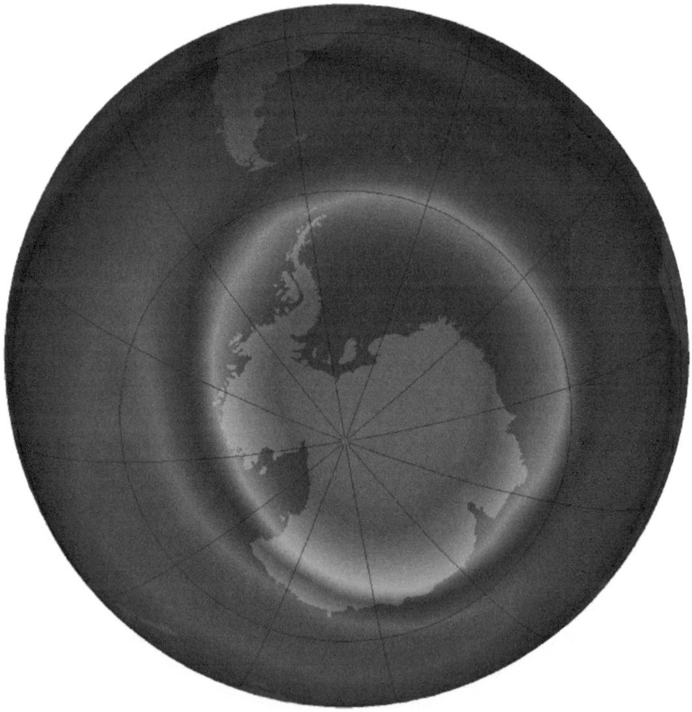

Fig. 13.1. Antarctic ozone hole in October 2017 recorded by satellite imaging. In the center the total ozone has fallen to below 200 Dobson units. From https://ozonewatch.gsfc.nasa.gov/monthly/SH.html with permission.

Rising Temperatures

The CO_2 concentration in the atmosphere has been rising from about 280 ppm before the Industrial revolution (Thomas et al. 2016) to the current level of about 400 ppm (Franks et al. 2013). Currently our atmosphere holds about 735 Gt carbon (not carbon dioxide). Each year about 7 Gt is released from fossil fuel burning and another 2 Gt from changed land usage, such as tropical deforestation. Both the terrestrial vegetation and the oceans take up about 50 Gt carbon each annually by photosynthetic activity, but this amount is released again to the atmosphere when the vegetation decays. However, a significant fraction of the anthropogenically released carbon is taken up by natural sinks. In the oceans decaying phytoplankton as well as organic material excreted by predators falls to the deep sea ocean; the sediments hold the largest amount of carbon (about 36,000 Gt compared to an estimated 7,500 Gt bound in fossil fuel). About 72% of the global dissolved organic carbon (DOC) is located in long-term deep ocean storage (Arrieta et al. 2015). This effective removal of part of the anthropogenically released carbon in the ocean is called biological pump (Meyer et al. 2016).

The surface temperature of our planet is kept at an elevated level by trace gases (greenhouse gases) in the atmosphere; without these the earth would be an inhospitable place with average temperatures below –30°C. Due to the anthropogenic emission of carbon dioxide from fossil fuel burning and changing land usage the mean global temperature has been rising from about 13°C before the onset of the Industrial revolution, because the earth absorbs about 0.85 ± 0.15 W/m^2 more energy from the Sun than it is emitting into space (Hansen et al. 2005). Global surface temperature in 2015 was the highest measured since the beginning of systematic recording and 0.87°C warmer than the average temperature in the period 1951–80 probably boosted by an exceptionally strong El Niño (Hansen and Sato 2016). However, there are large regional differences. While the mean temperature in central Europe, Northern Asia, and most of North America are about 2°C higher than the long-term average of 136 years, Alaska, Northern Canada and Siberia experienced temperature increases of about 4.5°C above average while part of the Antarctic coast were about 0.5°C colder than average. The temperature over most of the Atlantic, Indian, and the West Pacific Ocean was about 1 to 1.5°C above average. Even though future warming is predicted to be larger over land than over oceans (IPCC 2014) the mean ocean surface temperature has increased by about 1°C over the last 115 years; but the error bars are substantial (Fischetti 2013) (Fig. 13.2).

As on land there are large regional differences in the ocean temperatures. Especially in the Arctic increases of more than 4°C have been measured which is due to a feedback mechanism (Pithan and Mauritsen 2014): solar radiation is almost completely reflected into space by snow and ice surfaces (Lei et al. 2016). When these melt, the underlying soil and water absorb a much higher fraction of energy which effectively increases the temperature. Satellite images show that the September ice extent has decreased at the rate of 7% per decade from 1979 through 2001, and including 2013 it is about 14% per decade (Stroeve et al. 2014) resulting in a 50% loss of summer sea ice coverage. The possibility of an ice-free Arctic in the near future is being discussed (Msadek et al. 2014).

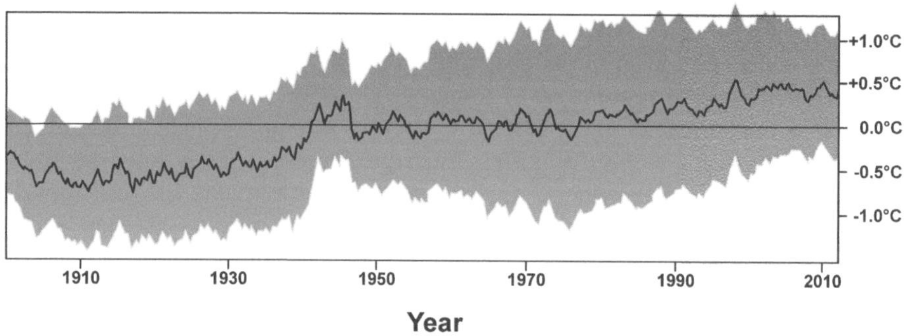

Fig. 13.2. Mean global surface water temperatures over the past 115 years. Modified after (Fischetti 2013).

Effects of Solar UV on Polar Macroalgae

The succession in polar marine benthic primary producers is affected by a number of biotic and abiotic factors (Campana et al. 2009). The first colonizers are usually diatoms and ephemeral macroalgae later followed by annual or perennial seaweeds. Ambient UV usually does not stress microalgae, such as diatoms very much. In contrast, early stages of macroalgal settlement are affected by solar short-wavelength radiation (Zacher 2014, de Oliveira et al. 2016). The mean photosynthetic optimum quantum yield of *Ascoseira mirabilis* propagules (gametangia, gametes and zygotes) was about 0.4. Irradiation with PAR decreased it to less than 0.2 after 1 h of exposure and to almost 0 after 8 h. Addition of UV-A and UV-B increasingly inhibited the yield. This inhibition was still noticeable after 48 h recovery in low white light irradiance (Fig. 13.3) (Roleda et al. 2007). Also the haploid tetraspores and diploid carpospores of the Antarctic *Gigartina skottsbergii* showed significant effects or UV radiation on photosynthetic performance, DNA damage, mortality and MAA concentrations but no differences between the ploidy levels (Roleda et al. 2008b).

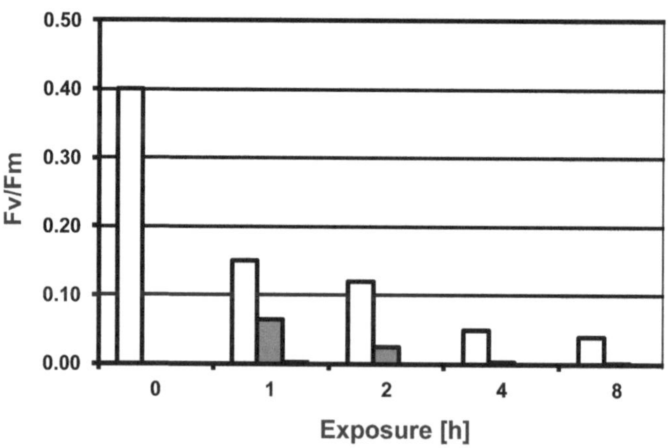

Fig. 13.3. Mean optimum quantum yield of *A. mirabilis* propagules during exposure to PAR (white bars), PAR + UV-A (gray bars) and PAR+UV-A + UV-B (black bars). Redrawn after (Roleda et al. 2007).

UVR causes cell biological and ultrastructural changes. Spores, gametes, and zygotes are the most vulnerable stages as has been shown in *Fucus*, several Laminariales and Gigartinales from Spitsbergen (Arctic) and King George Island (Antarctica) (Karsten et al. 2009). Germination, motility, nuclear division and photosynthesis are negatively affected (Huovinen et al. 2000, Makarov and Voskoboinikov 2001, Roleda et al. 2006b, Roleda et al. 2006a). UV radiation damages most cellular organelles, such as chloroplasts, mitochondria, nucleus, and the cytoplasm. In red algae the intrathylakoidal space is enlarged and the thylakoid membranes wrinkled after UV exposure, and the thylakoids become tubular as shown in *Bangia atropurpurea* (Poppe et al. 2003). In contrast, green algae, such as *Prasiola crispa* or *Urospora penicilliformis* showed no cytological damage after moderate UV exposure (Roleda et al. 2009b). However, gametophytes of *U. penicilliformis* showed increasing inhibition of the photosynthetic optimum quantum yield and an increase in CPDs by UV-B (Roleda et al. 2009a). Another study showed that chlorophylls, carotenoids, and xanthophyll cycle pigments (violaxanthin, antheraxanthin and zeaxanthin) were not affected in this alga after exposure to ambient and artificial radiation (Roleda et al. 2009b).

An important molecular target of UV radiation is the DNA resulting in single- or double-strand breaks, inter- or intra-strand crosslinks and the formation of cyclobutane pyrimidine dimers (CPD) and pyrimidine (6-4)-pyrimidone photoproducts which—if not repaired efficiently—can lead to errors in replication, mutations, and mortality (Karentz et al. 1991, Lois and Buchanan 1994). Different life stages of polar macroalgae show different susceptibility to UV-B damage of their DNA. Zoospores of the kelps *Alaria esculenta*, *Laminaria digitata,* and *Saccorhiza dermatodea* were found to be more susceptible to UV-B radiation than young sporophytes (Roleda et al. 2005, Roleda et al. 2006b, Roleda et al. 2006c, Roleda et al. 2006d). Low doses of UV-B < 72 J/m^2) were found to promote the growth of *Chondrus ocellatus* while higher doses inhibited it (Ju et al. 2015). Application of blue light promoted DNA photorepair while red light inhibited it. But both blue and red light reduced the synthesis of MAAs. Other major targets of solar UV radiation are proteins. The photosystem II (PS II) D1 protein and the Rubisco enzyme in the Calvin cycle are of significant importance (Bischof 2000, Bischof et al. 2000).

In addition to direct UV-B effects, the induction of reactive oxygen species (ROS), such as superoxide radicals, hydroxyl radicals and singlet oxygen ($^1O_2$), can result in cellular damage (Karsten et al. 2009, Bischof and Rautenberger 2012). In addition, ROS can be produced by photoactivation of dissolved organic matter (DOM) (Amado et al. 2015).

Mitigating strategies against UV-inflicted damage are avoidance by mat or turf formation, DNA repair, recovery of photosynthesis and shielding by UV-absorbing pigments, such as phlorotannins in brown algae (Roleda et al. 2005, Flores-Molina et al. 2016) and mycosporine-like amino acids in green and red algae (Sinha et al. 2007, Zacher and Campana 2008, Rastogi and Incharoensakdi 2013). DNA damage is repaired by photoreactivation, nucleotide and base excision repair (Britt 1996, Rastogi et al. 2010, Richa et al. 2015) and damage of the photosynthetic apparatus by resynthesis of the PS II reaction center D1 protein (Rautenberger et al. 2015). UV-B induced ROS are detoxified by enzymes, such as superoxide dismutase, catalase or glutathione peroxidase (Karsten and Holzinger 2014) or by non-enzymatic quenchers, such as ascorbate or carotenoids (Dummermuth et al. 2003, Athukorala et al. 2016). The concentration of UV-absorbing substances is linked to the growth depth, such as in the intertidal Rhodophyte *Devaleraea ramentacea*: much higher MAA concentrations were found in shallow-water thalli compared to deep-water isolates (Karsten et al. 1999). Similar results were found in other supra- and eulittoral red algae (Hoyer et al. 2001, Hoyer et al. 2003). Induction of phlorotannins by UV-B exposure was also demonstrated in *Laminaria hyperborea* (Halm et al. 2011). However, phlorotannins concentrations in zoospores of other Arctic brown algae (*Saccharina latissima, Laminaria digitata, Alaria esculenta*) depended on radiation and zoospore density (Müller et al. 2009).

Many deep-water Rhodophytes do not synthesize UV-absorbing MAAs since they are not exposed to strong solar UV-B. Likewise the concentration of MAAs depends on the seasonal exposure to UV-B. Young thalli of *Palmaria decipiens* collected during winter contained low MAA concentrations while older thalli collected during late spring and summer showed high concentrations (Post and Larkum 1993). Based on the available results red algae have been classified into three types: those which do not contain any MAAs (lower sublittoral), those with variable and inducible concentrations and those with permanently high MAA concentrations (supra- and eulittoral) (Hoyer et al. 2001).

One strategy to determine depth effects on benthic communities is using transplantation experiments (Fricke et al. 2008). The succession of benthic algae on ceramic tiles was studied after transplantation from 8 to 0.5 m water depth in Kongsfjord, Spitsbergen. Species composition of the communities changed showing a dominance of Ectocarpales (Phaeophyceae) at 8 m depth while green algae dominated at the near-surface site.

Effects of Temperature and Interaction with Other Stress Factors on Polar Macroalgae

In their habitat algae are exposed to a plethora of abiotic and biotic factors which do not operate independently but synergistically when the combined effects of several stressors are more pronounced than those to each individual stressor. If the response to two or more stress factors is smaller than that to the individual factors they operate antagonistically (Karsten et al. 2009). Global climate change has resulted in changes of several parallel stressors in polar regions. In addition to stratospheric ozone depletion and the resulting increase of solar UV-B, ocean warming and ocean acidification affects productivity, growth and development of macroalgae. In addition, pressure by predators and nutrient availability has to be considered. Therefore, it is mandatory to study biochemical, physiological, and ecological responses not to individual stress factors but to the array of factors in their habitat (Gao and Häder 2017).

A typical example for antagonistic effects are the responses of several Arctic and Antarctic Chlorophytes (*Enteromorpha bulbosa, Ulva clathrata, Alaria esculenta*) and the Rhodophytes (*Palmaria palmata, Coccotylus truncates, Phycodrys rubens*) which were less affected by UV-B radiation in the presence of elevated temperatures (van de Poll et al. 2002, Hoffman et al. 2003, Rautenberger and Bischof 2006, Fredersdorf et al. 2009).

In contrast, effects on zoospore germination and photosynthetic yield in sporophytes of the UV-B tolerant *Alaria esculenta* were basically due to elevated temperatures and the germination of UV-B sensitive zoospores of the Arctic Phaeophytes *Saccharina latissima* and *Laminaria digitata* was more affected at lower or higher temperatures outside their optimum (Müller et al. 2008).

Polar macroalgae are adapted to cold temperatures. Climate change might result in reduced sea ice cover and retreating glaciers which would alter irradiation, salinity, water temperature, and sedimentation and thereby change environmental conditions for macroalgal development and community structure. The recent break-off of a large fraction from the Larsen-C ice shelf in the Antarctic reduced the ice shelf area by about 10% (Hogg and Gudmundsson 2017). Increasing temperatures will augment invasion of temperate non-native species into polar regions (Hughes and Convey 2014). Rising temperatures may also modify the physiological responses in Antarctic marine macroalgae. Four brown algae (*Desmarestia menziesii, Desmarestia anceps, Ascoseira mirabilis, Himantothallus grandifolius*) and three red algae (*Iridaea cordata, Trematocarpus antarcticus, Palmaria decipiens*) were subjected to UV stress at two different temperatures (2 and 7°C) (Rautenberger et al. 2015). *D. menziesii* and *A. mirabilis* had the highest UV tolerance at 2°C. In contrast *H. grandifolius* was sensitive to UV at 2°C but had a higher tolerance at 7°C, which can be explained by a higher repair rate of the PS II D1 protein. However, higher temperatures did not increase UV tolerance in *D. anceps* indicating that the repair process itself was UV-sensitive. UV induction of phlorotannins could be responsible for the high UV tolerance in *Ascoseira*, because of their high radical scavenging potential. These detailed results indicate that increasing temperatures modulate UV sensitivity but to a different extent in different macroalgae. Another study on five macroalgal species collected in the eulittoral and the upper sublittoral from the Magellan Strait (*Ulva intestinalis*, *Porphyra columbina*, *Adenocystis utricularis*, *Desmarestia confervoides* and *D. ligulata*) showed that the Calvin cycle was not affected by 4 h exposure to UV-A and UV-B as indicated by the fact that ETRmax was not reduced (Rautenberger et al. 2009). Photosynthesis was not statistically significantly affected by PAR, but decreased in PAR+UV-A and even more so in PAR+UV-A+UV-B (Fig. 13.4) (Rautenberger et al. 2009). Combined effects of temperature, UV and PAR irradiances on phlorotannins content and zoospore germination were also studied in the Arctic brown alga *Saccorhiza dermatodea* (Steinhoff et al. 2011).

Since solar UV is a major stress factor for macroalgae productivity and development it is important to understand underwater optics in natural habitats (Huovinen et al. 2016). Penetration of solar radiation into the water body is remarkably variable in fjords, estuaries, channels, and open waters and subject to changes related to global climate change. Receding ice results in higher transparency and deeper UV-B penetration which spells higher risks for Antarctic macroalgal ecosystems. In contrast, increasing temperatures could mitigate UV stress by enhancing enzymatic photorepair.

Polar macroalgae in different taxa are remarkable since they can grow under extremely low irradiances and temperatures. Irradiances can be as low as 2 μmol photons m^{-2} s^{-1} for the compensation point and 10 μmol photons m^{-2} s^{-1} for saturated

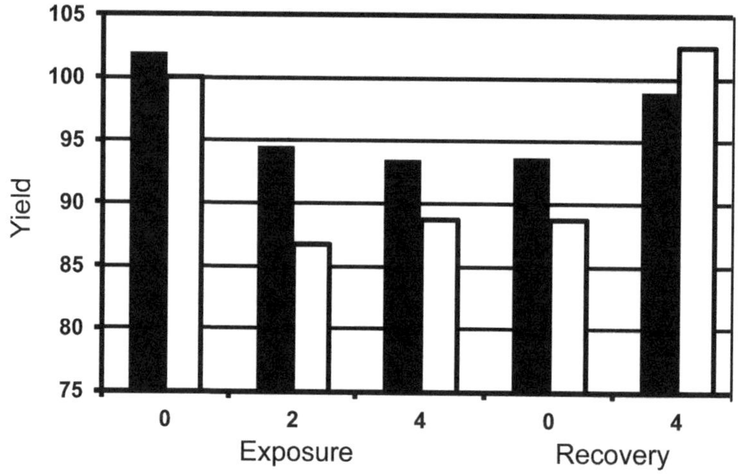

Fig. 13.4. Photosynthetic yield (ordinate, in percent of the control) in *Porphyra columbina* during exposure to PAR+UV-A (closed bars) or PAR + UV-A + UV-B (open bars) and during recovery. Redrawn after (Rautenberger et al. 2009).

photosynthesis (Gómez et al. 2009). With these values polar macroalgae can thrive under sea ice and at depths as low as 40 m. Some benthic microalgae can survive at even lower irradiances. The other remarkable characteristic is that macroalgae can complete their life cycle at temperatures close to 0°C year round. In contrast, Arctic algae require higher temperatures. With these abilities productivity of polar algae may equal or exceed productivity of temperature habitats. The interaction of several external factors, such as temperature, irradiance (PAR, UV-A and UV-B) and salinity were determined in two life stages of the Arctic kelp *Alaria esculenta* from Kongsfjord (Spitsbergen) by measuring variable chlorophyll fluorescence of PS II (Fredersdorf et al. 2009). Photosynthetic activity in sporophytes was strongly affected by temperature but hardly by radiation and salinity. In contrast to the adult macroscopic thalli, the germinating stages were more impaired: haploid zoospore germination was impaired by changes in temperature and/or salinity. Once the young thalli had developed, the kelps seem to be relatively tolerant to temperature (0°–21°C), UV stress (1 W/m^2) and reduced salinity (20, 28, 34) within species-specific limits which was supported by experiments in the Phaeophytes *Saccharina latissima* and *Laminaria solidungula* as well as the adult tetrasporophytes of the red alga *Devaleraea ramentacea* (Fredersdorf 2009). *Alaria esculenta* sporophytes were found to be affected by UV radiation but increasing temperature could mitigate the UV-induced damage. While most species tolerated hyposaline conditions, the deep-water *L. solidungula* was less tolerant. The chlorophyll *a* content was little affected by a combined temperature and salinity stress, but exposure to lower salinity resulted in higher concentrations of the photosynthetic antenna pigment fucoxanthin. The red alga *D. ramentacea* responded to increased UV stress by synthesising more MAAs, and in the Chlorophyte *Ulva lactuca* the concentration of the protective lutein was augmented by UV stress (Fredersdorf 2009). Germination of zoospores of the Phaeophytes *Laminaria digitata* and *Alaria esculenta* from Spitsbergen strongly depended on the temperature and light regime (Fig. 13.5) (Müller et al. 2008). Zoospores of both species germinated best at 7 and 12°V and hardly at 18°C. In *Alaria* additional UV-B had no marked effect, while in *Laminaria* it strongly reduced germination. These results show that environmental factors interact synergistically, but typically one factor prevails. The changing climate with increasing temperatures and reduced salinity, due to ice melting, has affected Arctic endemic species, such as *L. solidungula* more than temperate kelp species.

Rising temperatures in the Arctic leads to increased melting of ice and snow which also increases the load of sediments especially in fjords and coastal areas (Nygaard Markussen et al. 2014). This reduces the light availability for lower eulittoral and sublittoral macroalgae. On the other hand it protects algae, such as the kelp *Saccharina latissima*, from deleterious UV radiation especially in a high sediment-load environment (Roleda et al. 2008a).

With increasing temperatures salinity in polar regions decreases, due to increased melting, which is another stress factor which operates in conjunction with elevated temperatures and exposure to solar UV radiation. Macroalgae respond differently and species-specifically to this stress scenario. *Devaleraea ramentacea* tolerated large changes in salinity and was adapted to high UV radiation, while *Palmaria palmata* allowed only small changes in salinity and showed high mortality even at moderate hyposaline conditions; in addition, it is not very tolerant to excessive PAR and UV radiation (Karsten et al. 2003).

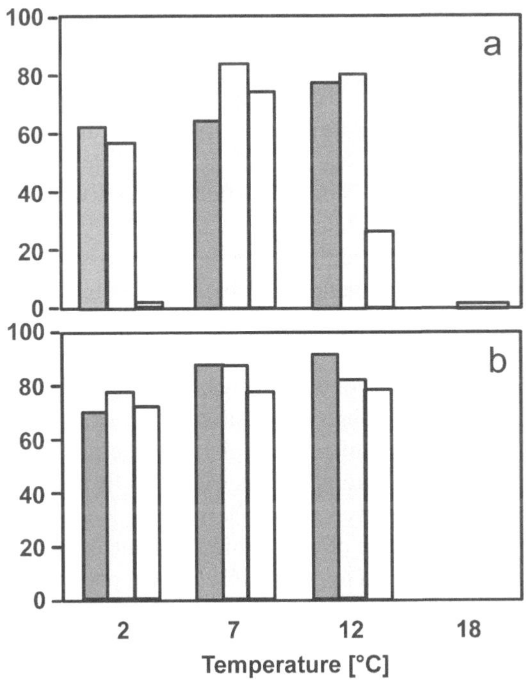

Fig. 13.5. Zoospore germination of *Laminaria digitata* (a) and *Alaria esculenta* (b) isolated from the Arctic Spitsbergen at four different temperatures (abscissa in °C) under PAR (left bars) PAR + UV-A (center bars) and PAR + UV-A and UV-B (right bars). Redrawn after (Müller et al. 2008).

Grazers show specific food preferences. Snails have been found to prefer UV-B sensitive Chlorophytes while they avoid leathery Rhodophytes which shifts the species composition in Antarctic phytobenthos communities (Zacher et al. 2007, Zacher and Campana 2008). In these communities grazing had a much more pronounced effect on community structure than UV radiation. Similarly grazers mitigated the effects of UV-B and temperature in the Chlorophyte *Ulva intestinalis* from Nova Scotia, Canada, while nutrient supplementation increased grazing pressure (Lotze and Worm 2002).

Conclusions and Ecological Consequences

Macroalgal communities are affected by several stress parameters linked to global climate change. While the individual effects of increasing temperature and UV exposure on algal growth, abundance, and community structure is fairly well understood (Müller et al. 2008), interactions with other abiotic factors and biological responses are substantially more complex (Harley et al. 2006). These factors include ocean circulation, changes in ocean chemistry and abundance of predators (Zacher et al. 2007, Molis et al. 2009, Browman 2017). For these reasons, it is important to study the combined effects of biotic and abiotic factors on algal communities. It is also of uttermost importance to do these studies in the field. Laboratory investigations are useful for understanding the underlying biochemical and physiological reactions of species to stress factors, but responses in the real habitat are often different, puzzling, and difficult to interpret (Häder et al. 2015, Gao and Häder 2017). This can be due to the wavelength distribution of artificial light sources which can be very different from natural solar radiation at the surface or filtered by the water column, including ratios of PAR:UV-A:UV-B. Early UV-B lamps and 'solar simulators' even contained UV-C radiation, while natural solar radiation at the earth surface never includes this wavelength band, which some researchers neglected to remove by long-pass cut-off filters (Rosette and Karin 1996). Ecological effects on algal communities are difficult to evaluate due to technical obstacles, limited accessibility of habitats and the vast areas of coastal ecosystems. In addition, strong effects of individual stress factors, such as increased solar UV-B radiation may be mitigated by adaptation over time or by species succession in microalgae and macroalgae (Wahl et al. 2004, Karsten et al. 2009). For this task the development has to be considered over several trophic levels in order to avoid misinterpretations of possible responses to key stress factors. An early and classical example was the observed positive effect of UV radiation on benthic diatoms in an artificial stream, which was difficult to interpret until it was found that UV had detrimental effects on natural grazers which finally resulted in the net growth of the primary producers (Bothwell et al. 1994). Considering several trophic levels are of importance since the effect of interacting stress factors in one level may affect the ecosystem at a different level by giving insensitive species an advantage (Xue et al. 2005, Bancroft et al. 2007). Changing environmental factors due to climate change may also alter zonation and colonization (Bischof et al. 2006) which would also affect epiphytes and grazers. While the existing results are a promising step towards understanding the effects of climate change with all its implied stress factors, more research is necessary to elucidate the scope of changes which may have negative or positive implications on the habitat, the food web and commercial exploitation of the aquatic ecosystems. However, it seems clear that increasing climate change is going to exert fundamental changes in polar regions because these are affected by the most extreme increases in temperature resulting in decreasing ice coverage, salinity, and increased sedimentation all of which will affect the macroalgae communities.

References

Abbasi, S. and T. Abbasi. The global ozone-depletion trends. pp. 101–111. *In*: 2017. Ozone Hole. Springer, New York.

Amado, A.M., J.B. Cotner, R.M. Cory, B.L. Edhlund and K. McNeill. 2015. Disentangling the interactions between photochemical and bacterial degradation of dissolved organic matter: amino acids play a central role. Microbial Ecology 69: 554–566.

Arrieta, J.M., E. Mayol, R.L. Hansman, G.J. Herndl, T. Dittmar and C.M. Duarte. 2015. Dilution limits dissolved organic carbon utilization in the deep ocean. Science 348: 331–333.

Athukorala, Y., S. Trang, C. Kwok and Y.V. Yuan. 2016. Antiproliferative and antioxidant activities and mycosporine-like amino acid profiles of wild-harvested and cultivated edible Canadian marine red macroalgae. Molecules 21: 119.

Bancroft, B.A., N.J. Baker and A.R. Blaustein. 2007. Effects of UVB radiation on marine and freshwater organisms: a synthesis through meta-analysis. Ecology Letters 10: 332–345.

Bischof, K. 2000. Effects of enhanced UV-radiation on photosynthesis of Arctic/cold-temperate macroalgae. Reports on Polar Research 375: 1–88.

Bischof, K., I. Gómez, M. Molis, D. Hanelt, U. Karsten, U. Lüder, M.Y. Roleda, K. Zacher and C. Wiencke. 2006. Ultraviolet radiation shapes seaweed communities. Reviews in Environmental Science and Biotechnology 5: 141–166.

Bischof, K., D. Hanelt and C. Wiencke. 2000. Effects of ultraviolet radiation on photosynthesis and related enzyme reactions of marine macroalgae. Planta 211: 555–562.

Bischof, K. and R. Rautenberger. 2012. Seaweed responses to environmental stress: reactive oxygen and antioxidative strategies. pp. 109–132 *In*: Wiencke, C. and K. Bischof. [eds.]. Seaweed Biology. Ecological Studies (Analysis and Synthesis). Springer, Berlin, Heidelberg.

Bothwell, M.L., D.M.J. Sherbot and C.M. Pollock. 1994. Ecosystem response to solar ultraviolet-B radiation: Influence of trophic-level interactions. Science 265: 97–100.

Britt, A.B. 1996. DNA damage and repair in plants. Annual Review of Plant Physiology and Plant Molecular Biology 47: 75–100.

Browman, H.I. 2017. Towards a Broader Perspective on Ocean Acidification Research. Oxford University Press.

Campana, G., K. Zacher, A. Fricke, M. Molis, A. Wulff, M.L. Quartino and C. Wiencke. 2009. Drivers of colonization and succession in polar benthic macro- and microalgal communities. Botanica Marina 52: 655–667.

de Oliveira, E.M., É.C. Schmidt, D.T. Pereira, Z.L. Bouzon and L.C. Ouriques. 2016. Effects of UV-B radiation on germlings of the red macroalga *Nemalion helminthoides* (Rhodophyta). Journal of Microscopy and Ultrastructure 4: 85–94.

DeSombre, E.R. 2000. The Experience of the Montreal Protocol: Particularly Remarkable, and Remarkably Particular. UCLA J. Envtl. L. & Pol'y 19: 49.

Douglass, A., R. Stolarski, M. Schoeberl, C. Jackman, M. Gupta, P. Newman, J. Nielsen and E. Fleming. 2008. Relationship of loss, mean age of air and the distribution of CFCs to stratospheric circulation and implications for atmospheric lifetimes. Journal of Geophysical Research: Atmospheres 113.

Dummermuth, A., U. Karsten, K. Fisch, G. König and C. Wiencke. 2003. Responses of marine macroalgae to hydrogen-peroxide stress. Journal of Experimental Marine Biology and Ecology 289: 103–121.

Fischetti, M. 2013. Deep heat threatens marine life. Scientific American 308: 92.

Flores-Molina, M.R., R. Rautenberger, P. Munoz, P. Huovinen and I. Gomez. 2016. Stress tolerance of the endemic antarctic brown alga *Desmarestia anceps* to UV radiation and temperature is mediated by high concentrations of phlorotannins. Photochemistry and Photobiology 92: 455–466.

Franks, P.J., M.A. Adams, J.S. Amthor, M.M. Barbour, J.A. Berry, D.S. Ellsworth, G.D. Farquhar, O. Ghannoum, J. Lloyd and N. McDowell. 2013. Sensitivity of plants to changing atmospheric CO_2 concentration: from the geological past to the next century. New Phytologist 197: 1077–1094.

Fredersdorf, J. 2009. Interactive abiotic stress effects on Arctic marine macroalgae - Physiological responses of adult sporophytes. PhD thesis, University of Bremen, Germany.

Fredersdorf, J., R. Müller, S. Becker, C. Wiencke and K. Bischof. 2009. Interactive effects of radiation, temperature and salinity on different life history stages of the Arctic kelp *Alaria esculenta* (Phaeophyceae). Oecologia 160: 483–492.

Fricke, A., M. Molis, C. Wiencke, N. Valdivia and A. Chapman. 2008. Natural succession of macroalgal-dominated epibenthic assemblages at different water depths and after transplantation from deep to shallow water on Spitsbergen. Polar Biology 31: 1191–1203.

Gao, K. and D.-P. Häder. Effects of ocean acidification and UV radiation on marine photosynthetic carbon fixation. pp. 235–250 *In*: Kumar, M. and P. J. Ralph [eds.] 2017. Systems Biology of Marine Ecosystems. Springer, Cham, Switzerland.

Gómez, I., A. Wulff, M.Y. Roleda, P. Huovinen, U. Karsten, M.L. Quartino, K. Dunton and C. Wiencke. 2009. Light and temperature demands of benthic algae in the polar regions. Botanica Marina 52.

Häder, D.-P., C.E. Williamson, S.-Å. Wängberg, M. Rautio, K.C. Rose, K. Gao, E.W. Helbling, R.P. Sinha and R. Worrest. 2015. Effects of UV radiation on aquatic ecosystems and interactions with other environmental factors. Photochemical & Photobiological Sciences 14: 108–126.

Halm, H., U.H. Lüder and C. Wiencke. 2011. Induction of phlorotannins through mechanical wounding and radiation conditions in the brown macroalga *Laminaria hyperborea*. European Journal of Phycology 46: 16–26.

Hansen, J., L. Nazarenko, R. Ruedy, M. Sato, J. Willis, A. Del Genio, D. Koch, A. Lacis, K. Lo and S. Menon. 2005. Earth's energy imbalance: Confirmation and implications. Science 308: 1431–1435.

Hansen, J. and M. Sato. 2016. Regional climate change and national responsibilities. Environmental Research Letters 11: 034009.

Harley, C.D., A. Randall Hughes, K.M. Hultgren, B.G. Miner, C.J. Sorte, C.S. Thornber, L.F. Rodriguez, L. Tomanek and S.L. Williams. 2006. The impacts of climate change in coastal marine systems. Ecology Letters 9: 228–241.

Hoffman, J.R., L.J. Hansen and T. Klinger. 2003. Interactions between UV radiation and temperature limit inferences from single-factor experiments. Journal of Phycology 39: 268–272.

Hofmann, D., S. Oltmans, W. Komhyr, J. Harris, J. Lathrop, A. Langford, T. Deshler, B. Johnson, A. Torres and W. Matthews. 1994. Ozone loss in the lower stratosphere over the United States in 1992–1993: Evidence for heterogeneous chemistry on the Pinatubo aerosol. Geophysical Research Letters 21: 65–68.

Hogg, A.E. and G.H. Gudmundsson. 2017. Impacts of the Larsen-C Ice Shelf calving event. Nature Climate Change 7: 540–542.

Hoyer, K., U. Karsten, T. Sawall and C. Wiencke. 2001. Photoprotective substances in Antarctic macroalgae and their variation with respect to depth distribution, different tissues and developmental stages. Marine Ecology Progress Series 211: 117–129.

Hoyer, K., U. Karsten and C. Wiencke. Inventory of UV-absorbing mycosporine-like amino acids in polar macroalgae and factors controlling their content. pp. 56–62 *In*: Huiskes, A. H. L., W. W. C. Gieskes, J. Rozema, R. M. L. Schorno, S. M. van der Vies and W. J. Wolff [eds.] 2003. Antarctic Biology in a Global Context. Backhuys Publ., Leiden, The Netherlands.

Hughes, K.A. and P. Convey. 2014. Alien invasions in Antarctica—is anyone liable? Polar Research 33: 22103.

Huovinen, P., J. Ramirez and I. Gomez. 2016. Underwater optics in sub-Antarctic and Antarctic coastal ecosystems. PLOS One 11: e0154887.

Huovinen, P.S., A.O.J. Oikari, M.R. Soimasuo and G.N. Cherr. 2000. Impact of UV radiation on the ealy development of the giant kelp (*Macrocystis pyrifera*) gametophytes. Photochemistry and Photobiology 72: 308–313.

IPCC. 2014. Climate Change 2014: Impacts, Adaptation, and Vulnerability. Part B: Regional Aspects. Contribution of Working Group II to the Fifth Assessment Report of the Intergovernmental Panel on Climate Change, 2014.

Ju, Q., H. Xiao, Y. Wang and X. Tang. 2015. Effects of UV-B radiation on tetraspores of *Chondrus ocellatus* Holm (Rhodophyta), and effects of red and blue light on repair of UV-B-induced damage. Chinese Journal of Oceanology and Limnology 33: 650.

Karentz, D., J.E. Cleaver and D.L. Mitchell. 1991. Cell survival characteristics and molecular responses of Antarctic phytoplankton to ultraviolet-B radiation. Journal of Phycology 27: 326–341.

Karsten, U., K. Bischof, D. Hanelt, H. Tüg and C. Wiencke. 1999. The effect of ultraviolet radiation on photosynthesis and ultraviolet-absorbing substances in the endemic Arctic macroalga *Devaleraea ramentacea* (Rhodophyta). Physiologia Plantarum 105: 58–66.

Karsten, U., A. Dummermuth, K. Hoyer and C. Wiencke. 2003. Interactive effects of ultraviolet radiation and salinity on the ecophysiology of two Arctic red algae from shallow waters. Polar Biology 26: 249–258.

Karsten, U. and A. Holzinger. 2014. Green algae in alpine biological soil crust communities: acclimation strategies against ultraviolet radiation and dehydration. Biodiversity and Conservation 23: 1845–1858.

Karsten, U., A. Wulff, M.Y. Roleda, R. Müller, F. Steinhoff, J. Fredersdorf and C. Wiencke. 2009. Physiological responses of polar benthic micro- and macroalgae to ultraviolet radiation. Botanica Marina 52: 639–654.

Lei, R., X. Tian-Kunze, M. Leppäranta, J. Wang, L. Kaleschke and Z. Zhang. 2016. Changes in summer sea ice, albedo, and portioning of surface solar radiation in the Pacific sector of Arctic Ocean during 1982–2009. Journal of Geophysical Research: Oceans 121: 5470–5486.

Lois, R. and B.B. Buchanan. 1994. Severe sensitivity to ultraviolet radiation in an *Arabidopsis* mutant deficient in flavonoid accumulation. II. Mechanisms of UV-resistance in *Arabidopsis*. Planta 194: 504–509.

Lotze, H.K. and B. Worm. 2002. Complex interactions of climatic and ecological controls on macroalgal recruitment. Limnology and Oceanography 47: 1734–1741.

Makarov, M. and G. Voskoboinikov. 2001. The influence of ultraviolet-B radiation on spore release and growth of the kelp *Laminaria saccharina*. Botanica Marina 44: 89–94.

McKenzie, R.L., P.J. Aucamp, A.F. Bais, L.O. Björn, M. Ilyas and S. Madronich. 2011. Ozone depletion and climate change: impacts on UV radiation. Photochemical & Photobiological Sciences 10: 182–198.

Meyer, K., A. Ridgwell and J. Payne. 2016. The influence of the biological pump on ocean chemistry: implications for long-term trends in marine redox chemistry, the global carbon cycle, and marine animal ecosystems. Geobiology 14: 207–219.

Molis, M., H. Wessels, W. Hagen, U. Karsten and C. Wiencke. 2009. Do sulphuric acid and the brown alga *Desmarestia viridis* support community structure in Arctic kelp patches by altering grazing, distribution patterns, and behaviour of sea urchins? Polar Biology 32: 71–82.

Msadek, R., G. Vecchi, M. Winton and R. Gudgel. 2014. Importance of initial conditions in seasonal predictions of Arctic sea ice extent. Geophysical Research Letters 41: 5208–5215.

Müller, R., C. Wiencke and K. Bischof. 2008. Interactive effects of UV radiation and temperature on microstages of Laminariales (Phaeophyceae) from the Arctic and North Sea. Climate Research 37: 203–213.

Müller, R., C. Wiencke, K. Bischof and B. Krock. 2009. Zoospores of three Arctic Laminariales under different UV radiation and temperature conditions: Exceptional spectral absorbance properties and lack of phlorotannin induction. Photochemistry and Photobiology 85: 970–977.

Nygaard Markussen, T., T. Joest Andersen, V. Brandbyge Ernstsen, M. Becker, C. Winter and B. Elberling. 2014. Redistribution and transport of melt-water supplied sediments and nutrients in Arctic fjords: The influence of flocculation in Disko Fjord, West Greenland. EGU General Assembly Conference Abstracts.

Pithan, F. and T. Mauritsen. 2014. Arctic amplification dominated by temperature feedbacks in contemporary climate models. Nature Geoscience 7: 181–184.

Poppe, F., R.A.M. Schmidt, D. Hanelt and C. Wiencke. 2003. Effects of UV radiation on the ultrastructure of several red algae. Phycological Research 51: 11–19.

Post, A. and A. Larkum. 1993. UV-absorbing pigments, photosynthesis and UV exposure in Antarctica: comparison of terrestrial and marine algae. Aquatic Botany 45: 231–243.

Rastogi, R.P. and A. Incharoensakdi. 2013. UV radiation-induced accumulation of photoprotective compounds in the green alga *Tetraspora* sp. CU2551. Plant Physiology and Biochemistry 70: 7–13.

Rastogi, R.P., Richa, A. Kumar, M.B. Tyagi and R.P. Sinha. 2010. Molecular mechanisms of ultraviolet radiation-induced DNA damage and repair. Journal of Nucleic Acids 2010: Article ID 592980.

Rautenberger, R. and K. Bischof. 2006. Impact of temperature on UV-susceptibility of two *Ulva* (Chlorophyta) species from Antarctic and Subantarctic regions. Polar Biology 29: 988–996.

Rautenberger, R., P. Huovinen and I. Gomez. 2015. Effects of increased seawater temperature on UV tolerance of Antarctic marine macroalgae. Marine Biology 162: 1087–1097.

Rautenberger, R., A. Mansilla, I. Gómez, C. Wiencke and K. Bischof. 2009. Photosynthetic responses to UV-radiation of intertidal macroalgae from the Strait of Magellan (Chile). Revista Chilena de Historia Natural 82: 43–61.

Richa, R.P. Sinha and D.-P. Häder. Physiological aspects of UV-excitation of DNA. pp. 203–248 *In*: Barbatto., M., A. C. Borin and S. Ullrich [eds.] 2015. Topics in Current Chemistry: Photoinduced Phenomena in Nucleic Acids II: DNA Fragments and Phenomenological Aspects. Springer, Berlin Heidelberg.

Roleda, M.Y., G. Campana, C. Wiencke, D. Hanelt, M.L. Quartino and A. Wulff. 2009a. Sensitivity of Antarctic *Urospora penicilliformis* (Ulotrichales, Chlorophyta) to ultraviolet radiation is life stage dependent. Journal of Phycology 45: 600–609.

Roleda, M.Y., D. Dethleff and C. Wiencke. 2008a. Transient sediment load on blades of Arctic *Saccharina latissima* can mitigate UV radiation effect on photosynthesis. Polar Biology 31: 765–769.

Roleda, M.Y., D. Hanelt and C. Wiencke. 2005. Growth kinetics related to physiological parameters in young *Saccorhiza dermatodea* and *Alaria esculenta* sporophytes exposed to UV radiation. Polar Biology 28: 539–549.

Roleda, M.Y., D. Hanelt and C. Wiencke. 2006a. Exposure to ultraviolet radiation delays photosynthetic recovery in Arctic kelp zoospores. Photosynthesis Research 88: 311–322.

Roleda, M.Y., D. Hanelt and C. Wiencke. 2006b. Growth and DNA damage in young *Laminaria* sporophytes exposed to ultraviolet radiation: implication for depth zonation of kelps on Helgoland (North Sea). Marine Biology 148: 1201–1211.

Roleda, M.Y., U. Lütz-Meindl, C. Wiencke and C. Lütz. 2009b. Physiological, biochemical, and ultrastructural responses of the green macroalga *Urospora penicilliformis* from Arctic Spitsbergen to UV radiation. Protoplasma 243: 105–116.

Roleda, M.Y., C. Wiencke and D. Hanelt. 2006c. Thallus morphology and optical characteristics affect growth and DNA damage by UV radiation in juvenile Arctic *Laminaria* sporophytes. Planta 223: 407–417.

Roleda, M.Y., C. Wiencke and U.H. Lüder. 2006d. Impact of ultraviolet radiation on cell structure, UV-absorbing compounds, photosynthesis, DNA damage, and germination in zoospores of Arctic *Saccorhiza dermatodea*. Journal of Experimental Botany 57: 3847–3856.

Roleda, M.Y., K. Zacher, A. Wulff, D. Hanelt and C. Wiencke. 2007. Photosynthetic performance, DNA damage and repair in gametes of the endemic Antarctic brown alga *Ascoseira mirabilis* exposed to ultraviolet radiation. Austral Ecology 32: 917–926.

Roleda, M.Y., K. Zacher, A. Wulff, D. Hanelt and C. Wiencke. 2008b. Susceptibility of spores of different ploidy levels from Antarctic *Gigartina skottsbergii* (Gigartinales, Rhodophyta) to ultraviolet radiation. Phycologia 47: 361–370.

Rosette, C. and M. Karin. 1996. Ultraviolet light and osmotic stress: activation of the JNK cascade through multiple growth factor and cytokine receptors. Science 274: 1194.

Sinha, R.P., S.P. Singh and D.-P. Häder. 2007. Database on mycosporines and mycosporine-like amino acids (MAAs) in fungi, cyanobacteria, macroalgae, phytoplankton and animals. Journal of Photochemistry and Photobiology B: Biology 89: 29–35.

Solomon, S. 1999. Stratospheric ozone depletion: A review of concepts and history. Reviews of Geophysics 37: 275–316.

Steinhoff, F.S., C. Wiencke, S. Wuttke and K. Bischof. 2011. Effects of water temperatures, UV radiation and low vs high PAR on phlorotannin content and germination in zoospores of *Saccorhiza dermatodea* (Tilopteridales, Phaeophyceae). Phycologia 50: 256–263.

Stroeve, J., T. Markus, L. Boisvert, J. Miller and A. Barrett. 2014. Changes in Arctic melt season and implications for sea ice loss. Geophysical Research Letters 41: 1216–1225.

Thomas, R., H. Graven, B. Hoskins and I.C. Prentice. 2016. What is meant by "balancing sources and sinks of greenhouse gases" to limit global temperature rise? Briefing note: 1–5.

Thompson, D.W.J., S. Solomon, P.J. Kushner, M.H. England, K.M. Grise and D.J. Karoly. 2011. Signatures of the Antarctic ozone hole in Southern Hemisphere surface climate change. Nature Geoscience 4: 741–749.

van de Poll, W.H., D. Hanelt, K. Hoyer, A.G.J. Buma and A.M. Breeman. 2002. Ultraviolet-B-induced cyclobutane-pyrimidine dimer formation and repair in Arctic marine macrophytes. Photochemistry and Photobiology 76: 493–500.

Wahl, M., M. Molis, A. Davis, S. Dobretsov, S.T. Dürr, J. Johansson, J. Kinley, D. Kirugara, M. Langer, H.K. Lotze, M. Thiel, J.C. Thomason, B. Worm and D.Z. Ben-Yosef. 2004. UV effects that come and go: a global comparison of marine benthic community level impacts. Global Change Biology 10: 1962–1972.

Waibel, A.E., T. Peter, K.S. Carslaw, H. Oelhaf, G. Wetzel, P.J. Crutzen, U. Pöschl, A. Tsias, E. Reimer and H. Fischer. 1999. Arctic ozone loss due to denitrification. Science 283: 2064–2069.

Williamson, C.E., R.G. Zepp, R.M. Lucas, S. Madronich, A.T. Austin, C.L. Ballaré, M. Norval, B. Sulzberger, A. Bais, R.L. McKenzie, S.A. Robinson, D.-P. Häder, N.D. Paul and J.F. Bornman. 2014. Solar ultraviolet radiation in a changing climate. Nature Climate Change 4: 434–441.

Xue, L., Y. Zhang, T. Zhang, L. An and X. Wang. 2005. Effects of enhanced ultraviolet-B radiation on algae and cyanobacteria. Critical Reviews in Microbiology 31: 79–89.

Zacher, K. 2014. The susceptibility of spores and propagules of Antarctic seaweeds to UV and photosynthetically active radiation—Field versus laboratory experiments. Journal of Experimental Marine Biology and Ecology 458: 57–63.

Zacher, K. and G. Campana. 2008. UV and grazing effects on an intertidal and subtidal algal assemblage: a comparative study. Ber. Polarforsch. Meeresforsch 571: 287–294.

Zacher, K., R. Rautenberger, D. Hanelt, A. Wulff and C. Wiencke. 2009. The abiotic environment of polar marine benthic algae. Botanica Marina 52: 483–490.

Zacher, K., A. Wulff, M. Molis, D. Hanelt and C. Wiencke. 2007. Ultraviolet radiation and consumer effects on a field-grown intertidal macroalgal assemblage in Antarctica. Global Change Biology 13: 1201–1215.

CHAPTER 14

Effects of Climate Change on Aquatic Bryophytes

*Javier Martínez-Abaigar** and *Encarnación Núñez-Olivera*

Aquatic Bryophytes

Bryophytes are 'true' plants, as pteridophytes and seed plants, because of the presence of an embryo and the alternation between gametophyte and sporophyte generations in the life cycle. Nevertheless, bryophytes are peculiar plants due to both structural and functional particularities. And within bryophytes, aquatic ones can also be considered peculiar bryophytes because of, again, special features affecting structure and functionality. The justification of this statement will be attempted in the following paragraphs.

Bryophytes are photosynthetic organisms important in plant evolution because they were the earliest diverging 'true' plants, having evolved from Zygnematophyceae green algae upon land colonization (Puttick et al. 2018). Several characteristics clearly differentiate bryophytes from the remaining plants (tracheophytes). The bryophyte life cycle consists of a dominant haploid green gametophyte, which is responsible for the vegetative functions and the production of gametes for sexual reproduction, and a diploid sporophyte growing attached to the gametophyte and specialized in spore production (Vanderpoorten and Goffinet 2009). An important ecological limitation for bryophyte sexual reproduction, and thus for genetic diversity, is that fertilization requires a continuous film of water between antheridia and archegonia. In many species, this limitation can be compensated by a great diversity of asexual vegetative propagation mechanisms (fragments of shoots or leaves, gemmae, bulbils, rhizoidal tubers, etc.). Bryophytes are structurally simple and lack the typical tracheophyte vascular tissues and organs (such as true roots, stems and leaves), but they are clearly

Facultad de Ciencia y Tecnología, Universidad de La Rioja, Madre de Dios 53, 26006 Logroño (La Rioja), Spain.
Emails: encarnacion.nunez@unirioja.es
* Corresponding author: javier.martinez@unirioja.es

separated from algae because of the presence of an embryo. Bryophytes are divided into three phylogenetic lineages: hornworts (always with a thalloid gametophyte), mosses (always with a gametophyte composed by rhizoids and primitive stems and leaves) and liverworts (with an either thalloid or leafy gametophyte). Sporophyte structure also differs between these three groups, whose evolutionary relationships remain to be completely elucidated (Morris et al. 2018, Puttick et al. 2018).

The main physiological features of bryophytes, which determine their small size, are poikilohydry, desiccation tolerance and limited transport of both mineral and organic substances within the gametophyte (Proctor 2014, Robinson and Waterman 2014). Their desiccation tolerance allows them to withstand continuous and frequent cycles (sometimes several cycles per day) of severe dehydration and subsequent rehydration. When exposed to the sun, they dry out and reach a dormant state with cessation of metabolism. Thus, their main growth occurs at low light levels when they are hydrated, responding immediately to intermittent favorable periods (Tuba et al. 2011a). Except the marine environment, bryophytes inhabit every kind of terrestrial and aquatic ecosystems, but they only generally prevail in tough habitats almost forbidden for other photosynthetic organisms, such as full sun-exposed bare rocks and soil crusts, heavily shaded forest soils, tree bark, peatlands, deep lakes and mountain streams.

Aquatic bryophytes show the same basic characteristics of bryophytes, but certain structural and ecophysiological features make them peculiar among bryophytes (Glime 2014). For example, aquatic bryophytes are frequently exposed to full sun in the hydrated state, because they occupy environments with a more stable access to water than those of terrestrial species. Therefore, the frequency of dehydration-rehydration cycles is much lower and desiccation periods can be from much longer (for example, in bryophytes from headwater streams) to non-existing (in bryophytes from deep lakes). The photosynthesis physiology, photoprotection mechanisms and growth rhythms of aquatic bryophytes must be adapted to these particular conditions.

The concept of 'aquatic bryophyte' is rather vague, although they can be broadly classified in a frame defined by water velocity and water depth (Vitt and Glime 1984). Regarding the first gradient, aquatic bryophytes range from limnophilous (living in standing waters) to rheophilous (living in running waters), whereas the second gradient classifies them into obligate aquatics, facultative aquatics and semiaquatic emergents.

Semiaquatic emergent bryophytes from wetlands, particularly species of the widely diversified *Sphagnum* genus that dominate peatlands, are crucial for the planet because these ecosystems sequestrate large quantities of carbon (Hájek 2014). Another important type of semiaquatic emergent bryophytes are those forming tufa, such as the mosses *Didymodon tophaceus*, *Eucladium verticillatum* and *Palustriella commutata*, and the liverworts *Southbya tophacea* and *Pellia endiviifolia* (Gimeno-Colera and Puche-Pinazo 1999, Pentecost and Zhang 2002). Bryophytes from wetlands and tufa are exposed to ambient sunlight, but at the same time they are permanently hydrated from the base of their shoots (apart from the water they can receive from precipitation), thus avoiding desiccation and adopting a quite different survival strategy to most bryophytes. Similar considerations can be applied to floating bryophytes (pleustophytes), such as the liverworts *Riccia fluitans* and *Ricciocarpus natans*. Facultative aquatic bryophytes is an ill-defined group of mainly mesophilous/hygrophilous species growing on soils or rocks that can be submerged during spates (for example, inhabiting stream banks), but otherwise are emerged and exposed to frequent cycles of desiccation-rehydration, like terrestrial species.

Finally, obligate aquatic bryophytes (OAB) are perennial species which stay submerged in flowing or stagnant water for most or the whole year. They play relevant roles in primary production, nutrient cycles and food webs, supporting periphyton and providing a refuge, and occasionally direct food, for a range of consumers (Bowden et al. 1999). In streams and rivers, OAB are represented in temperate European areas by different rheophilous mosses (mainly *Brachythecium rivulare*, *B. plumosum*, *Cinclidotus* spp., *Dichelyma falcatum*, *Fissidens crassipes*, *F. fontanus*, *F. grandifrons*, *Fontinalis* spp., *Hygroamblystegium tenax*, *H. fluviatile*, *Hygrohypnum duriusculum*, *H. ochraceum*, *Hyocomium armoricum*, *Leptodictyum riparium*, *Palustriella falcata*, *Platyhypnidium riparioides*, *Racomitrium aciculare*, *Schistidium agassizii* and *S. rivulare*) and liverworts (for example, *Aneura* spp., *Chiloscyphus polyanthos*, *Jungermannia atrovirens*, *J. exsertifolia* subsp. *cordifolia*, *J. pumila*, *Marsupella emarginata*, *Porella cordaeana*, *P. pinnata* and *Scapania undulata*) (Martínez-Abaigar and Ederra 1992, Muotka and Virtanen 1995, Vieira et al. 2012). Other species representing different biogeographic areas are mentioned by Vitt and Glime (1984), Suren (1996) and Shevock et al. (2017). OAB are almost absent in the tropics, mainly because of high water temperatures. Factors determining the bryophyte community composition in flowing waters are water chemistry (pH, alkalinity, inorganic carbon forms, nutrients and pollutants), temperature, current velocity, discharge, turbulence, water level fluctuation, light, turbidity and streambed stability. Bryophytes in these systems are exposed to important stress and disturbance factors, such as seasonal desiccation, abrasion by turbulent water and suspended solids, substratum movement, cold, nutrient limitation in soft waters, CO_2 limitation in the stagnant parts of alkaline streams, high photosynthetic and UV irradiances in unshaded high-altitude streams, and diaspore difficulties to attaching to new substrates (Glime 2014). This kind of bryophytes prefer small mountain streams to large rivers. On the other hand, dominant OAB in stagnant waters, mainly lakes, are usually limnophilous mosses belonging to the genera *Calliergon*, *Drepanocladus*, *Hamatocaulis*, *Sarmentypnum*, *Scorpidium*, *Straminergon*, *Warnstorfia* and *Fontinalis*, most of them included in the debated and complex families Calliergonaceae and Amblystegiaceae (Vanderpoorten et al. 2002). In these systems, once proved that alkalinity values are low and thus appropriate for bryophyte growth (Vestergaard and Sand-Jensen 2000), bryophytes typically grow permanently submerged, usually at medium or high depths (up to 140 m), to avoid desiccation (Ignatov and Kurbatova 1990, De Winton and Beever 2004, Wagner and Seppelt 2006). The main factors determining the composition of bryophyte communities from lakes are water chemistry (pH, conductivity, alkalinity), water transparency and human influence. Adverse factors for OAB in lakes are waterlogging, cold water, low irradiances in the deep zones, high hydrostatic pressure, and abrasion along the shores.

In general, rheophilous and limnophilous bryophytes may show different structural and physiological adaptations (Vitt and Glime 1984, Glime 2014, Shevock et al. 2017), although some species can grow in both flowing and stagnant waters. Rheophilous bryophytes often show stiff, wiry stems adapted to emersion and desiccation, being able to develop new shoots from apparently dead material. Leaves frequently are pluristratose, with differentiated margins and strong nerves, and shoots often form flow-resistant dense patches. On the other hand, limnophilous bryophytes usually have flexuose soft stems, and unistratose leaves frequently lacking nerve and without specialized margins. Despite these structural contrasts, the prevalence of bryophytes in streams and lakes is based on their economy in production, which in turn relies on their simple structure, and their physiological tolerance to adverse factors, such as cold and mineral nutrients scarcity. At the same time, permanently submerged bryophyte populations can take advantage of the buffering capacity of water with respect to the influence of environmental factors, avoiding for example the incidence of extreme temperatures.

Climate Change and Aquatic Systems

Climate change is a global-scale multi-faceted process which involves different anthropogenic environmental changes, such as global warming, increased CO_2 concentrations, enhanced solar UV-B radiation due to both stratospheric ozone depletion and reduced cloudiness, and increasing pollution of water, air and soils. Spatially and temporally heterogeneous changes in precipitation patterns and the frequency of severe weather events may also occur. All these aspects can interact producing multiplied effects on organisms. Despite being global, climate change manifests many times at more reduced regional or local scales, affecting naturally unstable ecosystems, such as rivers and streams.

The impacts of climate change on aquatic systems are more difficult to predict than those on terrestrial systems (Calow 1998). In this line, Häder et al. (2014) reviewed the effects of climate change on the productivity of aquatic primary producers, emphasizing that the responses are very variable in the different aquatic ecosystems across latitudes. For example, global warming can induce both positive (an increase in primary production) and negative (a decrease in photosynthesis due to melting of glaciers and accumulation of particles in the water column) effects in different types of polar waters, causing simultaneous negative effects (a decrease in primary production due to excess heat) in tropical waters. In addition, the interaction among climate change factors varies over the aquatic systems, and must be considered to make more reliable predictions on the effects on primary producers and the whole aquatic food web.

The main effects of climate change on freshwater aquatic systems affect water temperature, water level, solar radiation (including photosynthetic and UV wavelengths), CO_2 availability, and water nutrients and pollutants. The most frequently predicted trends regarding these variables are increases in water temperature, solar radiation across the water column, dissolved CO_2 (with the consequent water acidification), and diverse types of pollution (including eutrophication), and a decrease in water level. Obviously, exceptions to these general trends can occur, for example regarding water levels (see below). To assess the effects of climate change in a more global and realistic way, not only the responses of the different species to each individual factor should be considered, but also the interactions between the diverse factors, the shifts in the species composition of the communities, and the ecosystem effects (for example, food webs).

Climate Change and Bryophytes

The effects of climate change on bryophytes have been reviewed several times (Gignac 2001, 2011, Tuba et al. 2011b, He et al. 2016). In general, bryophytes are considered to be more responsive than vascular plants to climate change (Scarpitta et al. 2017), and have been identified as the taxonomic group most at risk from a reduction in distribution areas, although some species could have opportunities for range expansion (Pearce-Higgins et al. 2017). Thus, bryophytes would be good bioindicators of climate change because of their rapid responses (Gignac 2011), either in a positive or negative way, although this may not occur in all situations (Vanneste et al. 2017). In addition, they should not be treated as a homogeneous functional group because responses are species-specific. Much information produced on the topic of bryophytes and climate change is indirect, or was obtained in experiments performed under laboratory conditions or in short-term field studies. Thus, more specific and realistic long-term studies are badly needed. Despite this, important information obtained in terrestrial environments and wetlands can be highlighted.

Terrestrial bryophytes from different environments and geographic areas have been studied regarding the effects of climate change. In the Arctic tundra, summer warming will lead to a reduction in moss abundance (Callaghan et al. 2004). In alpine snowbeds, bryophytes will be replaced by adjacent shrubs and boreal species (Björk and Molau 2007). In subalpine Tibetan vegetation, the cover of bryophytes exposed to artificial nitrogen deposition and warming varied in a species-dependent manner (Sun et al. 2017). In Antarctica, Bramley-Alves et al. (2014) pointed out that the dominance of bryophytes may be threatened by climate change, but increasing temperature improved both moss growth (Amesbury et al. 2017) and sexual reproduction (Casanova-Katny et al. 2016). Antarctic moss archives show there exists a complex interaction between several factors sensitive to climate change (length of growing season, reduction in permanent snow cover, substrate stability, wind speed, nutrient availability and water supply), and thus predictions on bryophyte performance are difficult (Robinson and Erickson 2015, Royles and Griffiths 2015). Apparently, the extent of the hydrated (active) period was a more important influence on photosynthetic performance than temperature itself. In temperate broad-leaved forests, a dramatic bryophyte species turnover has taken place since 1900, primarily driven by the increased atmospheric nitrogen and temperature, leading to the introduction of more eutrophication-tolerant and thermophilic mosses and to a decline of liverworts (Dittrich et al. 2016). In boreal forests and temperate and tropical mountains, global warming is expected to shorten the periods of bryophyte metabolic activity (due to an increase in desiccation), which can negatively affect survival and bryophyte diversity (He et al. 2016). In dryland biocrust mosses, Coe et al. (2014) found restrictions on suitable microhabitats, limited sexual reproduction, reduced growth and establishment of young shoots, and replacement by shrubs. Overall, the effects of climate change on terrestrial bryophytes will depend on the species, the environment and the geographical area considered.

The effects of climate change on semiaquatic emergent *Sphagnum* species from peatlands have received much attention, due to their ecological relevance (Tuba et al. 2011b, Flanagan 2014, Hájek 2014). Overall, climate change factors (realistic increases in temperature, CO_2, UV radiation or N supply) may not cause substantial negative effects on photosynthesis and biomass of peatland bryophytes (Hájek 2014, Deane-Coe et al. 2015), and even some models show that, at least in North America, the suitable climatic area for *Sphagnum* peatland could expand because projected warming would be balanced by the increase in precipitation (Oke and Hager 2017). Model simulations taking into account several factors of climate change (temperature, precipitation, N-deposition and atmospheric CO_2) confirm predictions of little change in species composition, because the increase in vascular plants due to warming would be counteracted by the increase in *Sphagnum* due to the CO_2 increase (Heijmans et al. 2008). Nevertheless, some negative consequences of climate change on peatlands have also been predicted: a migration northward (Gignac et al. 1998), a shift in species abundances due to the different competition ability of each species (Breeuwer et al. 2008), and even *Sphagnum* death when exposed to high temperature (30°C) and moderate UV irradiance (1.30 W m^{-2}), at least under laboratory conditions (Cardona-Correa et al. 2015).

Terrestrial, emergent and facultative aquatic bryophytes mostly (or completely) respond to the consequences of climate change occurring in the atmospheric environment, because they are exposed to air temperature, atmospheric CO_2 (although in peatlands CO_2 can also be supplied from respiration in the underlying peat) and ambient sunlight (not filtered by the water column). Hence, they do not properly represent models to assess the effects of climate change on the hydrosphere. For these reasons, OAB (which stay submerged in streams and lakes for most or the whole year) will preferentially be considered in the present study. Indeed, these are the aquatic bryophytes that most reliably reflect the effects of climate change on water characteristics, such as water temperature and CO_2 levels, levels of water nutrients and pollutants, or the penetration of radiation in the water column. Thus, they deserve specific research in the context of climate change, because their responses to this process can be different to those shown by terrestrial or other types of aquatic bryophytes. In this sense, only Glime (2011) monographically studied the effects of climate change on, preferentially, OAB.

Climate Change and Obligate Aquatic Bryophytes (OAB)

Temperature

The relationship between air and water temperature will determine the responses of aquatic systems (and the organisms inhabiting them) to global warming. In streams, water temperature increases between 0.6–1.0°C for every 1°C increase in air temperature (Morrill et al. 2005). In lakes, temperature varies with depth in a more complex way than in streams, because of stratification and vertical mixing processes, and global warming can modify this thermal structure in different ways. The direct effects of increased temperature in aquatic systems will depend on the thermal tolerance of each species and the optimum temperatures for photosynthesis, respiration, and growth (Calow 1998). The levels of dissolved oxygen decrease with increasing water temperature, threatening the respiratory metabolism of aquatic species. Increased temperature can influence not only the basic physiology of the organisms, but also the distribution and abundance of the species (Vanderpoorten and Goffinet 2009).

OAB maintain a homogeneous temperature with the surrounding water, which acts as a temperature buffer, avoiding extreme temperatures and creating a much more stable environment for submerged bryophytes than that experienced by terrestrial bryophytes, which are exposed to air temperatures (but they desiccate when exposed to high temperatures, thus avoiding heat stress). Aquatic bryophytes are adapted to low temperatures, and show optimum photosynthesis and growth at around 15°C (Glime 2011, 2014). Under higher temperatures, respiration rapidly exceeds photosynthesis and growth ceases. In contrast, they show modest but still significant growth at low temperatures (down to 1.5°C) under field conditions (Beaucourt et al. 2001). Optimum temperatures depend on the species considered. In Icelandic spring-fed streams, with temperatures ranging from 7.1 to 21.6°C, the moss *Fontinalis antipyretica* dominated the bryophyte communities in the warmest streams, whereas the liverwort *Jungermannia exsertifolia* was only found in cold streams (Gudmundsdottir et al. 2011). As a methodological precaution, photosynthesis rates should not be used as a proxy of growth rates, because photosynthesis is usually measured, even in the field, using short incubation periods and sometimes unnatural conditions of light and carbon source, and these conditions are not adequate to evaluate real growth. Unfortunately, real growth measurements in OAB under field conditions are scarce (Kelly and Whitton 1987, Beaucourt et al. 2001, Li et al. 2009, Guo et al. 2013, Riis et al. 2014, 2016), mainly because these measurements require long-term experiments due to their slow growth. Using laboratory measurements, Davey and Rothery (1997) concluded that hydric Antarctic bryophytes would increase their photosynthesis and respiration rates, and their total productivity, under a climate-change-derived increase in temperature, because they are limited by cold.

Seasonality of OAB growth, measured under totally natural field conditions, may give clues about the real responses of these bryophytes to temperature and may help predict the effect of water warming. Unfortunately, there are few data of this kind. In temperate streams, maximum growth usually occurs in spring, but some growth can occur throughout the year (Kelly and Whitton 1987, Beaucourt et al. 2001). In Arctic pools, summer daily growth was 10-fold the winter daily growth (Guo et al. 2013), while in an Antarctic lake it was 5-fold (Li et al. 2009). Growth may be limited by low temperatures in aquatic systems of high latitudes and/or altitudes, and in these cases water warming (until around 15°C) could be positive. Any increase higher than this threshold would lead to negative consequences. In high-Arctic lakes, the aquatic mosses were not affected by temperature variation (with maximum water temperatures in summer of 10.8°C), probably due to the much lower temperature variation in water than in terrestrial ecosystems (Riis et al. 2014). Other factors were more influential on the control of bryophyte production in these ecosystems.

Regarding changes at a community level, higher temperatures might lead to the replacement of dominant bryophytes (*Scorpidium scorpioides*, *Drepanocladus trifarius* and *Sarmentypnum tundrae*) with southern species of higher plants in high-Arctic lakes (Riis et al. 2016).

In summary, the effects of increased water temperatures on OAB may be diverse, depending on the aquatic system (flowing vs. stagnant waters), the biogeographic region (in cold regions growth can increase but not always, and in temperate regions growth could decrease if temperatures go up beyond 15°C) and the species-specific tolerance to temperature. Given the buffering capacity of water with respect to temperature, and the consequent narrow thermic range that permanently submerged OAB experience throughout the year, the direct consequences of global warming on these organisms seem to be little dramatic. In this line, changes in species distribution (both expansions and reductions) attributed to global warming are restricted, in the vast majority of cases, to terrestrial species (Frahm and Klaus 2001, Bates and Preston 2011, Pócs 2011). Among OAB, only *Fissidens rivularis* has been found to change (expand) its range, probably because of the increase in winter temperature (Frahm and Klaus 2001). Some proven local extinctions of OAB are more probably due to habitat destruction than to climate change (Martínez-Abaigar and Núñez-Olivera 1996). On the other hand, variations in communities composition have been predicted but not fully confirmed in the field. Little long-term specific research has been performed on these matters, and thus much work remains to be done.

Water Level and Flow

The temperature increase associated to climate change will not only have direct but also indirect effects on OAB, mediated by changes in the hydrological regime (water level and water flow) derived from variations in precipitation, evaporation, and snow and ice melting. This indirect influence of global warming on aquatic systems may affect OAB more strongly than the mere increase in water temperature, and it should be taken into account to more adequately evaluate and predict the effects of climate change on these organisms.

In areas where less precipitation and more evaporation have been predicted (for example, Mediterranean climate zones), situations of low water levels and flows will be more frequent and long, impacting negatively on OAB. At the first moment, this will occur because of decreasing turbulence and free CO_2 availability, which is needed by OAB for photosynthesis (Glime 2011; see below). Then, longer periods of emersion and desiccation, with drastic reductions or complete cessation of growth, will impede OAB development. These processes would be associated to increasing temperature and levels of photosynthetic and UV irradiances while bryophytes desiccate. Even under these severely unfavorable conditions, some OAB (such as the mosses *Fontinalis antipyretica* and *Platyhypnidium riparioides*) can survive taking advantage of extremely desiccation-tolerant stems that, despite looking black and dead, can resprout and produce new healthy branches under subsequent rehydration. Additional physiological mechanisms of desiccation tolerance can be developed in this kind of species if dehydration rate is slow (Cruz de Carvalho et al. 2012). In contrast, less desiccation-tolerant species (like the softer aquatic liverworts of the genus *Jungermannia*) may die under these severe conditions. In the long-term, longer periods of low flow or complete dryness of water courses, or a higher frequency of desiccation-rehydration cycles of bryophytes (Glime 2011, 2014), can lead to changes in the species composition of the communities. Firstly, there could be an increasing predominance of aquatic but desiccation-tolerant mosses, which would have a higher competitive advantage than, for example, aquatic liverworts. Finally, aquatic communities would disappear and would be replaced by non-aquatic bryophytes typically growing on soils and even herbs or shrubs.

In contrast, water level can increase in lakes and rivers located in alpine and polar areas due to glaciers melting, with difficultly predictable consequences on water temperature, transparency and radiation penetration, and equally unpredictable effects on OAB. In both Mediterranean and alpine/polar environments, flooding events associated with climate change may be more frequent and destructive, with negative consequences for OAB due to breaking and drifting of bryophyte patches, habitat destruction, unstable substrata (remember the traditional saying 'rolling stones gather no moss') and more severe difficulties of diaspores to anchor on appropriate places. Nevertheless, seasonal changes in water flow in forested headwater streams, measured during two consecutive years, did not negatively affect the cover of two aquatic bryophytes (the moss *Leptodictyum riparium* and the liverwort *Porella pinnata*), and even the liverwort cover increased throughout the course of the study (Roberts et al. 2007). In German mountain streams, floods reduced biomass but did not affect the basic floristic structure (Tremp et al. 2012).

Photosynthetic Radiation

Photosynthetic radiation reaching organisms in aquatic systems is modified by diverse processes associated to climate change, such as variations in the riparian canopy, cloudiness, water level, turbulence, extent and duration of ice/snow cover, and concentration of suspended and dissolved materials (including organisms and dissolved organic carbon, DOC). Underwater radiation is much more easily measured in stagnant than in flowing waters, where turbulence can strongly alter the radiation climate (Frost et al. 2006).

Predictions of changes in photosynthetic radiation climate in freshwater systems are controversial. A decrease of radiation levels due to increases in precipitation, cloud cover and duration of snow cover periods has been anticipated in high-Arctic lakes (Riis et al. 2014). In this case, the annual net primary production of OAB would decrease. A similar response was predicted in a Danish lake due to increased DOC (Schwarz and Markager 1999). In addition, declining water clarity has been identified as a factor threatening OAB inhabiting deep lakes in New Zealand. However, increased radiation levels could take place in other lakes due to climate warming and consequent stronger DOC degradation and earlier disappearance of the ice and snow cover. Light-limited OAB could take advantage of these changes.

UV Radiation

The effects of UV radiation on underwater organisms depend on the dose received, which in turn is controlled by similar factors to those mentioned for photosynthetic wavelengths, especially the concentrations of dissolved and particulate organic matter, the proximity and extent of runoff from terrestrial ecosystems, and the depth and mixing processes in waters (Bais et al. 2018). The effects of UV-B (280–315 nm) and UV-A (315–400 nm), the two UV fractions reaching the biosphere, on photosynthetic organisms may be quite different. UV-A is the major component of UV solar radiation (around 95%), but its specific effects are much less known than those of UV-B (Verdaguer et al. 2017), probably because it is less harmful and research has been less profound due to the lack of relation with the stratospheric ozone depletion. In contrast, UV-B is specifically absorbed by stratospheric ozone and has traditionally been considered as a harmful environmental factor, although nowadays it is rather contemplated as a general regulator mostly inducing acclimation responses (Jansen and Bornman 2012). In aquatic systems, both UV-A and UV-B wavelengths are affected by other climate change factors than stratospheric ozone levels, such as variations in cloudiness, aerosols, DOC levels, surface reflectivity and duration of periods of snow/ice cover. Taking into account all these factors, some regions will likely receive enhanced UV levels (for example the tropics, where present UV levels are already intense) while others will receive reduced levels (polar latitudes, where current UV levels are generally less intense, and, to a lesser extent, mid-latitudes) (Williamson et al. 2014).

Within the context previously described, it is difficult to predict the effects of changing UV levels on OAB, because of the complex interaction between ozone reduction (and potential recovery) and other climate change factors. In addition, most studies to date were performed under laboratory conditions, and thus results cannot directly be extrapolated to the field (Martínez-Abaigar and Núñez-Olivera 2011). No study has been conducted on OAB from alpine lakes, which is surprising because these ecosystems have very transparent waters and thus organisms inhabiting them receive some of the highest UV levels in the world (Häder et al. 2011). Only Conde-Álvarez et al. (2002) studied the effects of UV radiation on OAB from stagnant waters (the thalloid liverwort *Riella helicophylla* from a shallow saline lake), concluding that solar UV merely caused some transitory and reversible damage to photosynthesis. The remaining studies have been performed in OAB from mountain streams. Field experimental studies are scarce and non-conclusive (Rader and Belish 1997), probably due to the inherent variability of the environmental conditions that OAB experience in these systems. Field observational studies conducted over three years in the aquatic liverwort *Jungermannia exsertifolia* subsp. *cordifolia*, showed that the main UV-absorbing compound (the hydroxycinnamic acid *p*-coumaroylmalic acid) varied seasonally in response to solar UV-B levels, protecting the liverwort from any significant physiological damage (assessed by chlorophyll fluorescence variables and the presence of thymine dimers in DNA). Thus, this liverwort is considered to be well-acclimated and tolerant to UV radiation. A similar conclusion was derived from another field study (Monforte et al. 2015a) in which 17 populations of the same species from mountain streams were analyzed. In addition, herbarium samples of this species have successfully been used to biomonitor the retrospective changes in UV radiation which occurred in northern and Mediterranean Europe (Otero et al. 2009, Monforte et al. 2015b).

Laboratory studies have been useful to characterize the responses of OAB to UV (mainly UV-B) radiation. UV stress may be preferentially indicated by a decrease in the maximum quantum yield of photosystem II (F_v/F_m), chlorophyll *a*/*b* ratio, and net photosynthesis rates. The effects depend not only on the species considered, but also on interacting environmental factors (temperature, presence of heavy metals) and the previous acclimation of the samples to sun or shade conditions (light history), low or high altitude, etc. (Martínez-Abaigar and Núñez-Olivera 2011). The basic responses to UV are similar in OAB and other photosynthetic organisms, from phytoplankton to seed plants. This is not surprising, because UV photoreception (at least in the green evolutionary lineage: Soriano et al. 2018), UV targets, UV damage, and protection and repair mechanisms, are quite similar. In general, the effects of realistic UV enhancements (analogous to those predicted as a consequence of ozone depletion) are not dramatic for most studied OAB, although they are structurally simple and lack the UV protections typical of higher plants, such as hairs, thick cuticles and epidermis. Due to this lack of structural protections, it is not surprising that the relative UV tolerance of OAB has been attributed to chemical protection, specifically the accumulation of UV-absorbing compounds, mainly of phenolic nature. In addition, the desiccation tolerance of many OAB may provide additional UV tolerance because both processes involve similar defense mechanisms (Takács et al. 1999).

Interestingly, the location of UV-absorbing compounds in different bryophyte cell compartments (mainly in vacuoles and cell walls) can influence their preferential protective role as antioxidants and/or UV screens (Fabón et al. 2010, 2012). In addition, location of these compounds serves to differentiate the two main evolutionary bryophyte lineages, since mosses show higher levels of cell wall-bound than vacuolar compounds, whereas liverworts show rather the contrary (Monforte et al. 2018). Given that the former are better UV protectants than the latter (Clarke and Robinson 2008, Robinson and Waterman 2014), mosses would better tolerate UV radiation than liverworts, which matches well with the general preference of mosses by more sun-exposed habitats than liverworts. Nevertheless, the UV-absorbing compounds of aquatic liverworts respond to UV more reliably than those of mosses, probably because the soluble vacuolar compounds are more UV-reactive than the relatively immobilized cell wall-bound compounds.

Overall, it is already clear that OAB are not as strongly UV-sensitive as could be anticipated considering their structural limitations, and that many of the species studied may suffer diverse physiological damage but can acclimate well to high UV levels through protection and repair mechanisms, mainly the accumulation of UV-absorbing compounds and the activity of DNA repairing systems. This would lead to a certain UV tolerance. Nevertheless, OAB responses to UV are still poorly understood, and thus further study is recommended. In particular, long-term realistic field manipulative experiments, together with experiments exploiting natural UV gradients along latitude, altitude or water depth, are needed to better understand the effects of current UV levels, and to predict the effects of potentially higher levels derived from the variations in stratospheric ozone and other factors of climate change. In addition, the effects of enhanced UV on the species composition of OAB communities and the ecosystem processes should be studied.

CO_2

Several forms of dissolved inorganic carbon (DIC) coexist in waters: free gaseous CO_2, bicarbonate (HCO_3^-), and carbonate (CO_3^{2-}). CO_2 reacts with water to form carbonic acid (H_2CO_3), which rapidly dissociates into bicarbonate (HCO_3^-) and carbonate (CO_3^{2-}) anions (cf. Chapter 3, this volume). Their relative proportions depend on chemical (pH and alkalinity), physical (type of water mass, depth, turbulence), and biological (photosynthesis and/or respiration of organisms) factors. Most epicontinental waters have pH values between 6.5 and 8.5, where HCO_3^- predominates. At pH 6.5, half of DIC is in the form of HCO_3^- and the other half is a mixture of CO_2 and H_2CO_3, whereas at pH 8.0 HCO_3^- represents 97% of the total DIC, while the remaining 3% is composed of H_2CO_3 and CO_3^{2-} (Margalef 1982). The increase in atmospheric CO_2 derived from climate change will shift the DIC equilibrium in waters mediated by a pH decrease, increasing the concentration of CO_2 and the relative proportion of CO_2 to HCO_3^- (Häder et al. 2011, Raven and Colmer 2016). Typically, plant responses to CO_2 enrichment include increases in photosynthesis, growth, total biomass and tissue C/N ratio, although long-term responses of photosynthesis may be both up- or down-regulated. Community and ecosystem responses can also be found, such as changes in species composition, nutrient cycling, and decomposition (Short and Neckles 1999).

OAB can use free CO_2 as a carbon source for photosynthesis, but most evidences show they can use HCO_3^- with low or no efficiency (Bain and Proctor 1980, Glime 2014). The low efficiency of HCO_3^- use would be the reason why OAB disappear in alkaline stagnant waters when depth increases, being replaced by other macrophytes able to use HCO_3^- (Martínez-Abaigar and Ederra 1992). In general, the access to free CO_2 is an important limiting factor for OAB photosynthesis, also constraining their competition ability. This is counteracted by the structural simplicity of bryophytes, whose mostly unistratose leaves may minimize internal diffusion resistance. Moreover, many OAB are rheophilous and inhabit fast and turbulent waters, which improves the access to free CO_2. Rheophilous species are exposed to stronger drag and mechanical damage, but adopt protective strategies, such as firm anchorage to the substrate, thick leaves and margins, and growth in dense patches (Vitt and Glime 1984). Limnophilous species can obtain CO_2 in acidic waters and/or from sediments (Maberly 1985), and most lakes worldwide are CO_2 supersaturated, indicating that lakes are sources rather than sinks of atmospheric CO_2 (Cole et al. 1994).

In summary, increasing atmospheric CO_2 would be expected to favor OAB due to increasing CO_2 levels and the relative proportion of CO_2 to HCO_3^- in freshwaters. Along this line, submerged *Sphagnum* species show an increase in growth rate in response to high CO_2 (Raven and Colmer 2016). However, long-term OAB growth under increased CO_2 can be higher, lower than or remain the same as that at present CO_2 (Raven and Colmer 2016). Thus, it would be an oversimplification to conclude that the CO_2 increase derived from climate change will directly be projected in an increased OAB growth, and other internal factors (Rubisco affinity, source-sink relationships, signaling and regulatory pathways of acclimation), and environmental limitations should be considered to make more reliable predictions.

Mineral Nutrients and Water Pollution

Freshwater systems are affected by diverse human activities leading to alterations in water quality, including increased levels of mineral nutrients, pollutants or acidity. OAB, especially those living permanently submerged, are good biomonitors of water chemistry and water pollution, including nitrogen, phosphorus, heavy metals, radioactive isotopes, and organic contaminants (for example, polycyclic aromatic hydrocarbons), because of their efficient absorption, capacity to integrate pollutant levels in the temporal scale, and easy manipulation (Martínez-Abaigar et al. 1993, 2002, García-Álvaro et al. 2000, Debén et al. 2015, 2017). In particular, OAB can absorb relatively high amounts of heavy metals because of the peculiar composition of their cell walls, enriched in uronic acids, which allow to immobilize these contaminants avoiding toxicity for the cells.

The relationships between climate change factors and the water levels of nutrients and pollutants may sometimes diffuse, but certainly exist. Higher temperatures may increase decomposition in terrestrial ecosystems, and increased nutrient run-off to streams and lakes can lead to a higher nutrient availability. This is predicted to increase moss primary production in high-Arctic lakes, which currently seems to be nutrient-limited (Riis et al. 2014, 2016). Similar findings were indicated in Arctic streams subjected to artificial phosphorus or nitrogen fertilization, although the responses were species-specific (Finlay and Bowden 1994, Arscott et al. 1998, Gudmundsdottir et al. 2011). However, slow-growing moss communities can be displaced in the long-term by other aquatic macrophytes more adapted to high nutrient levels. The increased atmospheric nitrogen deposition in the terrestrial environment, a phenomenon related to global change, also points in this direction (Arróniz-Crespo et al. 2008, Dittrich et al. 2016, Qu et al. 2016, Scarpitta et al. 2017). Indeed, deep water OAB from New Zealand lakes are particularly vulnerable to eutrophication (De Winton and Beever 2004).

Another different pollution problem is lake acidification caused by acid rain. This would improve the availability of free CO_2 and the consequent increase in OAB biomass, as Riis and Sand-Jensen (1998) have pointed out in Danish lakes after 40 years of observations.

Species Composition of OAB Communities

Combined responses to the diverse climate change factors will be reflected in changes in the distribution ranges of OAB, with consequent variations in the species composition of OAB communities. In general, species may be displaced to locations with more favorable temperature and precipitation conditions. For example, OAB could disappear from excessively hostile environments, such as hot waters, streams with markedly long dryness periods, etc., and could extend to (or take refuge in) higher latitudes and altitudes, if possible. These changes will especially occur in the long-term, but faster changes (occurring in a decade) have been pointed out (Glime 2011). The success or failure in colonizing new areas will depend on the ecophysiological adaptability of the different species and their dispersion ability, linked to the sexual and asexual diaspore production. Few case studies on the influence of climate change factors on OAB distribution ranges and communities structure are available. In 34 boreal springs, changes in the bryophyte communities were assessed from 1987 to 2015 in response to anthropogenic land drainage (Lehosmaa et al. 2017), finding a decline of the abundance and richness of most habitat specialist bryophytes (some of them OAB). Species extinctions were predicted to occur in one century. Surprisingly, a habitat specialist (*Brachythecium rivulare*) benefited from land drainage, clearly showing that the potential changes will be species-specific. Physiological tolerance to multiple adverse factors (increased temperatures, high solar radiation, frequent desiccation, etc.) may underlie the success of determined species.

Interactions between Climate Change Factors

Climate change involves different environmental factors and, under field conditions, species respond integrately to all of them. This complex interaction has been highlighted by several authors (Häder et al. 2003, Gignac 2011, Williamson et al. 2014). For example, trade-offs between high levels of photosynthetic and UV radiation, high temperature and desiccation may induce cross-tolerance. This would be important in the context of climate change because all these factors may simultaneously affect OAB during periods of low discharge and the consequent emersion. However, few studies exist on the interaction of several climate change factors on OAB. In two OAB (a moss and a liverwort), the influence of temperature on the effects of UV-B radiation depended on the species: the higher the UV-B tolerance, the lower the influence of temperature (Núñez-Olivera et al. 2004). Moreover, the adverse effects of cold and UV-B radiation were additive in the moss, probably because low temperatures inhibited the development of protection and repair mechanisms against UV-B, but this additiveness was lacking in the liverwort. In another study, UV-B-induced DNA damage in an aquatic liverwort was augmented in the presence of cadmium, which was attributed to the inhibition of DNA enzymatic repair mechanisms by the heavy metal (Otero et al. 2006). In high-Arctic lakes, two climate change factors, such as more prolonged snow cover (with concomitant reduction of primary production) and nutrient increase (with a consequent increase in primary production because of previous nutrient-limitation) would exert contrasting effects on OAB (Riis et al. 2014, 2016).

Ecosystem Processes

OAB play relevant ecological roles in aquatic ecosystems, where they interact with a constellation of producers, consumers, and decomposers (Bowden et al. 1999, Glime 2018). Thus, any change in OAB species or communities will have dramatic effects on the whole ecosystem. However, there is such a complex interaction between ecosystem components and climate change factors that consequences are difficult to predict, and specific case studies are needed.

Conclusion

The effects of climate change on OAB are greatly uncertain. Their preferential habitats, such as headwater streams and deep polar and alpine lakes, are predicted to be exposed to different changes, which will depend on the type of aquatic system and the geographical area considered: water temperature increase, stronger changes in water level, longer periods of low or no water flow, increase in solar radiation including UV (or decrease in lakes subjected to an enrichment in UV-absorbing matter), dissolved CO_2 increase, water acidification, eutrophication, and increase in pollution and extreme weather events (such as floods), etc. The influence of each individual factor on OAB may frequently counteract the effect of another factor, and the complex interactions between all those factors make OAB responses very difficult to predict. Positive or negative changes in photosynthesis and growth patterns can be expected. In addition, changes in competition between OAB themselves and with other organisms, the distribution ranges of species, the composition of the communities, and ecosystem processes, such as food webs and biogeochemical cycles, can be predicted. Some of these variations have already been assessed in the existing studies on the subject, but the species-specific responses (mainly depending on ecophysiological tolerance, adaptability and dispersal capacity) further complicate reliable predictions. Hence, neither bryophytes in general nor OAB in particular should be grouped as a single functional type regarding the effects of climate change.

As occurs in other organisms (Bais et al. 2018), impacts of climate change factors on OAB have been evaluated using individual species, whereas community-level processes have been more rarely analyzed. Thus, more observational and manipulative long-term field research, involving the different factors of climate change and their interactions, is needed. This is particularly required in headwater streams, a difficult environment to conduct field experiments due to their high spatial and temporal variability. Given that global predictions are notably difficult, and that extrapolations between different ecosystems and geographical areas are problematic, maybe regional predictions could be useful (Häder et al. 2014). In addition, the development of new tools to assess bryophyte vigor as influenced by environmental changes, including climate change (Malenovsky et al. 2017), can be essential to disentangle these challenging questions.

Overall, bryophytes are ideal plants for climate change bioindication, due to their sensitivity to ecological changes. Additionally, the responses of desiccation-tolerant bryophytes and non-tolerant higher plants to climate change are not comparable because of their basic physiological differences, and thus bryophytes should be separately studied in this context. Within bryophytes, OAB inhabit fragile ecosystems that are severely threatened nowadays (and will be so in the future) by climate change factors. Therefore, OAB are relevant tools to monitor and predict the effects of climate change on aquatic ecosystems.

Acknowledgments

This work was supported by the Ministerio de Economía y Competitividad of Spain (MINECO) and Fondo Europeo de Desarrollo Regional (FEDER) (Project CGL2014-54127-P).

References

Amesbury, M.J., T.P. Roland, J. Royles, D.A. Hodgson, P. Convey, H. Griffiths and D.J. Charman. 2017. Widespread biological response to rapid warming on the Antarctic Peninsula. Current Biology 27: 1616–1622.

Arróniz-Crespo, M., J.R. Leake, P. Horton and G.K. Phoenix. 2008. Bryophyte physiological responses to, and recovery from, long-term nitrogen deposition and phosphorus fertilisation in acidic grassland. New Phytologist 180: 864–874.

Arscott, D.B., W.B. Bowden and J.C. Finlay. 1998. Comparison of epilithic algal and bryophyte metabolism in an arctic tundra stream, Alaska. Journal of the North American Benthological Society 17: 210–227.

Bain, J.T. and M.C.F. Proctor. 1980. The requirement of aquatic bryophytes for free CO_2 as an inorganic carbon source: some experimental evidence. New Phytologist 86: 393–400.

Bais, A.F., R.M. Lucas, J.F. Bornman, C.E. Williamson, B. Sulzberger, A.T. Austin, S.R. Wilson, A.L. Andrady, G. Bernhard, R.L. McKenzie, P.J. Aucamp, S. Madronich, R.E. Neale, S. Yazar, A.R. Young, F.R. De Gruijl, M. Norval, Y. Takizawa, P.W. Barnes, T.M. Robson, S.A. Robinson, C.L. Ballaré, S.D. Flint, P.J. Neale, S. Hylander, K.C. Rose, S.A. Wängberg, D.P. Häder, R.C. Worrest, R.G. Zepp, N.D. Paul, R.M. Cory, K.R. Solomon, J. Longstreth, K.K. Pandey, H.H. Redhwi, A. Torikaiaj and A.M. Heikkiläak. 2018. Environmental effects of ozone depletion, UV radiation and interactions with climate change: UNEP Environmental Effects Assessment Panel, update 2017. Photochemical and Photobiological Sciences 17: 127–179.

Bates, J.W. and C.D. Preston. 2011. Can the effects of climate change on British bryophytes be distinguished from those resulting from other environmental changes? pp. 371–407. *In*: Tuba, Z., N.G. Slack and L.R. Stark (eds.). Bryophyte Ecology and Climate Change. Cambridge University Press, New York.

Beaucourt, N., E. Núñez-Olivera, J. Martínez-Abaigar, A. García-Alvaro, R. Tomás and M. Arróniz. 2001. Variaciones estacionales del crecimiento de *Fontinalis antipyretica* y *F. squamosa* en condiciones naturales. Boletín de la Sociedad Española de Briología 18/19: 37–44.

Björk, R.G. and U. Molau. 2007. Ecology of alpine snowbeds and the impact of global change. Arctic, Antarctic and Alpine Research 39: 34–43.

Bowden, W.B., D. Arscott, D. Pappathanasi, J. Finlay, J.M. Glime, J. LaCroix, C.L. Liao, A. Hershey, T. Lampella, B. Peterson, W. Wollheim, K. Slavik, B. Shelley, M.B. Chesterton, J.A. Lachance, R.M. LeBlanc, A. Steinman and A. Suren. 1999. Roles of bryophytes in stream ecosystems. Journal of the North American Benthological Society 18: 151–184.

Bramley-Alves, J., D.H. King, S.A. Robinson and R.E. Miller. 2014. Dominating the Antarctic environment: bryophytes in a time of change. pp. 309–324. *In*: Hanson, D.T. and S.K. Rice (eds.). Photosynthesis in Bryophytes and Early Land Plants. Springer, Dordrecht.

Breeuwer, A., M.M.P.D. Heijmans, B.J.M. Robroek and F. Berendse. 2008. The effect of temperature on growth and competition between *Sphagnum* species. Oecologia 156: 155–167.

Callaghan, T.V., L.O. Bjorn, Y. Chernov, T. Chapin, T.R. Christensen, B. Huntley, R.A. Ims, M. Johansson, D. Jolly, S. Jonasson, N. Matveyeva, N. Panikov, W. Oechel, G. Shaver, J. Elster, I.S. Jonsdottir, K. Laine, K. Taulavuori, E. Taulavuori and C. Zockler. 2004. Responses to projected changes in climate and UV-B at the species level. Ambio 33: 418–435.

Calow, P. 1998. The Encyclopedia of Ecology and Environmental Management. Blackwell Science, Oxford.

Cardona-Correa, C., J.M. Graham and L.E. Graham. 2015. Anatomical effects of temperature and UV-A plus UV-B treatments and temperature-UV interactions in the peatmoss *Sphagnum compactum*. International Journal of Plant Sciences 176: 159–169.

Casanova-Katny, A., G.A. Torres-Mellado and S.M. Eppley. 2016. Reproductive output of mosses under experimental warming on Fildes Peninsula, King George Island, Maritime Antarctica. Revista Chilena de Historia Natural 89: 13.

Clarke, L.J. and S.A. Robinson. 2008. Cell wall-bound ultraviolet-screening compounds explain the high ultraviolet tolerance of the Antarctic moss, *Ceratodon purpureus*. New Phytologist 179: 776–783.

Coe, K.K., J.P. Sparks and J. Belnap. 2014. Physiological ecology of dryland biocrust mosses. pp. 291–308. *In*: Hanson, D.T. and S.K. Rice (eds.). Photosynthesis in Bryophytes and Early Land Plants. Springer, Dordrecht.

Cole, J.J., N.F. Caraco, G.W. Kling and T.K. Kratz. 1994. Carbon dioxide supersaturation in the surface waters of lakes. Science 265: 1568–1570.

Conde-Álvarez, R.M., E. Pérez-Rodríguez, M. Altamirano, J.M. Nieto, R. Abdala, F.L. Figueroa and A. Flores-Moya. 2002. Photosynthetic performance and pigment content in the aquatic liverwort *Riella helicophylla* under natural solar irradiance and solar irradiance without ultraviolet light. Aquatic Botany 73: 47–61.

Cruz De Carvalho, R., M. Catalá, J.M. Da Silva, C. Branquinho and E. Barreno. 2012. The impact of dehydration rate on the production and cellular location of reactive oxygen species in an aquatic moss. Annals of Botany 110: 1007–1016.

Davey, M.C. and P. Rothery. 1997. Interspecific variation in respiratory and photosynthetic parameters in Antarctic bryophytes. New Phytologist 137: 231–240.

De Winton, M.D. and J.E. Beever. 2004. Deep-water bryophyte records from New Zealand lakes. New Zealand Journal of Marine and Freshwater Research 38: 329–340.

Deane-Coe, K.K., M. Mauritz, G. Celis, V. Salmon, K.G. Crummer, S.M. Natali and E.A.G. Schuur. 2015. Experimental warming alters productivity and isotopic signatures of tundra mosses. Ecosystems 18: 1070–1082.

Debén, S., J.R. Aboal, A. Carballeira, M. Cesa, C. Real and J.A. Fernandez. 2015. Inland water quality monitoring with native bryophytes: A methodological review. Ecological Indicators 53: 115–124.

Debén, S., J.R. Aboal, A. Carballeira, M. Cesa and J.A. Fernandez. 2017. Monitoring river water quality with transplanted bryophytes: A methodological review. Ecological Indicators 81: 461–470.

Dittrich, S., C. Leuschner and M. Hauck. 2016. Change in the bryophyte diversity and species composition of Central European temperate broad-leaved forests since the late nineteenth century. Biodiversity and Conservation 25: 2071–2091.

Fabón, G., J. Martínez-Abaigar, R. Tomás and E. Núñez-Olivera. 2010. Effects of enhanced UV-B radiation on hydroxycinnamic acid derivatives extracted from different cell compartments in the aquatic liverwort *Jungermannia exsertifolia* subsp. *cordifolia*. Physiologia Plantarum 140: 269–279.

Fabón, G., L. Monforte, R. Tomás-Las-Heras, J. Martínez-Abaigar and E. Núñez-Olivera. 2012. Cell compartmentation of UV-absorbing compounds in two aquatic mosses under enhanced UV-B. Cryptogamie Bryologie 33: 169–184.

Finlay, J.C. and W.B. Bowden. 1994. Controls on production of bryophytes in an arctic tundra stream. Freshwater Biology 32: 455–466.

Flanagan, L.B. 2014. Interacting controls on ecosystem photosynthesis and respiration in contrasting peatland ecosystems. pp. 253–268. *In*: Hanson, D.T. and S.K. Rice (eds.). Photosynthesis in Bryophytes and Early Land Plants. Springer, Dordrecht.

Frahm, J.P. and D. Klaus. 2001. Bryophytes as indicators of recent climate fluctuations in Central Europe. Lindbergia 26: 97–104.

Frost, P.C., A. Mack, J.H. Larson, S.D. Bridgham and G.A. Lamberti. 2006. Environmental controls of UV-B radiation in forested streams of Northern Michigan. Photochemistry and Photobiology 82: 781–786.

García-Álvaro, A., J. Martínez-Abaigar, E. Núñez-Olivera and N. Beaucourt. 2000. Element concentrations and enrichment ratios in the aquatic moss *Rhynchostegium riparioides* along the river Iregua (La Rioja, northern Spain). Bryologist 103: 518–533.

Gignac, L.D., B.J. Nicholson and S.E. Bayley. 1998. The utilization of bryophytes in bioclimatic modeling: predicted northward migration of peatlands in the Mackenzie river basin, Canada, as a result of global warming. Bryologist 101: 572–587.

Gignac, L.D. 2001. Bryophytes as indicators of climate change. Bryologist 104: 410–420.

Gignac, L.D. 2011. Bryophytes as predictors of climate change. pp. 461–482. *In*: Tuba, Z., N.G. Slack and L.R. Stark (eds.). Bryophyte Ecology and Climate Change. Cambridge University Press, New York.

Gimeno-Colera, C. and F. Puche-Pinazo. 1999. Flora y vegetación briofítica higro-hidrófila de la comunidad Valenciana (Este de España). Cryptogamie, Bryologie 20: 49–68.

Glime, J. 2014. Photosynthesis in aquatic bryophytes. pp. 201–232. *In*: Hanson, D.T. and S.K. Rice (eds.). Photosynthesis in Bryophytes and Early Land Plants. Springer, Dordrecht.

Glime, J.M. 2011. Ecological and physiological effects of changing climate on aquatic bryophytes. pp. 93–114. *In*: Tuba, Z., N.G. Slack and L.R. Stark (eds.). Bryophyte Ecology and Climate Change. Cambridge University Press, New York.

Glime, J.M. 2018. Bryophyte Ecology Volume 2: Bryological Interaction (electronic book). Michigan Technological University and the International Association of Bryologists, Houghton.

Gudmundsdottir, R., J.S. Olafsson, S. Palsson, G.M. Gislason and B. Moss. 2011. How will increased temperature and nutrient enrichment affect primary producers in sub-Arctic streams? Freshwater Biology 56: 2045–2058.

Guo, C.Q., R. Ochyra, P.C. Wu, R.D. Seppelt, Y.F. Yao, L.G. Bian, S.P. Li and C.S. Li. 2013. *Warnstorfia exannulata*, an aquatic moss in the Arctic: seasonal growth responses. Climatic Change 119: 407–419.

Häder, D.P., H.D. Kumar, R.C. Smith and R.C. Worrest. 2003. Aquatic ecosystems: effects of solar ultraviolet radiation and interactions with other climatic change factors. Photochemical and Photobiological Sciences 2: 39–50.

Häder, D.P., E.W. Helbling, C.E. Williamson and R.C. Worrest. 2011. Effects of UV radiation on aquatic ecosystems and interactions with climate change. Photochemical and Photobiological Sciences 10: 242–260.

Häder, D.P., V.E. Villafañe and W. Helbling. 2014. Productivity of aquatic primary producers under global climate change. Photochemical and Photobiological Sciences 13: 1370–1392.

Hájek, T. 2014. Physiological ecology of peatland bryophytes. pp. 233–252. *In*: Hanson, D.T. and S.K. Rice (eds.). Photosynthesis in Bryophytes and Early Land Plants. Springer, Dordrecht.

He, X., K.S. He and J. Hyvönen. 2016. Will bryophytes survive in a warming world? Perspectives in Plant Ecology, Evolution and Systematics 19: 49–60.

Heijmans, M.M.P.D., D. Mauquoy, B. Van Geel and F. Berendse. 2008. Long-term effects of climate change on vegetation and carbon dynamics in peat bogs. Journal of Vegetation Science 19: 307–320.

Ignatov, M.S. and N.B. Kurbatova. 1990. A review of deep-water bryophytes with new records from USSR. Hikobia 10: 393–401.

Jansen, M.A.K. and J.F. Bornman. 2012. UV-B radiation: from generic stressor to specific regulator. Physiologia Plantarum 145: 501–504.

Kelly, M.G. and B.A. Whitton. 1987. Growth rate of the aquatic moss *Rhynchostegium riparioides* in Northern England. Freshwater Biology 18: 461–468.

Lehosmaa, K., J. Jyvasjarvi, R. Virtanen, J. Ilmonen, J. Saastamoinen and T. Muotka. 2017. Anthropogenic habitat disturbance induces a major biodiversity change in habitat specialist bryophytes of boreal springs. Biological Conservation 215: 169–178.

Li, S.P., R. Ochyra, P.C. Wu, R.D. Seppelt, M.H. Cai, H.Y. Wang and C.S. Li. 2009. *Drepanocladus longifolius* (Amblystegiaceae), an addition to the moss flora of King George Island, South Shetland Islands, with a review of Antarctic benthic mosses. Polar Biology 32: 1415–1425.

Maberly, S.C. 1985. Photosynthesis by *Fontinalis antipyretica* II. Assessment of environmental factors limiting photosynthesis and production. New Phytologist 100: 141–155.

Malenovsky, Z., A. Lucieer, D.H. King, J.D. Turnbull and S.A. Robinson. 2017. Unmanned aircraft system advances health mapping of fragile polar vegetation. Methods in Ecology and Evolution 8: 1842–1857.

Margalef, R. 1982. Ecología. Omega, Barcelona.

Martínez-Abaigar, J. and A. Ederra. 1992. Brioflora del río Iregua (La Rioja, España). Cryptogamie, Bryologie Lichénologie 13: 47–69.

Martínez-Abaigar, J., E. Núñez-Olivera and M. Sánchez-Díaz. 1993. Effects of organic pollution on transplanted aquatic bryophytes. Journal of Bryology 17: 553–566.

Martínez-Abaigar, J. and E. Núñez-Olivera. 1996. The bryological work of Ildefonso Zubía Icazuriaga (1819–1891) in northern Spain. Nova Hedwigia 62: 255–266.

Martínez-Abaigar, J., E. Núñez-Olivera and N. Beaucourt. 2002. Short-term physiological responses of the aquatic liverwort *Jungermannia exsertifolia* subsp. *cordifolia* to KH_2PO_4 and anoxia. Bryologist 105: 86–95.

Martínez-Abaigar, J. and E. Núñez-Olivera. 2011. Aquatic bryophytes under ultraviolet radiation. pp. 115–146. *In*: Tuba, Z., N.G. Slack and L.R. Stark (eds.). Bryophyte Ecology and Climate Change. Cambridge University Press, New York.

Monforte, L., R. Tomás-Las-Heras, M.A. Del-Castillo-Alonso, J. Martínez-Abaigar and E. Núñez-Olivera. 2015a. Spatial variability of ultraviolet-absorbing compounds in an aquatic liverwort and their usefulness as biomarkers of current and past UV radiation: a case study in the Atlantic-Mediterranean transition. Science of the Total Environment 518–519: 248–257.

Monforte, L., E. Núñez-Olivera and J. Martínez-Abaigar. 2015b. UV radiation biomonitoring using cell compartmentation of UV-absorbing compounds in herbarium samples of a liverwort. Ecological Indicators 52: 48–56.

Monforte, L., G. Soriano, E. Núñez-Olivera and J. Martínez-Abaigar. 2018. Cell compartmentation of ultraviolet-absorbing compounds: an underexplored tool related to bryophyte ecology, phylogeny and evolution. Functional Ecology 32: 882–893.

Morrill, J.C., R.C. Bales and M.H. Conklin. 2005. Estimating stream temperature from air temperature: implications for future water quality. Journal of Environmental Engineering 131: 139–146.

Morris, J.L., M.N. Puttick, J.W. Clark, D. Edwards, P. Kenrick, S. Pressel, C.H. Wellman, Z.H. Yang, H. Schneider and P.C.J. Donoghue. 2018. The timescale of early land plant evolution. Proceedings of the National Academy of Sciences of the United States of America 115: 2274–2283.

Muotka, T. and R. Virtanen. 1995. The stream as a habitat templet for bryophytes: species' distributions along gradients in disturbance and substratum heterogeneity. Freshwater Biology 33: 141–160.

Núñez-Olivera, E., J. Martínez-Abaigar, R. Tomás, N. Beaucourt and M. Arróniz-Crespo. 2004. Influence of temperature on the effects of artificially enhanced UV-B radiation on aquatic bryophytes under laboratory conditions. Photosynthetica 42: 201–212.

Oke, T.A. and H.A. Hager. 2017. Assessing environmental attributes and effects of climate change on *Sphagnum* peatland distributions in North America using single- and multi-species models. PLoS One 12: e0175978.

Otero, S., E. Núñez-Olivera, J. Martínez-Abaigar, R. Tomás, M. Arróniz-Crespo and N. Beaucourt. 2006. Effects of cadmium and enhanced UV radiation on the physiology and the concentration of UV-absorbing compounds of the aquatic liverwort *Jungermannia exsertifolia* subsp. *cordifolia*. Photochemical and Photobiological Sciences 5: 760–769.

Otero, S., E. Núñez-Olivera, J. Martínez-Abaigar, R. Tomás and S. Huttunen. 2009. Retrospective bioindication of stratospheric ozone and ultraviolet radiation using hydroxycinnamic acid derivatives of herbarium samples of an aquatic liverwort. Environmental Pollution 157: 2335–2344.

Pearce-Higgins, J.W., C.M. Beale, T.H. Oliver, T.A. August, M. Carroll, D. Massimino, N. Ockendon, J. Savage, C.J. Wheatley, M.A. Ausden, R.B. Bradbury, S.J. Duffield, N.A. Macgregor, C.J. McClean, M.D. Morecroft, C.D. Thomas, O. Watts, B.C. Beckmann, R. Fox, H.E. Roy, P.G. Sutton, K.J. Walker and H.Q.P. Crick. 2017. A national-scale assessment of climate change impacts on species: assessing the balance of risks and opportunities for multiple taxa. Biological Conservation 213: 124–134.

Pentecost, A. and Z. Zhang. 2002. Bryophytes from some travertine-depositing sites in France and the U.K.: relationships with climate and water chemistry. Journal of Bryology 24: 233–241.

Pócs, T. 2011. Signs of climate change in the bryoflora of hungary. pp. 359–370. *In*: Tuba, Z., N.G. Slack and L.R. Stark (eds.). Bryophyte Ecology and Climate Change. Cambridge University Press, New York.

Proctor, M.C.F. 2014. The diversification of bryophytes and vascular plants in evolving terrestrial environments. pp. 59–78. *In*: Hanson, D.T. and S.K. Rice (eds.). Photosynthesis in Bryophytes and Early Land Plants. Springer, Dordrecht.

Puttick, M.N., J.L. Morris, T.A. Williams, C.J. Cox, D. Edwards, P. Kenrick, S. Pressel, C.H. Wellman, H. Schneider, D. Pisani and P.C.J. Donoghue. 2018. The interrelationships of land plants and the nature of the ancestral embryophyte. Current Biology 28: 733–745.

Qu, L.L., H.Y. Xiao, H. Guan, Z.Y. Zhang and Y. Xu. 2016. Total N content and delta N-15 signatures in moss tissue for indicating varying atmospheric nitrogen deposition in Guizhou Province, China. Atmospheric Environment 142: 145–151.

Rader, R.B. and T.A. Belish. 1997. Short-term effects of ambient and enhanced UV-B on moss (*Fontinalis neomexicana*) in a mountain stream. Journal of Freshwater Ecology 12: 395–403.

Raven, J.A. and T.D. Colmer. 2016. Life at the boundary: photosynthesis at the soil-fluid interface. A synthesis focusing on mosses. Journal of Experimental Botany 67: 1613–1623.

Riis, T. and K. Sand-Jensen. 1998. Development of vegetation and environmental conditions in an oligotrophic Danish lake over 40 years. Freshwater Biology 40: 123–134.

Riis, T., K.S. Christoffersen and A. Baattrup-Pedersen. 2014. Effects of warming on annual production and nutrient-use efficiency of aquatic mosses in a high Arctic lake. Freshwater Biology 59: 1622–1632.

Riis, T., K.S. Christoffersen and A. Baattrup-Pedersen. 2016. Mosses in high-Arctic lakes: *in situ* measurements of annual primary production and decomposition. Polar Biology 39: 543–552.

Roberts, B.J., P.J. Mulholland and W.R. Hill. 2007. Multiple scales of temporal variability in ecosystem metabolism rates: Results from 2 years of continuous monitoring in a forested headwater stream. Ecosystems 10: 588–606.

Robinson, S.A. and M.J. Waterman. 2014. Sunsafe bryophytes: Photoprotection from excess and damaging solar radiation. pp. 113–130. *In*: Hanson, D.T. and S.K. Rice (eds.). Photosynthesis in Bryophytes and Early Land Plants. Springer, Dordrecht.

Robinson, S.A. and D.J. Erickson III. 2015. Not just about sunburn—the ozone hole's profound effect on climate has significant implications for Southern Hemisphere ecosystems. Global Change Biology 21: 515–527.

Royles, J. and H. Griffiths. 2015. Climate change impacts in polar regions: lessons from Antarctic moss bank archives. Global Change Biology 21: 1041–1057.

Scarpitta, A.B., J. Bardat, A. Lalanne and M. Vellend. 2017. Long-term community change: bryophytes are more responsive than vascular plants to nitrogen deposition and warming. Journal of Vegetation Science 28: 1220–1229.

Schwarz, A.M. and S. Markager. 1999. Light absorption and photosynthesis of a benthic moss community: importance of spectral quality of light and implications of changing light attenuation in the water column. Freshwater Biology 42: 609–623.

Shevock, J.R., W.Z. Ma and H. Akiyama. 2017. Diversity of the rheophytic condition in bryophytes: field observations from multiple continents. Bryophyte Diversity and Evolution 39: 75–93.

Short, F.T. and H.A. Neckles. 1999. The effects of global climate change on seagrasses. Aquatic Botany 63: 169–196.

Soriano, G., C. Cloix, M. Heilmann, E. Núñez-Olivera, J. Martínez-Abaigar and G.I. Jenkins. 2018. Evolutionary conservation of structure and function of the UVR8 photoreceptor from the liverwort *Marchantia polymorpha* and the moss *Physcomitrella patens*. New Phytologist 217: 151–162.

Sun, S.Q., G.X. Wang, S.X. Chang, J.S. Bhatti, W.L. Tian and J. Luo. 2017. Warming and nitrogen addition effects on bryophytes are species- and plant community-specific on the eastern slope of the Tibetan Plateau. Journal of Vegetation Science 28: 128–138.

Suren, A.M. 1996. Bryophyte distribution patterns in relation to macro-, meso-, and micro-scale variables in South Island, New Zealand streams. New Zealand Journal of Marine and Freshwater Research 30: 501–523.

Takács, Z., Z. Csintalan, L. Sass, E. Laitat, I. Vass and Z. Tuba. 1999. UV-B tolerance of bryophyte species with different degrees of desiccation tolerance. Journal of Photochemistry and Photobiology B: Biology 48: 210–215.

Tremp, H., D. Kampmann and R. Schulz. 2012. Factors shaping submerged bryophyte communities: A conceptual model for small mountain streams in Germany. Limnologica 42: 242–250.

Tuba, Z. 2011a. Bryophyte physiological processes in a changing climate: an overview. pp. 13–32. *In*: Tuba, Z., N.G. Slack and L.R. Stark (eds.). Bryophyte Ecology and Climate Change. Cambridge University Press, New York.

Tuba, Z., N.G. Slack and L. Stark. 2011b. Bryophyte Ecology and Climate Change. Cambridge University Press, New York.

Vanderpoorten, A., L. Hedenäs, C.J. Cox and A.J. Shaw. 2002. Phylogeny and morphological evolution of the Amblystegiaceae (Bryopsida). Molecular Phylogenetics and Evolution 23: 1–21.

Vanderpoorten, A. and B. Goffinet. 2009. Introduction to Bryophytes. Cambridge University Press, Cambridge.

Vanneste, T., O. Michelsen, B.J. Graae, M.O. Kyrkjeeide, H. Holien, K. Hassel, S. Lindmo, R.E. Kapas and P. De Frenne. 2017. Impact of climate change on alpine vegetation of mountain summits in Norway. Ecological Research 32: 579–593.

Verdaguer, D., M.A.K. Jansen, L. Llorens, L.O. Morales and S. Neugart. 2017. UV-A radiation effects on higher plants: Exploring the known unknown. Plant Science 255: 72–81.

Vestergaard, O. and K. Sand-Jensen. 2000. Alkalinity and trophic state regulate aquatic plant distribution in Danish lakes. Aquatic Botany 67: 85–107.

Vieira, C., A. Seneca and C. Sergio. 2012. Floristic and ecological survey of bryophytes from Portuguese watercourses. Cryptogamie Bryologie 33: 113–134.

Vitt, D.H. and J.M. Glime. 1984. The structural adaptations of aquatic Musci. Lindbergia 10: 95–110.

Wagner, B. and R. Seppelt. 2006. Deep-water occurrence of the moss *Bryum pseudotriquetrum* in Radok Lake, Amery Oasis, East Antarctica. Polar Biology 29: 791–795.

Williamson, C.E., R.G. Zepp, R.M. Lucas, S. Madronich, A.T. Austin, C.L. Ballaré, M. Norval, B. Sulzberger, A.F. Bais, R.L. McKenzie, S.A. Robinson, D.P. Häder, N.D. Paul and J.F. Bornman. 2014. Solar ultraviolet radiation in a changing climate. Nature Climate Change 4: 434–441.

CHAPTER 15

Ecophysiological Responses of Mollusks to Oceanic Acidification

*Ting Wang and Youji Wang**

Introduction

The ocean, which plays an important role in buffering the rise of atmospheric CO_2, is a vital carbon reservoir absorbing approximately 24 million tons of CO_2 per day, accounting for approximately a quarter of the carbon dioxide produced by human activities (Sabine et al. 2004, Le Quere et al. 2010). The absorption is not entirely benign, and the 'ocean acidification' caused by the large amount of CO_2 entering the ocean has been considered to be "another CO_2 question brought by climate change" (Doney et al. 2009). This phenomenon has received much attention in contemporary ocean research. The rate of CO_2 entering the ocean is much greater than the rate of mixing, and the current absorption of CO_2 in the ocean mostly gathers on the ocean surface. Thus, acidification of the surface seawater causes relevant concern. The absorbed CO_2 gradually invades into the ocean, and excessive CO_2 incorporation into the seawater leads to not only a decrease in the pH of the seawater but also a decrease in carbonate and calcium carbonate saturation (Feely et al. 2004, Doney et al. 2009) (Fig. 15.1). Many invertebrates in the ocean are calcified animals, such as mollusks. In addition, when the carbonate content in seawater decreases the calcification capacity of calcified animals is directly affected. The decrease of seawater pH also affects the osmotic balance of marine animals and ultimately the growth and reproduction of marine calcified organisms. To date, acidification has a remarkable effect on the early development, calcification, metabolism, and other physiological processes of marine organisms, especially the calcified organisms. If the organisms cannot make

College of Fisheries and Life Science, Shanghai Ocean University, 999 Huchenghuan Road, Shanghai 201306, China.
Email: twang529@gmail.com
* Corresponding author: youjiwang2@gmail.com

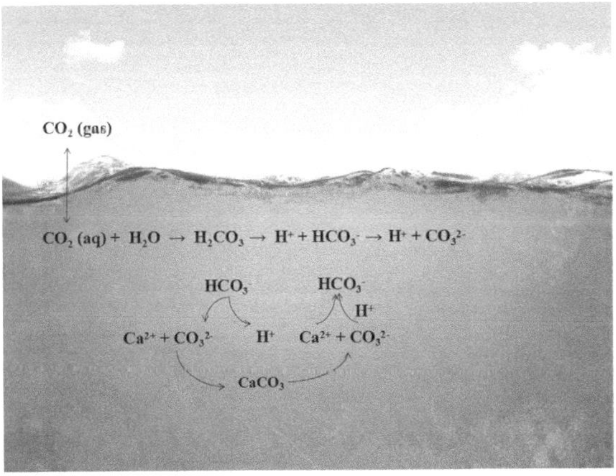

Fig. 15.1. Seawater acidification.

a corresponding physiological or behavioral adaptation to changes in seawater pH, their survival in the acid environment will be difficult.

Mollusks are connected to primary producers and secondary consumers. Mollusks are also an important link in the food chain and energy flow. Thus, mollusks play an important role in coastal ecosystem functions. Many bivalves are also of high economic value for fishing and aquaculture (Carrington et al. 2015). In addition, their long life cycle, sessile growth characteristics, filter feeding, *in vivo* toxin accumulation, and global distribution make mussels a popular model animal in marine environmental monitoring studies.

Physiological Effects of Acidification on Mollusks

Most mollusks have a stage of pelagic larvae, and their calcium carbonate shells usually begin to synthesize in the larval stage. Thus, ocean acidification can easily affect the early mollusk development (Fig. 15.2). The early life history stages (including gametes, embryos, and larvae) of estuarine and marine organisms are usually more sensitive than adults to increased CO_2 stress (Havenhand et al. 2008, Parker et al. 2009). Ocean acidification may inhibit the development and survival of mollusk larvae and subsequently the population reductions of a number of bivalve species globally. In addition, the larvae of the bay scallop *Argopecten irradians*, the hard clam *Mercenaria mercenaria*, and the American oyster *Crassostrea virginica* showed increased sensitivity to elevated pCO_2 (Talmage and Gobler 2009). The larval growth rate, survival rate, and lipid accumulation of mollusks also decreased (Talmage and Gobler 2010), and the shell length of *A. irradians* in a high-pCO_2 treatment was significantly shorter (White et al. 2013). Moreover, the survival of the flat oyster *Ostrea angasi* larvae decreased under acidification, but the larval development was not affected probably because *O. angasi* larvae can adapt to the high carbonate conditions in the hatchery (Cole et al. 2016). The abalone *Haliotis iris* juvenile shell length and weight are negatively affected by low pH, and large juveniles are less likely to respond

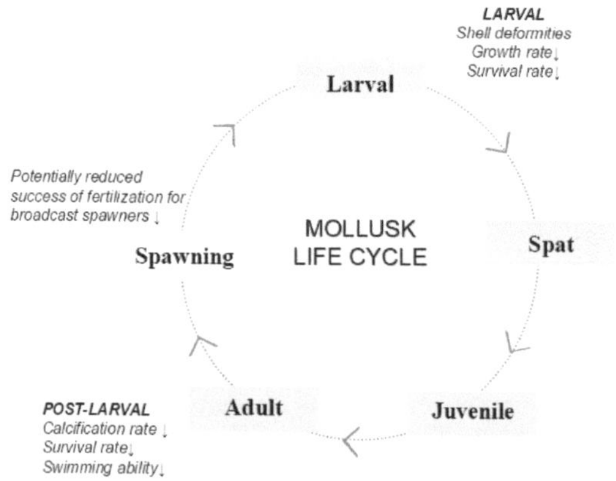

Fig. 15.2. Life history of mollusks and the effect of acidification on their life at different life stages.

to acidity than those of small individuals; their respiration rate is not significantly affected, suggesting that *H. iris* juveniles can maintain aerobic metabolism under acidification conditions (Cunningham et al. 2015). Under the condition of 2,000 ppm CO_2, 79.2% of the Pacific abalone *Haliotis discus hannai* larvae and 13.3% of the colored abalone *H. diversicolor* larvae were able to normally develop, whereas the oyster *Crassostrea angulata* larvae were the most tolerant to acidification, probably due to their adaptability to the estuarine acidification environment (Guo et al. 2015). Under elevated pCO_2, the average shell length of the clam *Mactra veneriformis* juveniles was significantly reduced, and the proportion of malformed individuals increased (Kim et al. 2016). The growth and survival of the scallop *Pecten maximus* under acidification were also adversely affected (Andersen et al. 2013). If the oyster *Ostrea lurida* experiences acidification during its early development, its subsequent growth, even under normal conditions, is slower than without experiencing acidification, confirming that acidification exerts a lasting effect on oysters (Hettinger et al. 2013). The above results indicated that acidification slows down or delays the development of most mollusks in the early life stages, increasing the risk of mollusk predation and the susceptibility to disease (Allen 2008).

Considering that acidification has a direct effect on the organisms with calcareous shells or bones, the calcification and mechanism of organisms have received great attention in the study of acidification (Fig. 15.3). Aragonite and calcite are the main forms of calcium carbonate in the ocean. With the continuous development of ocean acidification, undersaturation of the aragonite in the surface seawater may occur in the middle of this century, and calcite will be undersaturated at by the end of this century. Thus, this condition provides difficulty on calcification or maintaining their calcium carbonate shells and bones, resulting in threatened survival (Feely et al. 2009). During the embryonic period, the calcification of the Pacific oyster *Crassostrea gigas* was significantly negatively affected by the low pH and low $CaCO_3$ saturation state (Kurihara et al. 2007). Exposure to high CO_2 causes the Pacific oyster *C. gigas* larvae to calcify at a high proportion possibly due to the combined effects of an increased metabolic rate and energy resources, but growth retardation after three days

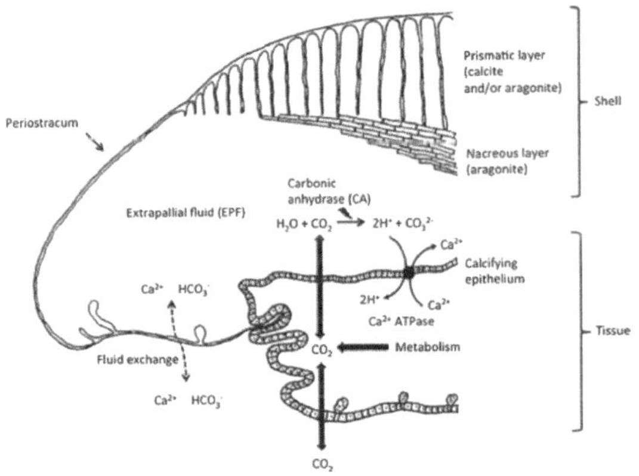

Fig. 15.3. Calcification of the mollusc shell. Modified from Gazeau (2013) with permission from Springer Nature Press.

of fertilization was observed, resulting in a small size of calcified juveniles (Timmins-Schiffman et al. 2013). The calcification rate of the blue mussel *Mytilus edulis* and the Pacific oyster *C. gigas* decreased linearly with decreasing pH, and the calcification rate is expected to decrease by 25% and 10%, respectively, by the end of this century (Gazeau et al. 2007). The pearl oyster *Pinctada fucata* was exposed to acidification conditions (pH values 7.8 and 7.6) for 28 days, and its calcification rate decreased by 25.9% and 26.8%, respectively, compared with the control group (pH 8.1–8.2), but the shell organic matter did not decrease under acidification conditions (Welladsen et al. 2010). The calcium deposition rate and total weight of the Chilean mussel *Mytilus chilensis* under acidic conditions were negatively affected by high CO_2 (Duarte et al. 2014). The basal shell diameter of the barnacle *Amphibalanus amphitrite* exposed at pH 7.4 increased during the growth period, whereas the ash content of the basement shell was predominantly calcium carbonate, confirming compensatory calcification of the barnacles under low pH conditions (Mcdonald et al. 2009). Shell weakening makes the barnacles susceptible to predator attack. Thus, the change of calcification under acidification is very important for the study of population dynamics. Even for the same species, a significant difference was observed in the sensitivity to ocean acidification. Parker et al. found that the daily growth of the shells of the cultured Sydney oyster *Saccostrea glomerata* decreased by 25%, whereas in the wild population it was reduced by 64% under elevated pCO_2 conditions (Parker et al. 2011). The larvae of the blue mussel *M. edulis* are still able to form shells in carbonate-unsaturated seawater, but the size of the shell was small at low pH (Bechmann et al. 2011), and hatching rate and shell thickness were reduced, resulting in a low settlement rate (Gazeau et al. 2010). Mollusk outer shells (calcite) are brittle under acidification conditions, and inner shells (aragonite) soften under acidification, indicating that acidification may cause changes in the physical properties of mollusk shells (Fitzer et al. 2015). When exposed to acidification conditions, the mollusks can still continue to maintain the shell calcification. However, excessive energy is consumed to resist environmental changes. This phenomenon reduces the energy supply for calcification. Thus, the new shell structure and strength are affected, and its protective effect is reduced.

Effects of Acidification on Immunity

Mollusks have different immune mechanisms to resist the invasion of outside pathogens compared with vertebrates, but ocean acidification may affect the immune process in mollusks. Beesley et al. found that acidification exhibited a significant negative effect on the health of the blue mussel *M.* edulis (Beesley et al. 2008). Calcium carbonate dissolution in shells resulted in an increased Ca^{2+} level in the hemolymph, and the phagocytosis was inhibited, resulting in a negative effect on the physiological function of the blood cells. The possible reason was the increased Ca^{2+} concentration caused by $CaCO_3$ dissolution followed by a cell signaling pathway that is dependent on Ca^{2+} (Bibby et al. 2008). The immune parameters of the blue mussel *M. edulis* (such as blood cell count and phagocytosis) are not significantly affected by acidification (Asplund et al. 2014). Antioxidant enzymes of the hemolymph in the thick shell mussel *Mytilus coruscus* exposed to pH 7.3 were higher than those under pH 8.1 (Hu et al. 2015). However, the lysosomal level decreased, and hemocyte mortality and reactive oxygen species (ROS) production increased (Wu et al. 2016). Acidification can cause blood cell apoptosis and increased ROS production in the Pacific oyster *C. gigas*, and elevated pCO_2 inhibits some antioxidant enzyme activities and decreases glutathione levels in the digestive gland (Wang et al. 2017). The total hemocyte count (THC) and phagocytosis of the bloody cockle *Tegillarca granosa* were significantly reduced, and the percentages of eosinophils and basophils were significantly reduced and increased, respectively, and the signal transduction was also negatively affected (Liu et al. 2016). Acidification has a significant effect on the immune system of mollusks and hinders the immune functions of bivalve mollusks. This condition may increase the susceptibility of mollusks to disease and make them susceptible to pathogens. To date, only a few bivalve species have been studied, and less is known about the effects of CO_2-driven ocean acidification on marine bivalve immunity. The underlying molecular mechanism remains unknown.

Effect of Acidification on Physiology

The decrease of pH in seawater affects the physiological metabolism of mollusks, such as food intake, absorption, oxygen consumption, and ammonia excretion. The clearance, ingestion, and the respiration rate of the clam *Ruditapes decussatus* have been found to be significantly reduced, whereas the ammonia excretion rate increased (Ma et al. 2011) under ocean acidification conditions. The clearance rate and absorption rate of the mussel *M. chilensis* were significantly lower after 35 days of exposure to highly acidic conditions (Navarro et al. 2013). Moreover, with the pH decrease, mortality increased, the health index (CI) decreased, and the physiological activities, such as clearance rate and oxygen consumption rate, were reduced in the mussel *M. edulis*, and decreases in calcification rate and carbonic anhydrase activity indicated that the carbon sink capacity of mussels is inhibited (Sun et al. 2016). Acidification reduces the thermal tolerance and aerobic scope of the scallop *P. maximus*, resulting in the loss of closure function and the impairment of adaptation to the extreme temperatures

(Schalkhausser et al. 2013). The shell thickness and daily growth increment of the mussel *Bathymodiolus brevior* in the northwest Eifuku volcano, Japan, living between pHs 5.36 and 7.29, is approximately half that of the mussels living at pH >7.8, indicating that low pH induces an inhibitory effect on physiological metabolism and growth (Tunnicliffe et al. 2009). The calcification rate of the Zhikong scallop *Chlamys farreri* was reduced by 33% at pH 7.9, the calcification rate was almost 0 at pH 7.3, and the respiration rate and oxygen consumption rate was reduced by 14% and 11%, respectively (Zhang et al. 2011). Respiration rate and clearance rate of the common cockle *Cerastoderma edule* significantly increased under low pH conditions (Ong et al., 2017). Wang et al. found that low pH induced a significant negative effect on respiration rate and ammonia excretion rate of the thick shell mussel *M. coruscus*, but its clearance rate and food absorption efficiency were almost unaffected (Wang et al. 2015). The reduction of food intake and the increase of excretion rates by mollusks will result in a reduction of energy for growth and, consequently, slowing the growth and the population increment. Compared with other invertebrates, the morphological and physiological differences between molluscs are significant. In both interspecies and intraspecies comparisons, the metabolic rates are significantly different. With the increased degree of ocean acidification, mollusk metabolic mechanism may change and that may affect their survival and population dynamics in the future.

Effects of Acidification on Acid–Base Regulation

As an important catalyst for carbohydrate reactions, carbonic anhydrase (CA) plays an important role in the acid–base regulation of vertebrates. However, the classification and regulation of CA in marine invertebrates, especially its responses to ocean acidification, have not been well clarified (Chegwidden et al. 1991). CA of some mollusks can respond to environmental stresses, such as salinity changes, pH reduction, and air exposure (Ren et al. 2014, Liu et al. 2015). Ocean acidification changes the seawater carbonate chemistry, thereby changing the body fluid balance in organisms (Parker et al. 2013). The CgCAII-1 protein of the Pacific oyster *C. gigas* has a conserved CA catalytic domain and is highly similar to alpha-CA of the invertebrate cytoplasm or the mitochondria, which converts CO_2 to H^+ and HCO_3^- in seawater. Under acidic conditions, the mRNA transcripts from the muscle, mantle, digestive gland, gill, and hemocyte increased significantly. CgCAII-1 protein can also interact with membrane-associated proteins, and this cytoplasmic CA physiological regulation mechanism may help explain the other physiological responses of marine mollusks to acidification (Wang et al. 2017). The blue mussel *M. edulis* showed increased levels of pCO_2 and bicarbonate in hemolymph under acidification conditions, indicating that mussels are not able to make fine acid–base regulation, and their ability to use bicarbonate to buffer the low pH in their hemolymph is limited (Lewis et al. 2016). Acid–base regulation is an important process responding to acidification under short-term acidification conditions. Although the mollusks have a certain ability of acid–base regulation, beyond the tolerance range, acidification can lead to disorders of acid–base balance, resulting in other abnormal physiological responses.

Effects of Acidification on Mollusk Behavior

Effects of Acidification on Sedimentation Behavior of Mollusks

Most marine invertebrates have a planktonic larval stage that sediments during metamorphosis, i.e., changing from planktonic to benthic life, significantly affecting the population dynamics and community structure (Caley et al. 1997). The settlement density of the oyster *Crassostrea madrasensis* larvae in the tropical estuaries is susceptible to pH changes, and settlement rate was half that at pH 7.31 compared to pH 7.49 (Kripa et al. 2014). Acidified sediments reduced the settlement behavior of the hard clam *M. mercenaria* (Green et al. 2013) and the excavation behavior of the clam *Mya arenaria* (Clements et al. 2017), causing negative effects on population supplementation and distribution (Clements and Hunt 2014). If ocean acidification is increasing, the deterioration of juvenile settlement habitats may lead to declined populations of marine mollusks.

Effects of Acidification on Feeding Behavior of Mollusks

The feeding rate of the abalone *Concholepas concholepas* larvae was reduced under acidification conditions, and when the pCO_2 increased, the food of the larvae consisted mostly of diatoms and flagellates, but not of cyanobacteria (Vargas et al. 2013). This study revealed a significant effect of low pH on larval feeding behavior, and ocean acidification may radically affect the feeding selectivity of *C. concholepas* larvae. The rock shell *Thais clavigera* prefers to prey on the large-size clam *Brachidontes variabilis*, with a significant reduction in prey treatment time under acidification conditions (Xu et al. 2017). The predation success rate of the predatory sea snail *Conus marmoreus* on the gastropod *Gibberulus gibbosus* decreased by 50% under acidification compared with the control group (Watson et al. 2017). Valve movement is essential for regulating water flow through the gills, providing food and promoting respiration. The ocean quahog *Arctica islandica* increased its valve motion when the pH was reduced to 6.2 (Bamber and Westerlund 2016). The above results showed that marine mollusks are able to make their own unique predation strategy, they prefer the optimum food which can increase the energy ingestion efficiency as much as possible. Thus, their survival and reproduction are guaranteed. However, such feeding behavior has been negatively affected by ocean acidification, most of which is manifested as decreases in prey preference and feeding rate.

Effect of Acidification on Mollusk Antipredation

Prey evolves a variety of strategies to avoid being hunted by predators in the natural environment, and preys and predators co-exist in a state of mutual balance. However, ocean acidification can cause a number of physiological responses, influencing energy metabolism and redistribution, thereby affecting the animal antipredation behavior. The herbivorous gastropod *Gibberos gibbosus* has a strong foot, providing the ability to escape from predation by jumping fast forward. Under the elevated CO_2 and facing threat, the number of jumping *G. gibbosus* individuals halved, and the jump latency increased, indicating that the antipredation performance was impaired by acidified conditions (Watson et al. 2014). The Chilean abalone *C. concholepas* can rapidly self-reset after being hit by waves and thus reduce the predation risk. However, when they were exposed to low pH and predators, the self-resetting time was approximately twice that under the absence of the predator (Manríquez et al. 2013). Moreover, the antipredation capacity of early developmental was lost (Manríquez et al. 2014). The ability of the black turban snail *Tegula funebralis* to resist the starfish predation under acidification conditions was reduced (Jellison et al. 2016). Acidification resulted in a decrease in the number of byssus in the thick shell mussel *M. coruscus* (Li et al. 2015a, Sui et al. 2015), and the decrease in the expression of the byssus secretion-related protein and the decrease in the adductor strength of the mussel at pH 7.3 indicated that the defense capacity is negatively affected (Sui et al. 2017). The intertidal gastropod *Littorina littorea* produces thick shells in response to crab predation, but this antipredation behavior is reduced at low pH, and a significant decrease of the metabolic rate upon predation risk and low pH was observed (Bibby et al. 2007). Notably, this snail can also evade predators by migration, reducing the predation risk and the feeding time. This condition resulted in slow growth and reproduction. The above results indicated that increasing CO_2 may impair the antipredation behavior of mollusks and alter the interaction between preys and predators. This phenomenon is not only reflected in the strategy and physiological state but also in the phenotype and life history.

Effects of Acidification on Reproduction

Unlike in *in vivo* fertilization organisms, most of the marine mollusks directly discharge gametes into the water, and the fertilization is finished in the sea. The breeding behavior of marine mollusks is susceptible to the effects of the environmental change, such as ocean acidification. Zhao et al. found that the growth rate of progeny from the parent Manila clam *Ruditapes philippinarum* exposed to low pH was significantly increased, demonstrating adaptation to acidification (Zhao et al. 2016). However, the fertilization rate of the clam *Macoma balthica* was reduced by 11%, and the hatching rate reduced by 87% under low pH 7.5 (Van et al. 2012). With the increase of pCO_2, the fertilization rate of the rock oyster *Gouldtea glomerata* was significantly reduced (Parker et al. 2009). The sperm motility, fertilization rate, and hatching rate of the rock oyster *C. gigas* were significantly affected by acidification (Barros et al. 2013), but no significant difference was observed in sperm motility between pH 7.8 and 8.15 (Schlegel and Havenhand 2009). Moreover, experiments have shown that acidification exhibits no significant effect on the fertilization of the pearl oyster *Pinctada martensii* (Liu and Lin 2012). The above studies showed that ocean acidification in the next 300 years will have a great effect on the reproduction of mollusks. However, some differences are observed on the effects on different species of mollusks.

Combined Effect of Acidification Associated with Other Factors

Carbon dioxide emissions lead to an increase in global mean temperature and ocean acidification, both occurring simultaneously. To clearly understand their combined effects on marine species, the interactive effects of multiple environmental stressors need to be elucidated. Environmental factors rarely act alone but with a variety of stress factors, especially in estuaries and coastal waters where human activities and environmental pressures are high, but most of the studies focus on a single environmental stress (Wernberg et al. 2012). Interactions of environmental factors include antagonistic effects, additive effects, and synergistic effects. The responses of marine organisms to ocean warming and acidification are quite different due to the differences in physiological and ecological characteristics (Dupont et al. 2008, Fabry 2008). Many marine organisms with calcium carbonate structures are very sensitive to ocean acidification, which can damage their ability to produce calcified bones. However, some calcifications can be achieved by buffering acidification through an acid–base compensation (Doney et al. 2009), changes in the metabolism (Widdicombe and Spicer 2008, Whiteley 2011) and energy redistribution (Wood et al. 2008). The negative impact of global warming is likely to stimulate rapid metabolism, growth, and development (Byrne et al. 2011) and even enhance the negative effects of acidification (Kleypas and Yates 2009).

The net calcium deposition rate and total weight of the Chilean mussel *Mytilid chilensis* were not significantly affected by temperature but high by CO_2, and the combination of temperature and CO_2 only affected the dissolution of the shell and displayed no significant effect on calcification (Duarte et al. 2014). The blue mussel *M. edulis* and the ocean quahog *A. islandica* are tolerant to a wide range of seawater temperature and pCO_2, with elevated lipofuscin concentrations under stress but declined shell growth (Hiebenthal et al. 2013). Thiobarbituric acid reactive substances and catalase of the clam *Chamelea gallina* were significantly decreased, but the superoxide dismutase was significantly increased under the reduced pH and increased temperature (Matozzo et al. 2013). The antioxidant activities of the mussel *Mytilus galloprovincialis* increased, whereas the THC, neutral red uptake, hemolymph lysozyme activity, and total protein levels were significantly inhibited (Matozzo et al. 2012). With the long-term exposure to elevated CO_2 and high temperature, the cell redox status of the oyster *C. virginica* and the clam *M. mercenaria* was altered (Matoo et al. 2013). The survival rate of the fluted clam *Tridacna squamosa* larvae decreased with the increase of pCO_2 and seawater temperature, and the survival rate was less than 20% under the highest combined stressed treatment, showing synergistic effects (Watson et al. 2012). Ocean acidification and warming have a potential impact on hemocyte immune parameters and calcium homeostasis. Li et al. found that acidification and warming reduced the hemolymph pH of the pearl oyster *P. fucata*, increasing the THC, the total protein content, and the percentage of large hyalinocytes and granulocytes but reducing the neutral red intake capacity (Li et al. 2015b).

Nardi et al. found that low pH and cadmium interactions caused permanent damage to hemocyte DNA of the mussel *M. galloprovincialis* and confirmed that the synergistic effect of temperature and acidification increased the toxic effects of cadmium (Nardi et al. 2016). Acidification and cadmium stress inhibit immune-related hemocyte functions of the hard clam *M. mercenaria* and the oyster *C. virginica*, and these effects are exacerbated by elevated pCO_2, resulting in reduced phagocytic activity, lectin, and heat shock protein HSP70 in hemocytes and the hemolysin activity of hemocytes and hemolymphs (Ivanina et al. 2014). However, acidification and copper stress resulted in an increase in the THC, a high phagocytosis, and increased hemocyte adhesion but reduced metabolic activity of hemocytes in the above mentioned two shellfish (Ivanina et al. 2016). Under the conditions of low pH and arsenic exposure, the activities of glutathione S-transferase (GST) in both *C. angulata* and *C. gigas* were reduced, and their antioxidant activities, immune defense, and biomineralisation were impaired (Moreira et al. 2016).

Under acidification and hypoxia stress, the growth rate of the oyster *C. virginica* decreased (Keppel et al. 2016), the respiratory function was inhibited (Steckbauer et al. 2015), the clearance rate, absorption efficiency, respiration rate, excretion rate, O:N ratio, and scope for growth of the mussel *M. coruscus* decreased (Sui et al. 2016). The growth of the Pacific oyster *C. gigas* before and after settlement was also inhibited (Ko et al. 2014). Under natural conditions, diel cycling pH and dissolved oxygen changes may occur in some bays. The biological effects of circadian acidification and hypoxia on mollusks have been rarely reported, and circadian hypoxia and acidification were only reported to affect larval development in the California mussel *Mytilus californianus* and the Mediterranean mussel *M. galloprovincialis* (Frieder et al. 2014). The mechanism of diel cycling hypoxia and acidification on the early development, energy metabolism, hemolymph immunity, and protein expression of mussels is unclear.

Conclusion

Seawater pH, temperature, hypoxia, and most of the heavy metal ions are important environmental factors that affect the survival of marine organisms. They can influence the growth, reproduction, and physiological aspects of marine life and then produce profound impacts on marine biodiversity. In future studies of mollusks, further considering the responses of interspecific differences and intraspecific variability of different mollusk species to multiple environmental stressors is necessary.

The existing understanding of the ecological effects of ocean acidification relies heavily on laboratory exposure experiments (simulating the chemical conditions of seawater acidification and then observing the responses of marine organisms) and mesocosm experiments (Jokiel et al. 2008, Breitbarth et al. 2010). Moreover, investigations on high CO_2 in the natural environments, such as the CO_2 that seeps from the volcano spout (Hall-Spencer et al. 2008) and areas affected by offshore upwellings (Feely et al. 2008), are all important research directions that have important environmental implications. To date, the physiological and ecological response processes and mechanisms of marine organisms to acidification are not understood very well, and less is known about the interaction between different species under acidification conditions and the response of the entire marine ecosystem to acidification. Therefore, we should strengthen the following research areas: (i) acidification and other factors that drive changes in marine ecosystems (such as temperature, dissolved oxygen, and light) and may produce complex interactions. Thus, to explore the ecological effects of multiple stresses from different levels of individuals to ecology is necessary; (ii) Many marine benthic invertebrates have different life stages and characteristics, showing different responses to acidification at different life stages, but most of the existing studies focus on the early life stage. The responses to acidification during the entire life history stage can reflect the adaptability of one species to acidification.

References

Allen, J.D. 2008. Size-specific predation on marine invertebrate larvae. The Biological Bulletin 214: 42-49.

Andersen, S., E. Grefsrud and T. Harboe. 2013. Effect of increased pCO_2 level on early shell development in great scallop (*Pecten maximus* Lamarck) larvae. Biogeosciences 10: 6161.

Asplund, M.E., S.P. Baden, S. Russ, R.P. Ellis, N. Gong and B.E. Hernroth. 2014. Ocean acidification and host–pathogen interactions: blue mussels, *Mytilus edulis*, encountering *Vibrio tubiashii*. Environmental Microbiology 16: 1029–1039.

Bamber, S.D. and S. Westerlund. 2016. Behavioral responses of *Arctica islandica* (Bivalvia: Arcticidae) to simulated leakages of carbon dioxide from sub-sea geological storage. Aquatic Toxicology 180: 295–305.

Barros, P., P. Sobral, P. Range, L. Chícharo and D. Matias. 2013. Effects of sea-water acidification on fertilization and larval development of the oyster *Crassostrea gigas*. Journal of Experimental Marine Biology and Ecology 440: 200–206.

Bechmann, R.K., I.C. Taban, S. Westerlund, B.F. Godal, M. Arnberg, S. Vingen, A. Ingvarsdottir and T. Baussant. 2011. Effects of ocean acidification on early life stages of shrimp (*Pandalus borealis*) and mussel (*Mytilus edulis*). Journal of Toxicology and Environmental Health, Part A 74: 424–438.

Beesley, A., D.M. Lowe, C.K. Pascoe and S. Widdicombe. 2008. Effects of CO—induced seawater acidification on the health of *Mytilus edulis*. Climate Research 37: 215–225.

Bibby, R., P. Cleall-Harding, S. Rundle, S. Widdicombe and J. Spicer. 2007. Ocean acidification disrupts induced defences in the intertidal gastropod *Littorina littorea*. Biology Letters 3: 699–701.

Bibby, R., S. Widdicombe, H. Parry, J. Spicer and R. Pipe. 2008. Effects of ocean acidification on the immune response of the blue mussel *Mytilus edulis*. Aquatic Biology 2: 67–74.

Breitbarth, E., R. Bellerby, C. Neill, M. Ardelan, M. Meyerhöfer, E. Zöllner, P. Croot and U. Riebesell. 2010. Ocean acidification affects iron speciation during a coastal seawater mesocosm experiment. Biogeosciences (BG) 7: 1065–1073.

Byrne, M., M. Ho, E. Wong, N.A. Soars, P. Selvakumaraswamy, H. Shepard-Brennand, S.A. Dworjanyn and A.R. Davis. 2011. Unshelled abalone and corrupted urchins: development of marine calcifiers in a changing ocean. Proceedings Biological Sciences 278: 2376.

Caley, M.J., M.H. Carr, M.A. Hixon, T.P. Hughes, G.P. Jones and B.A. Menge. 1997. Recruitment and the local dynamics of open marine populations. Annu Rev Ecol Syst 27: 477–500.

Carrington, E., J.H. Waite, G. Sarà and K.P. Sebens. 2015. Mussels as a model system for integrative ecomechanics. Annual Review of Marine Science 7: 443–469.

Chegwidden, W.R., N.D. Carter, Y.H. Edwards and R.P. Henry. 1991. The carbonic anhydrases: new horizons. Quarterly Review of Biology 30: 343–363.

Clements, J.C., M.M. Bishop and H.L. Hunt. 2017. Elevated temperature has adverse effects on GABA-mediated avoidance behaviour to sediment acidification in a wide-ranging marine bivalve. Marine Biology 164: 56.

Clements, J.C. and H.L. Hunt. 2014. Influence of sediment acidification and water flow on sediment acceptance and dispersal of juvenile soft-shell clams (*Mya arenaria* L.). Journal of Experimental Marine Biology & Ecology 453: 62–69.

Cole, V.J., L.M. Parker, S.J. O'Connor, W.A. O'Connor, E. Scanes, M. Byrne and P.M. Ross. 2016. Effects of multiple climate change stressors: ocean acidification interacts with warming, hyposalinity, and low food supply on the larvae of the brooding flat oyster *Ostrea angasi*. Marine Biology 163: 1–17.

Cunningham, S.C., A.M. Smith and M.D. Lamare. 2015. The effects of elevated pCO_2 on growth, shell production and metabolism of cultured juvenile abalone, *Haliotis iris*. Aquaculture Research 47.

Doney, S.C., V.J. Fabry, R.A. Feely and J.A. Kleypas. 2009. Ocean acidification: the other CO_2 problem. Annual Reviews of Marine Sciences 1: 169–192.

Duarte, C., J.M. Navarro, K. Acuña, R. Torres, P.H. Manríquez, M.A. Lardies, C.A. Vargas, N.A. Lagos and V. Aguilera. 2014. Combined effects of temperature and ocean acidification on the juvenile individuals of the mussel *Mytilus chilensis*. Journal of Sea Research 85: 308–314.

Dupont, S., J.N. Havenhand, Thorndyke, L.S, Peck and M.C. Thorndyke. 2008. CO_2-driven ocean acidification radically affect larval survival and development in the brittlestar *Ophiothrix fragilis*. Marine Biology 373: 285–294.

Fabry, V.J. 2008. Marine calcifiers in a high-CO_2 ocean. Science 320: 1020–1022.

Feely, R.A., S.C. Doney and S.R. Cooley. 2009. Ocean acidification: Present conditions and future changes in a high-CO_2 world. Oceangraphy 22: 36–47.

Feely, R.A., C.L. Sabine, J.M. Hernandezayon, D. Ianson and B. Hales. 2008. Evidence for upwelling of corrosive "acidified" water onto the continental shelf. Science 320: 1490–1492.

Feely, R.A., C.L. Sabine, K. Lee, W. Berelson, J. Kleypas, V.J. Fabry and F.J. Millero. 2004. Impact of anthropogenic CO_2 on the $CaCO_3$ system in the oceans. Science 305: 362–366.

Fitzer, S.C., W. Zhu, K.E. Tanner, V.R. Phoenix, N.A. Kamenos and M. Cusack. 2015. Ocean acidification alters the material properties of *Mytilus edulis* shells. Journal of the Royal Society Interface 12.

Frieder, C.A., J.P. Gonzalez, E.E. Bockmon, M.O. Navarro and L.A. Levin. 2014. Can variable pH and low oxygen moderate ocean acidification outcomes for mussel larvae? Global Change Biology 20: 754–764.

Gazeau, F., J.P. Gattuso, C. Dawber, A.E. Pronker, F. Peene, J. Peene, C.H.R. Heip and J.J. Middelburg. 2010. Effect of ocean acidification on the early life stages of the blue mussel *Mytilus edulis*. Biogeosciences Discussions 7: 2051–2060.

Gazeau, F., C. Quiblier, J.M. Jansen, J.P. Gattuso, J.J. Middelburg and C.H.R. Heip. 2007. Impact of elevated CO_2 on shellfish calcification. Geophysical Research Letters 34: L07603.

Green, M.A., G.G. Waldbusser, L. Hubazc, E. Cathcart and J. Hall. 2013. Carbonate mineral saturation state as the recruitment cue for settling bivalves in marine muds. Estuaries and Coasts 36: 18–27.

Guo, X., M. Huang, F. Pu, W. You and C. Ke. 2015. Effects of ocean acidification caused by rising CO_2 on the early development of three mollusks. Aquatic Biology 23: 147–157.

Hall-Spencer, J.M., R. Rodolfo-Metalpa, S. Martin, E. Ransome, M. Fine, S.M. Turner, S.J. Rowley, D. Tedesco and M.C. Buia. 2008. Volcanic carbon dioxide vents show ecosystem effects of ocean acidification. Nature 454: 96–99.

Havenhand, J.N., F.-R. Buttler, M.C. Thorndyke and J.E. Williamson. 2008. Near-future levels of ocean acidification reduce fertilization success in a sea urchin. Current Biology 18: R651–R652.

Hettinger, A., E. Sanford, T.M. Hill, E.A. Lenz, A.D. Russell and B. Gaylord. 2013. Larval carry-over effects from ocean acidification persist in the natural environment. Global Change Biology 19: 3317–3326.

Hiebenthal, C., E.E.R. Philipp, A. Eisenhauer and M. Wahl. 2013. Effects of seawater pCO_2 and temperature on shell growth, shell stability, condition and cellular stress of Western Baltic Sea *Mytilus edulis* (L.) and *Arctica islandica* (L.). Marine Biology 160: 2073–2087.

Hu, M., L. Li, Y. Sui, J. Li, Y. Wang, W. Lu and S. Dupont. 2015. Effect of pH and temperature on antioxidant responses of the thick shell mussel *Mytilus coruscus*. Fish & Shellfish Immunology 46: 573–583.

Ivanina, A.V., C. Hawkins and I.M. Sokolova. 2014. Immunomodulation by the interactive effects of cadmium and hypercapnia in marine bivalves *Crassostrea virginica* and *Mercenaria mercenaria*. Fish & Shellfish Immunology 37: 299–312.

Ivanina, A.V., C. Hawkins and I.M. Sokolova. 2016. Interactive effects of copper exposure and environmental hypercapnia on immune functions of marine bivalves *Crassostrea virginica* and *Mercenaria mercenaria*. Fish & Shellfish Immunology 49: 54–65.

Jellison, B.M., A.T. Ninokawa, T.M. Hill, E. Sanford and B. Gaylord. 2016. Ocean acidification alters the response of intertidal snails to a key sea star predator. Proc Biol Sci 283: 20160890.

Jokiel, P., K. Rodgers, I. Kuffner, A. Andersson, E. Cox and F. Mackenzie. 2008. Ocean acidification and calcifying reef organisms: a mesocosm investigation. Coral Reefs 27: 473–483.

Keppel, A.G., D.L. Breitburg and R.B. Burrell. 2016. Effects of co-varying diel-cycling hypoxia and pH on growth in the juvenile Eastern Oyster, *Crassostrea virginica*. Plos One 11: e0161088.

Kim, J.-H., O.H. Yu, E.J. Yang, S.-H. Kang, W. Kim and E.J. Choy. 2016. Effects of ocean acidification driven by elevated CO2 on larval shell growth and abnormal rates of the venerid clam, Mactra veneriformis. Chinese journal of oceanology and limnology 34: 1191–1198.

Kleypas, J.A. and K.K. Yates. 2009. Coral reefs and ocean acidification. Oceanography 22: 108–117.

Ko, G.W., R. Dineshram, C. Campanati, V.B. Chan, J. Havenhand and V. Thiyagarajan. 2014. Interactive effects of ocean acidification, elevated temperature, and reduced salinity on early-life stages of the pacific oyster. Environmental Science & Technology 48: 10079–10088.

Kripa, V., J.H. Sharma, S. Chinnadurai, L.R. Khambadkar, D. Prema and K.S. Mohamed. 2014. Effects of acidification on meroplanktonic oyster larval settlement in a tropical estuary. Indian Journal of Geo-Marine Science 43: 1675–1681.

Kurihara, H., S. Kato and A. Ishimatsu. 2007. Effects of increased seawater pCO_2 on early development of the oyster *Crassostrea gigas*. Aquatic Biology 1: 91–98.

Le Quere, C., T. Takahashi, E.T. Buitenhuis, C. Roedenbeck and S.C. Sutherland. 2010. Impact of climate change and variability on the global oceanic sink of CO_2. Global Biogeochemical Cycles 24.

Lewis, C., R.P. Ellis, E. Vernon, K. Elliot, S. Newbatt and R.W. Wilson. 2016. Ocean acidification increases copper toxicity differentially in two key marine invertebrates with distinct acid-base responses. Scientific Reports 6: 21554.

Li, L., W. Lu, Y. Sui, Y. Wang, Y. Gul and S. Dupont. 2015a. Conflicting effects of predator cue and ocean acidification on the mussel *Mytilus coruscus* byssus production. Journal of Shellfish Research 34: 393–400.

Li, S., Y. Liu, C. Liu, J. Huang, G. Zheng, L. Xie and R. Zhang. 2015b. Morphology and classification of hemocytes in *Pinctada fucata* and their responses to ocean acidification and warming. Fish & Shellfish Immunology 45: 194–202.

Liu, M., S. Liu, Y. Hu and L. Pan. 2015. Cloning and expression analysis of two carbonic anhydrase genes in white shrimp *Litopenaeus vannamei*, induced by pH and salinity stresses. Aquaculture 448: 391–400.

Liu, S., W. Shi, C. Guo, X. Zhao, Y. Han, C. Peng, X. Chai and G. Liu. 2016. Ocean acidification weakens the immune response of blood clam through hampering the NF-kappa β and toll-like receptor pathways. Fish & Shellfish Immunology 54: 322.

Liu, W.G. and J.S. Lin. 2012. Effect of ocean acidification on fertilization and early development of the pearl oyster *Pinctada martensii* Dunker. Marine Sciences 36: 19–23.

Ma, F.R., P. Range, X.A. Alvarez-Salgado and U. Labarta. 2011. Physiological energetics of juvenile clams *Ruditapes decussatus* in a high CO_2 coastal ocean. Marine Ecology Progress 433: 97–105.

Manríquez, P.H., M.E. Jara, M.L. Mardones, J.M. Navarro, R. Torres, M.A. Lardies, C.A. Vargas, C. Duarte, S. Widdicombe and J. Salisbury. 2013. Ocean acidification disrupts prey responses to predator cues but not net prey shell growth in *Concholepas concholepas* (loco). Plos One 8: e68643.

Manríquez, P.H., M.E. Jara, M.L. Mardones, R. Torres, J.M. Navarro, M.A. Lardies, C.A. Vargas, C. Duarte and N.A. Lagos. 2014. Ocean acidification affects predator avoidance behaviour but not prey detection in the early ontogeny of a keystone species. Marine Ecology Progress 502: 157–167.

Matoo, O.B., A.V. Ivanina, C. Ullstad, E. Beniash and I.M. Sokolova. 2013. Interactive effects of elevated temperature and CO_2 levels on metabolism and oxidative stress in two common marine bivalves (*Crassostrea virginica* and *Mercenaria mercenaria*). Comp Biochem Physiol A Mol Integr Physiol 164: 545–553.

Matozzo, V., A. Chinellato, M. Munari, M. Bressan and M.G. Marin. 2013. Can the combination of decreased pH and increased temperature values induce oxidative stress in the clam *Chamelea gallina* and the mussel *Mytilus galloprovincialis*? Marine Pollution Bulletin 72: 34–40.

Matozzo, V., A. Chinellato, M. Munari, L. Finos, M. Bressan and M.G. Marin. 2012. First evidence of immunomodulation in bivalves under seawater acidification and increased temperature. PLoS One 7: e33820.

Mcdonald, M.R., J.B. Mcclintock, C.D. Amsler, R. Dan, R.A. Angus, B. Orihuela and K. Lutostanski. 2009. Effect of ocean acidification over the life history of the barnacle *Amphibalanus amphitrite*. Marine Ecology Progress 385: 179–187.

Moreira, A., E. Figueira, A.M.V.M. Soares and R. Freitas. 2016. The effects of arsenic and seawater acidification on antioxidant and biomineralization responses in two closely related *Crassostrea* species. Science of the Total Environment 545–546: 569.

Nardi, A., L.F. Mincarelli, M. Benedetti, D. Fattorini, G. D'Errico and F. Regoli. 2016. Indirect effects of climate changes on cadmium bioavailability and biological effects in the Mediterranean mussel *Mytilus galloprovincialis*. Chemosphere 169: 493.

Navarro, J.M., R. Torres, K. Acuña, C. Duarte, P.H. Manriquez, M. Lardies, N.A. Lagos, C. Vargas and V. Aguilera. 2013. Impact of medium-term exposure to elevated pCO_2 levels on the physiological energetics of the mussel *Mytilus chilensis*. Chemosphere 90: 1242–1248.

Parker, L., P. Ross, W. O'Connor, H. Pörtner, E. Scanes and J. Wright. 2013. Predicting the response of molluscs to the impact of ocean acidification. Biology 2: 651–692.

Parker, L.M., P.M. Ross and W.A. O'CONNOR. 2009. The effect of ocean acidification and temperature on the fertilization and embryonic development of the Sydney rock oyster Saccostrea glomerata (Gould 1850). Global Change Biology 15: 2123–2136.

Parker, L.M., P.M. Ross and W.A. O'Connor. 2011. Populations of the Sydney rock oyster, *Saccostrea glomerata*, vary in response to ocean acidification. Marine Biology 158: 689–697.

Ren, G., Y. Wang, J. Qin, J. Tang, X. Zheng and Y. Li. 2014. Characterization of a novel carbonic anhydrase from freshwater pearl mussel *Hyriopsis cumingii* and the expression profile of its transcript in response to environmental conditions. Gene 546: 56–62.

Sabine, C.L., R.A. Feely, N. Gruber, R.M. Key, K. Lee, J.L. Bullister, R. Wanninkhof, C.S. Won, D.W.R. Wallace, B. Tilbrook, F.J. Millero, T.-H. Peng, A. Kozyr, T. Ono and A.F. Rios. 2004. The oceanic sink for anthropogenic CO_2. Science 305: 367–371.

Schalkhausser, B., C. Bock, K. Stemmer, T. Brey, H.O. Pörtner and G. Lannig. 2013. Impact of ocean acidification on escape performance of the king scallop, *Pecten maximus* , from Norway. Marine Biology 160: 1995–2006.

Schlegel, P. and J. Havenhand. 2009. Near-future levels of ocean acidification do not affect sperm motility and fertilization kinetics in the oyster *Crassostrea gigas*. Biogeosciences 6: 3009–3015.

Steckbauer, A., L. Ramajo, I.E. Hendriks, M. Fernandez, N.A. Lagos, L. Prado and C.M. Duarte. 2015. Synergistic effects of hypoxia and increasing $CO_{2\,o}$n benthic invertebrates of the central Chilean coast. Modares Mechanical Engineering 2: 8.

Sui, Y., M. Hu, X. Huang, Y. Wang and W. Lu. 2015. Anti-predatory responses of the thick shell mussel *Mytilus coruscus* exposed to seawater acidification and hypoxia. Marine Environmental Research 109: 159.

Sui, Y., H. Kong, X. Huang, S. Dupont, M. Hu, D. Storch, H.O. Pörtner, W. Lu and Y. Wang. 2016. Combined effects of short-term exposure to elevated CO_2 and decreased O_2 on the physiology and energy budget of the thick shell mussel *Mytilus coruscus*. Chemosphere 155: 207.

Sui, Y., Y. Liu, X. Zhao, S. Dupont, M. Hu, F. Wu, X. Huang, J. Li, W. Lu and Y. Wang. 2017. Defense responses to short-term hypoxia and seawater acidification in the thick shell mussel *Mytilus coruscus*. Frontiers in Physiology 8: 145.

Sun, T., X. Tang, B. Zhou and Y. Wang. 2016. Comparative studies on the effects of seawater acidification caused by CO2 and HCl enrichment on physiological changes in Mytilus edulis. Chemosphere 144: 2368–2376.

Talmage, S.C. and C.J. Gobler. 2009. The effects of elevated carbon dioxide concentrations on the metamorphosis, size, and survival of larval hard clams (*Mercenaria mercenaria*), bay scallops (*Argopecten irradians*), and Eastern oysters (*Crassostrea virginica*). Limnology & Oceanography 54: 2072–2080.

Talmage, S.C. and C.J. Gobler. 2010. Effects of past, present, and future ocean carbon dioxide concentrations on the growth and survival of larval shellfish. Proceedings of the National Academy of Sciences of the Unied States of America 107: 17246–17251.

Timmins-Schiffman, E., M.J. O'Donnell, C.S. Friedman and S.B. Roberts. 2013. Elevated pCO_2 causes developmental delay in early larval Pacific oysters, *Crassostrea gigas*. Marine Biology 160: 1973–1982.

Tunnicliffe, V., K.T.A. Davies, D.A. Butterfield, R.W. Embley, J.M. Rose and W.W. Chadwick, Jr. 2009. Survival of mussels in extremely acidic waters on a submarine volcano. Nature Geoscience 2: 344–348.

Van, C.C., E. Debusschere, U. Braeckman, G.D. Van and M. Vincx. 2012. The early life history of the clam *Macoma balthica* in a high CO_2 world. Plos One 7: e44655.

Vargas, C.A., M.D.L. Hoz, V. Aguilera, V.S. Martín, P.H. Manríquez, J.M. Navarro, R. Torres, M.A. Lardies and N.A. Lagos. 2013. CO_2-driven ocean acidification reduces larval feeding efficiency and changes food selectivity in the mollusk *Concholepas concholepas*. Journal of Plankton Research 35: 1059–1068.

Wang, X., M. Wang, Z. Jia, L. Qiu, L. Wang, A. Zhang and L. Song. 2017. A carbonic anhydrase serves as an important acid-base regulator in pacific oyster *Crassostrea gigas* exposed to elevated CO_2: Implication for physiological responses of mollusk to ocean acidification. Marine Biotechnology 19: 22–35.

Wang, Y., L. Li, M. Hu and W. Lu. 2015. Physiological energetics of the thick shell mussel *Mytilus coruscus* exposed to seawater acidification and thermal stress. Science of the Total Environment 514: 261–272.

Watson, S.-A., S. Lefevre, M.I. McCormick, P. Domenici, G.E. Nilsson and P.L. Munday. 2014. Marine mollusc predator-escape behaviour altered by near-future carbon dioxide levels. Proceedings of the Royal Society of London B: Biological Sciences 281: 20132377.

Watson, S.A., J.B. Fields and P.L. Munday. 2017. Ocean acidification alters predator behaviour and reduces predation rate. Biology letters 13.

Watson, S.A., P.C. Southgate, G.M. Miller, J.A. Moorhead and J. Knauer. 2012. Ocean acidification and warming reduce juvenile survival of the fluted giant clam,*Tridacna squamosa*. Molluscan Research 32: 177–180.

Welladsen, H.M., P.C. Southgate and K. Heimann. 2010. The effects of exposure to near-future levels of ocean acidification on shell characteristics of *Pinctada fucata* (Bivalvia: Pteriidae). Molluscan Research 30: 125–130.

Wernberg, T., D.A. Smale and M.S. Thomsen. 2012. A decade of climate change experiments on marine organisms: procedures, patterns and problems. Global Change Biology 18: 1491–1498.

White, M.M., D.C. McCorkle, L.S. Mullineaux and A.L. Cohen. 2013. Early exposure of bay scallops (*Argopecten irradians*) to high CO_2 causes a decrease in larval shell growth. PloS one 8: e61065.

Whiteley, N.M. 2011. Physiological and ecological responses of crustaceans to ocean acidification. Marine Ecology Progress 430: 257–271.

Widdicombe, S. and J.I. Spicer. 2008. Predicting the impact of ocean acidification on benthic biodiversity: What can animal physiology tell us? Journal of Experimental Marine Biology & Ecology 366: 187–197.

Wood, H.L., J.I. Spicer and S. Widdicombe. 2008. Ocean acidification may increase calcification rates, but at a cost. Proceedings Biological Sciences 275: 1767.

Wu, F., W. Lu, Y. Shang, H. Kong, L. Li, Y. Sui, M. Hu and Y. Wang. 2016. Combined effects of seawater acidification and high temperature on hemocyte parameters in the thick shell mussel *Mytilus coruscus*. Fish & Shellfish Immunology 56: 554–562.

Xu, X.Y., K.R. Yip, P.K.S. Shin and S.G. Cheung. 2017. Predator–prey interaction between muricid gastropods and mussels under ocean acidification. Marine Pollution Bulletin 124: 911–916.

Zhang, M.L., J. Zou, J.G. Fang, J.H. Zhang, M.R. Du, B. Li and L.H. Ren. 2011. Impacts of marine acidification on calcification, respiration and energy metabolism of Zhikong scallop *Chlamys farreri*. Progress in Fishery Sciences 32: 48–54.

Zhao, L., B.R. Schöne, R. Mertzkraus and F. Yang. 2016. Sodium provides unique insights into transgenerational effects of ocean acidification on bivalve shell formation. Science of the Total Environment 577: 360.

CHAPTER 16

Climate Change Effects on the Physiology and Ecology of Fish

Wang Xiaojie

Introduction

Since the industrial revolution the increasing amount of anthropogenic CO_2 emission in the air has caused the average surface temperature of the Earth to rise by about 0.7°C. One fourth of the CO_2 is absorbed by the ocean causing ocean acidification (Ciais et al. 2014). According to the CO_2 emission rate, by the end of the 21st century the Earth's surface temperature will rise by 2–4°C and the water pH value will drop by 0.3–0.4 units (IPCC 2007). In the freshwater and seawater ecosystems, the oxygen content is less than 2.0 mg/L, which is also a common ecological problem. Ocean warming together with increased stratification and changing circulation is leading to the decrease of the oxygen content in the sea. Therefore, hypoxia is considered to be the third largest climate change stress factor, along with ocean warming and ocean acidification (Levin and Breitburg 2015).

Temperature is the main non-biological determinant of fish physiology and ecology (Beitinger and Fitzpatrick 1979, Houde 1989). Ocean warming has changed the geographical distribution of ectothermic animals which has spread to poles to look for more comfortable temperatures (Sunday et al. 2012). Although fish have an ability of acid-base regulation, the effects of acidification on their sensory system and behavior of larval fish have been reported (Munday et al. 2009a, Bignami et al.

Institute for Marine Biosystem and Neurosciences Shanghai Ocean University Shanghai 201306, China; Key Laboratory of Exploration and Utilization of Aquatic Genetic Resources Shanghai Ocean University Ministry of Education, Shanghai 201306, China; and National Demonstration Center for Experimental Fisheries Science Education (Shanghai Ocean University), Shanghai 201306, China.
Email: xj-wang@shou.edu.cn

2013). Through affecting the behavior and physiology of fish, hypoxic events can induce shifts in spatial distribution and the structure of fish communities through direct mortality during extreme local events (Domenici et al. 2007). Currently, the research on the effects of multiple environmental stress factors on fish is attracting attention (Strobel et al. 2012, Enzor et al. 2013). In global fisheries production, fish play the most important role and account for 85% of the total. Therefore, the changes of the marine environment will seriously affect the global fisheries production (Branch et al. 2013). The effects of climate change on fish may also affect the service and output function of the entire marine ecosystem through the complex food web in the ocean.

Climate Change Effects on the Physiology of Fish

Ocean Warming Effects on the Oxygen Physiology of Fish

In aquatic ectotherms, such as fish, the body temperature closely follows the ambient water temperature. Changes in temperature directly affect the physiological processes of fish. A decreased capacity to perform aerobically (reduced aerobic scope) is hypothesized to be the key physiological mechanism that will determine the response of fishes to increased ocean temperature (Pörtner and Knust 2007, Pörtner and Farrell 2008). The aerobic scope sets the lower and higher thermal limits for an animal and is defined as the difference between the oxygen uptake needed to sustain basal metabolic functions and the maximal rate of oxygen uptake of the animal (Pörtner 2001) (Fig. 16.1). The limited capacity of the circulatory and ventilatory systems to keep pace with the increased oxygen demands of basal metabolism at higher temperatures causes a reduction in aerobic scope, allowing less energy to be devoted to, for example, feeding, growth and reproduction. Thus, a reduced capacity for aerobic function at higher temperatures affects all aspects of individual performance.

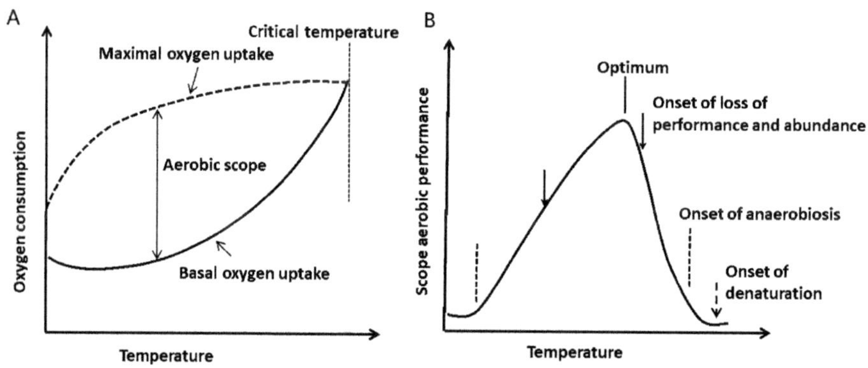

Fig. 16.1. Temperature effects on aquatic animals. (A) Aerobic scope is the difference between the minimum rate of oxygen consumption in unfed animals at rest and the maximal rate of oxygen consumption. (B) the thermal windows of aerobic performance display optima and limitations by pejus (pejus means 'turning worse'), critical, and denaturation temperatures, when tolerance becomes increasingly passive and time-limited (modified from Pörtner 2001, Pörtner and Farrell 2008).

Any imbalance between oxygen demand and supply will constrain the aerobic scope, thereby impairing individual performance. In warming environments, small-sized individuals are better able to balance demand and uptake because of their larger surface area to volume ratio (Atkinson 1994). This phenomenon, known as the temperature-size rule (TSR), has important consequences for global fisheries, whereby ocean warming is predicted to result in smaller fish and reduced biomass. In the Mediterranean Sea, the maximal size of 75 fish species along a steep temperature gradient supports the TSR. Moreover, size reduction in active fish species is dramatically larger than in more sedentary species (Van et al. 2017). Over a 40-year period, six out of eight commercial fish species examined in the North Sea underwent concomitant reductions in asymptotic body size with the synchronous component of the total variability coinciding with a 1–2°C increase in water temperature. Smaller body sizes decreased the yield-per-recruit of these stocks by 23% (Baudron et al. 2014). Changes in temperature, oxygen content and other ocean biogeochemical properties directly affect the ecophysiology of marine water-breathing organisms. Previous studies suggest that the most prominent biological responses are changes in distribution, phenology and productivity. Employing a model to examine the integrated biological response of over 600 species of marine fish due to changes in distribution, abundance and body size shows that assemblage-averaged maximum body weight is expected to shrink by 14–24% globally from 2000 to 2050 under a high-emission scenario. About half of this shrinkage is due to changes in distribution and abundance, the remainder to changes in physiology. The tropical and intermediate latitudinal areas will be heavily impacted with an average reduction of more than 20% (Cheung et al. 2013).

Oxygen supply has been shown to limit maximal body size of aquatic organisms (Hoefnagel and Verberk 2015). For fishes, oxygen uptake rate (i.e., supply) depends on gill surface area to body mass ratio. Therefore, individuals may approach a size where they can no longer acquire the oxygen needed for maintaining their metabolic demands (Pauly 1979). The theoretical underpinning for fish shrinking is that the oxygen supply to large-size fish cannot be met by their gills, whose surface area cannot keep up with the oxygen demand by their three-dimensional bodies (Pauly and Cheung 2018).

Ocean Acidification Effects on the Development and Sensory Systems

Marine fish have a strong tolerance to ocean acidification, because they can regulate the acid-base balance by the gills and kidneys (Brauner and Baker 2009, Esbaugh et al. 2012). However, larval fish differ from adults due to respiration and ion exchange patterns (Jonz and Nurse 2006, Pelster 2008), and have a higher surface area and volume ratio compared to adults (Kikkawa et al. 2003, Ishimatsu et al. 2004), so they are sensitive to ocean acidification. Acidification affects the embryonic development (Tseng et al. 2013) and the growth of larvae (Baumann et al. 2011), damages the structure of internal tissues and organs (Frommel et al. 2011) and reduces the survival of larvae (Baumann et al. 2011), etc. However, in other studies, it was found that ocean acidification has no effects on the fertilization of sperm and eggs (Franke and Clemmesen 2011), the incubation rate of fertilized eggs (Frommel et al. 2012), the growth and development of larvae (Munday et al. 2011, Frommel et al. 2011, 2012, Hurst et al. 2013, Bignami et al. 2013) and the swimming ability of larvae (Munday et al. 2009b).

Unlike internal organs, the sensory receptors of olfactory, visual, and auditory system are all on the fish's body surface, which are directly in contact with seawater. They are vulnerable to ocean acidification. Recent studies have shown that larval fish develop behavioral and sensory abnormalities when exposed to acidifying water. In 2009, Philip Munday et al. found that ocean acidification can reverse the loss of olfactory preference of coral reef fish (Munday et al. 2009a). Then more and more experiments and field observations found that acidification can also affect fish auditory sensitivity (Bignami et al. 2013), reduce the behavioral lateralization (Domenici et al. 2012) as well as influence larval visual identification of predators (Ferrari et al. 2012), showing that it affects the central nervous system of fish. And ocean acidification influences not only tropical fish, but also temperate fish, such as the Indian Ocean medaka (Wang et al. 2017) and California reef fish (Hamilton et al. 2014).

What is the underlying mechanism linking high CO_2 to these diverse responses? Nilsson et al. (2012) found that abnormal olfactory preferences and loss of behavioral lateralization exhibited by two species of larval coral reef fish exposed to high CO_2 can be rapidly and effectively reversed by treatment with an antagonist of the $GABA_A$ receptor. It suggests that ocean acidification changes the function of the $GABA_A$ receptor which affects the neural activity of the larvae. The $GABA_A$ receptor is an important inhibitory receptor in the central nervous system. After combining with agonists, the decoupling Cl^-, HCO_3^- channels open and the negative ions internal flow hyperpolarizes the neurons (decreased excitation). However, in the process of fetal brain development or in some epilepsy cases of human, following the change of Cl^-, HCO_3^- concentration gradient between the inside and outside of the membrane, there is an outflow of anions when $GABA_A$ receptor channels open. The neurons appear depolarized (increased excitation) (Farrant and Kaila 2007). In acidifying water, the fish will accumulate HCO_3^- and expel Cl^- to avoid the acidic effect (Ishimatsu et al. 2008, Brauner and Baker 2009). The acid-base adjustment of fish will change the anion flow direction after the $GABA_A$ receptor is activated. Its function will change from hyperpolarization to depolarization which will reverse the behavior of prey fish when it faces predators. The apparent link between CO_2-induced behavioral disturbances and the $GABA_A$ receptor has been supported by several other studies, using an array of sensory and behavioral assays as well as different $GABA_A$ receptor antagonists and agonists (Chung et al. 2014, Chivers et al. 2014). Examination of acid-base balance disturbances and associated compensatory mechanisms have also been performed at ocean acidification relevant scenarios. Spiny damselfish (*Acanthochromis polyacanthus*) exposed to 1,900 μatm CO_2 for 4 d exhibited significantly increased intracellular and extracellular HCO_3^- concentrations and elevated brain pH compared to control fish, providing evidence of CO_2 compensation. Moreover, high CO_2-exposed fish showed significantly olfactory abnormality, supporting a potential link between behavior disruption and CO_2 compensation (Heuer et al. 2016).

Hypoxia Effects on the Physiology of Fish

Both laboratory and field studies have demonstrated that hypoxia can slow down the growth of fish. It also can affect specific genes along the brain-pituitary-gonad axis, disrupt the synthesis and balance of sex hormones and lead to reproductive impairment and sex alteration in several fish species (Shang et al. 2006). The reproductive impairment by hypoxia can be transmitted to their offspring through epigenetic effects (Wang et al. 2016). Under a short-term low oxygen stress, fish can reduce the adverse effects through behaviors, such as adjusting activity intensity or finding relatively suitable habitats. For example, fish can increase the intake of oxygen through the gills, meanwhile reducing swimming activities to ensure that the intake of oxygen just meets the basic metabolic needs. If hypoxia persists, fish will reduce the range of aerobic activity by reducing the maximum metabolic rate (MMR), which is bound to affect growth, reproduction, and swimming behavior of fish (Claireaux et al. 2000). The amount of dissolved oxygen at different depths of water limits the distribution of fish in the ocean. The low-oxygen zones usually appear in the intermediate layer (100–200 m depth) up to the bottom (Diaz and Rosenberg 2008). Habitat compression is another negative effect of hypoxia on oceanic fish. For example, tuna is a large predator, and it often conducts extensive vertical movements during the feeding process. A 50% saturation of oxygen in the blood is an indicator of blood oxygen affinity, which can be used as a standard by animals for low oxygen tolerance. According to the 50% saturation of the oxygen pressure, due to a reduction of dissolved oxygen in the northern Pacific Ocean, the distribution of blue tuna in the Pacific and southern seas will be shallow (Mislan et al. 2017).

Multiple Stressors Effect on the Physiology of Fish

Today, marine life has been facing complex marine environmental changes, such as acidification, warming, and hypoxia. However, most studies have been done on the physiological effects of a single stress factor on fish, and there are few studies on the effects of multiple environmental factors. Understanding the combined effects of multiple stress factors on fish can better predict the impact of global climate change on fishery resources. The physiology and behavior of fish are more sensitive when exposed to both acidification and warming. In acidifying and warming water, feeding behavior (Nowicki et al. 2012), activity level (Faleiro et al. 2015) and aerobic scope of fish (Munday et al. 2009c) are affected. It is estimated that in the next century, the highest pH and temperature changes will occur at high latitudes (IPCC 2013). Under these double stressors, the metabolic rate of the polar fish will increase significantly (Enzor et al. 2013), and the mRNA expression associated with cell stress response will also increase (Huth and Place 2016). Especially in the early stage of development, the survival rate of eggs and larvae will be reduced. According to the SCREI (Simulator of Cod Recruitment under Environmental Influence) model, by 2100, under only ocean warming stress, the recruitment of very young Atlantic cod will fall to 35% of the long-term average (1983–2014). Under combined ocean warming and acidification, the average recruitment is reduced to 5% of the long-term average by the end of the century (Koenigstein et al. 2018).

Climate Change Effects on the Interspecies Relation of Fish

Previous studies of global climate change have focused on the direct physiological effects of ocean warming on fish. However, one recent meta-analysis (> 20 yr) on terrestrial and aquatic system found that, compared to the direct effects, the resulting new interspecies relationships are more likely to change the structure and function of ecosystems (Ockendon et al. 2014).

Climate Change Effects on the Distribution Areas of Fish

In some areas the temperature increase rate is higher than the global average temperature increase rate. And the isotherm is moving towards the poles (Burrows et al. 2011). As a way of phenotypic plasticity, behavioral thermoregulation can reduce the adverse effects of environmental temperature changes on species and populations. The main response of fish to the temperature rise is changing of the distribution area and swimming to high latitudes or deep water areas. The change of distribution in latitude makes tropical or subtropical fish the dominant species in temperate areas. The recombination of fish communities is called tropicalisation (Vergés et al. 2014). Fish migration to higher latitudes has become a global phenomenon. It appears at different latitudes, including the Australian eastern tropical to temperate transition zone (Vergés et al. 2016), the temperate area including Australia reef communities (Wernberg et al. 2016), European areas (Montero-Serra et al. 2015) and the northeast continental shelf sea area in America (Kleisner et al. 2017), and the Arctic waters, such as of Norway and the Barents sea (Koenigstein et al. 2018). In the context of global climate change, the change of species distribution brings about ecosystem recombination (ecosystem reconfigurations), which will have profound ecological, social, and economic impacts (Wernberg et al. 2016).

Using the fishing average temperature index (the mean temperature of the catch), fishery catches from 1970 to 2006 showed that marine fishery catches were significantly affected by the temperature change. More and more warm-water fish species become dominant at high latitudes. And subtropical species in tropical sea catches are lesser and lesser. The change of fishing composition has a great impact on the development of fisheries in coastal areas (Cheung et al. 2013). Tuna has an important economic value. These fish depend on a countercurrent heat exchange system to maintain their body temperature above the ambient temperature. Therefore, water temperature is an important factor in determining its distribution. Tropical tuna mainly distribute in water temperature higher than 18°C. According to long lines catches data, in subtropical areas of the Atlantic, the western Pacific and Indian Ocean the percentage of tropical tuna catch is on the rise. Thus, with global warming, tropical tuna are moving towards higher latitudes (Monllor-Hurtado et al. 2017). Fish move to higher latitudes and primary productivity increases with ocean warming. This may lead to an increase in the overall fishery income in the Arctic areas by 39% by 2050 as estimated by models. Compared with the tropical developing countries, the Arctic will likely benefit from climate change. However, the models predict that the potential benefits brought by warming will be offset by the reducing effects of ocean acidification in the Arctic Ocean (Lam et al. 2016).

Climate Change Effects on Fish Related Food Webs

Global climate change does not only directly but also indirectly affect the physiology and distribution of fish, through the mutual relationships between species, which further changes the population and community composition and ecosystem function. The indirect effects of climate change on fish include the following two aspects. First of all, species that have changed their distribution areas emerge in the new environment and generate new biological interactions. Secondly, some species migrate out of previous areas which results in the disappearance of biological interactions in these regions (Vergés et al. 2014). Even a moderate amount of temperature change, therefore, will produce a series of interaction processes in the food web. When an affected creature is the key species of the food web, climate change will have profound effects on the structure and function of the whole ecological system (Gilbert et al. 2014). Arctic creatures in the water of Norway and the Barents sea are thought to be the most influenced by global warming. With temperature increase, the fish in the north, such as the Atlantic cod move to the north (Drinkwater 2011). Meanwhile, as the seasonal ice cover area decreases and primary productivity increases, the pelagic species, such as blue and white cod (*Micromesistius poutassou*) and mackerel (*Scomber scombrus*) extend to the east and north (Loeng et al. 2005). How does the change of interspecies relationships affect the food web in Norway and the Barents sea? The increasing pelagic fish population will feed on the common species of the Norwegian sea, such as capelin, shrimp and red fish which are at the intermediate level of the food chain and are the main prey of cod. Furthermore, with increasing numbers, mackerel will also prey on more cod. Therefore, it is estimated that the cod population eventually migrates out of Norway and the Barents sea and gradually moves towards the Arctic ice-free zone (Bentley et al. 2017). Similarly, in the heavily impacted European North Sea ecosystem, a statistical non-deterministic model is used to analyze how climate and fishing interact between top-down and bottom-up control in food webs. Ocean warming pushes the bottom-up effects and mainly changes the community structure of zooplankton. Fishing brings the top-down effects and changes the biomass of economic fish. Planktivorous foraging fish (herring, Norway cod) are the key groups in the North Sea ecological system, linking bottom-up and top-down processes. The relationships between these fish (such as predation, competition, etc.), can affect the relative energy flow to the top consumers (whitting, pollack and seabirds) (Lynam et al. 2017).

When the ambient temperature changes the fish will seek an optimum temperature through migration. If the predatory fish at the top of the food chain is more sensitive to temperature, the transition habitat will change the structure and energy flow of the food web. For example, lake trout (*Salvelinus namaycush*) is a vulnerable, cold-adapted fish (optimum temperature 10–12°C) apex predator residing in many Canadian lakes. It is also the key species in the food web. The change of the average temperature from 15 to 20°C in the summer induces thermal stratification in lakes. In this scenario, salmon will leave the higher surface water temperature, and swim to the deep water area with colder temperature. Therefore, the biomass of small fish and invertebrates preyed upon by lake trout will decline. It will be more dependent on the phytoplankton productivity of deep water areas, which will change the structure of the food web (Tunney et al. 2014).

In both laboratory and mesoscale experiments, the double stressors of ocean warming and acidification affect food webs and key ecosystem processes through the mutual relationship between predator and prey. It was found that warming and acidification cumulatively effect the feeding rate of tropical coral reef fish, however, there is an antagonism between the selectivity of predator to prey (Ferrari et al. 2015). In another short-term experiment, exposure to high temperature led to an increase in attack and predation rates. And acidification had minor effects on the behavior of the predator. Ocean warming can significantly reduce the reaction of the prey to the predator, and acidification can have only a little effect on the escape behavior of the prey. The impact of acidification on the predator-prey interactions caused by warming only has a weak cumulative effect (Allan et al. 2017). In mesoscale experiments a food web of three trophic levels was simulated consisting of primary producers (algae), secondary producers (herbivorous invertebrates) and tertiary producer (predatory fish). Both increasing of the CO_2 concentration and warming of the water increased the primary productivity. Increasing the concentration of CO_2 influenced all three trophic levels because of the bottom-up effect. However, with simultaneous warming, the predators in the third trophic level increased the consumption of the prey in the second level due to the increased metabolic rate and the population of the second level decreased. These results suggested that anthropogenic CO_2 emissions as a carbon source increase the productivity of the food web. Ocean warming will reverse the positive effects of acidification through the nutritional relationship among species (Goldenberg et al. 2017).

Climate Change Effects on Biodiversity

As ocean temperature increases, many tropical fish invade temperate sea zones. From the point of invasive species, ocean warming will indirectly affect the relationship between the fish and other species, and then affect the community structure and diversity of marine organism. Invasive species mainly are the key vegetarian fish of coral reef systems. For example, the tropical vegetarian rabbit fish (*Siganus rivulatus*) colonized the Mediterranean through the Suez Canal as a result of ocean warming and established a large population. In the Mediterranean, they destroyed large kelp forests near the shore, and prevented the establishment of new algae farms, which profoundly changed the offshore reef system (Vergés et al. 2014). Eastern Australia is located in the tropical-temperate zone and is a key habitat for kelp forests. Using a 10-yr-old video dataset encompassing a 0.6°C warming period, the invasion of herbivore fish species can be seen to have increased as kelps gradually declined and then disappeared. Meanwhile, the previously abundant fish communities in this area have gradually disappeared, and they have been replaced by the dominant group of tropical herbivorous fish (Vergés et al. 2016). In some sea areas south of Japan, corals had invaded the temperate higher latitude areas. The interaction of corals and algae enhanced. In these areas the dominant species are changed from the algae to the corals (Mezaki and Kubota 2012). If large algae disappear and are not replaced, the biodiversity will decrease sharply. However, if they are replaced by corals, the species diversity may not be affected and even increase (Rossier and Kulbicki 2000).

What are the effects of ocean acidification on community structure and the diversity of fish species? The effects of ocean acidification on the diversity and structure of benthic fish community is studied at natural CO_2 vents. Because of elevated CO_2 concentration increase food resources and a decline of large predator fish by overfishing, the behaviorally dominant fish population in the middle trophic level of the food chain will increase, whereas the subordinate fish population will be suppressed. The fish diversity in local areas will decline. Reducing overfishing of predatory fish can delay the decline of fish diversity brought by acidification (Nagelkerken et al. 2017).

The Acclimation of Fish to Climate Change

In the evolutionary history of vertebrates, the rate at which animals adapt to changes in temperature is about 1°C every one million years (Quintero and Weins 2013). The current rate of warming in the ocean has exceeded the rate of change over the past 420,000 years (Collins et al. 2013). Global climate change is not only a sea surface temperature (SST) rise by 2.0 to 4.8°C in the 21st century; the intensity and frequency of the transient temperature fluctuations will also increase (IPCC 2013). Organisms living in extreme latitudes evolved within a narrow range of temperatures, such as those near the equator or the poles and will be more sensitive to global warming (Habary et al. 2017). Therefore, more studies on the adaptation ability of coral reef fish and polar fish to global climate change are necessary. When the water temperature is 1.5 to 3.0°C higher than the current summer average, it will significantly affect the aerobic scope, swimming, growth, and reproduction of most reef fish indicating that they are sensitive to climate change. Moreover, the optimum temperature of aerobic scope for rock fish on the equator and slightly higher latitudes are consistent. It suggests that fish have no significant local regional adaptation (Rummer and Munday 2017).

The impact of global climate change on fish populations depends on the adaptation potential of species, which depends on the genetic variability among individuals within the population. The effect of acidification on the behavior of different individuals in the fish population is different. And this difference can be transmitted across generations (Ferrari et al. 2011, Munday et al. 2012). According to olfactory behavior, the parents of coral reef fish (*Acanthochromis polyacanthus*) can be divided into CO_2-sensitive individuals and CO_2-tolerant individuals. Their offspring is exposed to acidified water. According to the genome, transcriptome and proteomics analysis of the brain, the gene and protein expression in offspring of tolerant and sensitive parents are different at high CO_2 concentration. This transgenerational molecular signature suggests that individual variation in CO_2 sensitivity could facilitate adaptation of fish populations to ocean acidification (Schunter et al. 2016). After culturing two generations of coral reef fish at a higher temperature, the key metabolic and individual size characteristics had significant heritability. By genotype and environment interaction analysis, it was found that there is significant genotype and environment interaction in the physiological metabolism. Different families of the same fish species have different tolerance to warming. The Northern Alaska red salmon (*Oncorhynchus nerka*) population has genetic plasticity to ocean warming. The individual genetic variation in the population supplies the probability for genetic selection under the global climate change (Munday et al. 2017, Sparks et al. 2017).

Acknowledgement

This work was supported by The National Science Foundation for Young Scientists of China (grant number 41306097).

References

Allan, B.J.M., P. Domenici, S.A. Watson, P.L. Munday and M.I. McCormick. 2017. Warming has a greater effect than elevated CO_2 on predator-prey interactions in coral reef fish. Proceedings of the Royal Society B, 284, 20170784. http://dx.doi.org/10.1098/rspb.2017.0784.

Atkinson, D. 1994. Temperature and organism size: A biological law for ectotherms? Advances in Ecological Research 25: 1–1.

Baudron, A.R., C.L. Needle, A.D. Rijnsdorp and C.T. Marshall. 2014. Warming temperatures and smaller body sizes: Synchronous changes in growth of North Sea fishes. Global Change Biology 20: 1023–1031.

Baumann, H., S.C. Talmage and C.J. Gobler. 2011. Reduced early life growth and survival in a fish in direct response to increased carbon dioxide. Nature Climate Change 2: 38–41.

Beitinger, T.L. and L.C. Fitzpatrick. 1979. Physiological and ecological correlates of preferred temperature in fish. Integrative and Comparative Biology 19: 319–329.

Bentley, J.W., N. Serpetti and J.J. Heymans. 2017. Investigating the potential impacts of ocean warming on the Norwegian and Barents seas ecosystem using a time-dynamic food-web model. Ecological Modeling 360: 94–107.

Bignami, S., I.C. Enochs, D.P. Manzello, S. Sponaugle and R.K. Cowen. 2013. Ocean acidification alters the otoliths of a pantropical fish species with implications for sensory function. Proceedings of the National Academy of Sciences of the United States of America 110: 7366–7370.

Branch, T.A., B.M. Dejoseph, L.J. Ray and C.A. Wagner. 2013. Impacts of ocean acidification on marine seafood. Trends in Ecology and Evolution 28: 178–186.

Brauner, C. and Baker, D. 2009. Patterns of Acid–Base Regulation During Exposure to Hypercarbia in Fishes. *In*: Glass, M. and Wood, S. (eds.). Cardio-Respiratory Control in Vertebrates. Springer, Berlin, Heidelberg.

Burrows, M.T., D.S. Shoeman, M.B. Buckley, P. Moore, E.S. Poloczanska, K.M. Brander, C. Brown, J.F. Bruno, C.M. Duarte, B.S. Halpern, J. Holding, C.V. Kappel, W. Kiessling, M.I. O'Connor, J.M. Pandolfi, C. Parmesan, F.B. Schwing, W.J. Sydeman and A.J. Richardson. 2011. The pace of shifting climate in marine and terrestrial ecosystems. Science 334: 652–655.

Cheung, W.W.L., J.L. Sarmiento, J. Dunne, T.L. Frolicher, V.W.Y. Lam, M.L.D. Palomares, R. Watson and D. Pauly. 2013. Shrinking of fishes exacerbates impacts of global ocean changes on marine ecosystems. Nature Climate Change 3: 254–258.

Chivers, D.P., M.I. Mccormick, G.E. Nilsson, P.L. Munday, S. Watson, M.G. Meekan, M.D. Mitchell, K.C. Corkill and M.C.O. Ferrari. 2014. Impaired learning of predators and lower prey survival under elevated CO_2: A consequence of neurotransmitter interference. Global Change Biology 20: 515–522.

Chung, W.S., N.J. Marshall, S.A. Watson, P.L. Munday and G.E. Nilsson. 2014. Ocean acidification slows retinal function in a damselfish through interference with $GABA_A$ receptors. The Journal of Experimental Biology 217: 323–326.

Ciais, P., C. Sabine, G. Bala, L. Bopp, V. Brovkin, J. Canadell, A. Chhabra, R. DeFries, J. Galloway, M. Heimann, C. Jones, C. Le Quere, R.B. Myneni, S. Piao and P. Thornton. 2014. Carbon and other biogeochemical cycles. pp. 465–570. *In*: Stocker, T.F., D. Qin, G-.K. Plattner, M. Tignor, S.K. Allen, J. Boschung, A. Nauels, Y. Xia, V. Bex and P.M. Midgley (eds.). Climate Change 2013: the Physical Science Basis. Contribution of Working Group I to the Fifth Assessment Report of the Intergovernmental Panel on Climate Change. Cambridge University Press, Cambridge, UK and New York, NY, USA,

Claireaux, G., D.M. Webber, J.P. Lagardere and S.R. Kerr. 2000. Influence of water temperature and oxygenation on the aerobic metabolic scope of Atlantic cod (*Gadus morhua*). Journal of Sea Resarch 44: 257–265.

Collins, M. and R. Knutti. 2013. Long-term climate change: projections, commitments and irreversibility. pp. 1029–1136. *In*: Stocker, T.F., D. Qin, G-.K. Plattner et al. (eds.). Climate Change: The Physical Science Basis. Contribution of Working Group I to the Fifth Assessment Report of the Intergovernmental Panel on Climate Change, Cambridge University Press, Cambridge.

Diaz, R.J. and R. Rosenberg. 2008. Spreading dead zones and consequences for marine ecosystems. Science 321: 926–929.

Domenici, P., C. Lefrancois and A. Shingles. 2007. Hypoxia and the anti-predator behaviour of fishes. Philos. Trans. R. Soc. B 362: 2105–2121.

Domenici, P., B. Allan, M.I. McCormick and P.L. Munday. 2012. Elevated carbon dioxide affects behavioural lateralization in a coral reef fish. Biology Letters 8: 78–81.

Drinkwater, K.F. 2011. The influence of climate variability and change on the ecosystems of the Barents Sea and adjacent waters: Review and synthesis of recent studies from the NESSAS Project. Progress in Oceanography 90: 47–61.

Enzor, L.A., M.L. Zippay and S.P. Place. 2013. High latitude fish in a high CO_2 world: Synergistic effects of elevated temperature and carbon dioxide on the metabolic rates of Antarctic notothenioids. Comparative Biochemistry and Physiology – Part A: Molecular & Integrative Physiology 164: 154–161.

Esbaugh, A.J., R. Heuer and M. Grosell. 2012. Impacts of ocean acidification on respiratory gas exchange and acid-base balance in a marine teleost, *Opsanus beta*. Journal of Comparative Physiology B 182: 921–934.

Faleiro F., M. Baptista, C. Santos, M.L. Aurelio, M. Pimentel, M.R. Pegado, J.R. Paula, R. Calado, T. Repolho and R. Rosa. 2015. Seahorses under a changing ocean: the impact of warming and acidification on the behaviour and physiology of a poor-swimming bony-armoured fish. Conservation Physiology 3: cov009. doi:10.1093/conphys/cov009.

Farrant, M. and K. Kaila. 2007. The cellular, molecular and ionic basis of GABAA receptor signaling. Progress in Brain Research 160: 59–87.

Ferrari, M.C.O., D.L. Dixson, P.L. Munday, M.I. McCormick, M.G. Meekan, A. Sih and D.P. Chivers. 2011. Intrageneric variation in antipredator responses of coral reef fishes affected by ocean acidification: implications for climate change projections on marine communities. Global Change Biology 17: 2980–2986.

Ferrari, M.C., R.P. Manassa, D.L. Dixson, P.L. Munday, M.I. McCormick, M.G. Meekan, A. Sih and D.P. Chivers. 2012. Effects of ocean acidification on learning in coral reef fishes. Plos One 7: e31478.

Ferrari, M.C.O., P.L. Munday, J.L. Rummer, M.I. Mccormick, K. Corkill, S. Watson, B.J.M. Allan, M.G. Meekan and D.P. Chivers. 2015. Interactive effects of ocean acidification and rising sea temperatures alter predation rate and predator selectivity in reef fish communities. Global Change Biology 21: 1848–1855.

Franke, A. and C. Clemmesen. 2011. Effect of ocean acidification on early life stages of Atlantic herring (*Clupea harengus* L.). Biogeoscience Discussions 8: 7097–7126.

Frommel, A.Y., R. Maneja, D. Lowe, A.M. Malzahn, A.J. Geffen, A. Folkvord, U. Piatkowski, T.B.H. Reusch and C. Clemmesen. 2011. Severe tissue damage in Atlantic cod larvae under increasing ocean acidification. Nature Climate Change 2: 42–46.

Frommel, A.Y., A. Schubert, U. Piatkowski and C. Clemmesen. 2012. Egg and early larval stages of Baltic cod, *Gadus morhua*, are robust to high levels of ocean acidification. Mar. Biol. 160: 1825–1834.

Gilbert, B., T.D. Tunney, K.S. Mccann, J.P. Delong, D.A. Vasseur, V. Savage, J.B. Shurin, A.I. Dell, B.T. Barton, C.D.G. Harley, H.M. Kharouba, P. Kratina, J.L. Blanchard, C. Clements, M. Winder, H.S. Greig and M.I. O'Connor. 2014. A bioenergetic framework for the temperature dependence of trophic interactions. Ecology Letters 17: 902.

Goldenberg S.U., I. Nagelkerken, C.M. Ferreira, H. Ullah and S.D. Connell. 2017. Boosted food web productivity through ocean acidification collapses under warming. Global Change Biology 23: 4177–4184.

Habary, A., J.L. Johansen, T.J. Nay, J.F. Steffensen and J.L. Rummer. 2017. Adapt, move, or die - how will tropical coral reef fishes cope with ocean warming? Global Change Biology 23: 566–577.

Hamilton T.J., A. Holcombe and M. Tresguerres. 2014. CO_2-induced ocean acidification increases anxiety in rockfish via alteration of GABAA receptor functioning. Proceedings of the Royal Society Biological Science 281: 1–7.

Heuer, R.M., M.J. Welch, J.L. Rummer, P.L. Munday and M. Grosell. 2016. Altered brain ion gradients following compensation for elevated CO_2 are linked to behavioural alterations in a coral reef fsh. Scientific Report, 6: 33216. doi:10.1038/srep33216.

Hoefnagel, K.N. and W.C.E.P. Verberk. 2015. Is the temperature-size rule mediated by oxygen in aquatic ectotherms? Journal of Thermal Biology 54: 56–65.

Houde, E. 1989. Comparative growth, mortality, and energetics of marine fish larvae: Temperature and implied latitudinal effects. Fish Bulletin 87: 471–495.

Huth, T.J. and S.P. Place. 2016. Transcriptome wide analyses reveal a sustained cellular stress response in the gill tissue of *Trematomus bernacchii* after acclimation to multiple stressors. BMC Genomics 17: 127.

Hurst, T.P., E.R. Fernandez and J.T. Mathis. 2013. Effects of ocean acidification on hatch size and larval growth of walleye pollock (*Theragra chalcogramma*). ICES Journal of Marine Science 70: 812–822.

IPCC. Summary for Policymakers . *In*: Solomon, S, Qin, D, Manning, M et al. (ed.). Climate Change 2007: The Physical Science Basis. Contribution of Working Group I to the Fourth Assessment Report of the Intergovernmental Panel on Climate Change. Cambridge: Cambridge University Press, 2007.

IPCC. Climate change 2013: The Physical Science Basis. Contribution Of Working Group I to the Fifth Assessment Report of the Intergovernmental Panel on Climate Change. Cambridge, UK: Cambridge University Press, 2013.

Ishimatsu A., T. Kikkawa, M. Hayashi, K. Lee and J. Kita. 2004. Effects of CO_2 on marine fish: larvae and adults. Journal of Oceanography 60: 731–741.

Ishimatsu, A., M. Hayashi and T. Kikkawa. 2008. Fishes in high-CO_2, acidified oceans. Marine Ecology Progress Series 373: 295–302.

Jonz, M.G. and C.A. Nurse. 2006. Ontogenesis of oxygen chemoreception in aquatic vertebrates. Respir. Physiol. Neurobiology 154: 139–152.

Kikkawa, T., A. Ishimatsu and J. Kita. 2003. Acute CO_2 tolerance during the early developmental stages of four marine teleosts. Environmental Toxicology 18: 375–382.

Kleisner, K.M., M.J. Fogarty, S. Mcgee, J.A. Hare, S. Moret, C.T. Perretti and V.S., Saba. 2017. Marine species distribution shifts on the U.S. northeast continental shelf under continued ocean warming. Progress in Oceanography 153: 24–36.

Koenigstein, S., F.T. Dahlke, M.H. Stiasny, D. Storch, C. Clemmesen and H.O. Pörtner. 2018. Forecasting future recruitment success for Atlantic cod in the warming and acidifying Barents sea. Global Change Biology 24: 526–535.

Lam, V.W.Y., W.W.L. Cheung and U.R. Sumaila. 2016. Marine capture fisheries in the Arctic: winners or losers under climate change and ocean acidification? Fish & Fisheries 17: 335–357.

Levin, L.A. and D.L. Breitburg. 2015. Linking coasts and seas to address ocean deoxygenation. Nat. Clim. Change 5: 401–403.

Loeng H., K. Brander, E. Carmack, S. Denisenko, K. Drinkwater, B. Hansen, K. Kovacs, P. Livingston, F. McLaughlin and E. Sakshaug. 2005. Marine systems. pp. 453–538. *In*: Symon, C., L. Arris and W. Heal (eds.). ACIA Arctic Climate Impact Assessment. Cambridge University Press.

Lynam, C.P., M. Llope, C. Möllmann, P. Helaouët, G.A. Bayliss-Brown and N.C. Stenseth. 2017. Interaction between top-down and bottom-up control in marine food webs. Proceedings of the National Academy of Sciences of the United States of America 114: 1952.

Mezaki, T. and S. Kubota. 2012. Changes of hermatypic coral community in coastal sea area of Kochi, high latitude, Japan. Aquabiology 201: 332–337.

Mislan, K.A.S., C.A. Deutsch, R.W. Brill, J.P. Dunne and J.L. Sarmiento. 2017. Projections of climate-driven changes in tuna vertical habitat based on species-specific differences in blood oxygen affinity. Global Change Biology 23: 4019–4028.

Monllor-hurtado, A., M.G. Pennino and J.L. Sanchezlizaso. 2017. Shift in tuna catches due to ocean warming. Plos One 12(6): e0178196.

Montero-Serra, I., M. Edwards and M.J. Genner. 2015. Warming shelf seas drive the subtropicalization of european pelagic fish communities. Global Change Biology 21: 144–53.

Munday, P.L., D.L. Dixson, J.M. Donelson, G.P. Jones, M.S. Pratchett, G.V. Devitsina and K.B. Doving. 2009a. Ocean acidification impairs olfactory discrimination and homing ability of a marine fish. Proceedings of the National Academy of Sciences of the United States of America 106: 1848–1852.

Munday, P. L., J.M. Donelson, D.L. Dixson and G.G. Endo. 2009b. Effects of ocean acidification on the early life history of a tropical marine fish. Proceedings Biological Sciences 276: 3275–3283.

Munday, P.L., N. Crawley and G. Nilsson. 2009c. Interacting effects of elevated temperature and ocean acidification on the aerobic performance of coral reef fishes. Marine Ecology Progress 388: 235–242.

Munday, P.L., M. Gagliano, J.M. Donelson, D.L. Dixson and S.R. Thorrold. 2011. Ocean acidification does not affect the early life history development of a tropical marine fish. Marine Ecology Progress Series 423: 211–221.

Munday, P. L. M.I. McCormick, M. Meekan, D.L. Dixson, S. Watson, D.P. Chivers and M.C.O. Ferrari. 2012. Selective mortality associated with variation in CO_2 tolerance in a marine fish. Ocean Acidification 1: 1–5.

Munday, P.L., J.M. Donelson and J.A. Domingos 2017. Potential for adaptation to climate change in a coral reef fish. Global Change Biology 23: 307–317.

Nagelkerken, I., S.U. Goldenberg, C.M. Ferreira, B.D. Russell and S.D. Connell. 2017. Species interactions drive fish biodiversity loss in a high-CO_2 world. Current Biology 27: 2177–2184.

Nilsson, G.E., D.L. Dixson, P. Domenici, M.I. McCormick, C. Sorensen, S. Watson and P.L. Munday. 2012. Near-future carbon dioxide levels alter fish behavior by interfering with neurotransmitter function. Nature Climate Change 2(3): 201–204.

Nowicki, J.P., G.M. Miller and P.L. Munday. 2012. Interactive effects of elevated temperature and CO_2 on foraging behavior of juvenile coral reef fish. J. Exp. Mar. Biol. Ecol. 412: 46–51.

Ockendon, N., D.J. Baker, J.A. Carr, E.C. White, R.E.A. Almond, T. Amano, E. Bertram, R.B. Bradbury, C. Bradley, S.H.M Butchart, N. Doswald, W. Foden, D.J.C. Gill, R.E. Green, W.J. Sutherland, E.V.J. Tanner and J.W. Pearce-Higgins. 2014. Mechanisms underpinning climatic impacts on natural populations: altered species interactions are more important than direct effects. Global Change Biology 20: 2221–2229.

Pauly, D. 1979. Gill size and temperature as governing factors in fish growth: a generalization of von Bertalanffy's growth formula. Institut fur Meereskunde an der Universitat Kiel 63: 156.

Pauly, D. and W.W.L. Cheung. 2018. Sound physiological knowledge and principles in modeling shrinking of fishes under climate change. Global Change Biology 24: e15–e26.

Pelster, B. 2008. Gas exchange. pp. 102–103. *In*: Finn, R.N. and B.G. Kapoor (eds.). Fish Larval Physiology, Part 2, Respiration & Homeostasis. Science Publishers, Einfield, NH.

Pörtner, H.O. 2001. Climate change and temperature-dependent biogeography: Oxygen limitation of thermal tolerance in animals. Nature 88: 137–146.

Pörtner, H.O. and R. Knust. 2007. Climate change affects marine fishes through oxygen limitation of thermal tolerance. Science 315: 95–97.

Pörtner, H.O. and A.P. Farrell. 2008. Physiology and climate change. Science 322: 690–692.

Quintero, I. and J.I. Weins. 2013. Rates of projected climate change dramatically exceed past rates of climatic niche evolution among vertebrate species. Ecology Letters 16: 1095–1103.

Rossier, O. and M. Kulbicki. 2000. A comparison of fish assemblages from two types of algal beds and coral reefs in the south-west lagoon of New Caledonia. Cybium 24: 3–26.

Rummer, J.L. and P.L. Munday. 2017. Climate change and the evolution of reef fishes: Past and future. Fish & Fisheries 18: 22–39.

Schunter, C., M.J. Welch, T. Ryu, H. Zhang, M.L. Berumen, G.E. Nilsson, P.L. Munday and T. Ravasi. 2016. Molecular signatures of transgenerational response to ocean acidification in a species of reef fish. Nature Climate Change 6: 1–5.

Shang, E.H., R.M. Yu and R.S. Wu. 2006. Hypoxia affects sex differentiation and development, leading to a male-dominated population in zebrafish (*Danio rerio*). Environmental Science and Technology 40: 3118–3122.

Sparks, M.M., P.A.H. Westley, J.A. Falke and T.P. Quinn. 2017. Thermal adaptation and phenotypic plasticity in a warming world: insights from common garden experiments on Alaskan sockeye salmon. Global Change Biology 23: 5203–5217.

Strobel, A., S. Bennecke, E. Leo, K. Mintenbeck, H.O. Portner and F.C. Mark. 2012. Metabolic shifts in the Antarctic fish *Notothenia rossii* in response to rising temperature and PCO_2. Frontiners in Zoology 9: 28.

Sunday, J.M., A.E. Bates and N.K Dulvy. 2012. Thermal tolerance and the global redistribution of animals. Nature Climate Change 2: 686–690.

Tseng, Y.C., M.H. Yu, M. Stumpp, L.Y. Lin, F. Melzner and P.P. Hwang. 2013. CO_2-driven seawater acidification differentially affects development and molecular plasticity along life history of fish (*Oryzias latipes*). Comparative Biochemistry and Physiology Part A 165: 119–130.

Tunney, T.D., K.S. Mccann, N.P. Lester and B.J. Shuter. 2014. Effects of differential habitat warming on complex communities. Proceedings of the National Academy of Sciences of the United States of America 111: 8077–8082.

Van, R.I., Y. Buba, J. Delong, M. Kiflawi and J. Belmaker, 2017. Large but uneven reduction in fish size across species in relation to changing sea temperatures. Global Change Biology 23: 3667–3674.

Vergés, A., P.D. Steinberg, M.E. Hay, A.G.B. Poore, A.H. Campbell, E. Ballesteros, K.L. Heck Jr, D.J. Booth, M.A. Coleman, D.A. Feary, W. Figueira, T. Langlois, E.M. Marzinelli, T. Mizerek, P.J. Mumby, Y. Nakamura, M. Roughan, E. Sebille, A.S. Gupta, D.A. Smale, F. Tomas, T. Wernber and S.K. Wilson. 2014. The tropicalization of temperate marine ecosystems: climate-mediated changes in herbivory and community phase shifts. Proceedings Biological Sciences 281: 20140846.

Vergés, A., C. Doropoulos, H.A. Malcolm, M. Skye, M. Garcia-Piza, E.M. Marzinelli, A.H. Campbell, E .Ballesteros, A.S. Hoey, A. Vila-Concejo, Y. Bozec and P.D. Steinberg. 2016. Long-term empirical evidence of ocean warming leading to tropicalization of fish communities, increased herbivory, and loss of kelp. Proceedings of the National Academy of Sciences of the United States of America 113: 13791.

Wang, S.Y., K. Lau, K. Lai, J. Zhang, A.C. Tse, J. Li, Y. Tong, T. Chan, C.K. Wong, J.M. Chiu, D.W. Au, A.S. Wong, R.Y. Kong and R.S. Wu. 2016. Hypoxia causes transgenerational impairments in reproduction. Nature Communication 7: 12114.

Wang, X., L. Song, Y. Chen, H. Ran and J. Song. 2017. Impact of ocean acidification on the early development and escape behavior of marine medaka (*Oryzias melastigma*). Marine Environmental Research 131: 10–18.

Wernberg, T., S. Bennett, R.C. Babcock, T.D. Bettignies, K. Cure, M. Depczynski, F. Dufois, J. Fromont, C.J. Fulton, R.K. Hovey, E.S. Harvey, T.H. Holmes, G.A. Kendrick, B. Radford, J. Santana-Garcon, B.J. Saunders, D.A. Smale, M.S. Thomsen, C.A. Tuckett, F. Tuya, M.A. Vanderklift and S. Wilson. 2016. Climate-driven regime shift of a temperate marine ecosystem. Science 353: 169.

Index

P

R

S

T

U

V

W

Z

9781138350052

9781138350052